REVIEWS IN MINERALOGY AND GEOCHEMISTRY

VOLUME 54   2003

BIOMINERALIZATION

EDITORS:

Patricia M. Dove   *Virginia Polytechnic Institute & State Univ. Blacksburg, VA, U.S.A.*

James J. De Yoreo   *Lawrence Livermore National Lab Livermore, CA, U.S.A.*

Steve Weiner   *Weizmann Institute Rehovot, Israel*

FRONT COVER: Coccolith calcite. Colorized high resolution scanning electron micrograph of *Calcidiscus leptoporus* ssp. *quadriperforatus* with heterococcoliths on left (blue) showing interlocking crystals with rhombic faces and holococcoliths on the right (violet) formed of minute rhombohedra (from Young and Henriksen, this volume; image by Markus Geisen).

Series Editor: ***Jodi J. Rosso***

MINERALOGICAL SOCIETY OF AMERICA
GEOCHEMICAL SOCIETY

COPYRIGHT 2003

MINERALOGICAL SOCIETY OF AMERICA

The appearance of the code at the bottom of the first page of each chapter in this volume indicates the copyright owner's consent that copies of the article can be made for personal use or internal use or for the personal use or internal use of specific clients, provided the original publication is cited. The consent is given on the condition, however, that the copier pay the stated per-copy fee through the Copyright Clearance Center, Inc. for copying beyond that permitted by Sections 107 or 108 of the U.S. Copyright Law. This consent does not extend to other types of copying for general distribution, for advertising or promotional purposes, for creating new collective works, or for resale. For permission to reprint entire articles in these cases and the like, consult the Administrator of the Mineralogical Society of America as to the royalty due to the Society.

REVIEWS IN MINERALOGY AND GEOCHEMISTRY

(Formerly: REVIEWS IN MINERALOGY)

ISSN 1529-6466

Volume 54

Biomineralization

ISBN 093995066-9

Additional copies of this volume as well as others in this series may be obtained at moderate cost from:

THE MINERALOGICAL SOCIETY OF AMERICA
1015 EIGHTEENTH STREET, NW, SUITE 601
WASHINGTON, DC 20036 U.S.A.

DEDICATION

Dr. William C. Luth has had a long and distinguished career in research, education and in the government. He was a leader in experimental petrology and in training graduate students at Stanford University. His efforts at Sandia National Laboratory and at the Department of Energy's headquarters resulted in the initiation and long-term support of many of the cutting edge research projects whose results form the foundations of these short courses. Bill's broad interest in understanding fundamental geochemical processes and their applications to national problems is a continuous thread through both his university and government career. He retired in 1996, but his efforts to foster excellent basic research, and to promote the development of advanced analytical capabilities gave a unique focus to the basic research portfolio in Geosciences at the Department of Energy. He has been, and continues to be, a friend and mentor to many of us. It is appropriate to celebrate his career in education and government service with this series of courses in cutting-edge geochemistry that have particular focus on Department of Energy-related science, at a time when he can still enjoy the recognition of his contributions.

BIOMINERALIZATION

Reviews in Mineralogy and Geochemistry

FOREWORD

In this volume, the editors Patricia Dove, James De Yoreo, and Steve Weiner have integrated a diverse group of contributors from the earth, biological and materials disciplines who introduce us to the concepts of biological mineralization, examine the major biomineralization processes, and describe their impact on earth history. This volume offers an excellent opportunity for both specialists and non-specialists to understand the basic principles of biomineralization. It also gives us a view into the future of this growing field of research. It was prepared in advance of a two-day short course (December 6–7, 2003) on Biomineralization, jointly sponsored by GS and MSA, and held at the Silverado Resort and Conference Center in Napa Valley, California prior to the fall AGU meeting in San Francisco.

Trish deserves a long vacation after the supreme effort she put into assembling this volume (and *on time* for the short course!). The many hours she spent reviewing, editing, revising, as well as her ability to keep things on-track made this volume possible. Steve's reviews also significantly improved many of the chapters. The authors are commended for their hard work. Thank you. And, as always, I thank my infinitely patient and supportive family, Kevin, Ethan and Natalie.

Jodi J. Rosso, Series Editor
West Richland, Washington
October 25, 2003

PREFACE AND ACKNOWLEDGMENTS

Since the dawn of life on earth, organisms have played roles in mineral formation in processes broadly known as biomineralization. This biologically-mediated organization of aqueous ions into amorphous and crystalline materials results in materials that are as simple as adventitious precipitates or as complex as exquisitely fabricated structures that meet specialized functionalities. The purpose of this volume of *Reviews in Mineralogy and Geochemistry* is to provide students and professionals in the earth sciences with a review that focuses upon the various processes by which organisms direct the formation of minerals. Our framework of examining biominerals from the viewpoints of major mineralization strategies distinguishes this volume from most previous reviews. The review begins by introducing the reader to over-arching principles that are needed to investigate biomineralization phenomena and shows the current state of knowledge regarding the major approaches to mineralization that organisms have developed over the course of Earth history. By exploring the complexities that underlie the "synthesis" of biogenic materials, and therefore the basis for how compositions and structures of biominerals are mediated (or not), we believe this volume will be instrumental in propelling studies of biomineralization to a new level of research questions that are grounded in an understanding of the underlying biological phenomena. To make this happen, the volume contains contributions from a special group of authors whose areas of expertise are as varied as the biominerals themselves. Special thanks are due to these scientists for meeting the deadlines with their comprehensive contributions.

Foreword, Preface & Acknowledgments

We also thank the individuals and organizations who have made possible the timely publication of this review. The Series editor, Jodi Rosso, cheerfully handled the production of the volume through our tight publication schedule. Alex Speer was helpful throughout the process and we thank him for sharing his experience regarding how to ensure a successful review and short course. Many other people have contributed to seeing the success of this volume and we thank Drs. Selim Elhadj, Nizhou Han, and Laura Wasylenki for carefully proofreading several chapters and helping with the many reference searches. Ellen Mathena and David Rugh helped to collect materials and to work with draft manuscripts. PMD also thanks the Department of Energy, Division of Chemical Sciences, Geosciences and Biosciences (DE-FG02-00ER15112) and the National Science Foundation (NSF-OCE-0083173) for supporting her time on this project.

We are very grateful to the U.S. Department of Energy, Office of Basic Energy Sciences, Chemical Sciences, Geosciences and Biosciences Division, for special funding in honor of Bill Luth (see dedication). We especially thank Dr. Nick Woodward for making this support possible. Additional funding was also provided by Lawrence Livermore National Laboratory. We also thank the College of Science and the Department of Geosciences at Virginia Tech for providing supplementary funds that allowed us to award registration fee waivers to 20 of the student participants. While we are all students of this fascinating field, it is a delight to see the high level of interest in the next generation of scientists.

Patricia M. Dove
Steve Weiner
James J. De Yoreo

Blacksburg, Virginia
September 2003

TABLE OF CONTENTS

1   An Overview of Biomineralization Processes and the Problem of the Vital Effect
Steve Weiner and Patricia M. Dove

INTRODUCTION	1
This chapter	2
OVERVIEW OF BIOMINERALS	2
What is a biomineral?	2
Some comments on the major groups of biominerals	3
Unique character of minerals formed by biological systems	7
BIOMINERALIZATION CONCEPTS	8
BASIC PROCESSES OF BIOMINERALIZATION	10
Biologically induced mineralization	10
Biologically controlled mineralization	10
Comments on biomineralization processes	17
VITAL EFFECTS	17
Two basic categories of vital effects	18
Equilibrium with the environment—the real puzzle	19
Concluding comment	23
ACKNOWLEDGMENTS	24
REFERENCES	24

2   Principles of Molecular Biology and Biomacromolecular Chemistry
John S. Evans

INTRODUCTION	31
GENERAL CELL ARCHITECTURE	31
Basic components of eukaryotic cells	31
Simpler in design: prokaryotic cells	33
MOLECULAR MANUFACTURING	33
Genetic code, gene organization: the protein "blueprints"	33
Protein synthesis: assembling the polypeptide chain	37
Post-manufacturing processes	40
Exporting proteins to the extracellular matrix	41
What about intracellular biomineralization?	41
Manufacturing of other macromolecules involved in the formation of biominerals	41
MOLECULAR PERFORMANCE	42
Protein structure	42
Biomineralization protein structure	44
Oligosaccharide and polysaccharide structure	44
Membrane assemblies	46

MOLECULAR MANIPULATION 47
    DNA manipulation.. 47
    Reverse transcription of RNA.. 50
    DNA amplification... 52
    Protein expression.. 52
    Chemical synthesis of biomacromolecules... 55
ACKNOWLEDGMENTS 56
REFERENCES 56

3   Principles of Crystal Nucleation and Growth
James J. De Yoreo and Peter G. Vekilov

INTRODUCTION 57
CRYSTAL NUCLEATION VS. CRYSTAL GROWTH.................................... 59
THERMODYNAMIC DRIVERS OF CRYSTALLIZATION 61
NUCLEATION... 63
    Nucleation on foreign surfaces 64
    Nucleation pathways and Ostwald's Law of phases..................................... 66
    The nucleus shape 68
    Crystal growth kinetics 71
    Step generation: 2D nucleation vs. growth at dislocations 76
MODIFYING THE SHAPES OF GROWTH HILLOCKS AND CRYSTALS 81
CONCLUSION... 89
ACKNOWLEDGMENTS 90
REFERENCES 90

4   Biologically Induced Mineralization by Bacteria
Richard B. Frankel and Dennis A. Bazylinski

INTRODUCTION 95
BIOLOGICALLY INDUCED MINERALIZATION ON ORGANIC SURFACES 96
    Bacterial surface properties.. 96
IRON AND MANGANESE MINERALIZATION PROCESSES..................... 98
    Iron and manganese oxidation 99
    Iron and manganese reduction 103
    Biologically induced mineralization of magnetite...................................... 104
    Magnetite dissolution.. 106
    Sulfate reduction 106
    Sulfide mineral oxidation... 107
INTRACELLULAR BIOLOGICALLY INDUCED MINERALIZATION ... 108
SIGNIFICANCE OF BIOLOGICALLY INDUCED MINERALIZATION ... 109
ACKNOWLEDGMENTS 110
REFERENCES 110

5 The Source of Ions for Biomineralization in Foraminifera and Their Implications for Paleoceanographic Proxies

Jonathan Erez

INTRODUCTION	115
Foraminifera, corals, and coccolithophores in the global carbon cycle	115
Importance of foraminifera and corals for paleoceanographic reconstructions	117
Subjects that will not be included	118
Geological aspects of rock-forming organisms	119
BIOMINERALIZATION IN FORAMINIFERA	119
Introduction	119
Light microscopy, TEM and SEM of test structure and function	120
Chamber formation in perforate foraminifera	122
Rates of calcification: radiotracers, weight increase and microsensor studies	122
Photosynthesis and calcification	124
Internal carbon and calcium pools	125
Incorporation of trace elements	129
Inhomogeneous distribution in the test	130
New observations and the role of seawater vacuolization	132
Observations on recovering individuals	133
Summary and working hypothesis	134
Relations between Ca^{2+} and CO_3^{2-} during $CaCO_3$ precipitation	137
Global CO_2 considerations	137
Concentration mechanism of ions inside seawater vacuoles and what prevents the precipitation of $CaCO_3$	138
THE ROLE OF Mg IN FORAMINIFERA	139
The carbon pathway and stable isotopes	141
SUMMARY AND CONCLUDING REMARKS	141
ACKNOWLEDGMENTS	143
REFERENCES	144

6 Geochemical Perspectives on Coral Mineralization

Anne L. Cohen and Ted A. McConnaughey

INTRODUCTION	151
THE SCLERACTINIAN SKELETON: MORPHOLOGY, MINERALOGY, GROWTH AND CHEMISTRY	153
The polyp	154
The sclerodermites	154
Centers of calcification	155
Sr/Ca geochemistry	158
Diurnal cycle of calcification	162
Models of skeletogenesis: physicochemical	163
Organic matrix models	166

CALCIFICATION MECHANISM 170
    Calcium ATPase and CO_2 based calcification... 170
CO_2 BASED CALCIFICATION AND STABLE ISOTOPES 173
    The kinetic model 173
    "Carbonate" scenarios... 174
    Respired CO_2 in the skeleton 177
    ^{13}C deficiencies, and $\delta^{18}O$ - $\delta^{13}C$ correlations in corals.................................. 177
WHY DO REEF CORALS CALCIFY SO FAST?... 178
    Photosynthesis does not cause rapid calcification 179
    Calcification relieves CO_2 stress.. 179
    Calcification may stimulate nutrient uptake 180
ACKNOWLEDGMENTS 181
REFERENCES 182

7 Biomineralization Within Vesicles: The Calcite of Coccoliths
Jeremy R. Young and Karen Henriksen

INTRODUCTION 189
    Biological affinities... 189
    Life cycles 190
HETEROCOCCOLITH BIOMINERALIZATION... 192
    Cytological aspects 192
    Biochemical aspects: organic molecules involved in coccolith
        biomineralization 194
    Morphological observations.. 195
    Nucleation 200
    Growth 201
HOLOCOCCOLITHS 205
    Cytological observations... 207
    Biochemical observations 207
    Morphological observations.. 207
THE EVOLUTION OF COCCOLITHOPHORE BIOMINERALIZATION
    MECHANISMS.. 209
    Summary... 211
    Crystallographic orientation—template control or a self-organizing system? ... 211
    Crystal growth regulation 212
    Genomic approaches... 212
ACKNOWLEDGMENTS 212
REFERENCES 212

8 Biologically Controlled Mineralization in Prokaryotes
Dennis A. Bazylinski and Richard B. Frankel

INTRODUCTION 217
THE MAGNETOTACTIC BACTERIA 218
    Classification and general features 218
    Ecology of magnetotactic bacteria 218
    Phylogeny of the magnetotactic bacteria 219
    Biogeochemical significance of the magnetotactic bacteria 220
THE MAGNETOSOME 220
    Composition of the magnetosome mineral phase 220
    Size of the magnetosome mineral phase 222
    Morphology of the magnetosome mineral phase 222
ARRANGEMENT AND EFFECT OF MAGNETOSOMES WITHIN
    THE CELL 225
    Function and physics of magnetotaxis 227
STUDYING BCM IN BACTERIA: SYNTHESIS OF THE BACTERIAL
    MAGNETOSOME 229
    Magnetite magnetosomes 230
    Environmental and physiological conditions that appear to support
        magnetosome synthesis 234
    Greigite magnetosomes 235
    Genetic systems in the magnetotactic bacteria 236
    Genomics and BCM in magnetotactic bacteria 237
SIGNIFICANCE OF BCM 238
    Magnetofossils and biomarkers 238
    Magnetic sensitivity in other organisms 240
ACKNOWLEDGMENTS 241
REFERENCES 241

9 Mineralization in Organic Matrix Frameworks
Arthur Veis

INTRODUCTION 249
BIOMINERALIZATION: GENERAL ASPECTS 249
THE COLLAGEN MATRIX 254
    Molecular assembly of the Type I collagen molecule 257
    Secretion 259
    Fibrillogenesis 260
    Fibril structure 261
    Cross-linking stabilization 264
    Cross-linkage chemistry 265
COLLAGEN MATRIX MINERALIZATION 268
    The SIBLING protein family 268
    Mineral placement 270

TOOTH ENAMEL	276
Cellular compartmentation	276
Enamel matrix proteins	277
ENAMEL MINERALIZATION	279
OTOLITHS	281
A mechanistic model	281
ACKNOWLEDGMENT	283
REFERENCES	283

10 Silicification: The Processes by Which Organisms Capture and Mineralize Silica

Carole C. Perry

INTRODUCTION	291
STRUCTURAL CHEMISTRY OF SILICA	291
SILICA CHEMISTRY IN AQUEOUS AND NON-AQUEOUS ENVIRONMENTS	293
THE STRUCTURAL CHEMISTRY OF BIOSILICAS	298
Techniques for the study of biosilica structure	299
SILICA FORMATION IN SPONGES	304
Introduction to sponges including structural chemistry	304
Silicateins	305
Collagen, spiculogenesis and the effect of silicon concentration	307
Model studies: primmorphs: studies of the effect of silicon and germanium on spiculogenesis	309
Diatoms: introduction and structural information	310
The first stage in silicification: transport of the raw ingredients into the cell!	312
Internal silicon pools	313
The silica deposition environment	314
What are the roles for the various extracted biopolymers (sponges and diatoms)?	317
How to go from molecular level chemical control to macroscopic control?	320
Progress towards understanding biosilicification at the molecular level in other systems	321
Unanswered questions, a selection	322
Where to from here?	323
ACKNOWLEDGMENTS	323
BIBLIOGRAPHY	323
REFERENCES	324

11 Biomineralization and Evolutionary History
Andrew H. Knoll

INTRODUCTION 329
THE PHYLOGENETIC DISTRIBUTION OF MINERALIZED SKELETONS 330
    Carbonate skeletons 330
    Silica skeletons 332
    Phosphate skeletons 333
THE GEOLOGIC RECORD OF SKELETONS 333
    Biomineralization in Precambrian oceans 333
    Biomineralization and the Cambrian explosion 337
    The Ordovician radiation of heavily calcified skeletons 341
    Permo-Triassic extinction and its aftermath 341
    Two Mesozoic revolutions 345
    The future 347
DISCUSSION AND CONCLUSIONS 347
ACKNOWLEDGMENTS 350
REFERENCES 350

12 Biomineralization and Global Biogeochemical Cycles
Philippe Van Cappellen

INTRODUCTION 357
BIOGEOCHEMICAL CYCLES 358
    Forcing mechanisms and time scales 358
    Models 362
    Carbon cycle 365
BIOMINERALIZATION IN A GLOBAL CONTEXT 368
    Biomineralization and biogeochemical cycles 368
    Biomineralization through time 370
MARINE BIOGEOCHEMICAL CYCLE OF SILICON 372
    Controls on biosiliceous production 372
    Weathering 374
    Preservation of biogenic silica 376
CONCLUSIONS 379
ACKNOWLEDGMENTS 379
REFERENCES 379

1 An Overview of Biomineralization Processes and the Problem of the Vital Effect

Steve Weiner
Department of Structural Biology
Weizmann Institute of Science
76100 Rehovot Israel

Patricia M. Dove
Department of GeoSciences
Virginia Tech
Blacksburg, Virginia 24061 U.S.A.

"Biomineralization links soft organic tissues, which are compositionally akin to the atmosphere and oceans, with the hard materials of the solid Earth. It provides organisms with skeletons and shells while they are alive, and when they die these are deposited as sediment in environments from river plains to the deep ocean floor. It is also these hard, resistant products of life which are mainly responsible for the Earth's fossil record. Consequently, biomineralization involves biologists, chemists, and geologists in interdisciplinary studies at one of the interfaces between Earth and life."

(Leadbeater and Riding 1986)

INTRODUCTION

Biomineralization refers to the processes by which organisms form minerals. The control exerted by many organisms over mineral formation distinguishes these processes from abiotic mineralization. The latter was the primary focus of earth scientists over the last century, but the emergence of biogeochemistry and the urgency of understanding the past and future evolution of the Earth are moving biological mineralization to the forefront of various fields of science, including the earth sciences.

The growth in biogeochemistry has led to a number of new exciting research areas where the distinctions between the biological, chemical, and earth sciences disciplines melt away. Of the wonderful topics that are receiving renewed attention, the study of biomineral formation is perhaps the most fascinating. Truly at the interface of earth and life, biomineralization is a discipline that is certain to see major advancements as a new generation of scientists brings cross-disciplinary training and new experimental and computational methods to the most daunting problems. It is, however, by no means a new field. The first book on biomineralization was published in 1924 in German by W.J. Schmidt (Schmidt 1924), and the subject has continued to intrigue a dedicated community of scientists for many years. Until the early 1980s the field was known as "calcification," reflecting the predominance of biologically formed calcium-containing minerals. As more and more biogenic minerals were discovered that contained other cations, the field became known as "biomineralization." An invaluable knowledge base has been established in a literature that is found at virtually every call number in the scientific library. With styles as varied as the biominerals themselves, the initiate will find that a number of authors have extensively assessed the state of knowledge in texts (e.g., Lowenstam and Weiner 1989; Simkiss and Wilbur 1989; Mann 2001) and specialized reviews (e.g., Westbroek 1983; Leadbeater and Riding 1986; Crick 1989; Bäuerlein 2000).

With this RiMG review, the foremost goal of the authors is to create a volume focused upon topics essential for scientists, and in particular earth scientists, entering the field. With this objective in mind, our approach is first to establish relevant aspects of molecular biology and protein chemistry (e.g., Evans 2003) and then the thermodynamic principles necessary for mineralization to occur (e.g., De Yoreo and Vekilov 2003). Equipped with this toolbox of essentials and Nature as our guide, we examine in some detail six major biomineralization processes. As much as possible, we attempt to de-emphasize specifics unique to some organisms and, instead focus on the major mineralizing strategies. Fundamental to this approach is an evolutionary perspective of the field. It is noteworthy that underlying mechanisms for controlling the biomineralization processes appear to be used again and again by members of many phyla.

The last part of this volume examines views of how biomineralization processes have been employed by organisms over Earth history and their intertwined relations to earth environments. By considering the impacts of these relations upon global biogeochemical cycles, studying the temporal and geochemical results of mineral-sequestering activities, and asking critical questions about the topic, we can arrive at a deeper understanding of biological mineralization and of life on earth.

This chapter

This introductory chapter is divided into two main sections: an overview of basic biomineralization strategies and processes and a discussion of how understanding more about biomineralization mechanisms may shed light on the manner in which environmental signals may or may not be embedded in the minerals produced by organisms.

Despite the fact that the hallmark of biomineralization is the control that organisms exert over the mineralization process, it has been noted by earth scientists for the last 50 years that biologically produced minerals often contain embedded within their compositions, signatures that reflect the external environment in which the animal lived. Thus many geochemists have focused on extracting the signal for past seawater temperatures, salinities, productivities, extent of sea water saturation, and more. The task has not been an easy one! In many cases, the control processes either completely eliminate the signals or shift them. Sorting out this so-called vital effect (Urey 1951) or physiological effect (Epstein et al. 1951) from the environmental signal remains a problem yet to be solved. In the second part of this chapter, we will discuss some of the principles of mineralization in terms of the vital effect—a subject of much current interest to the earth sciences community.

OVERVIEW OF BIOMINERALS

Over the last 3500 Myr or so, first prokaryotes and then eukaryotes developed the ability to form minerals. At the end of the Precambrian, and in particular at the base of the Cambrian some 540 Myr ago, organisms from many different phyla evolved the ability to form many of the 64 different minerals known to date (e.g., Knoll 2003). While the names and corresponding chemical compositions of minerals produced by organisms are given in Table 1 (Weiner and Addadi 2002), this list is unlikely to be complete, as new biologically produced minerals continue to be discovered.

What is a biomineral?

The term biomineral refers not only to a mineral produced by organisms, but also to the fact that almost all of these mineralized products are composite materials comprised of both mineral and organic components. Furthermore, having formed under controlled conditions, biomineral phases often have properties such as shape, size, crystallinity, isotopic and trace element compositions quite unlike its inorganically formed counterpart. The term "biomineral" reflects all this complexity. Figure 1 illustrates this by comparing part of a single calcite crystal formed by an echinoderm to synthetic single crystals of calcite.

Some comments on the major groups of biominerals

As indicated by Table 1, calcium is the cation of choice for most organisms. The calcium-bearing minerals comprise about 50% of known biominerals (Lowenstam and Weiner 1989). This comes as no surprise because calcium fulfills many fundamental functions in cellular metabolism (Lowenstam and Margulis 1980; Simkiss and Wilbur 1989; Berridge et al. 1998). This dominance of calcium-bearing minerals is what led to the widespread usage of the term calcification. However, reader beware! The term refers to the formation of calcium-containing phosphate, carbonate, oxalate and other mineral types. Table 1 also shows that about 25% of the biominerals are amorphous in that they do not diffract X-rays. Amorphous silica is commonly formed by organisms and has been investigated extensively (see Perry 2003). Another well studied biogenic amorphous mineral, is the granules of amorphous hydrous iron phosphate deposited as granules in the skin of the holothurian, *Molpadia* (Lowenstam and Rossman 1975), shown in Figure 2. Among these amorphous minerals are those that have the same chemical composition, but differ by degree of short range order (Addadi et al. 2003). Many of these have been discovered only in the last few years; hence they present a fascinating new area of research.

The calcium carbonate minerals are the most abundant biogenic minerals, both in terms of the quantities produced and their widespread distribution among many different taxa (Lowenstam and Weiner 1989). Of the eight known polymorphs of calcium carbonate, seven are crystalline and one is amorphous. Three of the polymorphs—calcite, aragonite and vaterite—are pure calcium carbonate, while two—monohydrocalcite and the stable forms of amorphous calcium carbonate—contain one water molecule per calcium carbonate (Addadi et al. 2003). Surprisingly, the transient forms of amorphous calcium carbonate do not contain water (Addadi et al. 2003). One of the major challenges in the field of biomineralization is to understand the mechanism(s) by which biological systems determine which polymorph will precipitate. This is genetically controlled and is almost always achieved with 100% fidelity.

Phosphates comprise about 25% of the biogenic mineral types. Except for struvite and brushite, most phosphate minerals are produced by controlled mineralization (see subsequent section). The most abundantly produced phosphate mineral is carbonated hydroxyapatite, also called dahllite (Lowenstam and Weiner 1989). It is the mineral present in vertebrate bones and teeth, as well as in the shells of inarticulate brachiopods. Note that the non-carbonated member of this family, hydroxyapatite, is not known to be formed biologically. Biogenic carbonate apatite crystals are usually plate-shaped and are exceedingly small (2–4 nm thick and some tens of nanometers long and wide; Weiner and Price 1986). It is interesting to note that synthetic carbonated hydroxyapatites precipitated under conditions similar to those found in vertebrate physiology are also plate-shaped and small (Moradian-Oldak et al. 1990). Therefore it is not the biological environment in which they form that gives them the plate shape which is most unusual for a mineral that crystallizes in the hexagonal crystallographic system. The fact that the biologically formed crystals are so small generally indicates that they are also rather unstable; hence they are typically more soluble than hydroxyapatite (Stumm 1992). They are also very difficult to characterize structurally because their high surface/volume results in many atoms being perturbed. With new spectroscopic applications designed to overcome the problems of working with small particles and the difficulties associated with hydrous phases (Waychunas 2001), current views of phosphate biomineral compositions and structures could still undergo significant revision.

Table 1. The names and chemical compositions of minerals produced by biologically induced and controlled mineralization processes

Name	Formula
Carbonates	
Calcite	$CaCO_3$
Mg-calcite	$(Mg_xCa_{1-x})CO_3$
Aragonite	$CaCO_3$
Vaterite	$CaCO_3$
Monohydrocalcite	$CaCO_3 \cdot H_2O$
Protodolomite	$CaMg(CO_3)_2$
Hydrocerussite	$Pb_3(CO_3)_2(OH)_2$
Amorphous Calcium Carbonate (at least 5 forms)	$CaCO_3 \cdot H_2O$ or $CaCO_3$
Phosphates	
Octacalcium phosphate	$Ca_8H_2(PO_4)_6$
Brushite	$CaHPO_4 \cdot 2H_2O$
Francolite	$Ca_{10}(PO_4)_6F_2$
Carbonated-hydroxylapatite (dahllite)	$Ca_5(PO_4,CO_3)_3(OH)$
Whitlockite	$Ca_{18}H_2(Mg,Fe)_2^{+2}(PO_4)_{14}$
Struvite	$Mg(NH_4)(PO_4) \cdot 6H_2O$
Vivianite	$Fe_3^{+2}(PO_4)_2 \cdot 8H_2O$
Amorphous Calcium Phosphate (at least 6 forms)	variable
Amorphous Calcium Pyrophosphate	$Ca_2P_2O_7 \cdot 2H_2O$
Sulfates	
Gypsum	$CaSO_4 \cdot 2H_2O$
Barite	$BaSO_4$
Celestite	$SrSO_4$
Jarosite	$KFe_3^{+3}(SO_4)_2(OH)_6$
Sulfides	
Pyrite	FeS_2
Hydrotroilite	$FeS \cdot nH_2O$
Sphalerite	ZnS
Wurtzite	ZnS
Galena	PbS
Greigite	Fe_3S_4
Mackinawite	$(Fe,Ni)_9S_8$
Amorphous Pyrrhotite	$Fe_{1-x}S$ $(x = 0-0.17)$
Acanthite	Ag_2S
Arsenates	
Orpiment	As_2S_3
Hydrated Silica	
Amorphous Silica	$SiO_2 \cdot nH_2O$
Chlorides	
Atacamite	$Cu_2Cl(OH)_3$
Fluorides	
Fluorite	CaF_2
Hieratite	K_2SiF_6
Metals	
Sulfur	S

Table 1 continued.

Name	Formula
Oxides	
Magnetite	Fe_3O_4
Amorphous Ilmenite	$Fe^{+2}TiO_3$
Amorphous Iron Oxide	Fe_2O_3
Amorphous Manganese Oxide	Mn_3O_4
Hydroxides & Hydrous Oxides	
Goethite	α-FeOOH
Lepidocrocite	γ-FeOOH
Ferrihydrite	$5Fe_2O_3 \cdot 9H_2O$
Todorokite	$(Mn^{+2}CaMg)Mn_3^{+4}O_7 \cdot H_2O$
Birnessite	$Na_4Mn_{14}O_{27} \cdot 9H_2O$
Organic Crystals*	
Earlandite	$Ca_3(C_6H_5O_2)_2 \cdot 4H_2O$
Whewellite	$CaC_2O_4 \cdot H_2O$
Weddelite	$CaC_2O_4 \cdot (2+X)H_2O$ ($X<0.5$)
Glushinskite	$MgC_2O_4 \cdot 4H_2O$
Manganese Oxalate (unnamed)	$Mn_2C_2O_4 \cdot 2H_2O$
Sodium urate	$C_5H_3N_4NaO_3$
Uric Acid	$C_5H_4N_4O_3$
Ca tartrate	$C_4H_4CaO_6$
Ca malate	$C_4H_4CaO_5$
Paraffin Hydrocarbon	
Guanine	$C_5H_3(NH_2)N_4O$

* by the convention of Lowenstam & Weiner (1989)
References: Lowenstam & Weiner (1989), Simkiss & Wilbur (1989), Mann (2001), Weiner and Addadi (2002)

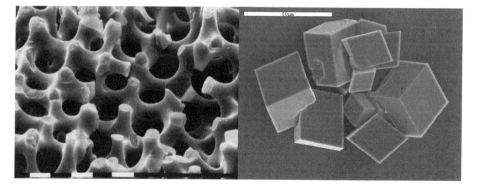

Figure 1. Comparison of calcite single crystals: (*left*) stereom of echinoderm and (*right*) synthetically produced rhombohedral forms.

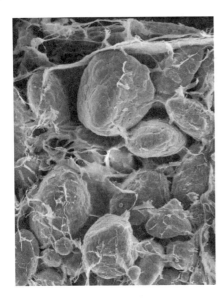

Figure 2. Granules of amorphous hydrous iron phosphate deposited in the skin of the holothurian, *Molpadia*. (Micrograph from the collection of the late H.A. Lowenstam). Diameter of largest granule about 200 microns.

It is noteworthy that each of the mineral classes includes one or more phases that contain water and/or hydroxyl groups (Table 1). These hydrated forms comprise about 60% of the biogenic minerals. In particular, all of the silica biominerals reported to date are hydrated (e.g., Lowenstam and Weiner 1989), and there is mounting evidence that many crystalline carbonate biominerals first form as hydrated phases (Beniash et al. 1997, 1999; Weiss et al. 2002; Addadi et al. 2003). The abundance of hydrated biominerals is no accident. Hydrated phases are favored over anhydrated counterparts by significantly lowering energetic barriers to nucleation and growth from aqueous solution (e.g., Stumm 1992). Organisms are metabolic misers; thus they use the Ostwald-Lussac rule to their advantage by favoring the precipitation of the lowest energy phases (Nancollas 1982).

The iron biominerals are not readily evaluated by mineral class because they have significant occurrences as oxides, hydroxides, and sulfides (Bazylinski and Moskovitz 1997; Konhauser 1997). Some iron sulfate and phosphate minerals are also reported (Konhauser 1998). When one considers the abundance of iron in the biosphere (particularly in the early earth) versus the very low solubility of most iron-bearing minerals, one expects this diversity and abundance. The iron biominerals are of particular significance because they comprise approximately 40% of all minerals formed by organisms (Lowenstam 1986; Bazylinski and Frankel 2003), and magnetite formation is believed to be the most ancient matrix-mediated biomineralizing system. As such, Kirschvink and Hagadorn (2000) have suggested that magnetic biominerals, and iron minerals in general, could contain clues to other aspects of controlled mineralization.

Table 1 also contains a group of so-called "organic" minerals. Despite the contradiction in terms, these are crystalline phases formed by organisms probably by the same underlying strategies used for "normal" mineral formation. We suspect that many of these minerals remain to be discovered, and exploring the functions they perform will be fascinating. Even DNA is known to transform into crystalline phases in bacteria subjected to stress (Minsky et al. 2002). The dense packing of the crystal provides protection from damage during periods of dormancy when much of the metabolic processes are nearly shut down.

Unique character of minerals formed by biological systems

Biominerals meet the criteria for being true minerals, but they can also possess other characteristics that distinguish them from their inorganically produced counterparts. The most obvious trait is that biogenic minerals have unusual external morphologies. It is perhaps the intricacy and diversity of bio-originated structures (e.g., Figure 3) that first attracts mineralogically-inclined persons into the field of biomineralization. For example, beautiful single crystals that express an unusual set of faces are found in magnetite (see Bazylinski and Frankel 2003). The SEM images presented throughout this volume capture the imagination by illustrating the astonishing ability of Nature to grow minerals of a complexity that is (for now) an impossible task for humankind to replicate! Implicit in this complexity is the intriguing ability of organisms to impose a "handedness" upon the external morphology of biominerals. Good examples are found from the microscopic (magnetosomes and coccoliths) to macroscopic (Nautilus) organisms. The question of how organisms use molecules to transfer chiral information to crystalline surfaces and thus induce asymmetrical biomineral structures is an on going one (e.g., Addadi and Weiner 2001; Orme 2001; De Yoreo and Vekilov 2003). Remarkably, genetic regulation runs the controlled-mineralization "program" to form biominerals with unusual morphologies using great fidelity again and again.

A second characteristic of biominerals is that many are actually composites or agglomerations of crystals separated by organic material. In many organisms, they exist as small bodies distributed within a complex framework of macromolecular frameworks such as collagen or chitin (Addadi et al. 2003). X-ray diffraction studies have led to an acceptance that the crystalline biominerals are typically single crystals. There are certainly examples of this, but higher resolution studies are showing this is not always the case (Wilt 1999). Rather, some "single crystal" biominerals are actually a mosaic of domains delimited by organic layers or they may contain significant occlusions of proteinaceous material (Wilt 2002). Yet, they exhibit many of the same diffraction properties as those of a single crystal (Simkiss 1986). As biominerals

Figure 3. Calcitic spicule formed by the ascidian *Bathypera ovoidea*. Length about 120 microns. (Micrograph from the collection of the late H.A. Lowenstam).

become increasingly characterized, a better understanding of both short-range and long-range structures and assembly of biominerals could change our interpretations of mineralization mechanisms.

BIOMINERALIZATION CONCEPTS

For nucleation and growth to occur, biomineral formation requires a localized zone that achieves and maintains a sufficient supersaturation. In most biological systems, the site of mineral deposition is isolated from the environment by a physical delimiting geometry. The actual size of that site or volume is sometimes ambiguous, but it is generally agreed that this region must limit diffusion into/out of the system or utilize a type of compartment. The extent of this isolation can be passive and minimal such as observed when bacteria cluster to form an intercellular zone that is diffusion-limited. At the other extreme, intracellular vesicles create compartmentalized environments where compositions can be precisely regulated. This compartment must be capable of modifying the activity of at least one biomineral constituent (usually the cation) as well as protons and possibly other ions. Any fluxes in ion chemistry must meet one constraint: the fluid must maintain electroneutrality. Ion supply (or removal) occurs by two means: Active pumping associated with organelles near the sites of mineralization or passive diffusion gradients. As will be shown throughout this volume, organisms use a great variety of anatomical arrangements to facilitate ion movement.

The chemical compositions of *in vivo* fluids at and adjacent to sites of biomineral formation have direct relevance to our understanding of the mineralization processes and of the consequent degree of control exerted by the organism on its internal environment. Later in this volume, we will be reminded that these fluid compositions reflect, to variable degrees, the marine roots of organisms in evolutionary history (e.g., Knoll 2003). Surprisingly, however, little information has been published concerning fluids for most organisms. Table 2 compares the major solute compositions of seawater and an average freshwater to the typical compositions of several biological fluids. One sees that the major constituents in the extrapallial fluid of marine mollusks reflect the higher salinity of its saltwater environment. (Extrapallial fluids are found in the space between the secretory epithelium of the mantle and the growth surface of the nacreous aragonite biomineral. See illustration in Zaremba et al. 1996). In contrast, extrapallial fluids of freshwater mollusks have much higher solute concentrations than their low salinity environment. A comparison of the two organisms reveals very different fluid compositions, yet both form aragonitic shells!

Table 2 reminds us that organisms possess fluids with significant ionic strengths. This means that mineralization studies must address the solution chemistry in terms of ion activity, not concentration. Careful use of activity coefficient models is necessary to estimate the supersaturation of growth environments. For example, calcium ion in a blood plasma with the physiological ionic strength of 0.15 molal (see Table 2) has an activity coefficient of approximately 0.3, compared to a value of 1.0 in an infinitely dilute solution (e.g., Langmuir 1997). In high salinity fluids associated with marine environments, the correction to activity becomes greater (and sometimes more ambiguous). In addition to affecting supersaturation, ionic strength mediates the charges of precursor molecules, thus affecting the stabilization of colloids and amorphous gels (e.g., Iler 1979; Perry 2003).

Investigations of cellular processes have led to general insights regarding cation concentrations in the tissues of higher organisms (da Silva and Williams 1991). In biology calcium is very highly controlled at levels in the range of 10^{-8} to 10^{-6} M, owing

Table 2. Summary of ion compositions in natural waters and biological fluids.

	Concentration (mmol/kg)							Concentration (µmol/kg)							(mmol/kg)	
	Na^+	K^+	Ca^{2+}	Mg^{2+}	HCO_3^-	Cl^-	SO_4^{2-}	Sr^{2+}	Li^+	Cu^{2+}	Zn^{2+}	Fe_T	Mn	$(PO_4)_T$	pH	I.S.
Seawater[1]	479	10.4	10.5	54.3	2.0	558	28.9	88-92	26-27	na	na	na	na	<0.1-3[1]	8.1	700
Marine Algae[3] (*Valonia*)	80	400	1.5	50	(10)?			na	na	na	na	na	na	5 (HPO_4^{2-})	na	na
Marine mollusk[4] (*Pinctada fucata*)	431	12.6	9.7	50.7	3.6	524	27.9	89	29	2.9	176	22.4	4.4	502	7.4	664
Freshwater[2]	0.31	0.04	0.37	0.15	0.87	0.22	0.12	na	na	na	na	na	na	na	4-7	
Freshwater mollusk[4] (*Hyriopsis schlegeli*)	22.1	0.57	4.1	0.63	10.5	15.0	5.2	4.3	na	5.2	22.3	56.4	54.2	41.1	8.15	44
Red blood cells[3]	11	92	10^{-4}	2.5	(10)?	50		na	na	na	na	na	na	3 (HPO_4^{2-})	na	94-106
Blood plasma[3]	160	10	2	2	30	100		na	na	na	na	na	na	3 (HPO_4^{2-})	7.4	158

Notes:
1. For a salinity of 35 parts per thousand, 25°C, P = 1 atm (Berner and Berner 1996)
2. Average "actual" world river composition. (Berner and Berner 1996)
3. da Silva and Williams (1991)
4. Wada and Fujinuki (1974)

to its key roles in signaling and metabolic processes. Similarly, free magnesium ion is reported to be approximately 10^{-3} M in all compartments except some vesicles. Incidentally, Simkiss (1986) noted the essential role of magnesium ion in mineralization processes because of its role in stabilizing carbonate and phosphate crystal formation. Other elements such as zinc are found at concentrations as low as 10^{-9} in the cytoplasm, while the free ion has concentrations as high as 10^{-3} M in some vesicles. In contrast, manganese has concentrations of approximately 10^{-8} M almost everywhere, whether in or out of the cell, but may be $<< 10^{-8}$ M in eukaryotic cells.

BASIC PROCESSES OF BIOMINERALIZATION

Biomineralization processes are divided into two fundamentally different groups based upon their degree of biological control. Lowenstam (1981) introduced these as "biologically induced" and "organic matrix-mediated," with the latter generalized by Mann (1983) to "biologically controlled" mineralization. Recognizing that the detailed processes of biomineralization within this convention are as varied as the organisms themselves, this section outlines basic mineralization strategies to help the reader place the information presented in subsequent chapters into a mechanistic framework. We emphasize "location, location" to show: 1) how/where biomineral constituents can be concentrated as ions or solid phases; 2) types of translocation that can occur; and 3) resting places and transformation of the end-products. Using this approach, we can furher define the variable degrees of biological control. It is the specific nature and degree of control that is central to understanding the extent of biological control of the elemental compositions of biominerals. Likewise, transport mechanisms and hydration environments are certain to also affect minor element chemistry. These are the roots of deciphering the vital effect.

Biologically induced mineralization

The secondary precipitation of minerals that occurs as a result of interactions between biological activity and the environment is termed "biologically induced" mineralization. In this situation, cell surfaces often act as causative agents for nucleation and subsequent mineral growth. The biological system has little control over the type and habit of minerals deposited, although the metabolic processes employed by the organism within its particular redox environment mediate pH, pCO_2 and the compositions of secretion products (e.g., McConnaughey 1989a; Fortin et al. 1997; Tebo et al. 1997; Frankel and Bazylinski and Frankel 2003). These chemical conditions favor particular mineral types in an indirect way (Figure 4). In some cases, biological surfaces are important in the induction stage because nucleation often occurs directly on the cell wall, and the resulting biominerals can remain firmly attached. In open waters, this epicellular mineralization can lead to encrustation so complete that gravitation overcomes buoyancy, and they settle through the water column. The sediment record attests to the extensive occurrence of this phenomenon (e.g., Knoll 2003; Van Cappellen 2003).

Heterogeneity is the hallmark of biologically induced minerals. Frankel and Bazylinski (2003) show that the compositions of minerals resulting from induced processes vary as greatly as the environments in which they form. This heterogeneity includes variable external morphology (typically poorly defined), water content, trace/minor element compositions, structure and particle size. Since these characteristics also typify inorganically precipitated minerals, unambiguous interpretations of the sediment and rock record continue to confound interpretations of earth environments.

Biologically controlled mineralization

In "biologically controlled" mineralization, the organism uses cellular activities to

Overview of Biomineralization Processes & Vital Effect Problem

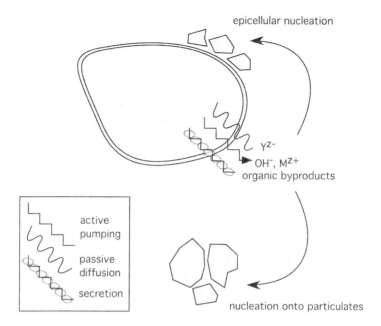

Figure 4. Schematic of biologically induced mineralization. Mineral precipitates form as a result of metabolic activities that affect pH, pCO_2, and secretion products. The cell is a causative agent only, without control over mineral type or habit.

direct the nucleation, growth, morphology and final location of the mineral that is deposited. While the degree of control varies across species, almost all controlled mineralization processes occur in an isolated environment. The results can be remarkably sophisticated, species-specific products that give the organism specialized biological functions.

Biologically controlled mineralization processes can be described as occurring extra-, inter- or intracellularly. These distinctions refer to the locations of the mineralization site with reference to the cells responsible for mineralization. However, not all mineralization processes can be classified in this simple manner. In some cases, mineral formation begins within the cell and then proceeds outside the cell. Identifying what are in essence, end members at least helps us understand the overall complexity. We will therefore discuss extra-, inter- and intracellular biomineralization separately.

Biologically controlled extracellular mineralization. In extracellular mineralization, the cell produces a macromolecular matrix outside the cell in an area that will become the site of mineralization. The term matrix refers to a group of macromolecules comprised of proteins, polysaccharides or glycoproteins that assemble to form a three-dimensional framework. The matrix composition is unique in that many of its proteins contain a high proportion of acidic amino acids (especially aspartate) and phosphorylated groups (Veis and Perry 1967; Weiner 1979; Weiner et al. 1983a,b; Swift and Wheeler 1992). The structures and compositions of these organic frameworks are genetically programmed to perform essential regulating and/or organizing functions that will result in the formation of composite biominerals.

There are two means by which the cell can transfer constituents to the matrix. In the first, the cell may actively pump cations through the membrane and into the surrounding region (e.g., Simkiss 1986). Once out of the cell, the supersaturation level of the fluid is established and maintained by ion diffusion over relatively large distances to the organic matrix (Figure 5a). In the second approach, the cations may be concentrated within the cell into cation-loaded vesicles, exported through the membrane and later broken down by precursor compounds at the organic matrix (Figure 5b). The latter mechanism is used for cartilage mineralization in the epiphyseal growth plate (Ali 1983). It is also probably used by sea urchin larvae to introduce amorphous calcium carbonate into spicule-forming vesicles, where it subsequently crystallizes into calcite (Beniash et al. 1999). Anion movement is typically the result of passive diffusion in response to the electroneutrality requirement and is ultimately driven by pH gradients created during cation transport (Simkiss 1976; McConnaughey 1989b). In both approaches, the cell works actively to supply cations to an external organic matrix for "on-site" nucleation and growth. This is distinguished from the epicellular nucleation and growth that occur during biologically induced mineralization.

Almost all structures that form by extracellular processes develop upon a pre-formed matrix derived from secretory products of multicellular epithelial tissues. Watabe and Kingsley (1989) suggest that these tissues have additional significance as extensive

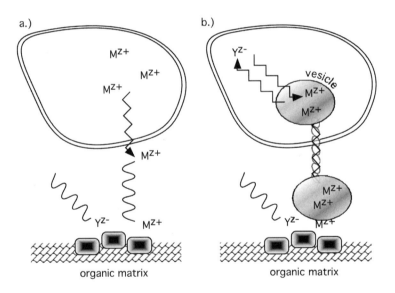

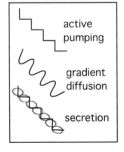

Figure 5. Illustrations of biologically controlled extracellular mineralization showing that this process is distinguished by nucleation outside of the cell. a.) Cations are pumped across the cell membrane and move by passive diffusion through extracellular fluids to the site of mineralization. b.) Cations are concentrated intracellularly as aqueous ions into a vesicle that is subsequently secreted. Compartment breakdown at site of mineralization releases cations for biomineral formation.

substrates that facilitate the formation of massive skeletons. Examples of organisms believed to mineralize primarily by extracellular biologically controlled processes include the external tests of certain foraminifera (Towe and Cifelli 1967; Zeebe and Sanyal 2002; Erez 2003), cephalopod statoliths (Bettencourt and Guerra 2000), shells of mollusks (Gregoire et al 1955; Crenshaw 1980; Weiner and Traub 1980, 1984; Falini et al. 1996; Pereira-Mouries et al. 2002; Weiss et al. 2002; Gotliv et al. 2003); exoskeletons of bryozoans (Rucker and Carver 1969), scleractinian corals (Constantz 1986; Constantz and Meike 1989), bones and teeth (e.g., discussion in Lowenstam and Weiner 1989; Veis 2003).

To understand mineralization mechanisms, one must possess knowledge regarding the structures of the organic matrix components and of the entire framework. One of the most thoroughly investigated matrices in this respect is the mollusk shell nacreous layer. X-ray and electron diffraction reveal that the most ordered component is β-chitin, and the most abundant component is silk fibroin, but it shows little evidence of order based on diffraction (Weiner and Traub 1984). It has also been shown that there is a well defined spatial relation between the chitin of the framework and the associated aragonite mineral. This implies that the nucleation mechanism is most likely epitaxial (Weiner and Traub 1984). The acidic components impart a chemical activity and template structure that direct the resulting mineralization process. Not much is known about their secondary structures. Mollusk acidic matrix proteins are thought to mainly adopt the β-sheet structure (Worms and Weiner 1986). More recent structural information on the nacreous layer raises the interesting possibilities that the silk fibroin component of the matrix is a gel (Levi-Kalisman et al. 2001) (Fig. 6) and that it does not form layers on either side of the chitin, as was proposed by Weiner and Traub (1984). Several mollusk shell matrix proteins have been sequenced. Most have been purified by gel electrophoresis and are not highly acidic.

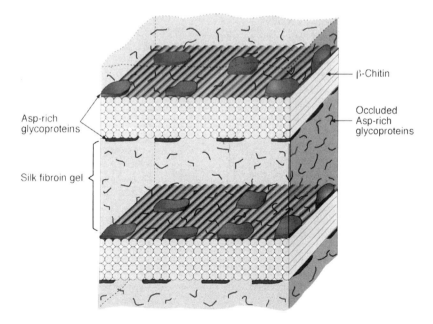

Figure 6. Model of the nacreous layer organic matrix, proposing that chitin and acidic macromolecules constitute the major framework constituents, and that the silk fibroin component forms a gel between the layers (Levi-Kalisman et al. 2001).

This contrasts with the overall amino acid composition of the soluble components in the matrix. This dichotomy may be due to the fact that only with massive and rapid fixation of the proteins in a gel after separation by electrophoresis can the highly acidic proteins be visualized (Gotliv et al. 2003). The one highly acidic protein that has been sequenced from mollusk shells has a fascinating domain-like structure that repeats itself many times (Sarashina and Endo 1998). It will be most interesting to discover its function.

Determining the functions of matrix proteins is the real bottleneck in understanding extracellular matrix mineralization. Several assays are used for demonstrating a protein's ability to modulate mineral nucleation and growth in various ways (Wheeler et al. 1981; Belcher et al. 1996). One that is reproducible and fairly specific was developed by Falini et al. (1996). It assays for aragonite-specific nucleation. It has been used to show that in mollusk nacreous layer matrices, some of the proteins are indeed able to specifically nucleate aragonite (Gotliv et al. 2003). Interestingly, in this *in vitro* assay, the aragonite is nucleated via a transient amorphous calcium carbonate (ACC) phase that also does not contain tightly-associated water molecules. It has been shown that mollusk larvae also produce their shells via an ACC transient phase (Weiss et al. 2002).

Biologically controlled intercellular mineralization. This type of mineralization is not widespread. It typically occurs in single-celled organisms that exist as a community. At first glance, intercellular mineral formation appears to be a variant of extracellular mineralization. In this case, however, the epidermis of the individual organisms serves as the primary means of isolating the site of mineralization (Mann 2001). As shown in Figure 7, the epithelial substrate reproducibly directs the nucleation and growth of

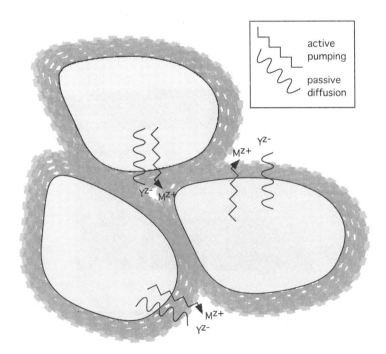

Figure 7. In biologically controlled intercellular mineralization, the epithelial surfaces of cells are used as the organic substrate for nucleation and growth with a preferred orientation. Cations are pumped out of the cell and fluid compositions are regulated to maintain control over the biomineral type and extent of growth.

specific biomineral phases over large areas of cell surfaces. Mineralization between cells can become so extensive as to completely fill the intercellular spaces, thus forming a type of exoskeleton. At first glance, this might appear to be a type of biologically induced mineralization, but studies have shown that the epidermis of individual organisms directs the polymorph and shape of the biomineral that forms. An example is found in calcareous algae that nucleate and grow calcite with a c-axis orientation that is perpendicular to the cell surface (Borowitzka et al. 1974; Borowitzka 1982).

Biologically controlled intracellular mineralization. Controlled mineralization can also occur also within specialized vesicles or vacuoles that direct the nucleation of biominerals *within* the cell. This is a widespread strategy. These compartmentalized crystallization environments govern the resulting biomineral composition and morphology. In this situation, the cell has a high degree of control upon the concentrations of cation and anion biomineral constituents in an environment where an organic matrix may also be active as a nucleating template. The compartment membrane also regulates the pH, pCO_2 and—at least to some extent—minor and trace element compositions. Indeed, the mineralized structures that develop from intracellular processes often exhibit highly intricate species-specific morphologies.

The schematics in Figure 8 show that the designation "intracellular mineralization"

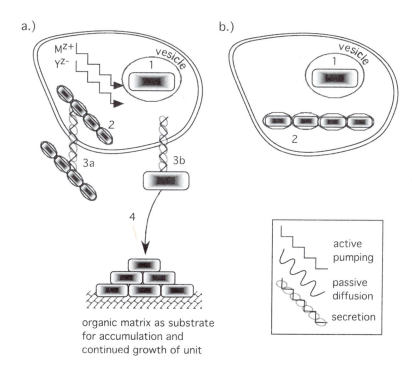

Figure 8. Schematics of biologically controlled intracellular mineralization shows that nucleation occurs within the cell in a specialized vesicle. a.) Biomineral is nucleated (1) within a compartment in the intracellular environment. These growth units may be assembled intracellularly (2) for subsequent secretion (3a) or secreted as an individual unit (3b) for subsequent organization into a higher order growth structures as a massive or organic-mineral composite. b.) In a less frequently used type of mineralization, the biomineral unit remains within the cell (1) as a single growth unit or is organized intracellularly (2) into a higher order structure.

is a broad concept that encompasses a number of fates for the initial compartment-based precipitate. Figure 8a illustrates the general scenario whereby biominerals form intracellularly before eventually becoming extracellular. These structures may leave the cell as individual units or be preassembled prior to extrusion through the membrane. An example of the latter is used by the Haptophyte algae. These organisms preassemble the single crystal segments of calcite intracellularly into a coccolith structure before passing the mature sheath through the cell membrane. Biomineral is transported through the cell membrane by two general means. First, the vesicle(s) or vacuole(s) may migrate to the membrane where the mature biomineral as an individual or pre-assembled structure is extruded by exocytosis. In an alternative process, the compartment membrane fuses with the plasma membrane, and a "premature" biomineral becomes exposed through a membrane breach (Watabe and Kingsley 1989).

Species-specific fates await the secreted crystals in the extracellular environment (Figure 8a). In a relatively straightforward usage, biominerals can be immediately employed without further growth or modification. For example, the mature coccolith structures extruded by Haptophyte algae are immediately used to enclose the cell (Brownlee et al. 1994; de Vrind-de Jong et al. 1994; de Vrind-de Jong and de Vrind 1997; Young and Henriksen 2003). Similarly, the silica mineralizing algae possess vesicles that appear to mold the amorphous silica into pre-fabricated scales prior to release (Watabe and Kingsley 1989).

The release of vesicular biominerals can also mark the beginning of a secondary assembly process whereby the biominerals interact with an extracellular organic matrix (Figure 8a) to become an ordered extracellular structure. Some miliolid foraminifera form bundles composed of an array of oriented crystals. Each crystal is enveloped by organic material, then bundles are passed through the cell membrane by exocytosis. Other miliolids accrete pre-formed crystals and matrix materials onto extracellular surfaces to form a loosely packed wall structure in a "stack-of-bricks" process (e.g., Berthold 1976; Hemleben, et al. 1986).

Echinoderms are perhaps the most surprising exploiters of an intracellular strategy to form huge mineralized products, some of which are centimeters long. These form within a vesicle that is the product of many cells fusing their membranes. The mineral is exposed to the environment only if and when the membrane is degraded (Markel 1986). Although the spicular structures of sea urchin larvae begin as amorphous calcium carbonate, the final products are typically single crystals. These biominerals often express a preferred crystallographic orientation and have smooth curved surfaces (Beniash et al. 1997; Beniash 1999; Wilt 2002).

Not all mineralized structures of intracellular origin become extracellular. Figure 8b shows that the biomineral-containing vesicles may form intracellularly and subsequently stay within the cell. While this mineralization process is not widely used by many phyla, for some organisms, it is a unique and important strategy. Perhaps the best example is exhibited by the magnetosome-producing bacteria. These structures are membrane-bound, euhedral crystals of magnetite or greigite that assemble via their magnetic fields to produce biomineral "chains" (Bazylinski 1996; Schüler and Frankel 1999; Bazylinski and Frankel 2003). Intracellular mineralization is also found in the radiolarians. These organisms form siliceous spicules within cytoplasmic extensions via numerous small vesicles that stream up to the cytoplasmic compartment (Simpson and Volcani 1981; Simkiss 1986). Plants also use intracellular mineralization processes to deposit calcium as oxalate crystals (Raven et al. 1982; Webb 1999; Skinner and Jahren 2003). Figure 9 illustrates an example of these unusually-shaped crystals of calcium oxalate monohydrate (whewellite) produced by the plant *Psychotria*.

Figure 9. Calcium oxalate monohydrate (whewellite) crystals produced by the plant *Psychotria.* Scale bar 10 microns. (From the collection of the late H.A. Lowenstam).

The walls, or frustrules, of diatoms are also the result of intracellular mineralization. Using a combination of two intracellular processes, diatoms first polymerize silica into specialized vesicles on an organic matrix of polypeptides (Swift and Wheeler 1991, 1992; Kroger et al. 1999, 2002). Silica is subsequently moved to the interior surface of the cell membrane to form an intracellular frame (Garrone et al. 1981). Matrix controls on frustrule patterning are discussed elsewhere (Noll et al. 2002; Pohnert 2002; Sumper 2002; Vrieling et al. 2002). For more details, see Perry (2003).

Comments on biomineralization processes

The framework of processes outlined above is useful to represent end members, but one must realize that as we examine the evolution and biochemistry of mineralization processes in this volume, these simple representations will be developed into a near-continuum of different strategies with considerable complexity. Part of this complexity lies in how organisms control their local environment to make a morphologically and functionally diverse group of minerals through changes in local chemistry and different shaping strategies. However, an additional question for earth scientists involves how biological mineralization processes determine the compositional signatures contained in biogenic minerals. With the increasing reliance upon compositional proxies as indicators of past environmental conditions, this is a problem that needs to be addressed. It is discussed in some detail below.

VITAL EFFECTS

The preservation of biogenic minerals in the fossil record offers wonderful opportunities for reconstructing the ancient environments in which organisms lived, and also for learning more about the evolution of their metabolism and physiology (reviewed in Chapter 11 of Lowenstam and Weiner 1989). What exactly can be learned, however, also depends upon how much we understand about the biomineralization processes involved in skeleton formation and the states of preservation of the fossil material being analyzed. Much attention has been paid to the latter when it comes to fossils extracted from strata now on the continent, but surprisingly little attention seems to be paid to the preservation states of shells of marine organisms in deep-sea sediments. We will focus, however, on the many open questions that concern the signals embedded in biogenic minerals and in particular questions related to the biological processes overriding the environmental signals—the so-called vital effect (Urey et al. 1951).

The first quantitative use of biogenic minerals for extracting information about the environment of formation was the demonstration that the stable oxygen isotopic composition of some taxa (mollusks in particular) reflects in part the temperature of the water in which the mollusks lived (Urey et al. 1951). Epstein et al. (1951) noted however that not all biogenic minerals are deposited in isotopic equilibrium and that "the presence of a physiological effect in the case of certain groups of animals such as the echinoderms and corals, and plants such as a coralline algae, has seemed probable …." (p. 424). This so-called "physiological effect" in time came to be referred to by the geochemical community as a "vital effect." Furthermore, because most geochemists investigated biogenic materials to reconstruct the paleoenvironments in which animals lived, the vital effect, just like post-burial diagenetic effects, represented a severe hindrance to their goals.

To this day, the most common approach used to resolve the problem involves identifying the taxa that faithfully record environmental parameters while avoiding the others. This empirical approach has the obvious advantage that while one uses these apparently "reliable" organisms, there is no need to worry about the other organisms with their associated complications. The problem is that in the absence of a deep understanding of the vital effects, it is well nigh impossible to know when the recording is really faithful, particularly when differentiating between equilibrium and small non-equilibrium effects. This problem has become acute in the last decade with the realization that the climate is changing and that one of our most promising means of assessing the significance of this change is to understand the fine details of past climatic changes. Understanding vital effects is therefore of considerable current importance.

At this point, we do not have the answers, and a perusal of this book describing the current state of the art in the field of biomineralization will probably not provide many answers. This is not because the subject matter is too difficult to resolve, but because the interests of most of the scientists investigating the field of biomineralization are elsewhere. Very few members of the geochemical community have made this their major interest, but those who have made very significant contributions (e.g., Erez 1978; McConnaughey 1989). A much more concerted effort, however, is needed.

Two basic categories of vital effects

Kinetic effects. McCrea (1950) investigated disequilibrium effects in an *in vitro* system in which calcium carbonate was precipitated under varying conditions. The greatest departures from equilibrium were found when the precipitates formed rapidly at temperatures below 15°C or above 60°C. Epstein et al. (1953) identified this potential problem in biogenic calcium carbonate when they noted that the calcium carbonate deposited by abalones, in an attempt to fill up holes that had been drilled in their shells, was not in oxygen isotopic equilibrium with the water in which the abalones lived (see also Epstein and Lowenstam 1954). Although the effect was small, it was clearly identifiable. They attributed it to the rapidity with which the material was laid down. Since then many different observations of diverse organisms have been reported in which mineral laid down rapidly appears to be out of isotopic equilibrium (Weber and Woodhead 1970; Land et al. 1975; Erez 1978; McConnaughey 1989a; Ziveri et al. 2003).

The mechanisms responsible for this may be due to varying kinetics of the uptake, transport and/or deposition stages of mineral formation. Clearly the issue is complex.

Taxonomic effects. Urey et al. (1951) also recognized that certain phyla may not deposit their skeletal material in isotopic equilibrium with the environment, and within a few years it was clearly demonstrated that some phyla completely control their isotopic compositions (Craig 1953) , as well as their Mg and Sr skeletal contents (Chave 1954; Blackmon and Todd 1959; Lowenstam 1963). Perhaps the best example is the Echinodermata (Weber and Raup 1966), although it has been noted in the echinoderms as well as other taxa that superimposed on the vital effect are variations that do reflect environmental changes (Weber and Raup 1966). At the same time other phyla were declared equilibrium depositers and hence useful for paleoenvironmental reconstructions. These were mainly the mollusks, the brachiopods and the planktonic foraminifera. It has subsequently been found in these phyla as well, that the situation is more complicated, with some members of the phylum being faithful recorders, and others not (Duplessey et al. 1970; Carpenter and Lohmann 1995; Letizia et al. 1997). Thus the concept of good and bad species was introduced. This designation, however, only applies to a given proxy. Many marine mollusks and planktonic foraminifera deposit their shells in isotopic equilibrium with the environment (Epstein et al. 1953; Erez and Luz 1983), but most do not have Sr or Mg contents that are in equilibrium with the environmental water in which they live (Elderfield et al. 1996). The mechanisms involved are therefore obviously different for different shell properties, and obtaining an in-depth understanding of this complicated phenomenon is by no means simple. In fact at this juncture, the key question to be addressed is probably how equilibrium is obtained, rather than how is non-equilibrium obtained? The latter is the rule, and the former is the exception to the rule.

Equilibrium with the environment—the real puzzle

It has been widely noted that in almost all organisms that control mineral formation, and the mollusks are no exception, the mineralization process takes place in isolated compartments within cells or between cells (Lowenstam and Weiner 1989; Simkiss and Wilbur 1989). Furthermore, all the different cells responsible for mineralization have to transport the raw materials to the site of deposition, which may be remote from the ultimate source of the ions, namely the environment. Often the ions are stored temporarily in membrane-bound vesicles and then re-dissolved. From this perspective, it is not surprising that equilibrium can actually be maintained between the shell and the environment. If we could understand how this is achieved, we would probably understand much more about the reasons for non-equilibrium.

In this section, we will try to analyze this issue, focusing more on the formation of the mineral phase itself rather than on the uptake, transport and temporary storage processes. See Erez (2003) for details on these important aspects, as well as for a discussion of the vital effect in this context.

Over the last two decades, the problem has become more compelling as more and more properties of biogenic minerals are being used as proxies for different aspects of the paleoenvironment. In the 1950s, the pioneering work was done on oxygen and carbon stable isotopic compositions, and soon afterwards the variations in minor elements (particularly strontium and magnesium contents) were also measured (Chave 1954; Lowenstam 1964). Today more than 10 proxies are known for biogenic minerals that are purported to monitor many different environmental parameters in addition to temperature. Each proxy may have its own vital effect story. We will approach the subject by discussing aspects of the biomineralization process in relation to possible vital effect mechanisms for different proxies.

Carbonic anhydrase. This enzyme essentially catalyses the hydration and dehydration of CO_2(aq), which in the absence of the enzyme is a rather slow reaction (Simkiss and Wilbur 1989). If it is present at the site of deposition and is working efficiently, it should not only overcome the slow hydration of aqueous CO_2, but also eliminate any kinetic isotopic fractionation at least with respect to oxygen (McConnaughey 1989a). Carbonic anhydrase is widely distributed (e.g., it is present in corals; Ip et al. 1991) and in the gland responsible for eggshell formation in birds (Gay and Mueller 1973). It is known indirectly to be involved in mineralization as an inhibitor of the enzyme to reduce the rate of mineralization (Goreau 1959). It has been identified in the membrane of the spicule forming vacuole in the gorgonian, *Leptogorgia* (Kingsley and Watabe 1987). In fact, a special form of carbonic anhydrase has been identified as a matrix component in mollusk shells (Miyamoto et al. 1996). This form is special in the sense that the carbonic anhydrase sequence is spliced into a shell matrix protein. So this enzyme really is at the site of mineral formation in mollusks. Is this the secret of equilibrium deposition of oxygen isotopes in mollusk shells?

The medium from which mineral forms (see also Erez 2003). There is little information on the chemistry of the medium from which minerals form in biological environments. Some studies involving mollusks analyzed the fluid between the organ that forms the shell (the mantle) and the shell itself and showed that it has a chemical composition close to seawater (Crenshaw 1972). It is not clear, however, whether this extrapallial fluid really is the medium from which mineral forms. There is much indirect evidence based on analyses of the trace elements in carbonate-containing biogenic minerals that the composition of the medium from which mineral forms does not generally reflect seawater composition. A survey of Mg and Sr contents in mollusk shells showed that members of almost all the 5 major classes control the amounts of these two elements that enter the shell (e.g., Table 2). Only the chitons have Mg and Sr contents that correspond to equilibrium deposition from seawater (Lowenstam 1963). Recently, Cohen et al. (2002) proposed that the reef corals changed the composition of the medium over the diurnal cycle. At night, this fluid has a composition that is approximately sea water but becomes significantly out of equilibrium during the day when symbiont photosynthesis is active (see Cohen and McConnaughey 2003).

It is always tacitly assumed that mineral deposition must occur from a saturated solution. Saturated it must be, although it is conceivable that a strategy for controlling mineral deposition at the appropriate site may be to localize saturation only at that site, and maintain under-saturated conditions elsewhere. There is no evidence for this, however. In fact a recent study indicates quite the opposite; there is no solution as such from which mineral forms! A cryo-TEM study of the forming spicule of sea urchin larvae showed that the membrane of the vesicle in which the spicule grows is juxtaposed to the forming spicule. Thus there is little or no space for a liquid phase (Beniash et al. 1999). An alternative is that the crystalline mineral forms from a transient amorphous precursor phase, and in echinoderm larvae at least, the latter appears to be formed in vesicles within cells adjacent to the location of spicule formation (Beniash et al. 1999). The vesicles presumably transfer their mineral content as a colloidal solid phase into the site where the spicule forms. Even though this is by no means proven to be a general strategy in biology, the possible implications it presents for understanding vital effects could be most significant.

Transient mineral precursor phases. It has been shown that both mollusk and echinoderm larvae form their aragonitic shells and calcitic spicules, respectively, via a transient amorphous calcium carbonate (ACC) precursor phase (Beniash et al. 1997; Weiss et al. 2002; Addadi et al. 2003). It has not as yet been demonstrated that adults also form their skeletons in this manner. Note that these phyla are on two completely different branches of the animal phylogenetic tree, and, as both have opted for the same strategy, the phenomenon may be widespread. Thus, even though it is premature to invoke this strategy as an important factor in explaining vital effects, it is worth examining briefly.

The formation of ACC occurs from a highly supersaturated solution. This can only be achieved if additives are present that prevent deposition of the crystalline phases. One such additive is Mg, and the concentrations required for ACC to form are similar to those found in sea water (Raz et al. 2000). It has also been shown that certain proteins are able to induce the formation and stabilization of ACC in the absence of Mg, which would otherwise almost instantaneously transform into a crystalline polymorph (Aizenberg et al. 1996). So at least in bivalve and echinoderm larval shells, Mg and proteins could play important roles in setting up the initial medium from which crystalline calcium carbonate forms. In fact *in vitro* experiments with proteins extracted from sea urchin larval spicules show that Mg has to be present for these proteins to induce ACC formation, although in lower amounts than is present in seawater (Raz et al. 2003). So in this case, a cooperative effect seems to exist between these specialized proteins and Mg.

If indeed the ACC precursor strategy turns out to be widespread in biology, this discovery will have many possible implications for vital effect mechanisms. Under what conditions can isotopic equilibrium be achieved during the transformation of ACC to aragonite or calcite, and what would be the differences in concentration of say, Mg, in the precursor versus mature phases? Interestingly, it has been noted that the transient forms of ACC are not hydrated in the two cases that have been studied (Addadi et al. 2003), as opposed to the stable forms of biogenic ACC. This lack of hydration would reduce the possibility of oxygen isotopic fractionation during the transformation process, as the major factor would be whether or not all of the ACC is converted into the mature crystalline form. If water is present and has to be removed during the transformation, then the efficiency of the process and the extent to which this water is in equilibrium with the environment, become important in terms of oxygen isotope fractionation. See Aizenberg et al. (2003a) for interesting observations and a discussion of this subject.

McCrea (1950) did show that when calcium carbonate forms very rapidly from seawater at temperatures below 15°C or above 60°C, variations in the oxygen isotopic composition of the precipitates occur of up to 5 ppm. He described these precipitates as gelatinous, as opposed to the coagulating precipitates formed within that temperature range. It would be interesting to repeat these type of experiments, while characterizing the atomic structure of the precipitates particularly in terms of atomic order. Could ACC be involved?

Adkins et al. (2003) studied the variations in oxygen and carbon isotopic compositions in deep sea corals. They noted that the most negative values for both oxygen and carbon isotopic compositions are at the trabecular centers and proposed an interesting mechanism to explain the vital effect related to the presence of a pH gradient. The assumption that they clearly state is that the mechanism of mineralization is the same for all parts of the trabecula. Is this really true? There has been considerable discussion about the precise mineral phase at these centers. An electron diffraction study indicates that calcite is present at the trabecular centers (Constantz and Meike 1989). To our knowledge, this observation has not been confirmed. Despite the real risk of speculating about speculations, it would be interesting to consider the implications if ACC functioned as a transient precursor phase of aragonite at the trabecular centers. Proving or disproving this will be a challenge because of the transient nature of ACC.

Biogenic mineral phase. Surprisingly, many open questions still remain regarding the detailed nature of the mineral phases formed by organisms. We submit that a better understanding of some of these questions will shed light on some vital effect mechanisms.

1. <u>The mineral phase itself</u>: The crystalline mineral phases produced by organisms are generally well known. If, however, a skeletal tissue contains both a crystalline phase and an amorphous phase, the latter may easily be overlooked, especially if only X-ray diffraction is used to identify the bulk mineral phase. Such juxtapositioning of crystalline and amorphous carbonate minerals has recently been reported in sponge and ascidian spicules (Aizenberg et al. 2003b). It is well known, for example, that within foraminiferal shells variations in Mg contents can differ markedly from one location to another (Bender et al. 1975; Duckworth 1977; Brown and Elderfield 1996). The structural properties of the high Mg calcite phase may be quite different from those of the low Mg phase. It is conceivable that even the relative proportions of the different phases may have a direct bearing on temperature-related effects, as they do for the proportions of calcite and aragonite in some mollusk shells (Lowenstam 1954).

2. <u>Surface to bulk ratio of individual crystals</u>: The surface of an individual crystal is a prime location for adsorption of trace metals. In biology the crystals are often enveloped in an organic layer (the matrix), which may or may not interfere with the adsorption of the metals. The surface areas of large crystals in relation to their bulk (volume) are small, and thus complications due to trace element adsorption on the crystal surfaces are less likely. This is not the case for small crystals. The extreme example is bone. Bone crystals have the shape of extremely thin plates, just a few nanometers thick (Weiner and Price 1986). Thus mature bone crystals are only some 15-20 atomic layers thick, and a major proportion of the atoms are disordered just because they are at, or close, to the surface. These are therefore preferred sites for metal adsorption. Biogenic aragonite crystals are almost always much smaller than biogenic calcite crystals, with cross-sectional dimensions of the usually needle-shaped crystals in the micron range (Treves et al. 2003). The one major exception is the aragonite of the nacreous layer of mollusk shells, where the crystals are much larger. Thus knowledge of the surface to bulk ratio of the biogenic tissue being used for proxy analysis may shed light on whether or not surface adsorption of trace metals is significant compared with incorporation within the bulk. Such variations may differ subtly between taxa.

3. <u>Extent of crystalline order and disorder within the bulk</u>: A single crystal contains defects over and above the defects present at or close to the surface of the crystal. For the trace elements that are included in the bulk of the crystal, it is of interest to differentiate between those that are located at grain boundaries and/or dislocations and those that substitute for a major ion in a lattice location. This has been checked for Mg in several biogenic calcites by analyzing the total Mg contents and the Mg content based on the peak shift in X-ray diffraction. The latter is only due to Mg present in lattice positions. It was found that in all cases the Mg is almost entirely in the lattice position (Raz et al. 2000). This need not always be the case. Knowing that for, say, foraminiferal calcite, it also is only in the lattice, will provide greater assurance that the trace element distribution can be characterized by a unique fractionation factor. Alternatively, the relative distributions of a specific trace element between the two sites may explain species variations in part.

4. <u>Occluded macromolecules</u>: Echinoderms, mollusks, brachiopods, calcareous sponges and ascidians are all known to occlude macromolecules within their

crystals (Berman et al. 1993). These are usually glycoproteins, and it has been shown by single-crystal, high-resolution X-ray diffraction analyses that they are often located along specific crystal planes. Despite the fact that they are very large compared to the atomic order within the crystal, the glycoproteins do not cause major dislocations. Some disorder does result, as can be detected by line width broadening. These sites (probably indistinguishable from grain boundaries) within the crystal bulk are also possible locations for trace metals. The presence of occluded macromolecules and ions influences the solubility of the mineral phase, and this may directly bear on the mechanisms of differential dissolution of different genera of foraminifera and coccolithophoridae observed in deep-sea sediments.

Crystal shape. The shape of a crystal reflects the nature of the environment in which it grows. In biology this environment, and hence crystal shapes, are generally very well controlled. Detailed analyses of crystal shapes and, in particular, the exact crystal faces expressed can shed much light on the mechanisms of growth. If a very unstable face is expressed, then some process must be responsible for its stabilization. The unusual face may be a nucleating plane, as has been suggested for the formation of certain types of calcium oxalate monohydrate crystals in plants (Bouropolous et al. 2001) and calcareous sponge spicules (Aizenberg et al. 1995). Another example of unstable faces being expressed occurs in the coccoliths (Young et al. 1999; Young and Henriksen 2003). These faces could in part explain different solubilities of coccolith calcite between species, as unstable faces are more soluble than stable ones. Curved surfaces of calcite are quite common, especially in cases where the crystal grows in a vesicle. These surfaces may be curved down to the atomic level. How such a curved surface is stabilized is not known, nor are its solubility characteristics. In general, detailed characterizations of crystal surfaces may be most helpful in understanding modes of formation and, in turn, vital effects, as well as solubility properties.

Ontogenetic variations. Variations in mineralization are known to occur in many different genera, including foraminifera. These too need to be understood in order to determine whether or not they can contribute to the complexities of equilibrium and non-equilibrium deposition. It may also be possible that the change in mineralization mechanism during growth is not a switch from one mechanism to another, but the addition of another mechanism in the juvenile and adult stages. Lowenstam and Weiner (1989) did raise this possibility for the scleractinian corals vis-a-vis the trabeculae centers of calcification.

Crystal maturation. Once formed, crystals may continue to change over long periods of time, especially if their surface to volume ratios are very high. This is the case in bone, where a sintering process occurs even within the lifetime of the animal and continues after death (Legeros et al. 1987). Thus diagenetic processes, even in the deep sea may not only involve dissolution, but also sintering. Obviously if in a closed system, this should have little effect on proxies. It can, however, complicate efforts to understand the basic mechanisms of deposition vis a vis proxies, as the mature mineral may not faithfully reflect the mineral phase formed initially. Chemical change may also occur during diagenesis. It has been shown, for example, that aragonite formed secondarily in coral skeletons even while the coral was still submerged (Spiro 1971).

Concluding comment

Given our present state of knowledge, the prospects for finding simple, robust explanations for different vital effects in terms of biomineralization mechanisms may seem hopeless at worst, and challenging at best. This, in our opinion, is most likely not the case. More than anything else, the apparent magnitude of the problem is due to so few studies having been carried out with this purpose in mind. We believe that once a critical mass of research gets underway, pieces of this puzzle will begin to fall into place. We concur with the opinions of some (Schrag and Linsley 2002) that as the importance of paleoceanography continues to increase because of global change effects, and with it the wide use of proxies for reconstructing the paleo-ocean environment, the need for understanding the basic mechanisms of mineralization that pertain to these problems is paramount. We hope that as our understanding of "vital effects" improves, this term which has so often been used as "a façade hiding our ignorance" (Lowenstam and Weiner 1989), will become instead a vehicle for understanding important aspects of the biomineralization process itself.

ACKNOWLEDGMENTS

We thank Lia Addadi for her insightful comments and discussions. SW is the incumbent of the Dr. Walter and Dr. Trude Burchardt Professorial Chair of Structural Biology. This work was supported in part by a U.S. Public Service Grant (DE06954) from the NIDCR. PD acknowledges generous support of the U.S. Department of Energy, Division of Chemical Sciences, Geosciences and Biosciences (DE-FG02-00ER15112) and the National Science Foundation (NSF-OCE-0083173).

REFERENCES

Addadi L, Weiner S (1992) Control and design principles in biological mineralization. Angew Chem Int Ed 31:153-169
Addadi L, Weiner S (2001) Crystals, asymmetry and life. Nature 411:753-755
Addadi L, Raz S, Weiner S (2003) Taking advantage of disorder: amorphous calcium carbonate and its roles in biomineralization. Adv Mat 15:959-970
Adkins JF, Boyle EA, Curry WB, Lutringer A (2003) Stable isotopes in deep-sea corals and a new mechanism for "vital effects." Geochim Cosmochim Acta 67:1129-1143
Aizenberg J, Grazul JL, Muller DA, Hamann DR (2003a) Direct fabrication of large micropatterned single crystals. Science 299:1205-1208
Aizenberg J, Hanson J, Koetzle TF, Leiserowitz L, Weiner S, Addadi L (1995) Biologically-induced reduction in symmetry: A study of crystal texture of calcitic sponge spicules. Chem Eur J 1(7):414-422
Aizenberg J, Lambert G, Addadi L, Weiner S (1996) Stabilization of amorphous calcium carbonate by specialized macromolecules in biological and synthetic precipitates. Adv Mat 8:222-226
Aizenberg J, Weiner S, Addadi L (2003b) Coexistence of amorphous and crystalline calcium carbonate in skeletal tissues. Conn Tissue Res 44:(in press)
Ali SY (1983) Calcification of cartilage. *In*: Cartilage, Structure, Function and Biochemistry. Hall BK (ed) Academic Press, New York, p 343-378
Bäuerlein E (ed) (2000) Biomineralization. Wiley-VCH Verlag GmbH, Weinheim Germany
Bazylinski DA (1996) Controlled biomineralization of magnetic minerals by magnetotactic bacteria. Chem Geo 132:191-198
Bazylinski DA, Moskowitz BM (1997) Microbial biomineralization of magnetic iron minerals: Microbiology, magnetism and environmental significance. Rev Mineral 35:181-223
Bazylinski DA, Frankel RB (2003) Biologically controlled mineralization in prokaryotes. Rev Mineral Geochem 54:217-247
Belcher AM, Wu XH, Christensen RJ, Hansma PK, Stucky GD, Morse DE (1996) Control of crystal phase switching and orientation by soluble mollusc-shell proteins. Nature 381(6577):56-58
Bender ML, Lorens RB, Williams DF (1975) Sodium, magnesium and strontium in the tests of planktonic foraminifera. Micropaleont 21:448-459
Beniash E, Addadi L, Weiner S (1999) Cellular control over spicule formation in sea urchin embryos: a structural approach. J Struct Biol 125:50-62
Beniash E, Aizenberg J, Addadi L, Weiner S (1997) Amorphous calcium carbonate transforms into calcite during sea-urchin larval spicule growth. Proc R Soc London B Ser 264:461-465
Berman A, Hanson J, Leiserowitz L, Koetzle TF, Weiner S, Addadi L (1993) Biological control of crystal texture: A widespread strategy for adapting crystal properties to function. Science 259:776-779

Berner EK, Berner RA (1996) Global Environment. Prentice Hall Inc, Upper Saddle River New Jersey
Berridge MJ, Bootman MD, Lipp P (1998) Calcium-A life and death signal. Nature 395:645-648
Berthold WU (1976) Biomineralisation bei milioliden Foraminiferen und die Matrizen-Hypothese. Naturwischenschaften 63:196
Bettencourt V, Guerra A (2000) Growth increments and biomineralization process in cephalopod statoliths. J Exper Mar Biol Ecol 248:191-205
Blackmon PD, Todd R (1959) Mineralogy of some foraminifera as related to their classification and ecology. J Paleont 33:1-15
Borowitzka MA, Larkum AWD, Nockolds CE (1974) A scanning electron microscope study of the structure and organization of the calcium carbonate deposits of algae. Phycologia 13:195-203
Borowitzka MA (1982) Morphological and cytological aspects of algal calcification. Intl Rev Cytology 74:127-160
Bouropolous N, Weiner S, Addadi L (2001) Calcium oxalate crystals in tobacco and tomato plants: morphology and *in vitro* interactions of crystal-associated macromolecules. Chem Eur J 7:1881-1888
Brown SJ, Elderfield H (1996) Variations in Mg/Ca and Sr/Ca ratios of the planktonic foraminifera caused by postdepositional dissolution: evidence of shallow Mg-dependent dissolution. Paleoceanography 11:543-551
Brownlee C, Nimer N, Dong LF, Merrett MJ (1994) Cellular regulation during calcification in Emiliania huxleyi. *In*: The Haptophyte Algae, Vol 51. Green JC, Leadbeater BSC (eds) Clarendon Press, New York, p 133-148
Carpenter SJ, Lohmann KC (1995) $\delta^{18}O$ and $\delta^{13}C$ values of modern brachiopod shells. Geochim Cosmochim Acta 59:3749-3764
Chave KE (1954) Aspects of the biogeochemistry of magnesium I. Calcareous marine organisms. J Geol 62:266-283
Cohen A, McConnaughey T (2003) Geochemical perspectives on coral mineralization. Rev Mineral Geochem 54:151-187
Cohen AL, Owens KE, Layne GD, Shimizu N (2002) The effect of algal symbiosis on the accuracy of Sr/Ca paleotemperatures from coral. Science 296(5566):331-333
Constantz BR (1986) Coral skeleton construction: A physiochemically dominated process. Palaios 1:152-157
Constantz BR, Meike A (1989) Calcite centers of calcification in *Mussa Angulosa* (Scleractinia). *In*: Origin, Evolution and Modern Aspects of Biomineralization in Plants and Animals. Crick RE (ed) Elsevier, Amsterdam, p 201-207
Craig H (1953) The geochemistry of the stable carbon isotopes. Geochim Cosmochim Acta 3:53-92
Crenshaw MA (1972) The inorganic composition of molluscan extrapallial fluid. Biol Bull 143:506-512
Crenshaw MA (1980) Mechanisms of shell formation and dissolution. *In*: Skeletal Growth of Aquatic Organisms. Rhoads DC, Lutz RA (eds) Plenum Publishing Corporation, New York, p 115-132
Crick RE (ed) (1989) Origin, Evolution, and Modern Aspects of Biomineralization in Plants and Animals. 5th Intl Symposium on Biomineralization 1986. Plenum Press, New York
da Silva JJRF, Williams RJP (1991) The Biological Chemistry of the Elements. Oxford University Press, New York
De Vrind-De Jong EW, Van Emburg PR, De Vrind JPM (1994) Mechanisms of calcification: Emiliania huxleyi as a model system. *In*: The Haptophyte Algae. Vol 51. Green JC, Leadbeater BSC (eds) Clarendon Press, New York, p 149-166
De Vrind-De Jong EW, De Vrind JPM (1997) Algal deposition of carbonates and silicates. Rev Mineral 35:267-307
De Yoreo, JJ, Vekilov PG (2003) Principles of crystal nucleation and growth. Rev Mineral Geochem 54:57-93
Deer WA, Howie RA, Zussman J (1966) An Introduction to the Rock-Forming Minerals. Longman, Essex, England
Duckworth DL (1977) Magnesium concentration in the tests of the planktonic foraminifer *Globoratalia truncatulinoides*. J Foraminiferal Res 7:304-312
Duplessey J-C, Lalou C, Vinot AC (1970) Differential isotopic fractionation in benthic foraminifera and paleotemperatures reassessed. Science 168:250-251
Elderfield H, Bertram CJ, Erez J (1996) A biomineralization model for the incorporation of trace elements into foraminiferal calcium carbonate. Earth Planet Sci Lett 142:409-423
Epstein S, Buchsbaum R, Lowenstam HA, Urey HC (1951) Carbonate-water isotopic temperature scale. Bull Geol Soc Am 62:417-426
Epstein S, Buchsbaum R, Lowenstam HA, Urey HC (1953) Revised carbonate-water isotopic temperature scale. Bull Geol Soc Am 64:1315-1326

Epstein S, Lowenstam HA (1954) Temperature-shell-growth relations of recent and interglacial Pleistocene shoal-water biota from Bermuda. J Geol 61(5):424-438

Erez J (1978) Vital effect on stable-isotope composition seen in foraminifera and coral skeletons. Nature 273:199-202

Erez J, Luz B (1983) Experimental paleotemperature equation for planktonic foraminifera. Geochim Cosmochim Acta 47:1025-1031

Erez J (2003) The source of ions for biomineralization in foraminifera and their implications for paleoceanographic proxies. Rev Mineral Geochem 54:115-149

Evans J (2003) Principles of molecular biology and biomacromolecular chemistry. Rev Mineral Geochem 54:31-56

Falini G, Albeck S, Weiner S, Addadi L (1996) Control of aragonite or calcite polymorphism by mollusk shell macromolecules. Science 271:67-69

Fortin D, Ferris FG, Beveridge TJ (1997) Surface-mediated mineral development by bacteria. Rev Mineral 35:161-180

Frankel RB, Bazylinski DA (2003) Biologically induced mineralization by bacteria. Rev Mineral Geochem 54:95-114

Garrone R, Simpson TL, Pottu-Boumendil J (1981) Ultrastructure and deposition of silica in sponges. *In*: Silicon and Siliceous Structures in Biological Systems. Simpson TL, Volcani BE (eds) Springer-Verlag, New York, p 495-525

Gay CV, Mueller WJ (1973) Cellular localization of carbonic anhydrase in avian tissues by labeled inhibitor autoradiography. J Histochem Cytochem 21:693-702

Goreau TF (1959) The physiology of skeletal formation in corals. I. A method for measuring the rate of formation of calcium deposition of corals under different conditions. Biol Bull 116:59-75

Gotliv B-A, Addadi, L, Weiner S (2003) Mollusk shell acidic proteins: in search of individual functions. Chem Bio Chem 4:522-529

Grégoire C, Duchateau GH, Florkin M (1955) La trame protidique des nacres et des perles. Ann Inst Oceanogr Monaco 31:1-36

Hemleben C, Anderson OR, Berthold W, Spindler M (1986) Calcification and chamber formation in foraminifera-a brief overview. *In*: Biomineralization in Lower Plants and Animals. Leadbeater BSC, Riding R (eds) Clarendon Press, Oxford, p 237-249

Iler RK (1979) The Chemistry of Silica. John Wiley & Sons, New York

Ip YK, Lim ALL, Lim RWL (1991) Some properties of calcium-activated adenosine triphosphatase from the hermatypic coral *Galaxea fascicularis*. Mar Biol 111:191-197

Kingsley R, Watabe N (1987) Role of carbonic anhydrase in calcification in the gorgonian *Leptogorgia virgulata* (Lamarck) (Coelenterata: Gorgonacea). J Exp Mar Biol Ecol 93:157-167

Kirschvink JL, Hagadorn JW (2000) A grand unified theory of biomineralization. *In*: Biomineralization. Bäuerlein E (ed) Wiley-VCH Verlag GmbH, Weinheim, Germany, p 139-149

Knoll A (2003) Biomineralization and evolutionary history. Rev Mineral Geochem 54:329-356

Konhauser KO (1997) Bacterial iron biomineralisation in nature. FEMS Microbiol Rev 20:315-326

Konhauser KO (1998) Diversity of bacterial iron mineralization. Earth-Science Rev 43:91-121

Kröger N, Deutzmann R, Sumper M (1999) Polycationic peptides from diatom biosilica that direct silica nanosphere formation. Science 286:1129-1132

Kröger N, Lorenz S, Brunner E, Sumper M (2002) Self-assembly of highly phosphorylated silaffins and their function in biosilica morphogenesis. Science 298:584-586

Land LS, Lang JC, Barnes DJ (1975) Extension rate: a primary control on the isotopic composition of the West Indian (Jamaican) scleractinian reef coral skeletons. Mar Biol 33:221-233

Langmuir D (1997) Aqueous Environmental Geochemistry. Prentice-Hall, Inc., Upper Saddle River, NJ

Leadbeater BSC, Riding R (eds) (1986) Biomineralization in Lower Plants and Animals. The Systematics Association Special Volume. Clarendon Press, Oxford

Legeros R, Balmain N, Bonel G (1987) Age-related changes in mineral of rat and bovine cortical bone. Calc Tissue Intl 41:137-144

Letizia ML, Moscariello A, Hunziker J (1997) Stable isotopes in Lake Geneva carbonate sediments and molluscs: review and new data. Eclog Geol Helv 90:199-210

Levi-Kalisman Y, Addadi L, Weiner S (2001) Structure of the nacreous organic matrix of a bivalve mollusk shell examined in the hydrated state using cryo-TEM. J Struct Biol 135:8-17

Levi-Kalisman Y, Raz S, Weiner S, Addadi L, Sagi I (2002) Structural differences between biogenic amorphous calcium carbonate phases using x-ray absorption spectroscopy. Advanced Functional Mat 12(1):43-48

Lowenstam HA (1954) Factors affecting the aragonite:calcite ratios in carbonate-secreting marine organisms. J Geol 62:284-322

Lowenstam HA (1963) Biological problems relating to the composition and diagenesis of sediments. *In*: The Earth Sciences: Problems and Progress in Current Research. Donnelly TW (ed) University of Chicago Press, Chicago, p 137-195
Lowenstam HA (1964) Sr/Ca ratio of skeletal aragonites from the recent marine biota at Palau and from fossil gastropods. *In*: Isotopic and Cosmic Chemistry. Craig H, Miller SL, Wasserburg GJ (eds), North Holland Publishing Co, Amsterdam, p 114-132
Lowenstam HA (1981) Minerals formed by organisms. Science 211:1126-1131
Lowenstam HA (1986) Mineralization processes in monerans and protoctists. *In:* Biomineralization in Lower Plants and Animals. Vol 30. Leadbeater BSC, Riding R (eds) Oxford University Press, New York, p 1-17
Lowenstam HA, Margulis L (1980) Calcium regulation and the appearance of calcareous skeletons in the fossil record. *In*: The Mechanisms of Biomineralization in Animals and Plants. Omori M, Watabe N (eds) Tokai University Press, Tokyo, p 289-300
Lowenstam HA, Rossman GR (1975) Amorphous, hydrous, ferric phosphatic dermal granules in Molpadia (Holothurodidea): Physical and chemical characterization and ecologic implications of the bioinorganic fraction. Chem Geol 15:15-51.
Lowenstam HA, Weiner S (1989) On Biomineralization. Oxford University Press, New York
Mann S (1983) Mineralization in biological systems. Struct Bonding 54:125-174
Mann S (2001) Biomineralization: Principles and Concepts in Bioinorganic Materials Chemistry. Oxford University Press, New York
Markel K (1986) Ultrastructural investigation of matrix-mediated biomineralization in echinoids (Echinodermata, Echinoidea). Zoomorphology 106:232-243
McConnaughey T (1989a) ^{13}C and ^{18}O isotopic disequilibrium in biological carbonates: I. Patterns. Geochim Cosmochim Acta 53:151-162
McConnaughey T (1989b) Biomineralization mechanisms. *In*: Origin, Evolution, and Modern Aspects of Biomineralization in Plants and Animals. Crick RE (ed) Plenum, New York, p 57-73
McCrea JM (1950) On the isotopic chemistry of carbonates and a paleotemperature scale. J Chem Phys 18:849-857
Minsky A, Shimoni E, Frenkel-Krispin D (2002) Stress, order and survival. Nature Rev Mol Cell Biology 3:50-60
Miyamoto H, Miyashita T, Okushima M, Nakano S, Morita T, Matsushiro A (1996) A carbonic anhydrase from the nacreous layer in oyster pearls. Proc Natl Acad Sci USA 93(18):9657-60
Moradian-Oldak J, Weiner S, Addadi L, Landis WJ, Traub W (1990) Electron diffraction study of individual crystals of bone, mineralized tendon and synthetic carbonate apatite. Conn Tissue Res 25:1-10
Nancollas GH (ed) (1982) Biological Mineralization and Demineralization. Life Sciences Research Report, Springer-Verlag, New York
Noll F, Sumper M, Hampp N (2002) Nanostructure of diatom silica surfaces and of biomimetic analogues. Nano Letters 2(2):91-95
Orme CA, Noy A, Wierzbicki A, McBride MT, Grantham M, Teng HH, Dove PM, DeYoreo JJ (2001) Formation of chiral morphologies through selective binding of amino acids to calcite surface steps. Nature 411:775-779
Pereira-Mouriès L, Almeida M-J, Ribeiro C, Peduzzi J, Barthélemy M, Milet C, Lopez E (2002) Soluble silk-like organic matrix in the nacreous layer of the bivalve Pinctada maxima. European J Biochem 269:4994-5003
Perry CC (2003) Silicification: the processes by which organisms capture and mineralize silica. Rev Mineral Geochem 54:291-327
Pohnert G (2002) Biomineralization in diatoms mediated through peptide- and polyamine-assisted condensation of silica. Angew Chem Intl E 41(17):3167-3169
Raven JA, Griffiths H, Glidewell SM, Preston T (1982) The mechanism of oxalate biosynthesis in higher plants: Investigating with the stable isotopes 18O and 13C. Proc R Soc London B, Biol Sci 216:87-101
Raz S, Hamilton P, Wilt F, Weiner S, Addadi L (2003) Proteins from sea urchin larval spicules mediate the transient formation of amorphous calcium carbonate on the way to calcite (to be submitted)
Raz S, Weiner S, Addadi L (2000) The formation of high magnesium calcite via a transient amorphous colloid phase. Adv Mater 12:38-42
Rucker JB, Carver RE (1969) A survey of the carbonate mineralogy of cheilostome bryozoa. J Paleont 43:791-799
Sarashina I, Endo K (1998) Primary structure of a soluble matrix protein of scallop shell: implications for calcium carbonate biomineralization. Am Mineral 83:1510-1515
Schlüter M, Rickert D (1998) Effect of pH on the measurement of biogenic silica. Mar Chem 63:81-92
Schmidt WJ (1924) Die Bausteine des Tierkorpers in Polarisiertem Lichte. F. Cohen Verlag, Bonn
Schrag DP, Linsley BK (2002) Corals, chemistry, and climate. Science 296:277-278

Schüler D, Frankel RB (1999) Bacterial magnetosomes: Microbiology, biomineralization and biotechnological applications. Applied Microbiol Biotech 52:464-473

Simkiss K (1976) Cellular aspects of calcification. *In*: The Mechanisms of Mineralization in the Invertebrates and Plants. Vol 5. Watabe N, Wilbur KM (eds) Univ. of South Carolina Press, Columbia, SC, p 1-31

Simkiss K (1986) The processes of biomineralization in lower plants and animals-an overview. *In*: Biomineralization in Lower Plants and Animals. Vol 30. Leadbeater BSC, Riding R (eds) Oxford University Press, NY, p 19-37

Simkiss K, Wilbur K (1989) Biomineralization. Cell Biology and Mineral Deposition. Academic Press, Inc., San Diego

Skinner HCW, Jahren AH (2003) Biomineralization. Biogeochemistry: Treatise on Geochemistry. W. H. Schlesinger, Elsevier Science. 8: (in press)

Spiro BF (1971) Diagenesis of some scleractinian corals from the Gulf of Elat, Israel. Bull Geol Soc Denmark 21:1-10

Stumm W (1992) Chemistry of the Solid-Water Interface. John Wiley & Sons Inc, New York

Sumper M (2002) A phase separation model for the nanopatterning of diatom biosilica. Science 295:2430-2433

Swift DM, Wheeler AP (1991) Some structural and functional properties of a possible organic matrix from the frustules of the freshwater diatom Cyclotella meneghiniana. *In*: Surface Reactive Peptides and Polymers. Vol 444. Sikes CS, Wheeler AP (eds) American Chemical Society, Washington, DC, p 340-353

Swift DM, Wheeler AP (1992) Evidence of an organic matrix from diatom biosilica. J Phycol 28:202-209

Tebo BM, Ghiorse WC, van Waasbergen LG, Siering PL, Caspi R (1997). Bacterially mediated mineral formation: Insights into manganese(II) oxidation from molecular genetic and biochemical studies. Rev Mineral 35:225-266

Towe KM, Cifelli R (1967) Wall ultrastructure in the calcareous foraminifera: Crystallographic aspects and a model for calcification. J Paleontol 41:742-762

Treves K, Traub W, Weiner S, Addadi L (2003) Aragonite formation in the chiton (Mollusca) girdle. Helv Chim Acta 86:1101-1112

Urey HC, Lowenstam HA, Epstein S, McKinney CR (1951) Measurement of paleotemperatures and temperatures of the Upper Cretaceous of England, Denmark, and the southeastern United States. Bull Geol Soc Am 62:399-416

Van Cappellen P (2003) Biomineralization and global biogeochemical cycles. Rev Mineral Geochem 54:357-381

Veis A, Perry CC (1967) The phosphoprotein of the dentin matrix. Biochemistry 6:2409-2416

Veis A (2003) Mineralization in organic matrix frameworks. Rev Mineral Geochem 54:249-289

Vrieling EG, Beelen TPM, van Santen RA, Gieskes WWC (2002) Mesophases of (bio)polymer-silica particles inspire a model for silica biomineralization in diatoms. Angew Chem Intl Ed 41(9):1543-1546

Wada K, Fujinuki T (1974) Biomineralization in bivalve molluscs with emphasis on the chemical composition of the extrapallial fluid. *In*: The Mechanisms of Mineralization in the Invertebrates and Plants. Vol 5. Watabe N, Wilbur KM (eds) Univ of SC Press, Columbia, SC, p 175-188

Watabe N, Kingsley RJ (1989) Extra-, inter-, and intracellular mineralization in invertebrates and algae. *In*: Origin, Evolution, and Modern Aspects of Biomineralization in Plants and Animals. Crick RE (ed) Plenum, New York, p 209-223

Waychunas GA (2001) Structure, aggregation and characterization of nanoparticles. Rev Mineral Geochem 44:105-166

Webb MA (1999) Cell-mediated crystallization of calcium oxalate in plants. Plant Cell 11:751-761

Weber JN, Raup DM (1966) Fractionation of the stable isotopes of carbon and oxygen in marine calcareous organisms—the Echinoidea. Part II. Environmental and genetic factors. Geochim Cosmochim Acta 30:705-736

Weber JN, Woodhead PM (1970) Carbon and oxygen isotope fractionation in the skeletal carbonate of reef-building corals. Chem Geol 6:93-117

Weiner S (1979) Aspartic acid-rich proteins: major components of the soluble organic matrix of mollusk shells. Calc Tissue Intl 29:163-167

Weiner S, Addadi L (1991) Acidic macromolecules of mineralized tissues: The controllers of crystal formation. Trends Biol Sc 16(7)

Weiner S, Addadi L (2002) At the cutting edge. Perspectives. Science 298:375-376

Weiner S, Price P (1986) Disaggregation of bone into crystals. Calcif Tissue Intl 39:365-375

Weiner S, Talmon Y, Traub W (1983a) Electron diffraction of mollusk shell organic matrices and their relationship to the mineral phase. Intl J Biol Macromol 5:325-328

Weiner S, Traub W (1980) X-ray diffraction study of the insoluble organic matrix of mollusk shells. FEBS Letters 111(2):311-316

Weiner S, Traub W, Lowenstam HA (1983) Organic matrix in calcified exoskeletons, D. Reidel Publishing Co, Dordrecht, Holland

Weiner S, Traub W (1984) Macromolecules in mollusk shells and their functions in biomineralization. Phil Trans R Soc London Ser. B 304:421-438

Weiner S, Traub W, Lowenstam HA (1983b) Organic matrix in calcified exoskeletons. *In*: Biomineralization and Biological Metal Accumulation. Westbroek P, de Jong EW (eds) Reidel Publishing Co, Dordrecht, Holland p 205-224

Weiner S, Wagner HD (1998) The material bone: Structure-mechanical function relations. Annual Rev Mat Sci 28:271-298

Weiss IM, Tuross N, Addadi L, Weiner S (2002) Mollusk larval shell formation: amorphous calcium carbonate is a precursor for aragonite. J Exp Zool 293:478-491

Westbroek P (1983) Biological metal accumulation and biomineralization in a geological perspective. *In*: Biomineralization and Biological Metal Accumulation. Westbroek P, de Jong EW (eds) D. Reidel Publishing Co, Dordrecht, Holland, p 1-11

Wheeler AP, George JW, Evans CR (1981) Control of calcium carbonate nucleation and crytal growth by soluble matrix of oyster shell. Science 212:1397-1398

Wilt FH (1999) Matrix and mineral in the sea urchin larval skeleton. J Struc Biol 126:216-226

Wilt FH (2002) Biomineralization of the spicules of sea urchin embryos. Zool Sci 19:253-261

Worms D, Weiner S (1986) Mollusk shell organic matrix: Fourier transform infrared study of the acidic macromolecules. J Exp Zool 237:11-20

Young JR, Davis SA, Brown PR, Mann S (1999) Coccolith ultrastructure and biomineralization. J Struct Biol 126:195-215

Young JR, Henriksen K (2003) Biomineralization within vesicles: the calcite of coccoliths. Rev Mineral Geochem 54:189-215

Zaremba CM, Belcher AM, Fritz M, Li Y, Mann S, Hansma PK, Morse DE, Speck JS, Stucky GD (1996) Critical transitions in the biofabrication of abalone shells and flat pearls. Chem Mat 8:679-690

Zeebe RE, Sanyal A (2002) Comparison of two potential strategies of planktonic foraminifera for house building: Mg^{2+} or H^+ removal? Geochim Cosmochim Acta 66(7):1159-1169

Ziveri P, Stoll H, Probert I, Klaas C, Geisen M, Ganssen G, Young J (2003) Stable isotope "vital effects" in coccolith calcite. Earth Planet Sci Lett 210:137-149

2 Principles of Molecular Biology and Biomacromolecular Chemistry

John S. Evans
Laboratory for Chemical Physics
New York University
345 E. 24th Street
New York, New York 10010, U.S.A.

INTRODUCTION

Crucial events in biomineral formation—such as compartmentalization, supersaturation, precipitation, export of macromolecules, and cessation (Lowenstam and Weiner 1989)—require a "referee" who can control these events with precision and fidelity. This job falls to the cell, and in particular, a specialized cell, such as an osteoblast, odontoblast, mantle epithelium, or bacterium who has evolved or differentiated into a "molecular factory" that generates and controls the biomineralization process. Thus, to understand how biominerals form, we must first revisit basic concepts in biology that explain how cells function at the molecular level. This information will provide us with a basis for understanding subsequent chapters in this Review.

To provide a basic understanding for the non-specialist, this chapter will focus on four topics. <u>General Cell Architecture</u> will introduce the basic components of a cell and their function. <u>Molecular Manufacturing</u> will describe the major biomineralization-related macromolecules and how they are created and exported for use in biomineral formation. <u>Molecular Performance</u> will present structural concepts of cell-generated macromolecules which function in the biomineralization process. Finally, in <u>Molecular Manipulation</u>, we will briefly review contemporary techniques in molecular biology and chemical synthesis that may be mentioned in other chapters of this Review.

The reader should be aware that no single chapter or overview can cover all of the relevant information regarding cellular biology, molecular biology, and biochemistry, the three disciplines which outline not only the structure and function of cells and macromolecules, but also the biomineralization process itself. In fact, I would argue that it is difficult to condense decades of research and technology into a single chapter. Obviously, a lot will be left out in the process. For this reason, I strongly recommended that the reader consult basic college-level biochemistry, cell biology, and molecular biology textbooks to fill in the gaps with regard to knowledge, and/or to clarify the topics discussed in this overview (Alberts et al. 1983; Jendrisak et al. 1987; Sambrook et al. 1989; Singer and Berg 1991; Stryer 1995; Baynes and Dominicak 1999).

GENERAL CELL ARCHITECTURE

Basic components of eukaryotic cells

The predominant cell type that produces biominerals is the eukaryotic cell, although a number of bacterial strains also generate mineral deposits (Lowenstam and Weiner 1989). The eukaryotic cell can be characterized by a high degree of internal organization, as evidenced by the presence of intracellular organelles, i.e., compartmentalized structures which perform specialized functions for cellular function. Figure 1 presents the general picture of an eukaryotic cell and its associated organelle structures. The cell boundary itself is delineated by a cell membrane, consisting of a protein-containing lipid

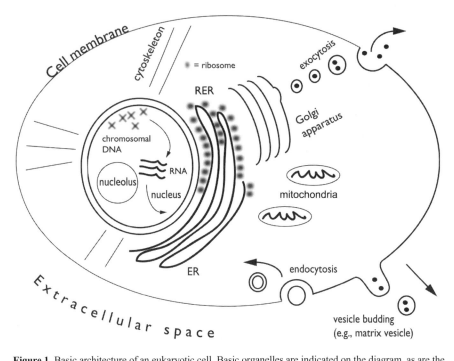

Figure 1. Basic architecture of an eukaryotic cell. Basic organelles are indicated on the diagram, as are the processes of endo- and exocytosis. Note that, with the exception of the nucleus, all organelles are bound by a lipid bilayer membrane; in the case of the nucleus, a double bilayer surrounds the nuclear compartment. For simplicity, the ER is arbitrarily represented as having "smooth" (i.e., ribosome-free) and "rough" (i.e., ribosome-studded ER membranes) regions. Please refer to the text for more description.

bilayer—the area within the cell is called the cytoplasm or cytosol. The endoplasmic reticulum, Golgi apparatus, vesicles, and mitochondria are also bound by a membrane bilayer. However, the nucleus is actually bound by a double membrane bilayer. Membrane bilayers compartmentalize each organelle such that important functions are isolated and regulated.

Within the nucleus are the chromosomes, i.e., protein-bound deoxyribonucleic acid (DNA) complexes which contain genetic information (genes). The chromosomal DNA gives rise to heterogeneous nuclear ribonucleic acid RNA (hnRNA) via a process called transcription. The hnRNA, in turn, is processed into messenger RNA (mRNA), which is exported out to the cytoplasm for translation of the genetic code into proteins (i.e., protein synthesis). The nucleolus is a large diffuse structure within the nucleus; it is the site of ribosome-specific RNA synthesis (denoted as rRNA) and the assembly of ribosomes. DNA, RNA, and ribosome components will be discussed later in this chapter.

Within the cytoplasm, the cell generates energy via metabolic conversion of molecules. Part of the metabolic cycle concerned with oxygen reduction to water (aerobic metabolism) takes place within membrane bound organelles called mitochondria. Mitochondria also possess their own DNA and RNA. Other organelles, such as the endoplasmic reticulum and Golgi apparatus, are sites for protein synthesis for export, oligosaccharide and lipid synthesis, membrane growth and post-translational modification of proteins. Note that the cytoplasm membrane itself is highly dynamic, and experiences budding off to form membrane vesicles (sometimes termed matrix vesicles) as well as fusion of intracellular membrane vesicles and subsequent release of vesicle contents to the extracellular space (exocytosis), or, engulfment of extracellular materials and their transport into the cytoplasm (endocytosis) (Fig. 1). Specialized vesicles, such as peroxisomes and lysosomes, are membrane-bound compartments that form from either the Golgi apparatus or from the cell membrane, and these contain hydrolytic and oxidative enzymes.

The cell is motile, i.e., it can move through media or on surfaces. To do this, the cell relies on a cytoskeletal network of contractile and support proteins that connect the cell membrane internally (Fig. 1). Examples of cytoskeletal proteins include actin (actin filaments), tubulin (microtubules), and intermediate filament proteins. The cytoskeletal proteins attach to the cell membrane and are concentrated near the nuclear membrane. These contractile proteins provide shape to the cell and allow the membrane edges of the cell to extend/retract at various points, providing the cell with locomotion. Moreover, the cytoskeletal network also facilitates cytoplasmic streaming, i.e., movement or flow of the cytoplasm throughout the cell interior. The exterior of the cell membrane itself contains receptor proteins which allow the cell to adhere to surfaces and/or detect extracellular molecules.

One type of mineral-producing eukaryotic cell that features a different cell architecture are plant cells (Lowenstam and Weiner 1989). The plant cell possesses a rigid polysaccharide-containing coating called the cell wall which surrounds the cell membrane bilayer. Additionally, plant cells possess intracellular membrane-bound organelles known as chloroplasts. These organelles contain the chlorophyll-based photosynthetic apparatus utilized in O_2 production.

Simpler in design: prokaryotic cells

By contrast, prokaryotic cells, or bacteria, are simpler in their organization. They possess few if any internal membrane-bound organelles and do not have a cytoskeletal network. Thus, in those prokayotes with no internal organelles, all intracellular processes are combined together within cell cytoplasm and are not partitioned. Prokaryotes are bounded by a cell wall, and in some cases, a cell membrane which is surrounded by the cell wall.

MOLECULAR MANUFACTURING

Within the confines of the cell, there is substantial biosynthetic and metabolic activity that maintains the viability of the cell and its replication, and provides functional activity which supports the host organism. In this section, we will introduce the basic building blocks of biochemistry and molecular biology and describe how proteins—one of the major players in biomineralization—are generated. Given the complexity of the genetic information in a cell and the numerous pathways available for protein production, perhaps the best way to present this material is to tell the story of a hypothetical biomineralization protein, which I will call "hybiopro." This hypothetical protein contains 135 amino acids and is produced by a eukaryotic cell which creates extracellular mineral deposits. Hybiopro is synthesized within the cell and is released into the extracellular space, where it regulates the nucleation of mineral crystals. What now follows is a description of how the DNA gene coding for hybiopro is manipulated to generate the actual protein.

Genetic code, gene organization: the protein "blueprints"

Proteins are basically polymers consisting of linear combinations of 20 amino acids, and it is the sequential linear combination of amino acids that defines any protein sequence. Amino acids are typically represented by a 3-letter abbreviation or by a single letter code. Every protein has an amino terminus, called the N-terminal end or N-terminus, which is considered the start of the protein sequence, and, a carboxylate terminus, called the C-terminal end or C-terminus, which is considered the end of the protein sequence. All proteins are coded for by DNA genes which are stored within the cell nucleus. DNA itself is folded and packed into complexes called chromosomes, i.e, highly compact assemblies of DNA and DNA-specific proteins (histones) that are arranged into nucleosome complexes that are further organized into higher-order assemblies known as chromatin (Fig. 2). The cell which produces hybiopro has a hypothetical chromosome, 18, which contains a gene region on its DNA which contains the complete amino acid sequence for this protein. When the cell becomes activated for mineral production, chromosome 18 is targeted by the cell for the production of hybiopro.

The DNA, in turn, is double stranded (abbreviated as dsDNA) and consists of two antiparallel chains, one running 5' to 3', the other, in an antiparallel direction, 3' to 5', where the 3' and 5' designators indicate the position of the sugar ring –OH group. In DNA, the deoxyribose sugar units are linked by phosphodiester bonds wherein a phosphate ester linkage is formed between the 5' –OH of one deoxyribose with the 3' –OH of the other deoxyribose (Fig. 2). Each deoxyribose sugar, in turn has one of 4 possible nitrogen-containing base groups [adenine (A), guanine (G), thymine (T), and cytosine(C)] covalently linked via a C–N bond at the C1' carbon. The sugar + base unit represents a nucleoside, and when a phosphate group is attached at either the 3' or 5' end, it is termed a nucleotide. These nucleotide bases form complementary hydrogen bonds with corresponding bases on the opposite strand called base pairs. The rule of thumb is: A forms a base pair with T via two hydrogen bonds, and G forms a base pair with C using three hydrogen bonds. Thus, if one DNA strand has a hypothetical nucleotide sequence, 5'-ATTATTGCGCTA-3', the complementary, opposite antiparallel strand would read 3'-TAATAACGCGAT-5'.

Each amino acid is coded for by one or more three-nucleotide stretch termed the codon. Rather than reproduce the codon list here, we refer the reader to any standard biochemistry textbook for the complete listing. For example, the DNA codon CTT codes for the amino acid, leucine, and the corresponding RNA codon would be CUU. Thus, for a 135 amino acid protein like hybiopro, we would expect 135 codons to make up the gene, which would result in a DNA gene consisting of 3 × 135 = 405 nucleotide base pairs. Note that of the two antiparallel strands of DNA, only one strand codes for a protein. This is called the coding or sense strand. The other complementary strand is called the template or antisense strand. In eukaryotic cells, a protein gene may be organized as several dsDNA coding fragments called exons that are separated from one another by non-coding dsDNA fragments called introns (Fig. 3). In our hypothetical hybiopro protein gene, there are 3 coding exons, separated by 2 non-coding intron regions. The reason for this non-contiguous arrangement of the DNA coding sequence will be explained shortly.

To understand what comes next, let's imagine a factory that has an area for manufacturing and a reference room which contains blueprints for the manufacture of certain items. According to factory rules, the original blueprints must remain in the reference room for security. However, the factory does allow us to make copies of any blueprint we want to take with us to the production floor. The cell is like a factory, too, and the reference room is the nucleus where the chromosomes and DNA are kept. To

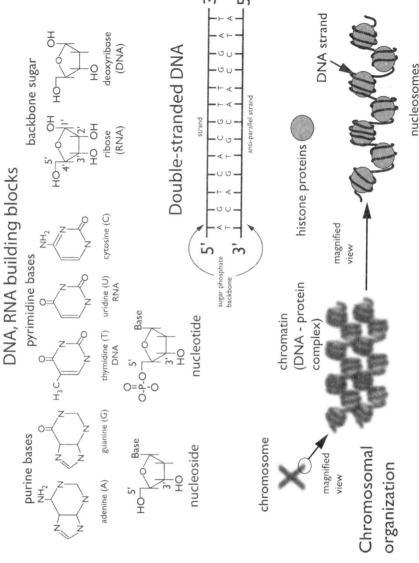

Figure 2. Chromosomal DNA, DNA double helix, genes. The top part of this figure gives the basic "building blocks" of nucleic acids, i.e., their elementary components, the nucleotides. The ribose (or deoxyribose) sugar units and purine/pyrimidine bases. The carbon numbering scheme for ribose and deoxyribose sugars is given as well. A simplified representation of double-stranded DNA(dsDNA) is likewise presented. Although not shown in this drawing, A–T base pairs form 2 hydrogen bonds, G–C base pairs form 3 hydrogen bonds, and the DNA helix itself exhibits a twisted helical structure. Lastly, chromosomes are DNA-protein complexes that are extremely condensed and supercoiled. The chromosome contains numerous genes, and if we were to magnify or enlarge a small region of the chromosome, we would observe that the chromosome is organized into chromatin and nucleosome DNA-protein superstructures. Such compact packing allows an incredible amount of genetic information to be stored in a very small volume within the nucleus.

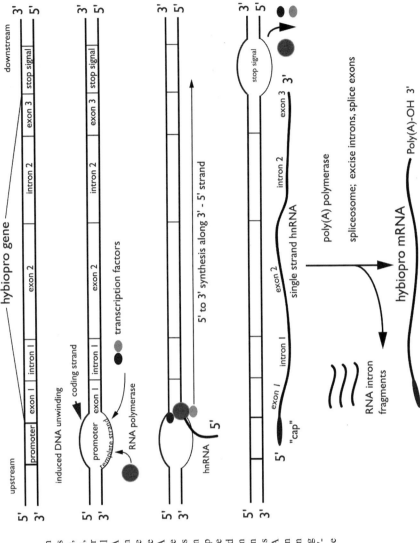

Figure 3. RNA transcription and exon gene splicing. This diagram depicts dsDNA that contains the hybiopro gene, consisting of 3 exons and 2 introns, along with the upstream RNA promoter region and the downstream stop signal region. With the assistance of DNA binding proteins called transcription factors, RNA polymerase binds to the promoter dsDNA region and unwinds the DNA into single strands. The RNA polymerase then advances along the template strand (3'–5') and synthesizes the hnRNA in a 5'–3' direction. When the RNA polymerase reaches the stop signal region of the template DNA, the polymerase, transcription factors, and single stranded hnRNA dissociate from the template DNA and the process can begin anew. Subsequently, the hnRNA is processed within the nucleus by the RNA spliceosome, which removes the intron regions and ligates or joins the exon regions together. This process, along with the addition of the 5' cap and the 3'-poly(A) tail region, result in the formation of mRNA.

make a protein, we need to generate a copy of the original protein blueprint, or DNA gene, and take that copy out to the cytoplasm, or production floor, where we can manufacture the protein. For the cell, that copy of DNA is called heterogeneous nuclear ribonucleic acid, or hnRNA for short. RNA differs from DNA in two respects. First, RNA uses uracil (U) in place of thymidine (T). Second, the sugar backbone of RNA consists of ribose units instead of deoxyribose (Fig. 2).

To generate the hybiopro hnRNA, the cell utilizes enzymes, or proteins which catalyze chemical reactions. The enzymes responsible for hnRNA synthesis are RNA polymerases (Fig. 3), and these enzymes synthesize RNA in a 5' to 3' direction, using the 3' to 5' DNA template strand to make this complementary RNA strand. The RNA polymerase binds to dsDNA region upstream from the hybiopro gene called the promoter region. Other proteins, called transcription factors, also bind to the RNA polymerase and this complex, called the basal transcription apparatus, begins to synthesize the coding hnRNA strand from the unwound, single stranded DNA (ssDNA) template strand. This process of hnRNA synthesis from DNA is called transcription. During transcription, the nascent hnRNA chain will be chemically modified at the 5' end by the addition of a "capping" terminus. The RNA polymerase/transcription factor complex terminates the transcription of the hnRNA once it reaches a specific DNA region downstream of the hybiopro gene (Fig. 3). This termination region or stop signal causes the RNA polymerase to come off the template ssDNA strand. Simultaneously, the hnRNA strand dissociates from the DNA template strand and the ssDNA refolds back into a double-strand again. Once the hnRNA has dissociated from the DNA template strand, specific endonucleases cleave the 3' end, and a poly(A) polymerase adds a polyadenylate "tail" to the 3' end of the hnRNA. The 5' capping and 3' poly(A) tail modifications permit the mRNA to be recognized by the cell for the process of protein synthesis, and, protect the mRNA from degradation within the cell for a certain period of time.

The transcription of a gene is not a random event; no factory could survive by randomly producing products without a production order, timetable, and appropriate scheduling. The cell takes all of this into account by highly regulating the transcription process. This can be achieved by responding to external molecular signals received outside the cell at the cell membrane using a process called receptor-ligand binding and activation. Here, the external cell membrane surface contains membrane proteins which recognize only certain molecules, such as hormones or growth factors. Once these molecules bind to a receptor protein, the receptor becomes activated and sends a chemical message to the nucleus to initiate the transcription of a particular protein, or a set of proteins. Another form of transcription activation can arise internally. Here, certain proteins or molecules that are produced within the cell can travel to the nucleus, where they activate RNA polymerase-promoter binding and transcription. In either instance, protein genes are transcribed on demand.

The hnRNA now becomes the next focus in the process of hybiopro production. Remember that the exon regions contain the DNA sequence that codes for hybiopro, but these coding regions are not contiguous or immediately adjacent to one another. Rather, they are separated by inserted intron regions. Thus, we need to dispense with the non-coding intron regions that currently reside in the hnRNA transcript and join the coding exon regions together. This is achieved by removal of intron sequences from the hnRNA and rejoining of hnRNA exon segments by RNA-protein complexes called spliceosomes (Fig. 3). The resulting spliced hybiopro RNA is now called messenger RNA (mRNA).

Protein synthesis: assembling the polypeptide chain

With the production of mRNA within the nucleus, the cell now has the correct "blueprint copy" from which hybiopro can be made. The process of protein synthesis takes place in the cell cytoplasm, and involves the interaction of the mRNA with two key cellular machines: a two-piece RNA-protein complex called the ribosome and a series of small folded RNAs called transfer RNAs (tRNAs) to which individual amino acids are enzymatically attached via an ester linkage to the adenosine group at either the 2′ or 3′ end of the tRNA (Fig. 4). The tRNA-amino acid covalent complex is called aminoacyl-tRNA or tRNA-AA, and each amino acid has a specific tRNA. Each tRNA has a 3-base sequence near its end called the anticodon loop. This anticodon is complementary to one of the existing amino acid codons on the mRNA. For example, in our earlier example, CUU is the RNA codon for leucine, and the corresponding tRNA anticodon would be GAA. Hence, when a mRNA codon for amino acid such as serine becomes available, the tRNA - Ser complex will bind at the Ser codon using the anticodon region of the tRNA. This logic then extends to the other amino acids. In this way, the gene sequence encoded on mRNA can be "read" by appropriate tRNA-amino acid complexes and the correct amino acid is now positioned in its proper place with regard to the intended protein sequence.

The process of synthesizing a protein is very much like the assembly of a link chain: you sequentially join one link with another in a linear fashion, head-to-tail (Fig. 4). The ribosome, consisting of a small and large subunit, assembles onto the mRNA near its 5′ end with the help of proteins called initiation factors. Two sites within the ribosome—"A" and "P"—are available for tRNAs to come and bind to two contiguous codons on the mRNA. In all eukaryotic mRNAs, the first codon at the 5′ end of mRNA codes for methionine (AUG). This is called the initiation codon. Hence, tRNA-Met will come into site "P" and use its anticodon region to form base pairs with the Met initiation codon on mRNA, and this process is called initiation. Subsequently, individual tRNAs are recognized by proteins called elongation factors and brought to the ribosome "A" site. If the tRNA-amino acid complex contains an RNA anticodon sequence which will base pair or bind to the exposed mRNA codon at site "A", then this tRNA will bind there. Now, with two tRNA-amino acid complexes residing next to one another in the ribosomal complex, an enzyme called peptidyl transferase will take the two amino acids and join them together, releasing the Met-specific tRNA molecule in the process and creating a tRNA-dipeptide complex. This dipeptide complex now moves from the "A" site into the "P" site in a process called translocation. Translocation occurs via the assistance of elongation factor proteins. Now, we have the first two amino acids of the N-terminal sequence of hybiopro joined together.

In the elongation process, the ribosome now advances along the mRNA in a 5′ to 3′ direction, with initiation factors filling the empty "A" site with a new tRNA-amino acid complex whose anticodon recognizes the next available mRNA codon. A peptide bond is formed between the new "A" site tRNA-AA and the previous tRNA-peptide chain residing in the "P" site, the old tRNA is removed, and the nascent tRNA-polypeptide complex is translocated to the "P" site (Fig. 4). This cycle is repeated over and over as each individual amino acid complex is added to the growing protein chain, proceeding from the N-terminus of the sequence. Once the last amino acid of the sequence is added, the next codon on mRNA, called the stop codon, is exposed. This codon, which takes the form of –UGA–, –UAA– or –UAG–, does not code for any amino acid. Rather, proteins called releasing factors recognize this stop codon and cause the release of the entire hybiopro polypeptide chain from the last tRNA and the ribosomal complex. Subsequently, the ribosome and mRNA dissociate from one another, and the entire translation process repeats itself anew, starting from the initiation stage. We have now successfully produced a molecule of hybiopro! Note that a single mRNA molecule can have several ribosomal complexes simultaneously synthesizing polypeptide chains. This assembly is called a polyribosomal complex.

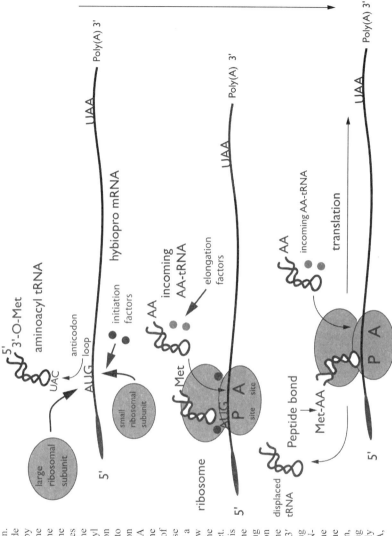

Figure 4. Overview of protein translation. The scheme for building the polypeptide chain using the codons provided by mRNA starts with the assembly of the small and large ribosomal subunits at the mRNA AUG initiation codon to form the ribosomal complex. This assembly takes place at the AUG initiation codon of the mRNA, and requires the aminoacyl tRNA-Met and proteins called initiation factors to allow ribosomal assembly to proceed. Next, with the help of elongation factor proteins, the aminoacyl tRNA which has an anticodon that matches the next mRNA codon enters the "A" site of the ribosome. The peptidyl transferase enzyme will catalyze the formation of a peptide bond between Met and the new amino acid, leading to the release of the tRNA that was originally linked to Met. The new Met-AA-tRNA complex is translocated to the "P" site, leaving the "A" site vacant for the next incoming tRNA-AA aminoacyl whose anticodon matches the next mRNA codon. The ribosomal complex proceeds in a 5' to 3' direction along the mRNA, synthesizing the nascent peptide chain from the N-terminal to the C-terminal end, using the repetitive process just described. The arrival of the ribosome at the stop codon, combined with the action of releasing factor proteins, leads to the disassembly of the ribosomal complex, the last tRNA, and the release of the polypeptide chain.

Since the translation process gives rise to functional polypeptide chains, it is obviously a process that the cell must exert control over. The most advantageous means of regulating translation is to control the availability of mRNAs that enter and remain in the cytoplasm. Thus, by regulating the transcription of hnRNA, splicing of hnRNA into mRNA and by the degradation or "chopping up" of cytoplasmic mRNA by RNA enzymes, the cell can regulate how many "blueprints" enter the production floor of the cellular factory.

One final note regarding protein expression from DNA. Each amino acid possesses one or more codons, and, some of these codons differ from one another by only 1 base. Thus, if a codon region in DNA were to change via a mutation at one nucleotide base, the resulting amino acid coded for could be radically different from the native amino acid originally specified, and likewise, this change in amino acid usage could have an impact on the final function and structure of the protein that is ultimately produced.

Post-manufacturing processes

In some cases, the production of a protein by translation is only the first step in manufacturing. Some proteins require chemical groups, such as sugars, hydroxy groups, phosphate or sulfate anions, to be added to amino acid sidechains. This process, called post-translational modification, is similar to the final packaging of a product in a factory prior to shipment. In most cases, for efficiency, the packaging and shipping facilities of a factory are kept separate from the actual manufacturing floor. The cell has evolved two organelles for protein intracellular transport packaging: the endoplasmic reticulum (ER) and the Golgi complex or apparatus. These membrane-bound organelles are located around the cell nucleus (see Fig. 1).

The cell performs much of the post-translational modification of proteins in the ER. For hybiopro and other proteins which are destined to be exported from the cell, the actual translation process itself takes place on the outer surface of the ER membrane. Here, ribosomal complexes dock and bind to the ER outer membrane receptors called the signal recognition particle docking protein, and the elongating polypeptide chain emerging from the ribosome is inserted directly into the ER compartment itself. Regions of the ER which bind to ribosomes physically appear rough, and are named the rough endoplasmic reticulum or RER (Fig. 1). Now, the newly synthesized hybiopro polypeptide chain resides within the ER and is isolated from other enzymes in the cytoplasm. Subsequently, this polypeptide can be conveniently modified within via the action of ER-specific post-translational enzyme complexes.

With regard to biomineralization proteins, there are several important post-translational modifications that can occur in eukaryotic cells. For example, within the ER the Met amino acid at the first position of each protein sequence is removed. Esterification of amino acid sidechains also takes place in the ER. These include phosphorylation of Thr, Ser, and Tyr sidechain –OH groups, and the sulfate ester formation of Tyr–OH. Cysteine residues within the polypeptide chain can be joined together to form disulfide bridging groups that physically unite parts of a polypeptide chain together. Finally, a number of biomineralization proteins have covalently attached sugar units called oligosaccharides. These oligosaccharide-modified proteins, called glycoproteins, are assembled and modified in either the ER or the Golgi complex via sugar-transferring enzymes which covalently link sugar units to either Asn (N-linked glycoprotein) or Ser/Thr (O-linked glycoprotein) sidechain groups. Oligosaccharides provide additional molecular information for a protein, and, permit unique molecular recognition of a glycoprotein via the placement of different oligosaccharide chains on the surface of the protein.

Exporting proteins to the extracellular matrix

So far, hybiopro has been manufactured at levels required to induce mineral formation outside our hypothetical cell. If necessary, hybiopro has been chemically modified as well. Now, the last hurdle in this manufacturing process is shipment or export of the protein to the extracellular matrix region outside the cell.

The challenge that confronts the cell "factory" is to get the protein, unscathed, outside the factory to its "customers" in the extracellular matrix. Luckily, the cell factory has adapted the membrane-bound Golgi apparatus to be a packaging facility and a shipping dock as well (Fig. 1). The Golgi apparatus consist of stacks of disc-shaped membrane chambers that have small vesicles or membrane-bilayer particles associated with it. The hybiopro enters the Golgi apparatus via a transport vesicle which buds off the ER and travels to the Golgi, where it fuses with the Golgi membrane. Once inside the Golgi, post-translational modifications can occur, such as glycosylation or modification of oligosaccharide chains. The hybiopro travels through the membrane compartments of the Golgi until it reaches the region closest to the cytoplasmic membrane of the cell. This region of the Golgi, termed the maturing face, is where secretory vesicles are formed by budding off from the Golgi membrane. Hybiopro is then packaged into a secretory vesicle at this region. Subsequently, the secretory vesicles migrate from the Golgi and reach the cytoplasmic membrane, where they fuse with the membrane bilayer and the contents of each vesicle are emptied into the extracellular matrix region outside the cell (Fig. 1). At this point, the cell has finished with the task of creating, modifying, and exporting our hybiopro biomineralization protein to the region outside the host cell. The job is finished!

What about intracellular biomineralization?

In the foregoing, we have only considered genomic synthesis events for proteins which participate in mineral formation outside the parent cells. However, the reader should be aware that intracellular biomineralization also takes place in Nature (Lowenstam and Weiner 1989), with magnetite-containing bacteria being a prime example of this phenomenon. Despite their eventual destination, both intra- and extracellular-based proteins are still generated from the genome as described above. Where the differences occur is in what happens after polypeptide translation: biomacromolecules which participate in intracellular mineral formation obviously forego the Golgi-based processing and vesicle exocytosis that we associate with extracellular biomacromolecules. Instead, like other cytoplasmic-based proteins, intracellular biomineralization proteins are most likely earmarked for intracellular retention, possibly as a result of primary amino acid sequence features that allow the protein to be recognized and permit bypassing of the extracellular trafficking route that we described above. There may also be different transcriptional activation pathways for extracellular and intracellular proteins. The reader is urged to consult the primary literature regarding specific intracellular biomineralization-based macromolecules for further details regarding post-translational events or other processes that impact the fate of intracellular-based biomacromolecules.

Manufacturing of other macromolecules involved in the formation of biominerals

In the last section, we alluded to the synthesis of other macromolecules, such as oligosaccharides and lipids. These also play a role in the biomineralization process, either on an intra- or extracellular level, and we will briefly review how these macromolecules are created and exported to the extracellular space.

Oligosaccharides and polysaccharides. Extracellular polymers of sugar or monosaccharide units, known as oligosaccharides (<30 monosaccharides/chain) or polysaccharides (>30 monosaccharides/chain) are created intracellularly by polymerizing enzymes known as transferases. Some important polysaccharides which play a role in the biomineralization process include β-chitin—a scaffolding polysaccharide found in exoskeleton and in invertebrate mineralized tissues—and glycosaminoglycans, sugar units which comprise proteoglycan complexes found in bone, cartilage, and tooth dentine (Lowenstam and Weiner 1989). The formation of sugar polymers takes place in both the ER and in the Golgi apparatus, where the extracellular sugar polymers are packaged into secretory vesicles and exported outside the cell.

Lipids and membrane assemblies. Lipids are macromolecules consisting of a glycerol backbone, two fatty acids which are linked to –OH groups of the glycerol molecule via ester bonds, and, and polar headgroup (i.e., phosphate, or phosphate + organic molecule) which is covalently linked to the remaining –OH group of the glycerol molecule (Fig. 7). The synthesis of lipids takes place on the surface of the ER membrane via a set of enzymes which links the individual components together. Eventually, the synthesized lipids assemble together and form membrane fragments which fuse with the ER membrane. In turn, the ER membrane buds off transport vesicles which travel to the Golgi apparatus. However, there are some intracellular vesicles which also form (e.g., peroxisomes, lysosomes, etc.), and these are earmarked for specific functions within the cell. Eventually, the fusion of secretory vesicles with the cytoplasmic membrane allow newly synthesized lipids to regenerate the cytoplasmic membrane.

MOLECULAR PERFORMANCE

Now that we understand how proteins are produced from the genome, we will now explore macromolecular structure and function. There are three categories of macromolecules that have been implicated in the biomineralization process: proteins, polysaccharides, and membrane assemblies. These will be discussed in the following sections.

Protein structure

Essentially, the amino acid sequence of every protein dictates the structure of each protein, its shape, and its function. A protein is basically a polymer comprised of amino acids linked together by chemical bonds (Fig. 5). There are twenty naturally-occurring amino acids in Nature, and, there are also artificial or non-natural amino acids that are made in the laboratory, plus natural modifications (phosphorylation, sulfation, hydroxylation, etc.) that occur to amino acids during post-translational processing. Each amino acid has an a central carbon atom (α–carbon) to which four substituents are attached: (1) α–amino group, (2) α–carboxylate group, (3) α–hydrogen atom, (4) sidechain chemical group (R). The polypeptide chain amide bond, also referred to as the peptide bond, consists of a head-to-tail condensation of the α–carboxylate group with the α–amino group of another amino acid, with loss of a water molecule. The most important feature of the amino acids are the 20 natural sidechain groups, which can be categorized into hydrophobic, anionic, cationic, and polar. The reader is urged to consult standard biochemistry textbooks for a complete listing of amino acid sidechains and their chemistry. Thus, depending on the amino acid composition of a polypeptide, it may possess a net charge based upon the summation of all sidechain charges at a given pH, including the free amino and carboxylate termini. The linear ordering of a polypeptide chain is referred to as the <u>primary sequence</u> or <u>primary structure</u>, and it is directly derived from the linear codon usage in the DNA which codes for the protein. Moreover, it is the amino acid sequence that allows us to define similarities and differences in structure and function between different proteins.

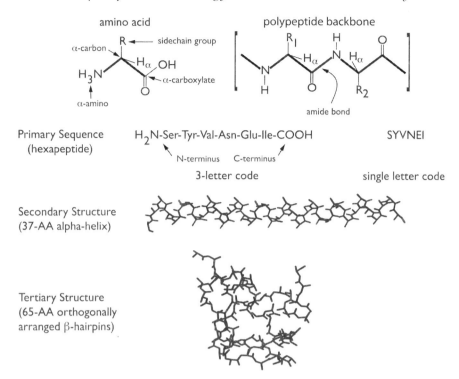

Figure 5. Overview of polypeptide structure. Presented here are the basic structures of an amino acid and an example of a polypeptide, with the amide bond (CONH) indicated by the arrow. Note that the peptide amide bond has partial p bond character, and adopts a planar structure. Examples of primary sequence (6-residue polypeptide presented in 3- and 1-letter amino acid code formats), secondary structure (37-residue alpha-helix), and tertiary structure (65-residues orthogonally arranged beta-hairpin structure) are given. Although not shown, quaternary structure is defined as the assemblage of individual protein subunits which are stabilized as a complex by non-covalent bonds between subunits.

Protein structure is hierarchical. The single polypeptide chain can be ordered at one level where there exist local folding and amide NH–carbonyl C=O intrastrand hydrogen bonding along the peptide backbone. The spatial arrangement of sequential amino acids is called the secondary structure (Fig. 5), and it is directly defined by the primary sequence and by solution conditions. In globular proteins which adopt a compact, folded structure in solution, there are several categories of secondary structure: for example, α–helix (Fig. 5), β–sheet, 3_{10} helix, β–helix, β–turn, polyproline type II. These structures, in turn, can be organized into supersecondary structures, which are clusters of individual secondary structure units. For example, two contiguous α–helices that are clustered together are referred to as coiled coils; two β–strands which possess an intervening α–helix are referred to as "βαβ" unit, and so on.

The next level of organization is the ordering of secondary structure units that are non-contiguous. This level is called tertiary structure, and in globular proteins it defines the three-dimensional fold of the polypeptide chain into a compact unit that has an interior region that is shielded from solvent and an exterior surface that is exposed to solvent (Fig. 5). Typically, the interior region of a protein tertiary fold is stabilized by intrastrand non-covalent sidechain-sidechain interactions; for example, hydrophobic-

hydrophobic interactions between apolar amino acids, complementary charge-charge interactions (also known as salt-bridging), and hydrogen bonding are among the possible tertiary stabilizing interactions that occur within the protein interior. Another type of internal stabilization of a protein fold can be had via the use of covalent disulfide –S–S– bond formation between cysteine residues in the polypeptide chain. The formation of disulfide bonds within a protein occurs post-translationally within the ER via enzyme catalysis, and, these bonds can be reversibly cleaved by the use of chemical agents which reduce the –S–S– linkage to two individual –SH thiol groups.

The final level of protein organization is quaternary structure, where a protein consists of more than one separate polypeptide chain. In this case, each folded polypeptide chain is considered a subunit, and quaternary structure refers to the spatial organization of subunits within the overall protein.

Biomineralization protein structure

From the foregoing, it would be expected that biomineralization proteins would obey the hierarchical rules of protein structure in the same way that globular or folded proteins do. And in some cases, this may be true. However, recent structure studies of biomineralization protein sequences are beginning to paint a different picture, particularly for mineral-associated proteins (Wustman et al. 2002; Evans 2003; Michenfelder et al. 2003; Wustman et al. 2004). Many biomineral-associated proteins possess primary structure or sequences that do not match those of folded, globular proteins (Xu and Evans 1999; Zhang et al. 2000, Wustman et al. 2002a,b, 2004; Zhang et al. 2002; Michenfelder et al. 2003). These unusual sequences adopt unfolded structures in solution, featuring extended, repeating β–turn, loop, or random coil conformations with little if any evidence for tertiary structure (Xu and Evans 1999; Zhang et al. 2000, Wustman et al. 2002a,b, 2004; Zhang et al. 2002; Michenfelder et al. 2003;). Although the exact reasons for these conformational preferences have not been fully explored, it is likely that the genes which code for mineral-associated proteins possess sequences that generate unfolded polypeptide structures in solution. Unfolded, conformationally labile polypeptides are reactive, and, in the presence of mineral and/or macromolecular interfaces, will attempt stabilize their conformational state by binding to exposed interfaces (Evans 2003; Michenfelder et al. 2003). In addition, conformationally labile sequences are more likely to adapt to irregular mineral surface topologies that one might expect to find in the extracellular matrix, compared to a more rigid α–helix or β–hairpin structure (Evans 2003).

Oligosaccharide and polysaccharide structure

We will briefly review carbohydrate chemistry in terms of monosaccharide units, monosaccharide chemical modifications, glycosidic bond, and oligosaccharide/polysaccharide structure. A monosaccharide is the building block of oligosaccharides and polysaccharides, and exist in either 5-(furanose) or 6-(pyranose) atom ring structures (Fig. 6). Ring structures are more rigid than linear chains, but it should be noted that 5- and 6-member ring structures can adopt a number of different conformations. Ring structures can open into a chain structure in solution, generating a reactive aldehyde or ketone group. The carbon of this reactive group is called the anomeric carbon, and it is the site of glycosidic bond formation (see below). Monosaccharides normally are hydroxylated. However, chemical modifications mediated by enzymes can generate substitutions of –OH groups with hydride, carboxylate, phosphate, sulfate, amino, and N-acetyl groups. These substitutions create a wide variety of possible interaction sites, and in the case of phosphate, carboyxlate, and sulfate groups, complementary charge interactions with metal ions, other macromolecules, or mineral surfaces.

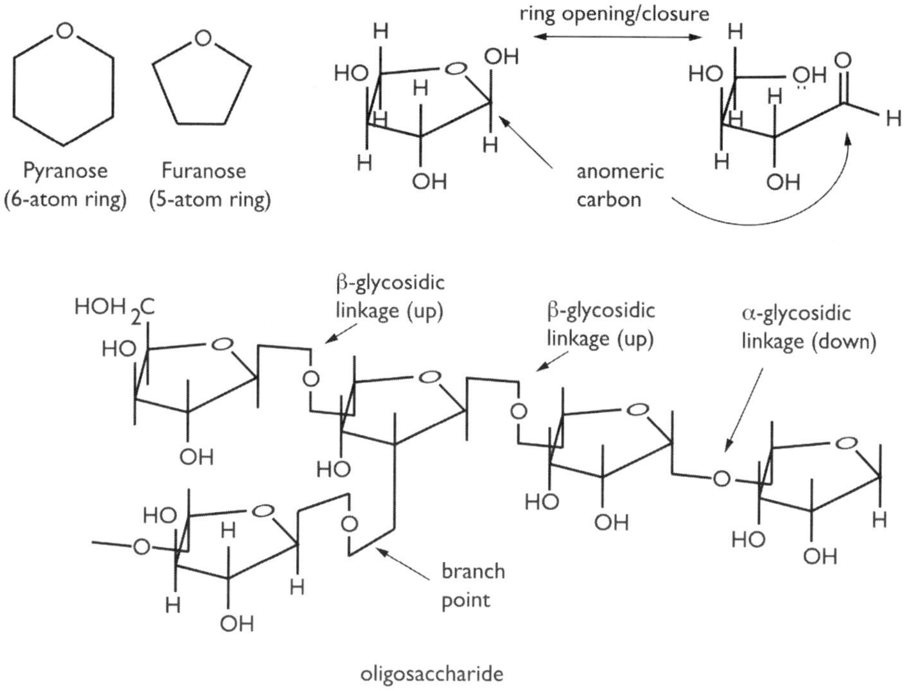

Figure 6. Overview of carbohydrates. The most common form of monosaccharides in biology are 5- or 6- atom ring structures, which are termed furanose and pyranose, derived from the elemental organic ring structures furan and pyridine. The structures presented here are in a common format referred to as Haworth projections; other structural representations may also be used, and the reader is referred to contemporary biochemistry texts for these alternate formats. The most important feature of any ring monosaccharide is the presence of an anomeric carbon, which represents the site of O atom attachment in the ring, and, is the starting point for numbering of ring carbon atoms. The ring structure can undergo reversible opening and closure in solution; the anomeric carbon in the linear form is either an aldehyde or ketone group, which undergoes nucleophilic attack by one of the monosaccharide –OH groups to reform the ring structure. This reversible reaction involves the loss of a water molecule, which is not shown above. The glycosidic bond is the covalent bond that joins the anomeric carbon of one monosaccharide to the –OH or other strong nucleophilic group of another monosaccharide. The type of glycosidic bond, a or b, is defined by the position of the glycosidic bond (above, beta; below, alpha) with respect to the ring structure of the first monosaccharide. The covalent assemblage of monosaccharides leads either to the formation of oligosaccharide or polysaccharide chains. These chains can be purely linear, feature branching points for additional chain attachment, or, can possess any combination of linear and branching bonds.

Oligosaccharides and polysaccharides are created via polymerization of individual monosaccharide units, using linkages called glycosidic bonds (Fig. 6), which involve bond formation between the anomeric carbon of a monosaccharide with the –OH group of another monosaccharide, with loss of a water molecule. This is called an O-glycosidic bond. A glycosidic bond is usually designated as being either α– or β–, depending on the position of the glycosidic bond (up, down) with respect to the ring structure of the monosaccharide bearing the anomeric carbon. The interesting feature of a glycosidic bond is that the donor –OH group can be from nearly any position on the monosaccharide ring, which can lead to linear chains or branching chains. Any number of branch points can occur within an oligosaccharide or polysaccharide. Add to this the number of permissible ring conformations for each monosaccharide in the chain and the potential for chemical group modification, and it becomes clear that oligo- and polysaccharides can generate incredible molecular structural diversity for recognition purposes.

In the previous section, oligosaccharide-modified proteins, or glycoproteins, were introduced. The attachment of oligosaccharide chains to a protein is usually via an O–glycosidic linkage with either serine–OH or threonine–OH sidechain groups, or, a N–glycosidic linkage with the amide nitrogen of asparagine. Most glycoproteins are either O- or N-linked, but there are some which feature both attachment types. The addition of oligosaccharide chains to a protein conveys an additional layer of molecular recognition capabilities, in that the oligosaccharide chains can be recognized by other proteins, and, they themselves may recognize and bind to other macromolecules, charged groups or ions, and mineral interfaces.

Polysaccharides which are involved in the formation of mineralized tissues belong to the family of structural polysaccharides. Structural polysaccharides typically feature linear chains with very little branching, and, possess β–O–glycosidic bonds which allow individual polysaccharide chains to adopt an open structure and form hydrogen bonds with other linear polysaccharide chains. Two structural polysaccharides are worth noting: β-chitin, a linear chain polymer of N-acetylglucosamine monosaccharides found in the extracellular matrices of invertebrates, and glycosaminoglycans (GAGs), linear chain polymers comprised of amino disaccharide repeat units that comprise proteoglycans of vertebrate connective tissues (Lowenstam and Weiner 1989).

Membrane assemblies

Lipids, consisting of two fatty acid chains, a glycerol group, and a phosphate group, will assemble into a monolayer or bilayer structure in aqueous solutions, in which the amphipathic or polar/hydrophobic nature of each lipid drives the assembly process, i.e., polar headgroups are oriented towards the aqueous environment, and the hydrophobic fatty acid chains are oriented away from the solvent and towards each other in the interior of the assemblage (Fig. 7). The simplest assembly is the micelle, a spherical monolayer assembly where all fatty acid chains are oriented inwards. This assembly is also adopted by detergents and long chain alcohols in water. Small assemblies of lipids can form continuous bilayer structures called liposomes or vesicles, compartmentalized structures which contain an aqueous interior. Micelles, liposomes, and vesicles can fuse together or with other membrane bilayer assemblies.

Because of the amphipathic nature of membrane bilayers, other amphipathic molecules, such as cholesterol, proteins, or detergents, can insert within the bilayer structure and become part of the overall assembly. However, polar molecules cannot easily transfer across the bilayer, and require the assistance of polypeptide-forming channels within the bilayer, or induction of small ruptures of the bilayer via chemical methods, to traverse the bilayer. The fact that the polar headgroups of the micelle (exterior) or bilayer (both interior and exterior) are available for interaction with metal ions and anions permits the use of membrane surfaces to nucleate inorganic solids (Lowenstam and Weiner 1989). In addition, membrane vesicles known as matrix vesicles (see Fig. 1) are produced by biomineral-competent cells to synthesize and deposit inorganic solids in the extracellular matrix (Lowenstam and Weiner 1989), and, magnetotactic bacteria use intracellular vesicles to compartmentalize magnetite deposits within the cell for magnetic field detection (Lowenstam and Weiner 1989).

Principles of Molecular Biology & Biomacromolecular Chemistry

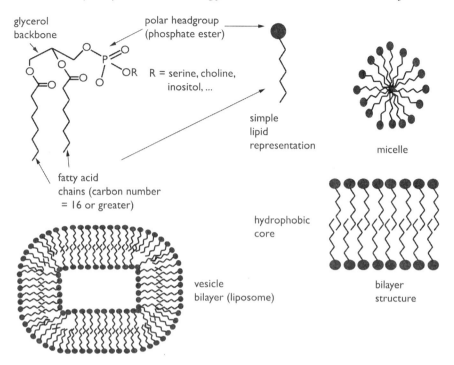

Figure 7. Lipids and membrane assemblies. The general structure of membrane lipids is presented, with the glycerol molecule serving as a "backbone" for fatty acid esterification and the addition of a phosphate ester group. Lipids can assemble into a number of amphipathic structures, the simplest being the micelle, which is essentially a spherical monolayer, up to bilayer structures.

MOLECULAR MANIPULATION

From the foregoing, we now have an understanding of how proteins are generated from the genome. More than 20 years ago, scientists began devising recombinant DNA techniques that would permit the manipulation of DNA genes, and ultimately allow the creation of artificial proteins and facilitate the amplified production of native proteins. In addition to molecular biology technology, chemical methods for synthesizing proteins and polypeptides have advanced to the point where complete proteins <150 amino acids in length can be created using solid-phase chemistries. In this chapter, we will discuss some of the more common DNA-and protein-based molecular biology and chemical synthesis methods that are essential to modern molecular biology and macromolecular chemistry. In the chapters which follow, many of the concepts and techniques that are introduced are based upon one or more of the methods described in this section. It should be noted that additional DNA and chemical synthesis techniques exist and/or are currently being developed, and it is recommended that the reader consult contemporary molecular biology, protein chemistry textbooks or the primary literature for additional information.

DNA manipulation

There are several central techniques that are crucial to modern DNA biotechnology. These techniques are often used in combination as a strategy for extracting gene organizational information, for identifying specific genes that code for proteins, for assaying the transcriptional and translational processes, and for identifying cells which express certain genes. One technique is restriction endonuclease fragmentation, which allows us to cut dsDNA into specific sized lengths for analyses (Fig. 8). This is accomplished by the use of bacterial DNA cutting enzymes (restriction endonucleases) that recognize specific base pair sequences within double stranded DNA and cut specific phosphodiester bonds within both strands, leaving either cut, flush-ended double stranded DNA, or double stranded DNA with short, single stranded ends (i.e., "sticky ends"). Figure 8 shows an example of each type of enzyme (Bam H1, Hae III); note that the name of each enzyme is derived from the abbreviated name of the parent bacterial strain. Restriction endonucleases are used to cleave dsDNA molecules into specific fragments that are more readily analyzed and manipulated. By the use of multiple restriction endonucleases, we can generate a fragment map or "fingerprint" of a specific DNA. Among other things, these enzymes allow us to excise a specific gene region and splice or insert a different one in its place, or, to obtain a specific DNA fragment that can be used to screen other genomic or RNA libraries.

A technique which complements restriction fragmentation is DNA ligation. A DNA enzyme called DNA ligase catalyzes the formation of a phosphodiester bond between dsDNA molecules containing free 3' –OH and 5' phosphate groups. (Fig. 8). Ligation can occur between flush-ended DNA molecules, or between DNA molecules that possess "sticky ends." This technique allows the joining of different DNA fragments together to form new DNA molecules.

We also have the ability to "tag" or label any piece of DNA, such that it can be identified and tracked in any subsequent experiment. DNA labeling (Fig. 8) can involve the use of radioactive (e.g., ^{32}P, ^{35}S, represented by the black circle in Fig. 8) or molecular fluorescent labels. Tagging or labeling can be accomplished in one of three ways. The end labeling technique involves the use of the enzyme, polynucleotide kinase, which exchanges the unlabeled 5' phosphate group at the end of a target ssDNA or dsDNA molecule for the terminal ^{32}P of labeled adenosine triphosphate (ATP), resulting in ^{32}P-labeled DNA and unlabeled adeosine diphosphate (ADP) (Fig. 8). The polymerase-based method utilizes DNA polymerase I Klenow subunit enzyme, which adds deoxyribonucleotides (A, T, G, C) and creates a complementary strand from a ssDNA template. If the nucleotides are labeled with either a radioactive or fluorescent label, then the label will be incorporated into the newly synthesized strand of DNA and the resulting dsDNA will be labeled on one strand. Finally, the nick translation method utilizes dsDNA, DNAase I endonuclease, and DNA polymerase I. Here, the DNAase I endonuclease introduces a limited number of single-stranded breaks ("nicks," shown as triangles in Fig. 8) in the DNA molecule, leaving exposed 5' phosphate and 3' hydroxyl groups. One then adds DNA polymerase I Klenow subunit enzyme, which then adds labeled nucleotides to the nicked regions and fills in the gap in the DNA strand. Thus, with any of these methods, we can produce a DNA fragment that we can trace throughout isolation and manipulation processes via the incorporated label.

Using above-ambient temperatures, we can induce dsDNA in solution to partially unwind or unfold ("denature"), exposing its single strands. These exposed single strands can now form base pairs with other DNA or RNA single strands if we first introduce the "new" DNA or RNA at above ambient temperatures, then slowly lower the temperature and allow the new + old strands to bind or anneal together during the cooling stage. This process of annealing or hybridizing DNA and/or RNA strands together is called nucleic acid hybridization and this technique permits us to identify new pieces of DNA or RNA which have base pair complementarity to the target DNA or RNA. This method exploits

Principles of Molecular Biology & Biomacromolecular Chemistry 49

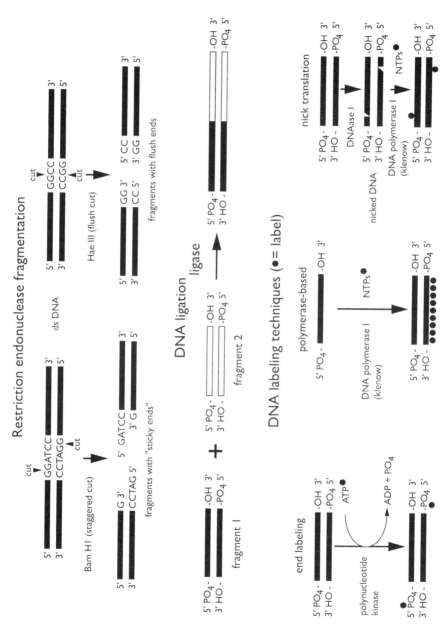

Figure 8. Molecular biology techniques: Restriction enzymes, DNA ligation, DNA labeling.

the fact that nucleotides which have complementary bases (A with T, G with C) will form double strands via hydrogen bonding between bases. Thus, if one takes ssDNA and mixes it with either ssDNA or RNA, there will be reactions that occur between the strands, depending on base pair matching. The larger the degree of nucleotide sequence complementarity between the single strands, the greater the number of hydrogen bonds will be between both strands and, subsequently, the stronger the interaction between the two strands. Hybridization can be used to define the complementarity or degree of nucleotide sequence homology between nucleic acid strands; in turn, this allows one to identify RNA or DNA strands which possess complementary nucleotide sequences to a target DNA sequence probe. Using a slightly different approach, hybridization can also be carried out in specific cells, wherein a DNA or RNA fragment is introduced within a cell by various techniques, then monitored for its binding with the cellular genome or RNA pool.

The determination of base pair sequence in a fragment of DNA can be readily performed via DNA sequencing using the chain termination/dideoxy method (Fig. 9). To sequence a given piece of DNA, we need to know a partial protein sequence that exists within the protein gene. From that primary sequence, we work backwards and generate a DNA oligonucleotide single stranded probe that will bind within the target protein DNA sequence. Using annealing techniques, we introduce the DNA probe to the single stranded target DNA molecule that codes for the protein; the primer forms base-pairs with the target DNA and creates a duplex double strand. When we introduce DNA polymerase I, this primer sequence will serve as a starting point for creating a complementary DNA strand of the protein gene. To make the complementary DNA strand, DNA polymerase I needs deoxyribonucleotides (dNTPs, where N = C, A, G, T). We provide these, but we also introduce chain terminating dideoxyribonucleotides (ddNTPs), which lack a 3' –OH group. Thus, as the DNA polymerase synthesizes the complementary strand from dNTPs, a ddNTP will eventually get incorporated instead. When this happens, there is no 3' –OH group available for strand extension, and DNA polymerase I cannot add any more dNTPs or ddNTPs to this complementary strand and chain termination occurs. Now, realize that the actual length of DNA that is synthesized will depend on the random incorporation of ddNTPs; in other words, we will end up with a series of different DNA molecules, each of which is random in length. Thus, if we conduct 4 parallel reactions like this, where labeled ddATP, ddCTP, ddGTP, or ddTTP only are introduced in one of 4 reaction mixtures that include DNA polymerase I and the DNA-primer complex, then we will end up with a mixture of DNA molecules labeled at the 3' end with ddA, ddG, ddC, or ddT. Using polyacrylamide gel electrophoresis (PAGE), which can separate DNA molecules that differ by only 1 nucleotide in length, we get a series of labeled molecules for each reaction mixture. Comparing each reaction mixture in terms of DNA chain length (5' end of sequence: smallest DNA fragment; 3' end of sequence: largest DNA fragment) and ddN termination allows us to deduce the DNA sequence (Fig. 9). This powerful technique has allowed scientists to determine entire genomic DNA sequences from eukaryotes and prokaryotes. In recent years, much of the sequencing tasks have now become automated, and, are utilizing novel technologies, such as DNA sequencing on chip arrays, to accomplish the task.

Reverse transcription of RNA

Reverse transciptase is a viral enzyme which takes 5'–3' mRNA and reverse transcribes a 3'–5' complementary DNA (cDNA) from the mRNA. Thus, if we provide a DNA primer that hybridizes to the target mRNA, we can synthesize a cDNA via extension from the primer to the 5' end of the mRNA molecule (Fig. 9). In the case of eukaryotic mRNA, we know that the 3' end of the mRNA contains a poly-A tail, so it becomes rather straightforward to hybridize a poly-dT probe to the mRNA and use this to

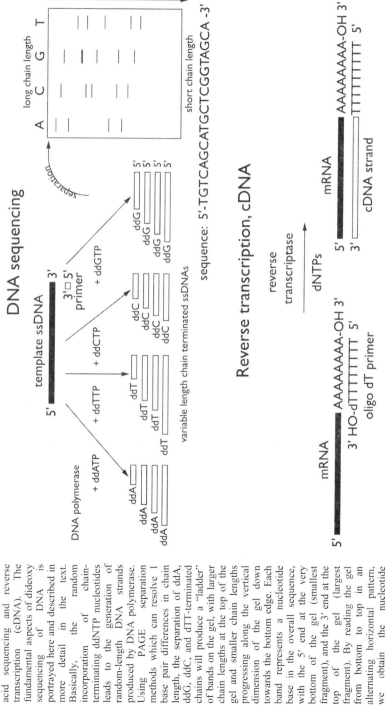

Figure 9. Molecular biology techniques: Dideoxy nucleic acid sequencing and reverse transcription (cDNA). The fundamental aspects of dideoxy sequencing of DNA is portrayed here and described in more detail in the text. Basically, the random incorporation of chain-terminating ddNTP nucleotides leads to the generation of random-length DNA strands produced by DNA polymerase. Using PAGE separation methods which can resolve 1 base pair differences in chain length, the separation of ddA, ddG, ddC, and dTT-terminated chains will produce a "ladder" of bands on the gel, with larger chain lengths at the top of the gel and smaller chain lengths progressing along the vertical dimension of the gel down towards the bottom edge. Each band represents a nucleotide base in the overall sequence, with the 5' end at the very bottom of the gel (smallest fragment), and the 3' end at the top of the gel (largest fragment). By reading the gel from bottom to top in an alternating horizontal pattern, we obtain the nucleotide sequence from the 5' end to the 3' end, as shown below the gel.

permit 5' to 3' cDNA synthesis from the mRNA strand. The resulting cDNA can then be used for sequencing purposes to determine the protein sequence of the coding mRNA, or, can be used in other analyses such as chromosome gene mapping, protein expression, etc.

DNA amplification

Oftentimes, in organisms, the quantity of a specific DNA gene is very limited, and thus this low quantity represents an experimental challenge with regard to isolation, sequencing, and further manipulation of the DNA as described above. However, one technique, polymerase chain reaction (PCR) (Fig. 10), has made it possible to use even minute levels of DNA for analyses. This technique relies on the ability of a specific DNA polymerase, Taq I, to catalyze the elongation of a DNA strand complementary to a template DNA. By using primers specific for the extreme 3' end of the sense and antisense strands, and, repetitive annealing/denaturing cycles, we can perform chain elongation with Taq I, then denature the dsDNA and separate off the newly synthesized DNA single strands. By repeating this denaturation-annealing-polymerase elongation/denaturing cycle over and over, we utilize each newly synthesized DNA strands as a "template" for additional DNA synthesis. The result is that each new DNA that is synthesized becomes smaller and smaller in size until only the gene region is represented in the new DNA pool. In this manner, we can repetitively generate numerous DNA molecules from a single DNA template strand. These PCR-generated DNA copies can then be recovered and used for further analyses.

Another route to DNA amplification is cell-based plasmid DNA cloning (Fig. 11). This method takes advantage of the fact that bacteria can accept foreign DNA, usually in the form of a plasmid (i.e., closed circular dsDNA) and make multiple copies of the DNA via a process called DNA replication. The use of host cells which undergo rapid cell division and replication ensures that a significant amount of DNA is produced in a very short period of time (i.e., hours). To amplify the target DNA, we insert the target DNA within the expressible gene using restriction endonucleases and DNA ligase. The modified plasmid DNA is then inserted into the bacteria, and its foreign DNA insert will be replicated numerous times as the bacterial cells undergo cell division. Multiple copies of the target DNA can then be purified and isolated from plasmid DNA isolated from the expanded bacterial cell population.

Protein expression

Oftentimes, quantities of a protein may be limited, and thus performing characterization experiments becomes limited as well. Biomineralization proteins often are limited in quantities, either due to the low expression within mineralized tissues, or, due to the difficulty in obtaining the protein from the mineralized matrix itself. Thus, it becomes necessary to seek additional methods which can amplify the amounts of protein. For this reason, genetic overexpression of proteins in host cells such as bacteria, yeast, mammalian or insect cells has become an important area of biotechnology. The central idea is to use the cell-based DNA cloning procedure similar to the one described above, but instead of isolating the target DNA, we simply allow the host cell to express the target DNA and make the protein using the host cell machinery (Fig. 11). Using restriction enzymes and DNA ligase, the foreign protein DNA gene is inserted within a host protein gene whose expression can be induced and amplified by the addition of a chemical agent to the cell media. This gene insertion creates a "fusion" protein which, during the protein isolation phase, can be cleaved away using a specific polypeptide-cleaving enzyme (protease) to free the foreign protein. This foreign-host "fusion" protein will be regulated by an RNA polymerase promoter transcription site, and the fusion gene will be transcribed upon addition of a chemical agent to the host cells that activates the

Principles of Molecular Biology & Biomacromolecular Chemistry

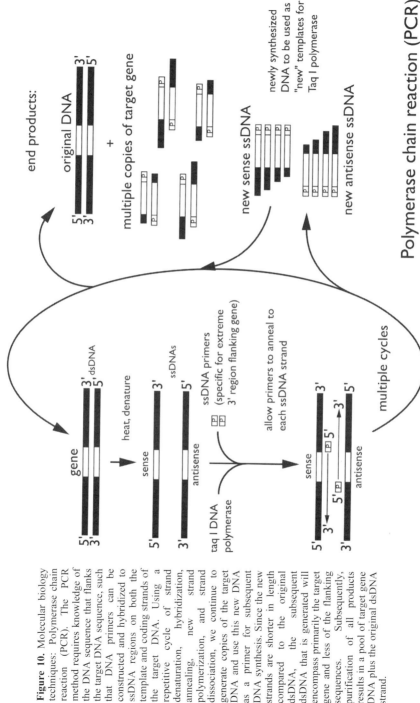

Figure 10. Molecular biology techniques: Polymerase chain reaction (PCR). The PCR method requires knowledge of the DNA sequence that flanks the target DNA sequence, such that DNA primers can be constructed and hybridized to ssDNA regions on both the template and coding strands of the target DNA. Using a repetitive cycle of strand denaturation, hybridization, annealing, new strand polymerization, and strand dissociation, we continue to generate copies of the target DNA and use this new DNA as a primer for subsequent DNA synthesis. Since the new strands are shorter in length compared to the original dsDNA, the subsequent dsDNA that is generated will encompass primarily the target gene and less of the flanking sequences. Subsequently, purification of all products results in a pool of target gene DNA plus the original dsDNA strand.

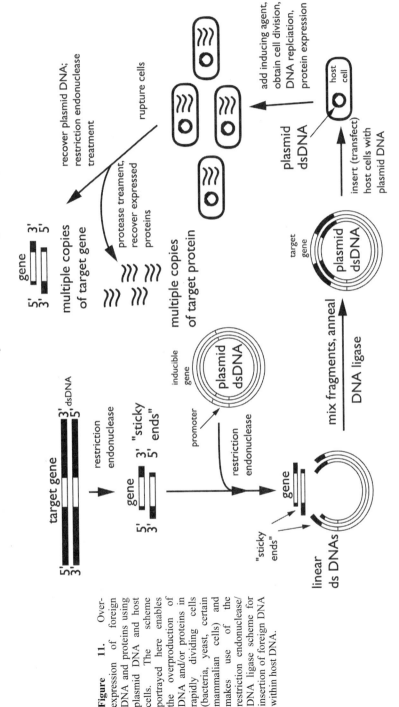

Figure 11. Overexpression of foreign DNA and proteins using plasmid DNA and host cells. The scheme portrayed here enables the overproduction of DNA and/or proteins in rapidly dividing cells (bacteria, yeast, certain mammalian cells) and makes use of the restriction endonuclease/DNA ligase scheme for insertion of foreign DNA within host DNA.

RNA polymerase. Using this approach, large quantities of host cells can be grown in culture, with each cell expressing multiple copies of the fusion protein. By rupturing the host cells, recovering the overexpressed fusion protein using standard biochemical isolation techniques, and then cleaving the fusion protein away from the foreign protein via a site-specific protease, one can obtain the necessary amounts of foreign protein needed for subsequent experiments.

Chemical synthesis of biomacromolecules

The ability to chemically synthesize a polypeptide, oligonucleotide, or oligosaccharide chain is based upon the use of a solid support, i.e., a polymer resin to which individual monomer units (i.e., amino acids, monosaccharides, nucleotides) can be sequentially attached one at a time (Fig. 12). This scheme, termed solid-phase synthesis, allows the selective formation of a head-to-tail covalent bond between the resin-attached monomer and an incoming free monomer in solution. The covalent addition of one monomer at a time is accomplished by outfitting different protecting or blocking groups to each free monomer unit to chain synthesis, then selectively cleaving off these protecting groups using chemical agents when the coupling of a new monomer to the nascent resin-bound chain is desired (Fig. 12). The joining or coupling of a new monomer utilizes a coupling agent which activates one chemical "end" of the incoming monomer group, such that it will readily react and form a bond with the exposed, unprotected group

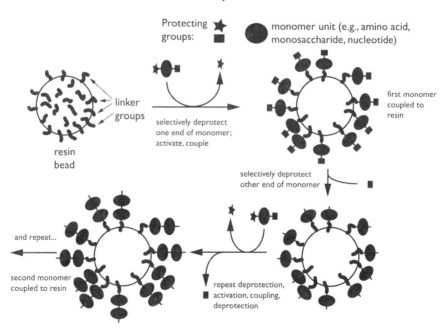

Figure 12. Chemical synthesis of biomacromolecules. This scheme has a general application to the syntheses of polypeptides, carbohydrates, and oligonucleotides and has been used recently to generate defined-length polymers and other organic molecules. The protecting groups at both ends of the monomer differ in terms of the reagents required for their removal; this insures that only one specific end become deprotected at any given step in the scheme.

at the end of the nascent resin-bound chain. Using this strategy, we insure that only the monomer at that chain position is added at that step. This entire procedure is conducted in solvent and the beads allow filtration and washing to be performed, such that displaced protecting groups, cleavage reagents, and excess monomer can be washed away. Once the entire chain is completed, all protecting groups can be chemically removed and the chain is cleaved from the resin support, yielding a free biomacromolecule that can be further purified using standard biochemical purification techniques. This solid-support method has also been utilized to synthesize discrete-length polymer chains and complex organic molecules as well.

ACKNOWLEDGMENTS

I would like to thank the Office of Army Research and the Department of Energy for their current support of our research in the area of biomineralization proteins and biomimetic polypeptides.

REFERENCES

Alberts B, Bray D, Lewis J, Raff M, Roberts K, Watson JD (1983) Molecular Biology of the Cell. Garland Publishing, Inc., New York, NY
Baynes J, Dominicak MH (1999) Medical Biochemistry. Harcourt Brace and Company, London, England.
Evans JS (2003) Apples and oranges. Structural aspects of biomineral- and ice-associated proteins. Curr Opin Colloidal Interace Sci 8:48-54
Lowenstam HA, Weiner S (1989) On Biomineralization. Oxford University Press, Oxford, England
Michenfelder M, Fu G, Lawrence C, Weaver JC, Wustman BA, Taranto L, Evans JS, Morse DE (2003) Characterization of two molluscan crystal-modulating biomineralization proteins and identification of putative mineral binding domains. Biopolymers, in press.
Jendrisak J, Young RA, Engel J (1987) Guide to Molecular Cloning Techniques. Berger S, Kimmel A (eds) Academic Press, San Diego, CA
Sambrook J, Fritsch EF, Maniatis T (1989) Molecular cloning: A Laboratory Manual. Cold Springs Harbor Laboratory, Cold Springs Harbor, NY
Singer M, Berg P (1991) Genes and Genomes: A Changing Perspective. University Science Books, Mill Valley, CA
Stryer L (1995) Biochemistry. W.H. Freeman and Company, New York, NY
Wustman B, Morse DE, Evans JS (2002a) Strucutral analyses of polyelectrolyte sequence domains within the adhesive elastomeric biomineralization protein Lustrin A. Langmuir 18:9901-9906
Wustman B, Santos R, Zhang B, Evans JS (2002b) Identification of a glycine loop-like coiled structure in the 34-AA Pro, Gly, Met repeat domain of the biomineral-associated protein, PM27. Biopolymers 65:1305-1318
Wustman BA, Weaver JC, Morse DE, Evans JS (2004) Characterization of a Ca (II)-, mineral-interactive polyelectrolyte sequence from the adhesive elastomeric biomineralization protein, Lustrin A. Langmuir, in press
Xu G, Evans JS (1999) Model peptide studies of sequence repeats derived from the intracrystalline biomineralization protein, SM50. I. GVGGR and GMGGQ repeats. Biopolymers 49:303-312
Zhang B, Wustman B, Morse D, Evans JS (2002) Model peptide studies of sequence regions in the elastomeric biomineralization protein, Lustrin A. I. The C-domain consensus -PG-, -NVNCT- motif. Biopolymers 64:358-369
Zhang B, Xu G, Evans JS (2000) Model peptide studies of sequence repeats derived from the intracrystalline biomineralization protein, SM50. II. Pro, Asn-rich tandem repeats. Biopolymers 54:464-475

3 Principles of Crystal Nucleation and Growth

James J. De Yoreo
Chemistry and Materials Science Directorate
Lawrence Livermore National Laboratory
Livermore, California 94551 U.S.A.

Peter G. Vekilov
Department of Chemical Engineering
University of Houston
Houston, Texas 77204 U.S.A.

INTRODUCTION

In the most general sense, biomineralization is a process by which organisms produce materials solutions for their own functional requirements. Because so many biomineral products are derived from an initial solution phase and are either completely crystalline or include crystalline components, an understanding of the physical principles of crystallization from solutions is an important tool for students of biomineralization. However, crystal growth is a science of great breadth and depth, about which many extensive texts have been written. In addition, there are already other thorough reviews that specifically address the crystal growth field of study as it relates to biomineral formation. Consequently, the goals of this chapter are both modest and specific. It is intended to provide: 1) a simple narrative explaining the physical principles behind crystallization for those who are completely new to the topic, 2) a few basic equations governing nucleation and growth for those who wish to apply those principles—at least in a semi-quantitative fashion—to experimental observations of mineralization, and 3) an overview of some recent molecular-scale studies that have revealed new insights into the control of crystal growth by small molecules, both organic and inorganic.

This last topic gets to the heart of what makes crystallization in biological systems unique. Every day, many tons of crystals are produced synthetically in non-biological processes, but by-and-large, the degree of control over nucleation and growth achieved by deterministic additions of growth modifiers or the presence of a controlling matrix is very minor. More commonly, crystal growers view modifying agents as unwanted impurities and work extremely hard to eliminate them from the starting materials. Indeed, the degree to which living organisms are able to control the crystallization process is most striking when contrasted to the products of such synthetic crystallization processes. This contrast applies to both the compositional differences that result (e.g., Weiner and Dove 2003) and the external morphology. Figure 1a shows a typical photomicrograph of a crystal of calcium carbonate grown in a closed beaker from a pure solution. It has a simple rhombohedral shape and consists of calcite, the most stable form of calcium carbonate. Moreover, crystals grown in this way tend to nucleate randomly throughout the beaker or on its surfaces with no preferred orientation and grow at a rate determined by the solution conditions as well as various inherent materials parameters. Figure 1b-d shows examples of biological calcite in the form of shaped single crystals, intricate crystal composites, and regular crystal laminates. The calcite crystals, as coccolith plates (Fig. 1b), lie in specific locations and are oriented with respect to the crystallographic axes of calcite; whereas the mollusk shell crystallites (Fig. 1c) are aragonite, which in standard laboratory conditions is a less stable form of calcium carbonate than is calcite. The calcite crystals from the human inner ear (Fig. 1d) exhibit a spindle-like shape that is

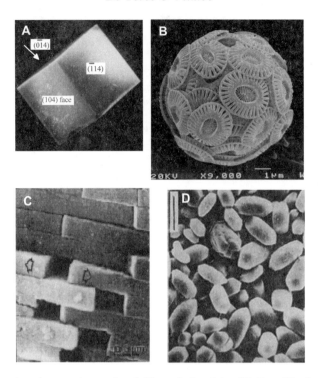

Figure 1. Scanning electron micrographs of (A) synthetic calcite, (B) Coccolithophorid, *Emiliania huxleyi* (from International Nannoplankton Association website), (C) Aragonitic layers from Mollusc shell (From Calvert 1992 with permission from Materials Research Society), and (D) calcite crystals from the human inner ear (From Mann 2001 with permission from Oxford University Press).

quite different from that of synthetic calcite. These examples demonstrate that organisms are able to control the location and crystallographic orientation of nucleation, the shape of the growing crystallites, and even the phase of the resulting material.

One of the major challenges in crystal growth science today is to understand the physical mechanisms by which this level of control is achieved. A very general and useful construct for thinking about the problem is the "energy landscape," illustrated in Figure 2. This landscape highlights the fact that crystallization is, first and foremost, a phase transition through which matter is transformed from a state of high free energy in a solvated state to one of low free energy in the crystal lattice. All aspects of a crystal, including its phase, habit, and growth rate, are determined by the shape of this landscape. Equilibrium crystal habit and phase are controlled by the depths and shapes of the energy minima. By varying the heights of the barriers, the growth kinetics can be controlled, and non-equilibrium final or intermediate states can be selected. It stands to reason that living organisms modulate crystal growth by manipulating the energy landscapes. A complete physical picture of biomineral growth requires a description of the geometry and stereochemistry of the interaction between the crystal lattice and the organic modifiers, the magnitude of the interaction energy, the effect of that interaction on the energy landscape, and the impact of the change in the landscape on crystallization.

Finally, we note that this physiochemical picture provides only a small piece to the

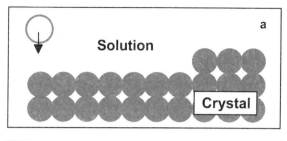
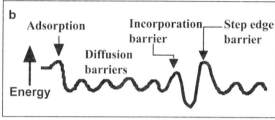

Figure 2. Schematic of (a) physical landscape and (b) energy landscape seen by a solute molecule as it becomes part of the crystal. The vertical axis in (b) can either refer to the actual potential energy if tied directly to the physical landscape, or the free energy if referring to processes encountered in the transformation to the solid state, such as dehydration or incorporation.

puzzle of biomineralization. As Figure 3 illustrates, there are many levels of regulation at work that influence these physical processes (Mann 2001). While they include other physical parameters such as spatial constraints and controls over ion fluxes, these are, in turn, influenced by a myriad of biological machines that shuttle reactants and products to desired locations, undoubtedly utilizing biochemical feedback. These processes are further influenced by a combination of genetic and environmental factors.

CRYSTAL NUCLEATION VS. CRYSTAL GROWTH

The association of organic compounds with biomineralized structures (Wada and Fujinuki 1976; Crenshaw 1972; Collins et al. 1992) and their observed effect on crystallization kinetics (Sikes et al. 1990; Wheeler et al. 1990) strongly suggest that these compounds modify the growth stage of minerals. This conclusion is reinforced by observations of faceted crystal surfaces in a number of cases, including the calcite crystals shown in Figure 1d as well as numerous examples of calcium oxalate monohydrate crystals, both functional—as in plants—and pathogenic—as in humans (Fig. 4). On the other hand, many biomineralized structures in nature suggest that organic components also control nucleation. In particular, the growth of biominerals in precise locations within complex crystal composites and the generation of crystallites with specific crystallographic orientations are difficult to explain without appealing to active controls during the nucleation stage. Moreover, there is a substantial body of evidence to suggest that proteins and other organic molecules serve as "templates," providing preferential sites for nucleation and controlling the orientation of the resulting crystals. Many organic-inorganic composites, including the shells of birds' eggs (Fink et al. 1992) and mollusks (Calvert 1992), exhibit a layered structure in which the inorganic component grows at the organic-solution interface. One of the most remarkable examples is that of bone. The carbonated apatite crystals that give bone its stiffness grow amongst

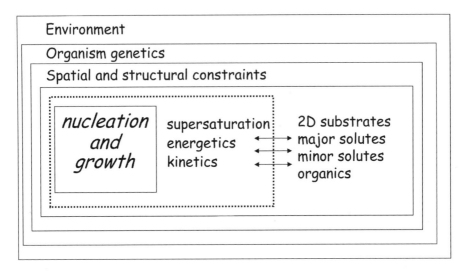

Figure 3. Levels of regulation in biomineralization. (from Dove et al. 2004, based on Mann 2001).

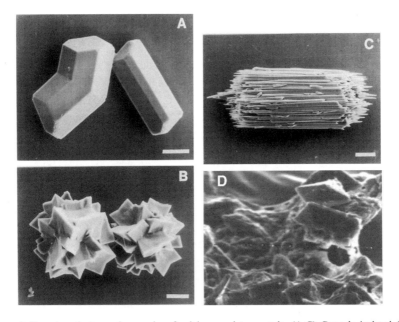

Figure 4. Scanning electron micrographs of calcium oxalate crystals. (A-C) Crystals isolated from selected plants. Scale bar = 5 μm. (A) Twinned, prismatic crystals from the seed coat of bean (*Phaseolus vulgaris*). (B) Druse crystals isolated from velvet leaf (*Abutilon theophrasti*), a common weed. (C) Isolated bundle of raphides, needle-shaped crystals, from leaves of grape (*Vitis labrusca*). (From Webb 1999 with permission from the American Society of Plant Biologists.) (D) Crystals grown in vivo on uretal stent. Crystals are about 50 mm in size. (Used with kind permission of Professor Heywood).

molecular fibers of collagen. But the location and orientation of the crystallites is not random. The collagen fibers are believed to pack into bundles with a periodicity that leaves rows of small gaps known as "hole zones" 40 nm in length and 5 nm in width, each of which provides an identical molecular-scale environment for mineralization, as illustrated in Figure 5 (Katz and Li 1973; Miller 1984; Mann 2001). The crystallites themselves grow with a specific geometric relationship between the crystal lattice and the collagen fibers, with the {001} axis lying along the length of the fiber and the {110} axis pointing along the rows of hole zones. This implies molecular-scale control by the collagen molecules over both the orientation and location of the crystal nuclei.

THERMODYNAMIC DRIVERS OF CRYSTALLIZATION

Whether considering nucleation or growth, the reason for the transformation from solution to solid is the same, namely the free energy of the initial solution phase is greater than the sum of the free energies of the crystalline phase plus the final solution phase (Gibbs 1876, 1878). In terms of solution activities (which are often well approximated by solution concentrations), an equivalent statement is that the actual activity product of the reactants, AP, exceeds the equilibrium activity product of those reactants, the latter being simply the equilibrium constant, K_{sp}. In cases where a single chemical component crystallizes, the driving force has been expressed not in terms of total free energy change during crystallization, but rather as the change in chemical potential of the crystallizing species, $\Delta\mu$. This $\Delta\mu$ measures the free energy response to molecules transferring from one phase to the other. The larger $\Delta\mu$ becomes, the greater is the driving force for crystallization (Mullin 1992).

Not surprisingly, both the change in free energy and the change in chemical potential are directly related to the activity products (Johnson 1982). For the precipitation reaction:

$$a\text{A} + b\text{B} + \ldots + n\text{N} \rightarrow \text{A}_a\text{B}_b\cdots\text{N}_n \qquad (1)$$

the activity product of the reactants, AP, and the value of K_{sp} are given by:

$$AP = [\text{A}]^a[\text{B}]^b[\text{C}]^c\cdots[\text{N}]^n \qquad (2)$$

$$K_{sp} = [\text{A}]_e^a[\text{B}]_e^b[\text{C}]_e^c\cdots[\text{N}]_e^n \qquad (3)$$

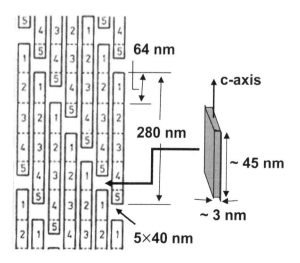

Figure 5. Schematic of collagen fibrils showing the stagger of the hole zones and the size and orientation of the carbonated hydroxyapatite crystals relative to the fibril direction. (Based on Mann 2001).

where the subscript "e" refers to the activity at equilibrium. In turn, the free energy of solution per molecule, Δg_{sol}, and the change in chemical potential are:

$$\Delta g_{sol} = -k_B T \ln K_{sp} \tag{4}$$

$$\Delta \mu = k_B T \ln AP - \Delta g_{sol} \tag{5a}$$

$$= k_B T \ln(AP/K_{sp}) \tag{5b}$$

where k_B is the Boltzmann constant, and T is the absolute temperature. Rather than use $\Delta \mu$, most crystal growth analyses refer to the supersaturation, σ, which is related to $\Delta \mu$ by:

$$\Delta \mu = k_B T \ln \sigma \tag{6}$$

$$\sigma \equiv \ln\{AP/K_{sp}\} \tag{7}$$

Readers may often see σ defined as $(AP/K_{sp}) - 1$ or even $(C/C_e) - 1$ where C and C_e are the equilibrium reactant concentrations. These are referred to as absolute supersaturations. They are valid approximations only at small values of σ and, in the case of the expression that utilizes concentrations, only useful when either the activity coefficients, $\chi = [N]/C$ and $\chi_e = [N_e]/C_e$ for the reactants are close to unity, or when $\chi = \chi_e$. Nonetheless the absolute supersaturation is important when considering the kinetics of growth. This will become clear later in the chapter.

In the sections that follow, the way in which σ and $(AP - K_{sp})$ influence nucleation and growth will be presented in detail, but it is instructive at this point to discuss, in a general sense, the external factors that can influence these parameters, since crystal growth can be modulated through their manipulation. The values of K_{sp} and C_e depend on solution composition and temperature as, to some extent, do the values of the activity coefficients, but presumably temperature variation is not an important tool for biological systems. Consequently, in order to manipulate these parameters, organisms must rely primarily on variations in solution chemistry, including ionic strength, pH, and impurity content, whether inorganic or organic.

One other important factor worth discussing is the role of water. It has recently been realized that the structuring of water molecules around ions, molecules, and colloid-sized particles in dissolved state is an important factor in the thermodynamics of phase transitions in solution. In some cases, it was found that the structured water molecules are released when the ions or molecules join a growing crystal (Yau et al. 2000a), while in other cases, additional water molecules are trapped during crystal formation (Vekilov et al. 2002a). The release or trapping of water molecules has significant enthalpic and entropic consequences. The enthalpy effect depends on the strength of the bonds between water and solute molecules and the strength of the bonds forming in the crystals. Hence, it varies significantly from system to system. The entropy effect, however, appears to be consistently ~20 J K^{-1} per mole of water released or trapped, with a positive sign in the cases of release and negative sign in the cases of trapping. Not surprisingly, the magnitude of the entropy effect above is comparable to the entropy loss upon freezing of water (Tanford 1994; Dunitz 1994).

Especially with larger species such as proteins, the entropy contribution due to solvent trapping or release can be very significant and can even be the determinant of the free energy change for crystallization. For example, it was found that the enthalpy effect of crystallization of the protein apoferritin is negligible, and crystallization occurs only because of the significant entropy gain due to the release of just one or two water molecules in each of the 12 intermolecular contacts in the crystal (Yau et al. 2000a). This entropy gain adds up to positive (+160 J K^{-1}) per mole of apoferritin, with a free energy contribution of about -47 kJ mol^{-1}. An even more striking example is the human mutant hemoglobin C, which has a significant positive enthalpy of crystallization of 160 kJ mol^{-1}, i.e., crystallization is only possible because of the release of about 10 water molecules when a crystal contact is formed. In this case, the total entropy gain due to this release is about 600 J K^{-1} per mole of hemoglobin, and its contribution to the free energy of crystallization is -180 kJ mol^{-1}, making the net gain of free energy upon crystallization about -20 kJ mol^{-1} (Vekilov et al. 2002a).

Note that this discussion of the effects of water structuring focuses only on the contributions to the initial and final states on the free energy landscape. As discussed below in relation to the kinetics of crystallization, water structuring is also the most important contributor to the free energy barrier for crystallization, and the kinetics of restructuring of the water associated with a solute molecule seems to determine the overall crystallization rate.

NUCLEATION

Nucleation is one of the two major mechanisms of the first order phase transition, the process of generating a new phase from an old phase whose free energy has become higher than that of the emerging new phase (Hohenberg and Halperin 1977; Chaikin and Lubensky 1995). Nucleation occurs via the formation of small embryos of the new phase inside the large volume of the old phase. Another prominent feature of nucleation is metastability of the old phase, i.e., the transformation requires passage over a free energy barrier (Kashchiev 1999). This is easily understood by considering the free energy changes associated with the formation of the nucleus. The statement that the free energy per molecule of the new phase is less than that of the solvated phase only applies to the bulk of the new phase. The surface is a different matter. Because the surface molecules are less well bound to their neighbors than are those in the bulk, their contribution to the free energy of the new phase is greater. The difference between the free energy per molecule of the bulk and that of the surface is referred to as the interfacial free energy. (It is sometimes called the surface free energy, but, strictly speaking, this term should be reserved for surfaces in contact with vacuum.) The interfacial free energy is always a positive term and acts to destabilize the nucleus. As a consequence, at very small size when many of the molecules reside at the surface, the nucleus is unstable. Adding even one more molecule just increases the free energy of the system. On average, such a nucleus will dissolve rather than grow. But once the nucleus gets large enough, the drop in free energy associated with formation of the bulk phase becomes sufficiently high that the surface free energy is unimportant, and every addition of a molecule to the lattice lowers the free energy of the system. There is an intermediate size at which the free energy of the system is decreased whether the nucleus grows or dissolves, and this is known as the critical size. This phenomenon is referred to as the Gibbs-Thomson effect. Of course, if the supersaturation is high enough, the critical size can be reduced to less than one growth unit. Then the barrier vanishes and the old phase becomes unstable so that an infinitesimal fluctuation of an order parameter, such as density, can lead to the appearance of the new phase. The rate of generation and growth of the new phase is then only limited by the rate of transport of mass or energy. This second process is referred to as spinodal decomposition, and the boundary between the regions of metastability and instability of the old phase is called a spinodal line (Binder and Fratzl 2001; Kashchiev 2003).

The existence of a critical size has a number of implications. First and foremost, because nucleation of the new phase is the result of fluctuations that bring together sufficient numbers of molecules to exceed the critical size, the probability of nucleation will be strongly affected by the value of the critical size. This means that nucleation can be controlled, to some extent, by modulating the critical size, which is in turn a function of the interfacial energy. The smaller the interfacial energy, the smaller the critical size and the more likely nucleation becomes for any given supersaturation. As a consequence, by varying either the solution composition or the supersaturation, the probability of nucleation can be manipulated.

Nucleation on foreign surfaces

The presence of a foreign surface can be used to exert even greater control over nucleation because, quite often, the interfacial energy between a crystal nucleus and a solid substrate is lower than that of the crystal in contact with the solution (Neilsen 1964; Abraham 1974; Chernov 1984; Mullin 1992; Mutaftschiev 1993). This is because the molecules in the crystal can form bonds with those in the substrate that are stronger than the bonds of solvation. Because the enthalpic contribution to the free energy comes primarily from chemical bonding, stronger bonds lead to a smaller interfacial free energy. This may well be the key physical phenomenon that allows living organisms to delineate both location and orientation of crystallites. Clearly, the strength of bonding at the interface is strongly dependent on the structure and chemistry of the substrate surface. If the atomic structure of the substrate surface closely matches a particular plane of the nucleating phase so that lattice strain is minimized and, in addition, the substrate presents a set of chemical functionalities that promote strong bonding to the nucleus, then the enthalpic contribution to the interfacial free energy becomes small, and nucleation occurs preferentially on that crystal plane.

As discussed in the introduction, many lines of evidence suggest that organisms utilize this principle to control the location and orientation of nuclei. In addition, several experimental studies (Mann et al 1992; Heywood and Mann 1994; Wong et al. 1994; Berman et al. 1995; Archibald et al. 1996; Addadi et al. 1998; Aizenberg et al. 1999a,b, 2003; Travaille et al. 2002) have used successfully used organic films to mimic the process. For example, Addadi et al. (1997) showed that β-sheet polyaspartate adsorbed onto sulfonated polystyrene substrates induced nucleation of (001) oriented calcite crystals. Acidic glycoproteins from mollusk shells had similar effects *in vitro*. Aizenberg et al. (1999a,b) showed that varying the structure of terminal groups on self-assembled monolayers (SAMs) of alkane thiols on gold precisely controlled the plane of nucleation of calcite and Travaille et al. (2002) demonstrated a 1:1 relationship between the underlying gold structure and the in-plane calcite orientation. Based on the above discussion, it is difficult to interpret these observations in any way other than the result of varying the interfacial energy. Nonetheless, little or no work has been done to quantify the relative importance of thermodynamic control through variations in interfacial energy, local increases in supersaturation due to changes in pH or ion concentration near the surface films, or even kinetic controls on adsorption/desorption rates that affect surface adsorption lifetimes and thus rates of nucleation.

We can quantify these concepts by considering the dependence of the change in free energy during precipitation on the size of the nucleus (Kashchiev 1999; Mullin 1992). For simplicity, we consider a spherical nucleus of radius r, as illustrated in Figure 6a, nucleating within the bulk solution. (This is referred to as homogeneous nucleation.) The free energy change per molecule (Δg) is given by the sum of bulk (Δg_b) and surface terms (Δg_s), namely:

$$\Delta g = \Delta g_b + \Delta g_s \qquad (8a)$$

$$= -\{[(4/3)\pi r^3]/\Omega\}\Delta\mu + 4\pi r^2 \alpha \qquad (8b)$$

where Ω is the volume per molecule, and α is the interfacial free energy. The first term is

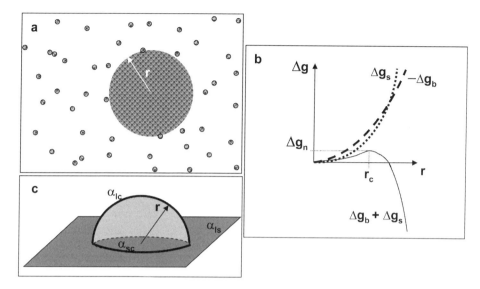

Figure 6. (a) Formation of a spherical nucleus of radius r from a solution leads to the free energy changes shown in (b). The cross-over of the bulk and surface terms combined with their opposing signs leads to a free energy barrier. (c) Heterogeneous formation of a hemispherical nucleus at a foreign substrate.

negative and varies as the cube of r, while the second term is positive and varying as r^2. As shown in Figure 6b, the sum of the two terms has a maximum that occurs when $d\Delta g/dr = 0$. The value of r at this point can be found by taking the derivative of (8b) and setting it equal to zero. This is known as the critical radius, r_c, and is given by:

$$r_c = 2\Omega\alpha/\Delta\mu \quad (9a)$$

$$= 2\Omega\alpha/kT\sigma \quad (9b)$$

The same analysis can be performed for heterogeneous nucleation, i.e., nucleation at a foreign surface. There are now two interfacial energies to consider, one between the crystal and solution and the other between the crystal and the substrate, as illustrated in Figure 6c. If we assume, for simplicity, that the nucleus is a hemisphere of radius r, we have:

$$\Delta g = -\{[(2/3)\pi r^3]/\Omega\}\Delta\mu + \pi r^2(2\alpha_{lc} + \alpha_{sc} - \alpha_{ls}) \quad (10)$$

where the subscripts "sc", "lc", and "ls" refer to substrate-crystal, liquid-crystal, and liquid-substrate respectively. The new expression for r_c becomes:

$$r_c = 2\Omega\alpha'/k_BT\sigma \quad (11a)$$

$$\alpha' = \alpha_{lc}\{1-(\alpha_{ls} - \alpha_{sc})/2\alpha_{lc}\} \quad (11b)$$

The term in the brackets is always less than one, provided the free energy of the crystal-substrate interface is less than that of the substrate-liquid interface. Thus the value of r_c at the substrate is reduced below that for nucleation in free solution. Figure 7 shows an example of the control over nucleation by surface structure in case of calcite nucleation on alkane thiol self-assembled monolayers.

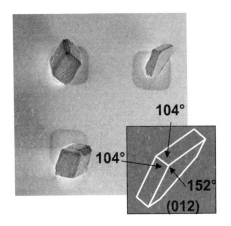

Figure 7. Nucleation of calcite on a template. The substrate consists of gold-coated mica which has been patterned with mercaptohexadecanoic acid (MHA)—a carboxyl terminated alkane thiol—using a micro-stamping method. The surrounding region was then coated with octadecane thiol. The crystals all nucleated on the MHA terminated regions and almost always with the (012) face as the plane of nucleation.

Nucleation pathways and Ostwald's Law of phases

If a solution is supersaturated, then regardless of the critical size or the presence of a foreign surface, the solution will eventually crystallize. What is significant about the critical size is that it controls the probability of a nucleus forming on any given timescale. In other words, it determines the kinetics of nucleation. To understand and quantify this, again consider Figure 6b. It shows that along with a critical nucleus size comes a nucleation barrier whose magnitude, Δg_n can be determined by substituting (9b) or (11a) into (8b) to get:

$$\Delta g_n = (16/3)\pi\alpha^3(\Omega/k_B T\sigma)^2 \tag{12a}$$

$$\propto \alpha^3/\sigma^2 \tag{12b}$$

It is this barrier that determines the kinetics of nucleation. As with any kinetically-limited chemical process, the nucleation probability is proportional to the exponential of the barrier height divided by $k_B T$. Thus the nucleation rate (Neilsen 1964; Abraham 1974) is given by:

$$J_n = A\exp(-\Delta g_n/k_B T) \tag{13}$$

where A is a factor that depends on many parameters. Substituting in the expression for Δg_n from (12) gives:

$$J_n = A\exp(-B\alpha^3/\sigma^2) \tag{14}$$

where we have grouped all of the factors other than interfacial energy and supersaturation into the coefficient B. Equation (14) shows just how strongly the nucleation rate depends on the supersaturation and the interfacial energy. They come in as the 2nd and 3rd powers respectively in the argument of an exponential! Figure 8 illustrates how strongly this function varies with supersaturation by showing the reciprocal of J_n, which is a measure of the induction time until nucleation at any given supersaturation.

An inherent assumption in the above discussion is that the pathway of nucleation goes directly from the solution to the formation of nuclei with ordered crystalline structure identical to that of the eventual bulk crystal. When the main concepts concerning equilibrium between large and small phases and the generation of nuclei of a new phase from an old were introduced by J.W. Gibbs, he was careful to point out that the composition of the "globulae," as he called them (i.e., the nuclei), would likely differ

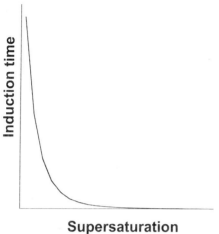

Figure 8. The induction time for nucleation is a strong function of supersaturation.

from the composition of a large sample of the new phase (i.e., the bulk) (Gibbs 1876, 1878). Since the surface tension of the "globulae," the main factor that determines the height of the barrier for nucleation, is not necessarily related to the surface tension of the interface between bulk samples of the two phases and cannot be measured independently, Gibbs stated that surface tension should be viewed as an adjustable parameter.

In further developments, some of the Gibbs's assumptions regarding the thermodynamics of nucleation of a fluid phase were transferred to the nucleation of ordered solids, such as crystals, from dilute or condensed fluids or from other solid states (Katz and Ostermier 1967; Neilsen 1967; Kahlweit 1969; Walton 1969). The classical nucleation theory emerged, in which the so called "capillary approximation" was widely used. In application to crystals, the approximation means that the molecular arrangement in a crystal's embryo is identical to that in a large crystal and hence, the surface free energy of the nuclei will equal the one of the crystal interface. However, there are many instances in which this is unlikely to be the case, simply because the energy barrier leading to a more disordered, less stable state is less than the one leading to the most stable state. This is the basis of the Ostwald-Lussac law of phases, which maintains that the pathway to the final crystalline state will pass through all less stable states in order of increasing stability (Nancollas 1982). If this law were strictly true, then, for example, the crystallization of $CaCO_3$ at room temperature from a pure solution would proceed by nucleation of amorphous material, which would then transform into vaterite, then aragonite, and finally calcite. Indeed, new evidence shows some calcitic biominerals begin as an amorphous calcium carbonate precursor that later transforms to its lower energy crystalline counterpart (Addadi et al. 2003).

Whether or not the Ostwald-Lussac law generally holds is unknown, as nucleation events are so difficult to study, both because they occur spontaneously and because the critical nucleus size is very small, typically in the 1 to 100 nm range. Nonetheless, one can construct a reasonable physical basis for the law. Taking Equation (14) as the rate of nucleation of a given phase, the only way a less stable phase can nucleate more rapidly than the most stable phase is if α^3/σ^2 is reduced by nucleating the less stable phase. But to say that this phase is less stable implies that the K_{sp} is larger, which in turn implies that σ is smaller. This means that, in order for the ratio α^3/σ^2 to become smaller, α must be reduced by an even greater factor for the less stable phase. Recall that α is proportional to the difference between the change in free energy in forming a crystal with a surface (Δg) and that of forming an infinite crystal (Δg_b). In fact, starting with Equations (8a) and (8b), we have

$$\alpha = (\Delta g - \Delta g_b)/4\pi r^2 \qquad (15a)$$

$$= \Delta g_s/4\pi r^2 \qquad (15b)$$

The free energy terms, Δg_i, are themselves comprised of two terms: the change in enthalpy, Δh_i, and the change in entropy, Δs_i, i.e., $\Delta g_i = \Delta h_i - T\Delta s_i$). Now let's examine each of the terms. The statement that a phase is less stable is equivalent to one stating that Δg_b is smaller. That this should be so makes sense because the poorer bonding of the more disordered phase ensures that Δh_b will be smaller for its formation, and the higher level of disorder has the same impact on Δs_b, as illustrated in Figure 9. Therefore, the only way that α can become smaller is for the total change of free energy to decrease by even more, i.e., Δg_s must decrease as well. Equation (15b) reflects this. As with the bulk, since the surface of the less stable phase is likely to be more disordered than the surface of the most stable phase, both Δh_s and Δs_s are likely to be smaller. Nonetheless, this is not a proof, and the Ostwald-Lussac law of phases should only be viewed as a guiding principle rather than a statement of fact.

The nucleus shape

Equations (8)–(14) above were derived under the assumption of a spherical shape of the nucleus. If the nucleus has any other shape, obviously, the coefficients included in the parameters A and B in Equation (14) will take different values, but the overall conclusions on nucleation kinetics, it appears, will not change significantly. There is an important caveat in the latter statement, and it is in the assumption that we have a means to predict the nucleus shape in advance. In some cases, this is indeed so. Thus, thinking of a fluid phase nucleating within another fluid, Gibbs suggested that the shape of the nuclei is the one ensuring the lowest free energy of the nuclei, i.e., the sphere (Gibbs 1876, 1878). Accounting for crystal anisotropy and applying the free energy minimization selection criterion, one comes up with cubic or other faceted shapes and only slight modifications to Equations (8)–(14).

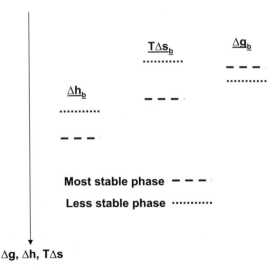

Figure 9. Schematic illustrating the relative contributions of enthalpy and entropy changes to the free energy change upon crystallization into the most stable and a less stable phase.

However, free energy minimization is not an absolute criterion for nucleus shape selection. In the simple model system considered by Gibbs, minimal free energy of the nucleus provides for a minimal free energy barrier for nucleation and, hence, for a nucleation pathway of fastest kinetics. Thinking of formation of structured crystalline nuclei, one is forced to realize that the one-dimensional picture, with the nucleus size as the only variable, is perhaps oversimplified. The free energy landscape then appears as a surface in a multi-dimensional space. In this space, multiple pathways exist between the initial state, that of single molecules suspended in the solvent, and the final state, crystals of sizes larger than the critical that grow, driven by $\Delta\mu$. Obviously, the observed rate of nucleation depends on the pathway that was selected or, in other words, we only see the products of the fastest nucleation pathway. Thus, we come up with a more general criterion for nucleus shape selection—which ensures the fastest nucleation rate.

With this in mind, one can understand the two experimental observations of crystalline nuclei that have yielded shapes far from the expected compact, smooth and polyhedral. It was found that the nuclei of the protein apoferritin are raft-like and consist of several rows of molecules in a single crystalline layer as shown in Figure 10 (Yau and Vekilov 2000, 2001). Obviously, due to the low number of intermolecular contacts this shape has a free energy significantly higher than a conceivable compact shape. However, it was found that for this protein, the formation of intermolecular contacts is significantly hampered (Chen and Vekilov 2002; Petsev et al. 2003) by a free energy barrier linked to the water structuring around the polar surface residues (see Figs. 11 and 12) (Petsev and Vekilov 2000a,b). It has been suggested that the quasi-planar shape ensures faster

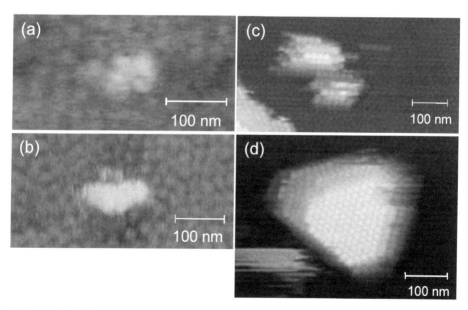

Figure 10. AFM images of sub-critical, near-critical clusters, and super-critical crystallites in the nucleation pathway of the apoferritin crystals. (a) A cluster consisting of two molecules with apoferritin concentration, $C = 0.23$ mg/mL and $\sigma = 2.3$. (b) A cluster consisting of six molecules in two rods with four and two molecules in each of the rods with $C = 0.04$ mg/mL and $\sigma = 0.5$. In both (a) and (b), the clusters have landed on the bottom of an AFM cell covered with a single layer of apoferritin molecules. (c) A near critical cluster that has landed on the (111) face of a apoferritin crystal at $\sigma = 1.1$. (d) A crystallite of ~ 150–180 molecules on a large crystal at $\sigma = 1.1$.

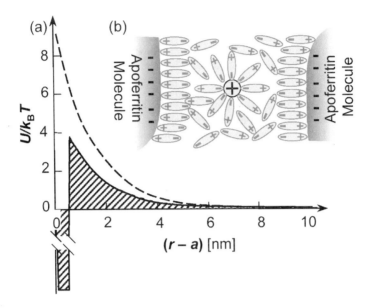

Figure 11. (a) The potential of pair interactions between apoferritin molecules. *Dashed line*: in the presence of $[Na^+] = 0.20$ M according to a hydration force model. *Solid line*: schematic representation of the interaction potential after the addition of Cd^{2+}. (b) Schematic representation of the build-up of counterions at the surface of the protein molecules that leads to the hydration repulsion reflected in (a).

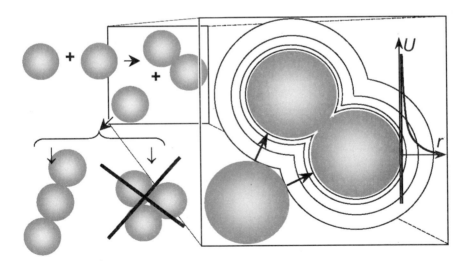

Figure 12. Schematic of formation of a trimer of molecules with an interaction potential consisting of long-range repulsion and short-range attraction, highlighted on $U(r)$ plot at far right. Contours around zoomed-in dimer represent repulsive potential lines in the plane of the image.

kinetics of nucleation because of the low number of intermolecular contacts that have to form and the low number of barriers that have to be overcome (Vekilov and Chernov 2002). Note that the presence of similar free energy barriers for the formation of crystalline contacts cannot be excluded for any material crystallizing from solution. Hence, non-equilibrium nucleus shapes maybe the rule, rather than the exception, for solution-grown crystals.

Another example of nucleus visualization comes from studies of a crystallizing suspension of colloid particles in a specially selected organic solvent (Gasser et al. 2001). At the relatively low supersaturations involved, the nuclei were quite large, consisting of around 100 particles, and had the shape of oblate ellipsoids. This shape deviates from the expected sphere and the surface was found to be rough, with many protrusions of three or four particles. The surface roughness is associated with high surface free energy, and a likely explanation for its presence is that it facilitates the attachment of particles to the nucleus and in this way contributes to faster kinetics of nucleation.

Crystal growth kinetics

Figure 13a schematically illustrates atomic processes occurring at a crystal surface. The surface consists of flat regions called terraces and raised partial layers called steps (Chernov 1961, 1984, 1989). The steps themselves are also incomplete, containing kinks. The kink sites are very important because molecules that attach there make more bonds to neighboring molecules than the ones that attach to the terraces or to flat step edges. Consequently they are more likely to stick. Conversely, when molecules leave the crystal, they can do so more easily by detaching from kinks than from either complete step edges

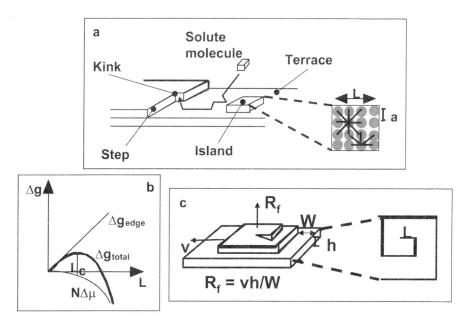

Figure 13. (a) Illustration of the atomic processes during crystal growth. Solute molecules enter kinks either directly from solution or after adsorbing and diffusing across terraces. Islands can nucleate on the terraces provided they overcome the free energy barrier to 2D nucleation illustrated in (b). (c) shows the geometry of a dislocation hillock. The existence of a critical size leads to the formation of the spiral structure, since the new segment of the step cannot move until it reaches that size.

or from embedded sites in the terraces. As a result, the rate at which molecules can be added to a crystal, for a given solute concentration, scales with the kink density. This means that the growth rates of crystals can be altered by either blocking kink sites or by roughening steps. We will come back to this later.

Even at equilibrium, the steps have kinks due to thermally activated detachment of molecules from the steps onto either the step edges or the terraces or even back into solution (Frenkel 1932; Burton et al. 1951; Chernov 1984). Consequently the step edges are not static; molecules are constantly attaching and detaching, even at equilibrium (Williams and Bartelt 1991; Barabasi and Stanley 1996). Figure 14a shows an example of a step edge at high resolution showing how this process makes the edge "fuzzy." If the crystal has weak bonds in all directions (think of non-interacting colloids as an extreme example), the step is perfectly rough and has an equilibrium kink density of about 1/2, i.e., every other site is a kink. Of course in most crystals, the bonds are anisotropic with some strong and some weak. The strengths of the bonds and the anisotropy in bonding determine the equilibrium kink density. For example, if a crystal possesses strong bonds parallel to a step edge but weak bonds perpendicular to the edge, the energy penalty for creating kinks is large, and the step will have a low kink density. If the converse is true, then the energy penalty is small, and the step will have a high kink density. All other factors being equal, these two types of steps will advance at very different rates. Figure 14b shows an example of steps on a crystal that has high anisotropy in bonding and exhibits rough steps in one direction and smooth steps in the other. Not surprisingly, at

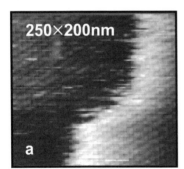
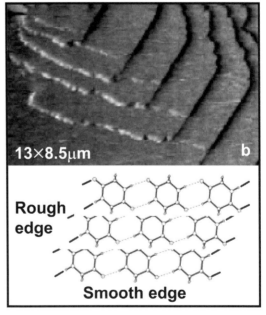

Figure 14. AFM images of (a) a step on a crystal of the protein canavalin showing the fuzziness of the step due to attachment and detachment of molecules, and (b) two types of steps on the surface of a glycine derivative of diketopiperazine (Gly-DKP) along with the packing geometry of the Gly-DKP molecules. The strong amine bonds along one axis and weak inter-ring bonds perpendicular to the plane of the rings leads to smooth steps along the strong-bonding direction and rough steps along the other.

the same supersaturation, the rate at which the rough step advances is ten times the rate for the smooth steps.

Growth from a supersaturated solution occurs because the flux of molecules attaching to the crystal surface exceeds the flux of molecules detaching from the surface. The probability that a molecule will detach from the crystal is solely determined by the strength of its bonds to its neighbors. Since the strength of bonding is a function of temperature rather than the flux to the surface, the total flux from the surface is nearly independent of concentration. In contrast, the flux to the surface is proportional to the bathing concentration. The solubility is then the concentration at which the two fluxes are equal. This raises an important aspect of crystal growth kinetics. Since the flux of molecules to the surface depends on the actual solute concentration (or activity), then, other factors being equal, highly soluble crystals will grow faster than sparingly soluble crystals, *even at equivalent supersaturations*. Shifting the solubility automatically changes the growth kinetics, providing yet another means of altering crystal growth rates (Davis et al. 2000).

The kinetics of the attachment and detachment processes at step edges are determined by the energy barriers seen by the molecules. It has been suggested that the barrier to desolvation, i.e., the breaking of bonds to solvent molecules, is the dominant barrier to attachment of a solute molecule to the step (Chernov 1961, 1984). While the basic idea that the solvation water is a major part of the barrier seems to have withstood the tests of time, recent simulations and modeling have modified the original scenario. Thus molecular dynamics simulations have shown that the life time of the hydrogen bonds in the structured water layer on the surface of protein molecules is of the order of nanoseconds (Makarov et al. 2000, 2002)—many orders of magnitude shorter than the characteristic time scales of the molecular attachment. In other words, the bonds between solvent molecules are broken and restored many times while a molecule is *en route* to an incorporation site, and the breaking of such a bond cannot be the rate-limiting stage. While an exact picture of the rearrangement of the water structure upon incorporation is still missing, there is significant evidence that the kinetic barrier is related to the energy characteristics of this rearrangement, and the rate of molecular attachment is determined by its rate (Petsev et al. 2003).

Clearly, all processes related to the solvent structure are influenced by solution composition. In particular, pH and ionic strength strongly influence step advancement rates. As an example, the dependence of step advancement rates on solute concentration at three different pH values for crystals of the protein canavalin is shown in Figure 15 (Land et al. 1997).

The primary barrier to detaching a molecule from a step are its bonds to adjacent molecules in the crystal. These barriers are difficult to influence just by altering solution composition because they are controlled by the crystal itself. However, introduction of impurities that incorporate into the crystal at sufficiently high concentrations can alter these detachment rates. As an example, Figure 16 shows the dependence of step advancement rates on Ca^{2+} activity for calcite grown in the presence of magnesium (Davis et al. 2000). With increasing magnesium levels, the activity at which the step speed goes to zero moves to higher values, i.e., the solubility increases, and, as discussed above, the solubility is a direct measure of equilibrium detachment rates.

The equations that govern growth kinetics depend on many assumptions about the pathways of mass transfer, but a very general set of considerations leads to some useful results. Since growth occurs on steps, for now we will ignore the source of the steps and just analyze the step kinetics. We will assume that the steps are rough and that the

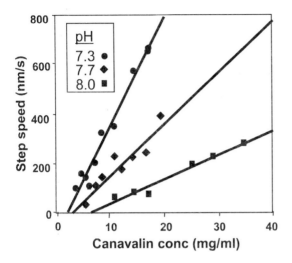

Figure 15. Dependence of step speed on concentration as a function of pH for the protein canavalin (From Land et al. 1997 with permission of Elsevier).

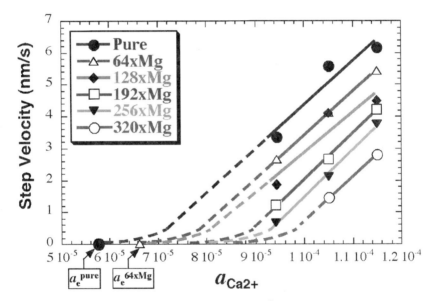

Figure 16. Measured dependence of step velocity on Ca^{2+}, activity, a_{Ca2+} as a function of Mg^{2+} concentration. The solubility shifts to higher activities, while the kinetic coefficient remains roughly constant. The solution activity of magnesium is expressed in a shorthand form where $320 \times Mg = 4.81 \times 10^{-4}$ M (Modified after Davis et al. 2000).

concentration and activity are equal. The step advancement rate, R_s, is given by the difference between the attachment rate, R_a and detachment rate, R_d, (in numbers of molecules per unit of time). The step speed is just $R_s b$ where b is the depth of a kink (units of length). The attachment rate is given by the flux, F, of molecules to the step, times the area, ch, of a site on the step, times the probability of attachment, P_a, where c is the distance between molecules along the step and h is the step height. The detachment rate from any given site is the attempt frequency, ν, times the probability of detachment, P_d. In the case of attachment, we will assume that the attachment probability is controlled by the interaction with the surrounding solvent and ignore site-to-site differences. However, because different kinds of sites have very different detachment rates, we must sum over the detachment probabilities for each type of site. The step speed becomes:

$$v/b = R_a - R_d \qquad (16c)$$

$$= chFP_a - \nu \, \Sigma n_i P_{d,i} \qquad (16b)$$

The flux is proportional to the concentration, and the detachment probability for a given site is $\exp(-E_i/kT)$. Equation (16) becomes:

$$v/b = chBC - \nu \, \Sigma n_i \exp(-E_i/kT) \qquad (17)$$

where B is a concentration-independent coefficient that contains geometric factors, temperature, and mass terms describing the dynamics of solute diffusion near the surface. Also, in the rough step limit, the second term is independent of concentration. At equilibrium v is zero and $C = C_e$; consequently, we can replace the detachment term with $chBC_e$. Now the step speed becomes:

$$v = cbhB(C - C_e) \qquad (18a)$$

$$= \Omega\beta(C - C_e) \qquad (18b)$$

where Ω is the volume per molecule. The parameter β is commonly referred to as the kinetic coefficient (Chernov 1961, 1984). Table 1 gives derived kinetic coefficients for a number of crystal systems and shows that it has an inverse scaling with molecular size (Land and De Yoreo 1999; Vekilov 2003).

The above expression for the rate of the step propagation as the product of molecular size, kink density and net flux into a kink has been subjected to critical tests. Not surprisingly—it is essentially a reformulation of the mass preservation law—it was found to be exactly followed during growth of ferritin and apoferritin crystals (Yau et al. 2000a,b; Chen and Vekilov 2002; Petsev et al. 2003).

Table 1. Measured kinetic coefficients normalized by molecular size

System	β (cm/s)	$b(Å)$ [c]	β/b (rows/s)
Calcite	0.3–0.5	3.2	$0.9–2\times10^7$
KH_2PO_4	0.1–0.4	3.7	$2–8\times10^6$
CdI_2	9×10^{-3}	4.2	2.1×10^5
Brushite	1.8×10^{-2}	5.8	3.1×10^5
GlyDKP [a]	3×10^{-3}	3.98	7.5×10^4
canavalin	1.2×10^{-3}	83	1.4×10^3
thaumatin	2.4×10^{-4}	58.6	4.1×10^2
catalase	3.2×10^{-5}	87.8	3.6×10^1
STMV [b]	4×10^{-6}	160	2.5×10^0

Notes:
    a glycine derivative of diketopiperazine
    b Satellite Tobacco Mosaic Virus
    c kink depth

Derivation of the step speed dependence on concentration under many different assumptions leads to a similar relationship, the difference lying in the physical parameters of which β is comprised. One important exception is in the case of smooth steps. Then the density of kink sites depends on concentration increasing as $(C/C_e)^{1/2}$ for sufficiently small σ and one can no longer replace the expression for the detachment rate with the equilibrium expression. Then Equation (18) becomes:

$$v = \Omega \beta [(C - C_e) - C_e f(n_1)] \quad (19)$$

where n_1 is the density of nucleated kinks on a step edge and the function $f(n_1)$ rises linearly at low σ and approaches a constant at high concentrations. In other words, the step speed is non-linear at low concentrations, but becomes linear at high concentrations and then just looks like the rough limit with a lateral offset, as shown in Figure 17. Note the similarity of v vs. C for the smooth limit to the data on calcite shown in Figure 16.

One of the important consequences of Equations (18) and (19) is that the step speed scales with the <u>absolute</u> supersaturation, not the <u>actual</u> supersaturation. This means that if two types of crystals are placed in solutions of the same supersaturation, the crystal that is more soluble will grow faster than the other, simply because there is a larger flux of molecules to the surface. In other words, *step speed scales with solubility* and one cannot assume that faster growth rates imply faster kinetics at the kink sites.

Step generation: 2D nucleation vs. growth at dislocations

So far we have assumed that steps are pre-existing on a crystal surface. But without a new source of steps, any pre-existing steps would rapidly grow out to the edge of the crystal leaving a featureless terrace and no source for growth (For a demonstration of this phenomenon see Rashkovich 1991). One way to generate new steps is to generate two-dimensional islands of molecules that then spread outward (see Fig. 13a). But as in the case of three-dimensional nucleation, there is a critical size and a free energy barrier associated with this process, although in this case they are due to the excess free energy of creation of the new step edge rather than creation of a new surface (see Fig. 13b). (We will derive the equivalent 2D nucleation expressions later.) As Figure 18 shows, at sufficiently high supersaturations, crystal surfaces do grow from solutions by this mechanism (Teng et al. 2000; Land et al. 1997; Land and De Yoreo 1999). (Table 2

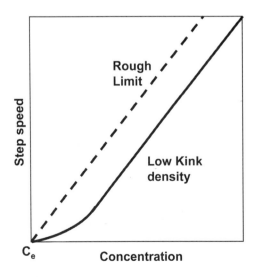

Figure 17. Dependence of step speed on concentration in the rough limit (Eqn. 18) and when steps are smooth due to low kink density (Eqn. 19).

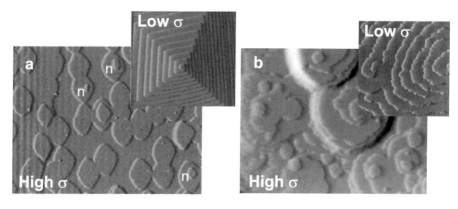

Figure 18. AFM images showing examples of 2D nucleation at high supersaturation for (a) calcite and (b) canavalin. N - locations where islands have nucleated on top of other islands.

Table 2. Measured step edge energies and critical sizes

System	$\gamma \cdot a$ (ergs) [a]	σ_{2D} [b]	N_c [c]
KH$_2$PO$_4$	4.2×10^{-14}	$\sigma \geq 0.1$	650
Catalase	4.2×10^{-14}	$\sigma < 0.2$	<2000
Thaumatin	1.3×10^{-13}	$\sigma \geq 0.2$	800
CaCO$_3$	1.3×10^{-12}	$\sigma < 3$	<600
Canavalin	1.5×10^{-12}	$\sigma \geq 2.5$	800

Notes:
  [a] where a equals the lattice spacing
  [b] the supersaturation at which 2D nucleation was observed.
  [c] number of molecules in the critical island at σ_{2D}

shows the supersaturation at which 2D nucleation is observed and the critical size at that supersaturation, based on measured step edge energies.) But at low supersaturations, the critical size is large and the probability of achieving it becomes prohibitively small. As an example, the critical island size for KH$_2$PO$_4$—the canonical solution-grown crystal—at an absolute supersaturation of 5% is about 10 nm or roughly 10,000 KH$_2$PO$_4$ molecules. The odds of obtaining an island of this size through a fluctuation that moves up a free energy gradient are understandably small. (In fact, when nucleation theory was applied to the growth of crystal surfaces back in the 1950's, researchers quickly realized that at the low supersaturations often used in crystal growth, 2D nucleation could never account for the observed growth rates.)

Fortunately, crystals are not perfect. They contain dislocations, i.e., breaks in the crystal lattice that generate permanent sources of steps (Frank 1949). Figure 13c shows a schematic of a step generated at a dislocation, and Figure 19 shows some examples of dislocation growth sources on crystal surfaces. Of course one is immediately struck by the fact that all of these sources generate spiral arrangements of steps. These features are referred to as dislocation spirals or dislocation hillocks or just growth hillocks. Besides the step speed, which we discussed above, the characteristics of the growth spiral are the distance between steps (i.e., terrace width) and shape. The terrace width, W is important in determining the growth rate, R_f, of a crystal face. As Figure 13c shows, R_f is equal to

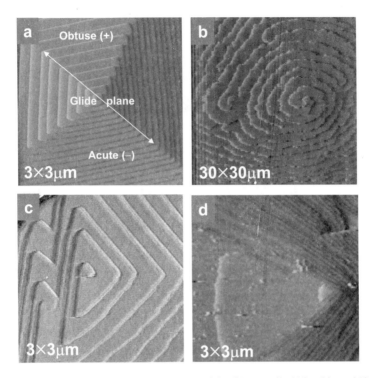

Figure 19. AFM images of dislocation hillocks on (a) calcite, (b) canavalin (c) brushite and (d) calcium oxalate monohydrate.

the hillock slope, p, times the step speed, v. But the hillock slope is just h/W where h is the step height, which is fixed by the crystal structure. So controls on terrace width are also important factors in determining growth rates (Burton et al. 1951).

To quantify growth on dislocation hillocks, we use the example of calcite (Teng et al. 1998, 2000). Growth of calcite on the {104} faces occurs on 3.1 Å monomolecular steps generated by dislocations. The advance of these steps leads to the formation of polygonal growth hillocks with steps parallel to the $<\overline{4}81>$ and $<\overline{4}41>$ directions as illustrated in Figure 19a. The presence of a c-glide plane generates two distinct pairs of crystallographically identical steps denoted as the positive and negative steps due to the angles that the step risers make with the terraces (Paquette and Reeder 1995). The terrace widths of the two step types, $W_\pm$ are related to the step speeds, $v_\pm$ through $W_+/W_- = v_+/v_-$.

The birth of a new spiral segment is shown in Figure 20. If this new step segment remains in equilibrium with the adjacent reservoir of growth units, then it will only advance when the change in free energy, Δg, associated with the addition of a new row of growth units is negative. Taking into account the anisotropy in calcite step structure, one can show that Δg for a straight step is:

$$\Delta g_\pm = -(L/b)\Delta\mu + 2c<\gamma>_\pm \quad (20a)$$

$$<\gamma>_+ = (1/4)[2(\gamma_+ + \gamma_-) + (\gamma_{++} + \gamma_{+-})] \quad (20b)$$

$$<\gamma>_- = (1/4)[2(\gamma_+ + \gamma_-) + (\gamma_{--} + \gamma_{+-})] \quad (20c)$$

Figure 20. AFM images showing the birth of a new step segment on calcite. In (a), the smallest step is below the critical length and is not moving. In (b) it has exceed the critical length and begun to move. A new sub-critical step segment now appears. (Modified after Teng et al. 1998)

where L is the length of the step, b is the 6.4 Å intermolecular distance along the step, c is the 3.1 Å distance between rows, γ_+ and γ_- are the step edge free energies along the + and − steps, and γ_{++}, γ_{--} and γ_{+-} are contributions to the step edge free energy from the corner sites as illustrated in Figure 1c. (When step curvature is included, the terms on the right hand side of Eqns. 20b and 20c become integrals of $\gamma_\pm$ as a function of orientation.) Setting Δg to zero and substituting $kT\sigma$ for $\Delta\mu$ shows that free energy only decreases if the length of the step exceeds a critical value, L_c, given by:

$$L_{c\pm} = 2bc{<}\gamma{>}_\pm/k_bT\sigma \qquad (21)$$

Alternatively, setting $\Delta g = 0$ for a step of arbitrary length gives the length dependent equilibrium activity, $a_e(L)$:

$$a_e(L) = a_{e,\infty}\exp(2bc{<}\gamma{>}/Lk_bT) \qquad (22a)$$

$$= a_{e,\infty}\exp(\sigma L_c/L) \qquad (22b)$$

where the subscript ∞ refers to the infinitely long step and a and $a_{e,\infty}$ are related by $a = a_{e,\infty}\exp(\sigma)$. Equations (22a) and (22b) are statements of the Gibbs-Thomson effect. It predicts that, even for the growth spiral, the critical length should scale inversely with the supersaturation as in Equation (21). Figure 21 shows that the results for calcite agree with the predicted dependence. The slopes of the lines give the values of $<\gamma>_\pm$.

Figure 21 also shows that $L_{c\pm}$ vs. $1/\sigma$ exhibits a non-zero intercept. It marks the supersaturation, $\sigma_{c\pm}$ where $L_{c\pm}$ goes to zero. For $\sigma > \sigma_{c\pm}$, 1D nucleation along the step edge lowers the free energy of the steps. This supersaturation gives the approximate free energy barrier, $g_{1D\pm}$, to formation of a stable dimer on the step edge through $g_{1D\pm} = kT\sigma_{c\pm}$ (Chernov 1998).

The Gibbs-Thomson effect also affects step speeds for short step segments. Combining Equations (6), (18), and (22) leads to a length dependent step speed:

$$v_\pm(L) = v_{\pm\infty}\{1-[e^{(\sigma L_{c\pm}/L)}-1]/[e^\sigma-1]\} \qquad (23)$$

where $v_{\pm\infty}$ is given by Equation (18) for $a_e(L\rightarrow\infty)$. When $\sigma \ll 1$, Equation (23) reduces to a commonly used approximation: $v = v_\infty(1-L_c/L)$ (Rashkovich, 1991). Unfortunately, in the few cases that have been investigated, the prediction of Equation (23) has turned out to be incorrect (See for example, Teng et al. 1998). Figure 22 shows the predicted dependence of v^+/v_∞ on L/L_c for calcite along with the measured dependence. The measured speed is independent of supersaturation or step direction and rises much more rapidly than that predicted by Equation (23).

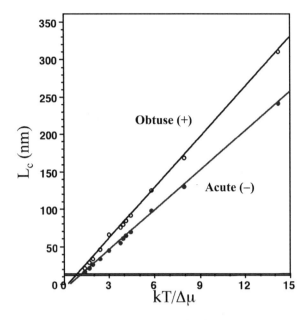

Figure 21. Dependence of critical length on supersaturation for calcite based on images like that in Figure 20. (Modified from Teng et al. 1998)

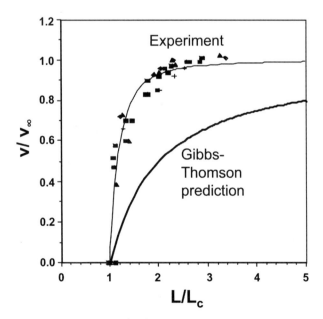

Figure 22. Dependence of step speed on step length for calcite showing the strong deviation from that expected based on the Gibbs-Thomson effect. (Modified from Teng et al. 1998)

Equation (23) was originally derived for spirals with isotropic step edge energies and kinetics, and assumes a high kink site density such that the rate of attachment is not limited by the availability of kinks. Voronkov (1973) proposed that for highly polygonized spirals, the distance between kink sites along a step is large so that the attachment rate is limited by their availability. He pointed out that the equilibrium shape of a step should be curved towards the corners and that the straight, central portion—which determines the speed—comprises only a small fraction of the step. (This prediction of step shape is verified in Fig. 20.) When the neighboring step advances, there is a rapid increase in the relative size of the straight portion and hence in the number of available kink sites. As a result, when L exceeds L_c, the step speed rises rapidly to its limiting value as its length increases from L_c to a length L' whose order of magnitude is given by $L_c + 2\Omega/ch\sigma$. Substituting in the constants of Ω, h and c specific to calcite (see Teng et al. 1998), we obtain $L'/L_c \sim 1.1$, a prediction which is consistent with the results in Figure 22. This result calls into question the use of data like those of Figure 21 to derive step edge energies by applying the Gibbs-Thomson formalism. Unfortunately, this is currently an unresolved area of crystal growth science.

Both the existence of a critical length and the length dependence of the step speed impact the terrace width and hence the growth rate. Consider an isotropic square spiral. If the step speed is zero for $L < L_c$ and jumps discontinuously to its maximum step speed, v_∞ at $L = L_c$, then the terrace width would be given by $4L_c$ (Burton et al. 1951; Rashkovich 1991). However, because the step speed rises gradually from zero to v_∞, the terrace width is larger by a factor that we will refer to as the Gibbs factor, G. Equation (23) leads to $G = 2.4$ for small σ, giving a terrace width of $9.6L_c$, which decreases to $G = 1$ in the limit of large σ.

For the rhombohedral spiral of calcite, the terrace widths for the two step directions become:

$$W_+ = 2G(1+B)<L_c>/\sin\theta \qquad (24a)$$

$$W_- = 2G(1+1/B)<L_c>/\sin\theta \qquad (24b)$$

where B is v^+/v^-, $<L_c>$ is the average value of L_c for the two step directions and θ is the angle between adjacent turns of the spiral. Figure 23 shows the measured dependence of $W_\pm$ on σ as well as $W = 9.6<L_c>$ obtained using Equation (24) at small σ. As predicted, the terrace widths scale inversely with σ, but due to the anomalously rapid rise in $v(L)$, the measured G factor is close to one. As a consequence, the growth rate of the calcite surface is about 2.5 times that predicted from classical growth theory (Burton et al. 1951; Chernov 1961).

MODIFYING THE SHAPES OF GROWTH HILLOCKS AND CRYSTALS

The shapes of crystals are controlled by a combination of energetic and kinetic factors (Burton et al. 1951; Chernov 1961). The equilibrium shape is that which minimizes the total surface free energy of the crystal, which is in turn the sum of the individual products of surface area times interfacial energy (see Fig. 24a). Thus low energy faces are preferentially expressed. On the other hand, crystal shape is rarely achieved by equilibration in a solution at equilibrium conditions, but rather during exposure to growth conditions. The slowest growing faces become the largest, and rapidly growing faces either become small or disappear altogether. Not surprisingly, the faces which are the low energy faces tend to also be those that grow slowly. Thus faces expressed during growth tend to be those expressed at equilibrium as well, although the sizes of the faces depend more critically on kinetic factors as well as the details of the

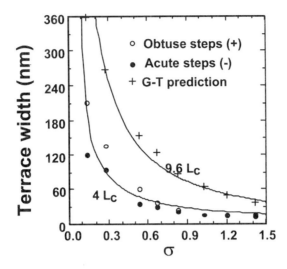

Figure 23. Dependence of terrace width on supersaturation on calcite. The solid and open symbols are the measured values. The solid line fit to those values is just four times the measured critical length. The crosses give the prediction using the Gibbs-Thomson equation and the fit is just 9.6 times the measured critical length. (Modified from Teng et al. 1998)

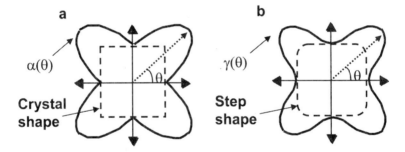

Figure 24. Schematic illustrating the relationship between (a) interfacial free energy and crystal shape, and (b) step edge free energy and step shape for simple cases.

growth sources, such as the number steps generated at a dislocation source. Because the chemical and structural factors that lead to low energy faces are also present at the step edges, the shapes of growth hillocks as well as the equilibrium shapes of the critical nucleus also tend to mimic the shapes of the growth faces, although, once again, kinetic factors lead to preferential expression of certain steps. Strictly speaking, the orientation dependence of the step edge energy, which is what determines the equilibrium step shape, is not the same as that of the surface energy in the same plane. For one thing, crystal facets are the result of true cusps in the dependence of surface energy on orientation, which is a 2D feature (Herring 1951; Chernov 1961). In contrast, while the minima in the dependence of step edge energy on orientation lead to the preferential expression of certain step directions, due to the presence of kinks, in 1D there are no true cusps (see Fig. 24b) (Chernov 1961, 1984). Calcite again provides an excellent example. As Figure 25 shows, the shape of the crystal face mimics the hillock shape, which in turn mimics the shape of the critical step segment.

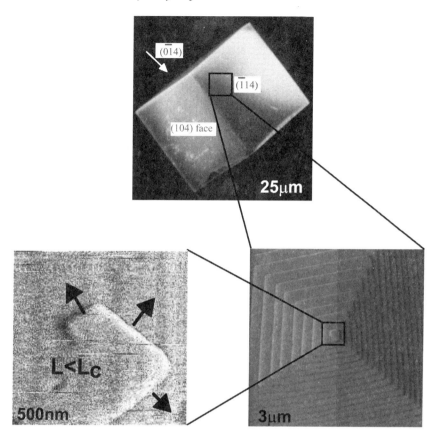

Figure 25. For many systems, there is a close similarity between step shape, hillock shape and facet shape. This example is from calcite.

As a result of these relationships, it is possible to modify the shapes of crystals by introducing ions or molecules that alter the shapes of growth hillocks through interactions with the step edges. These growth modifiers can be organic and include both peptides and large proteins, but even small inorganic modifiers may play a role in this process, given their presence in naturally occurring waters and cellular fluids (Weiner and Dove 2003). There are four mechanisms by which ions or molecules can modify growth hillocks, either by changing the step speed or by altering the step edge energy. They are: 1) step pinning, 2) incorporation, 3) kink blocking, and 4) step edge adsorption. Each of these major mechanisms for growth inhibition exhibits a characteristic dependence of step speed on supersaturation and impurity concentration as illustrated in Figure 26 (Dove et al. 2004).

1) Step pinning: Certain types of impurities block the attachment of molecules to the step edges (see Fig. 26a). The only way the step can continue to advance is by growing around those blocking sites (Cabrera and Vermileya 1958). As long as the average spacing between the impurities is much greater than the critical radius of curvature for the step, the step can continue to advance unimpeded. But when the spacing becomes comparable to the critical curvature, once again the Gibbs-Thomson effect

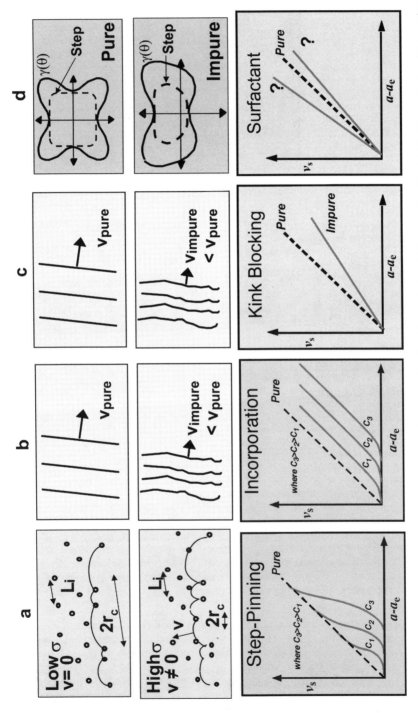

Figure 26. The four models for impurity interactions and their effect on step kinetics: (a) step pinning, (b) incorporation, (c) kink blocking, and (d) surfactant action (Dove et al. 2004).

influences growth. Moreover, if the average impurity spacing becomes so small that the step curvature must exceed the critical radius in order to pass between them, the step stops moving altogether. Although the step kinetics appear to be altered, this is actually a thermodynamic effect. It isn't the impurity *per se* that stops the step, but rather the fact that, according to the Gibbs-Thomson equation, the true supersaturation is a function of curvature and goes to zero at the critical curvature. If the radius of curvature should fall below r_c due, for example, to step fluctuations, the solution actually becomes undersaturated and the step retreats until the radius of curvature increases to r_c. In terms of supersaturation, for a given impurity content, there is a critical supersaturation, σ_d below which the crystal surface enters a so-called "dead zone" where no growth occurs (Cabrera and Vermileya 1958; Chernov 1961, 1984; Rashkovich 1991; Voronkov and Rashkovich 1994). As the supersaturation is increased, the crystal begins to grow with a speed given by Equation (23), with L_c and L replaced by r_c and r, eventually recovering the speed of the step in pure solution as shown in Figure 26a. Because the average impurity spacing decreases as the impurity level increases, the value of σ_d and width of the dead zone both become larger.

The relationship between σ_d and the impurity concentration can be derived by comparing the average impurity spacing with the critical radius (Cabrera and Vermileya 1958; Potapenko 1993; van Enckevort and van der Berg 1998). If the solution impurity concentration is C_i, then the surface density is $n_i = \Gamma C_i$ where Γ is derived from the Langmuir isotherm for the impurity. The average impurity spacing is equal to $1/(\kappa n_i^{0.5})$ where κ is a number of order unity that depends on the lattice geometry. Equating this to twice the critical radius at σ_d leads to:

$$1/(\kappa n_i^{0.5}) = 2r_c = 2\Omega\alpha/kT\sigma_d \qquad 25a$$

Rearranging gives the result that $\sigma_d \propto C_i^{0.5}$. Specifically,

$$\sigma_d = (2\kappa\Omega\alpha)n_i^{0.5} = (2\kappa\Gamma\Omega\alpha)C_i^{0.5} \qquad 25b$$

$$\sigma_d \propto C_i^{0.5} \qquad 25c$$

Interestingly, one example for which the data clearly demonstrates this square root dependence is the growth of ice in the presence of antifreeze proteins. Other systems spectacularly fail to display this behavior. In some cases, the reason for the discrepancy is clear (Land et al. 1999), while in other highly relevant cases for biomineralization, such as calcite growing in the presence of Sr^{3+}, the source of the observed anomalous dependence is unknown (Wilson et al. 2002).

Step pinning is highly dependent upon the details of the impurity-step interactions. Consequently, the same impurities that may block one type of step can leave steps on adjacent faces—or even other types of steps on the same face—unimpeded. In this way step pinning generally leads to a change in hillock and crystal shape (For examples, see De Yoreo et al. 2002 and Qiu et al. 2003). Figure 27 shows the example of the growth of calcium oxalate monohydrate (COM), the main mineral forming phase of kidney stones, in the presence of citrate, a naturally occurring modulator of stone formation (Qiu et al. 2003). The images were collected in the regime just above σ_d. Citrate has no effect on the {010} face but dramatically slows growth on the {$\bar{1}$01} face. In fact, one particular step (the [101]) is most strongly affected. In pure solution, its speed is more than an order of magnitude higher than that of the adjacent [120] steps. But at the citrate level used in this experiment, it slows by a factor of 12 and loses lateral stability, while the <120> steps only slow by a factor of two and becomes rough and rounded. The result is that the step speed on that face is nearly isotropic, and the hillocks become rounded. The crystal shape reflects these changes. Crystals become rounded plates with large {$\bar{1}$01} faces. Molecular

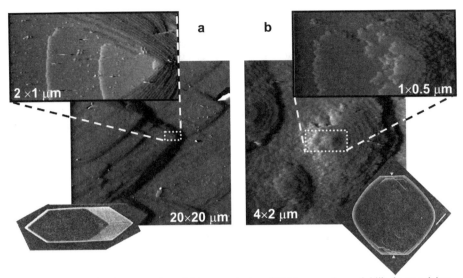

Figure 27. Example of a system that exhibits step pinning. AFM images of growth hillocks on calcium oxalate monohydrate growing (a) pure solution and (b) solution containing sodium citrate. The lower insets show the resulting crystal morphologies.

simulations provide an explanation for these observations by predicting citrate-step binding energies that scale with the observed impacts on step propagation. The naturally occurring protein osteopontin has the inverse effect: it pins steps on the {010} face but not the {$\bar{1}$01} face. In combination, the two modifiers can poison growth on both sets of faces.

2) Incorporation: Impurity incorporation occurs when foreign ions or molecules become captured by advancing steps or otherwise incorporate at kink sites along a step edge to become part of the growing crystal. Typically the impurity molecules distort the crystal structure, thereby increasing the internal energy of the solid through an enthalpic contribution (van Enckevort and van der Berg 1998; Davis et al. 2000). The resulting increase in free energy is manifested as an increase in the solubility (K_{sp}) of the crystal, leading to a lower effective supersaturation. Since σ is defined as the natural logarithm of the ion activity product divided by the solubility constant, an increase in the solubility of the crystal reduces the effective supersaturation (σ_{eff}). As shown in Figure 26b, this result shifts step velocity curves to higher equilibrium activities, resulting in what appears as a dead zone at low supersaturation ($\sigma < \sigma_d$). But this is a misinterpretation, because the supersaturation needs to be recalculated using the value of K_{sp} for the impurity-bearing phase. Above σ_d, the linear relationship between step velocity and concentration characteristic of the pure system is regained. However, the absolute magnitude of the step velocity always remains below that of the pure system at the same supersaturation. Moreover the slope of v versus C remains unchanged. Recalling from Equation (18) that dv/dC gives the kinetic coefficient, β this implies that there is no impact on kinetics of attachment processes at the step edges. The example of Mg incorporation in calcite was shown in Figure 16.

There is an important caveat to discuss here. Incorporation does not have to result in decreasing growth rates and can even increase the growth rate at sufficiently low impurity concentrations. This is because the incorporation of the impurity *always* increases the entropy of the solid, making it more stable and decreasing its solubility

(Astilleros et al. 2002). This type of behavior is seen for low Sr concentrations in calcite (Wilson et al. 2003).

Incorporation mechanisms can also change crystal shape. In particular, when the impurities incorporate at different rates into adjacent steps, the result is a crystal with sharp changes in impurity content near the boundaries of the two step directions (e.g., the positive and negative step directions in calcite). Those spatial variations lead to strain near the boundary, which lowers the supersaturation and inhibits growth. In calcite grown with Mg, this effect has been proposed as the source of crystal elongation along the {001} axis (Davis et al. 2003).

3) Kink blocking: When impurities adsorb to kink sites for very short residence times, they lead to an effective reduction in kink site density (Bliznakow 1958; Chernov 1961), with or without incorporation. That is not to say that they permanently block step advancement at that site, just kink propagation. Thus there is no dead zone, and the solubility is unaffected. Instead, the kinetic coefficient—and hence the slope of v vs. C is reduced as shown in Figure 26c. As with step pinning, because this effect is highly dependent on specific impurity-step interactions, kink blocking can result in a change in crystal shape. Figure 28 shows an example of a system that appears to exhibit this behavior, though the only proof is the kinetic data presented in the figure, along with the observation that the steps, though roughened slightly, do not change shape.

4) Surfactants: Impurities that lower the interfacial energy by adsorbing to surfaces modify many aspects of the surface dynamics and are known as surfactants. Impurities that adsorb to step edges can have similar effects, modifying growth by lowering the step edge energy. Although there are a number of variants on this process, here we consider the end-member situation whereby the modification is a purely thermodynamic effect that changes the orientation dependence of the step edge energy, γ. As a result, the minimum energy step shape undergoes a transition to a new lowest energy form when the impurity concentration exceeds a threshold value. Figure 26d illustrates this evolution by a "2D Wulff construction" of step edge free energy for the case of a growth hillock that develops in a pure and impurity-modified system. The concepts behind this type of impurity modification were worked out for changes in 3D crystal shape due to true 2D surfactants (Hartman and Kern 1964; Kern 1969), but the extension to 1D is straightforward.

Evidence for this type of impurity effect comes from changes in microscopic (step edge direction) and macroscopic (expression of crystal facets) morphology in the absence of a kinetic effect. Indeed, impurity interactions by step edge adsorption are unique because step flow rates can exhibit little deviation from the pure system. As an example of this type of behavior, we again look to the calcite system. Figure 29 shows the effect of right-handed and left-handed aspartic acid on the shapes of growth hillocks and the resulting macroscopic crystals (Orme et al. 2001). The steps do not exhibit the pinning observed for citrate on COM, yet the shapes of the hillocks are dramatically altered. Moreover, the symmetry about the calcite glide plane is broken such that L-aspartic acid gives one chirality while D-aspartic acid gives the opposite chirality. There are new step directions that switch from one side of the glide plane to the other when the amino acid enantiomer is switched from L to D. These shapes were analyzed in terms of changes in step edge energetics with the modifier acting as a 1D surfactant. In fact, the overall reduction in terrace width demonstrates a true reduction in average step edge energy. But the relative degree to which the orientation dependence of step edge free energy is changed compared to the kinetic coefficient is unclear. What is clear from these experiments, however, is that, as in the case of COM with citrate and calcite with Mg, the overall shapes of the resulting crystals are being controlled by the shapes of the growth hillocks on existing faces.

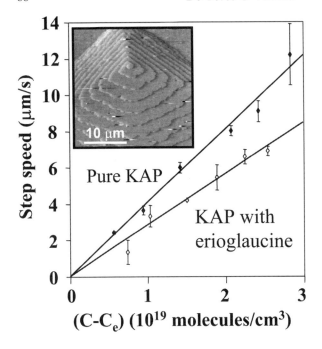

Figure 28. Example of system that exhibits behavior expected for kink blocking. AFM image and dependence of step speed on concentration for potassium acid phthalate with and without the organic dye, erioglaucine. The steps roughen slightly and the kinetic coefficient is reduced, but there is no dead zone, little or no shift in equilibrium solubility, and no change in shape.

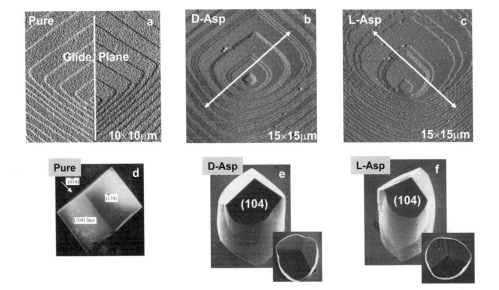

Figure 29. Example of system that exhibits behavior expected for addition of a surfactant. AFM images of calcite grown in (a) pure solution, (b) Solution containing D-aspartic acid and (c) solution containing L-aspartic acid. The shape changes dramatically and even shows a left-right shape dependence that corresponds to that of the additive. (d-f) show that the resulting crystal shape reflects these changes. (After Orme et al. 2001).

CONCLUSION

The physical processes of nucleation and crystal growth are inherent to the biological process of mineralization. Deciphering the control mechanisms exerted by organisms over nucleation and growth is one of the central challenges facing us as we try to understand how biominerals are formed. The concept of the energy landscape, though quite detached from the biology, provides a framework for relating the interactions between the controlling agents and the physical changes in shape, structure, and phase, regardless of whether those agents act as members of solid molecular scaffolds or ions in solutions. In this chapter, we have attempted to show how the principles of crystallization relate to these physical properties.

The single most important concept for understanding nucleation is that of the critical size and its relationship to the main external control parameter, the supersaturation, and the primary materials control parameter, the interfacial energy. Most importantly, the magnitude of the interfacial energy depends on the atomic-scale at the interface and, consequently, surfaces can be used by organisms to control both the location and orientation of biominerals. Secondarily, the details of the free energy landscape that separates the solvated phase from the final bulk crystal has a large impact on the pathway of nucleation and its manipulation may allow organisms to precipitate inorganic compounds into an easily shaped but metastable amorphous phase, which is then driven to convert to the final crystalline phase. Unfortunately, while these concepts are based on a firm physical foundation, clear and quantifiable links to real systems have yet to be established.

Critical size and free energy barriers continue to be central concepts during the growth phase of crystals, but the potential mechanisms for controlling the process appear to be greater in number and diversity. The critical parameters continue to include supersaturation and free energy, but the free energy of the interface is supplanted by that of the step. Moreover, the kinetics of attachment and detachment become of equal importance in determining growth rates and crystal shapes. These processes are typically quantified by a single parameter, the kinetic coefficient, which unfortunately masks the physics behind solute incorporation. Surprisingly, while many processes including desolvation, adsorption, surface diffusion, chemical reaction, and eventual incorporation can, in principle, both impact this coefficient and depend on the details of system chemistry, AFM studies of recent years are pointing towards rearrangement of waters to allow the solute to access the surface as the dominant factor. Consequently, the kinetic coefficient shows a clear scaling with molecular size. Once size is taken into account, the apparent energy barrier to step motion is surprisingly similar for all systems analyzed.

Due to the combined importance of both the thermodynamics of the step edge and the kinetics of solute attachment and detachment, there are a number of opportunities for organisms to modulate the growth phase. Unlike the situation with nucleation of biominerals, a widely accepted model for shape modification has been developed based on the concept of stereochemical recognition. Within this model, the modulators of growth exhibit stereochemical recognition for otherwise unexpressed faces of the mineral. Attachment to those faces then stabilizes them, producing a new crystal habit. Proposed examples include calcium oxalate monohydrate with citrate, ice with antifreeze glycoproteins, and calcite with inorganic impurities such as Li and Mg, as well as organic modifiers containing carboxyl groups such as aspartic acid, glutamic acid, and acidic peptides. In principle, this mechanism can and probably does occur for some systems. However, as shown in this chapter, many of the classic systems for which this model was proposed clearly display behavior better explained through specific impurity-step interactions on existing faces. The differences between these two models of shape modification have yet to be reconciled and highlight an inescapable conclusion: While the past century has led to a deep understanding of the physical principles of nucleation and growth, we are far from understanding how living organisms utilize those principles to control the formation of biomineral structures.

ACKNOWLEDGMENTS

This work was performed under the auspices of the U.S. Department of Energy by the University of California, Lawrence Livermore National Laboratory under contract No. W-7405-Eng-48

REFERENCES

Abraham FF (1974) Homogeneous Nucleation Theory. Academic Press, New York
Addadi L, Moradian J, Shay E, Maroudas NG, Weiner S (1987) A chemical model for the cooperation of sulfates and carboxylates in calcite crystal nucleation: relevance to biomineralization. Proc Nat Acad Sci 84:2732-2736
Addadi L, Raz S, Weiner S (2003) Taking advantage of disorder: amorphous calcium carbonate and its roles in biomineralization. Adv Mat 15:959-970
Aizenberg J, Black AJ, Whitesides GM (1999a) Control of nucleation by patterned self-assembled monolayers. Nature 398:495-498
Aizenberg J, Black AJ, Whitesides GM (1999b) Oriented growth of calcite controlled by self-assembled monolayers of functionalized alkanethiols supported on gold and silver. J Am Chem Soc 121:4500
Aizenberg J, Grazul JL, Muller DA, Hamann DR (2003) Direct fabrication of large micropatterned single crystals. Science 299:1205-1208
Archibald DD, Qadri SB, Gaber BP (1996) Modified calcite deposition due to ultrathin organic films on silicon substrates. Langmuir 12:538–546
Astilleros JM, Pina CM, Fernndez-daz L, Putnis A (2002) Molecular-scale surface processes during the growth of calcite in the presence of manganese. Geochim Cosmochim Acta 66:3177-3189
Barabási A-L, Stanley EH (1995) Fractal Concepts in Surface Growth. Cambridge University Press, Cambridge
Berman A, Ahn DJ, Lio A, Salmeron M, Reichert A, Charych D (1995) Total alignment of calcite at acidic polydiacetylene films: Cooperativity at the organic-inorganic interface. Science 259:515-518
Binder K, Fratzl P (2001) Spinodal decomposition. In: Phase Transformation in Materials. Kostorz G (ed) Wiley, New York
Bliznakow G (1958) Crystal habit and adsorption of cosolutes. Fortsch Min 36:149
Burton WK, Cabrera N, Frank FC (1951) The growth of crystals and the equilibrium structure of their surfaces. Royal Soc London Philos Trans A243:299-358
Cabrera N, Vermileya DA (1958) Growth and Perfection of Crystals. John Wiley & Sons, New York; Chapman and Hall, London
Calvert P (1992) Biomimetic ceramics and composites. MRS Bulletin XVII:37-44
Chaikin PM, Lubensky TC (1995) Principles of condensed matter physics. Cambridge University Press, Cambridge
Chen K, Vekilov PG (2002) Evidence for the surface diffusion mechanism of solution crystallization from molecular-level observations with ferritin. Phys Rev E 66:21606
Chernov AA (1961) The spiral growth of crystals. Sov Phys Uspekhi 4:116-148
Chernov AA (1984) Modern Crystallography III: Crystal Growth. Springer, Berlin
Chernov AA (1989) Formation of crystals in solutions. Contemp Phys 30:251-276
Chernov AA (1998) Theoretical and Technological Aspects of Crystal Growth. In: Materials Science Forum. Fornari R, Paorichi C (eds) TransTech Publications, USA, p 71-78
Collins MJ, Westbroek P, Muyzer G, DeLeeuw JW (1992) Experimental evidence for condensation reactions between sugars and proteins in carbonate skeletons. Geochim Cosmochim Acta 56:1539-1544
Crenshaw MA (1972) The soluble matrix from Mercenaria mercenaria shell. In: International Symposium on Problems in Biomineralization. Schattauer FK (ed) Verlag, New York
Davis KJ, Dove PM, De Yoreo JJ (2000) Resolving the controversial role of Mg^2 in calcite biomineral formation. Science 290:1134-1137
Davis KJ, Dove PM, Wasylenki LE, De Yoreo JJ (2004) Morphological consequences of differential Mg^{2+} incorporation at structurally distinct steps on calcite. Am Min (in press)

DeYoreo JJ, Orme CA, Land TA (2001) Using atomic force microscopy to investigate solution crystal growth. *In*: Advances in crystal growth research. Sato K, Nakajima K, Furukawa Y (eds) Elsevier Science, New York, p 361-380

DeYoreo JJ (2001) Eight Years of AFM: What has it taught us about solution crystal growth., *In:* 13th International Conference on Crystal Growth. Hibiya T, Mullin JB, Uwaha M (eds) Elsevier, Kyoto, Japan

De Yoreo JJ, Burnham A, Whitman PK (2002) Developing KDP and DKDP crystals for the world's most powerful laser. Int Mat Rev 47:113-152

Dove PM, Davis KJ, DeYoreo (2004) Inhibition of $CaCO_3$ crystallization by small molecules: the magnesium example. *In:* Solid-Fluid Interfaces to Nanostructural Engineering, Vol. II. Liu XY, De Yoreo JJ (eds) Kluewer/Plenum Academic Press, New York (in press).

Dunitz JD (1994) The entropic cost of bound water in crystals and biomolecules. Nature 264:670-670

Fink DJ, Arnold IC, Heuer AH (1992) Eggshell mineralization: A case study of a bioprocessing strategy. MRS Bulletin XVII:27-31

Frank FC (1949) The influence of dislocations on crystal growth. Discussions Faraday Soc 5:48

Frenkel J (1932) Note on a relation between the speed of crystallization and viscosity. Phys J USSR 1:498-510

Gasser U, Weeks E, Schofield A, Pusey P, Weitz D (2001) Real-space imaging of nucleation and growth in colloidal crystallization. Science 292:258-262

Gibbs JW (1876) On the equilibrium of heterogeneous substances. Trans Connect Acad Sci 3:108-248

Gibbs JW (1878) On the equilibrium of heterogeneous substances. Trans Connect Acad Sci 16:343-524

Hartman P, Kern R (1964) Le changement de facies par adsorption et al theorie dea "PBC." Acad Sci Paris 258:4591-4593

Herring C (1951) The use of classical macroscopic concepts in surface tension problems. *In:* Collection: Structure and properties of solid surfaces. McGraw-Hill, New York, p 143

Heywood BR, Mann S (1994) Template-directed nucleation and growth of inorganic materials. Adv Mater 6:9-20

Hohenberg PC, Halperin BI (1977) Theory of dynamic critical phenomena. Rev Mod Phys 49:435-479

Johnson DA (1982) Some thermodynamic aspects of inorganic chemistry. Cambridge University Press, Cambridge

Kahlweit M (1969) Nucleation in Liquid solutions. *In*: Physical Chemistry, Vol. VII. Eyring H (ed) Academic Press, New York, p 675-698

Katz JL, Ostermier BJ (1967) Diffusion cloud chamber investigation of homogeneous nucleation. J Chem Phys 47:478-487

Katz EP, Li S (1973) The intermolecular space of reconstituted collagen fibrils. J Mol Biol 21:149-158

Kashchiev D (1999) Nucleation: Basic Theory with Applications. Butterworths, Heinemann, Oxford

Kashchiev D (2003) Thermodynamically consistent description of the work to form a nucleus of any size. J Chem Phys 118:1837-1851

Kern R (1969) Crystal growth and adsorption. *In*: Growth of Crystals. Sheftal' NN (ed) Consultants Bureau, New York, p 3-23

Lacmann R, Schmidt F (1977) Nucleation and equilibrium forms of mixed crystals. *In:* Current Topics in Materials Science 2. Kaldis E, Scheel HJ (eds) North–Holland, Amsterdam, p 301–325

Land TA, De Yoreo JJ, Lee JJ (1997) *In situ* AFM investigation of canavalin crystallization kinetics. Surf Sci 384:136

Land TA, De Yoreo JJ (1999) *In situ* AFM investigation of growth source activity on single crystals of canavalin. J Cryst Growth 208:623

Land TA, Martin TL, Potapenko S, Palmore GT, De Yoreo JJ (1999) Recovery of surfaces from impurity poisoning during crystal growth. Nature 399:442

Makarov VA, Andrews BK, Smith PA, Pettitt BM (2000) Residence times of water molecules in the hydration sites of myoglobin. Biophys J 79:2966-2974

Makarov VA, Pettitt BM, Feig M (2002) Solvation and hydration of proteins and nucleic acids: a theoretical view of simulation and experiment. Acc Chem Res 35:376-384

Mann S, Archibald DD, Didymus JM, Heywood BR, Meldrum FC, Wade VJ (1992) Biomineralization: Biomimetic potential at the inorganic-organic interface. MRS Bulletin XVII:32-36

Mann S (2001) Biomineralization: Principles and Concepts in Bioinorganic Materials Chemistry. Oxford University Press, New York

Miller A (1984) Collagen: The organic matrix of bone. Philos Trans R Soc London B 304:455-477

Mullin JW (1992) Crystallization, 3rd edition. Butterworths, Oxford

Mutaftschiev B (1993) Nucleation. *In:* Handbook on Crystal Growth. Hurle DTJ (ed) North–Holland, Amsterdam, p 187-248

Nancollas GH (ed) (1982) Biological Mineralization and Demineralization. Life Sciences Research Report. Springer-Verlag, New York

Neilsen AE (1964) Kinetics of Precipitation. Pergamon, Oxford

Neilsen AE (1967) Nucleation in aqueous solutions. *In:* Crystal Growth. Peiser S (ed) Pergamon, Oxford, p 419-426

Orme CA, Noy A, Wierzbicki A, McBride MY, Grantham M, Dove PM, DeYoreo JJ (2001) Selective binding of chiral amino acids to the atomic steps of calcite. Nature 411:775-779

Paquette J, Reeder RJ (1995) Relationship between surface structure, growth mechanism, and trace element incorporation in calcite. Geochim Cosmochim Acta 59:735-749

Petsev DN, Vekilov PG (2000a) Evidence for non-DLVO hydration interactions in solutions of the protein apoferritin. Phys Rev Lett 84:1339-1342

Petsev DN, Thomas BR, Yau S-T, Vekilov PG (2000b) Interactions and aggregation of apoferritin molecules in solution: Effects of added electrolytes. Biophysical J 78:2060-2069

Petsev DN, Chen K, Gliko O, Vekilov PG, (2003) Diffusion-limited kinetics of the solution-solid phase transition of molecular substances. Proc Natl Acad Sci USA, 100:792-796

Potapenko SY (1993) Moving of steps through an impurity fence. J Cryst Growth 133:147-154

Qiu SR, Wierzbicki A, Orme CA, Cody AM, Hoyer JR, Nancollas GH, Zepeda S, De Yoreo JJ (2003) Molecular Modulation of Calcium Oxalate Crystallization by Osteopontin and Citrate (submitted)

Rashkovich LN (1991) KDP Family of Crystals. Adam-Hilger, New York

Sikes CS, Yeung ML, Wheeler AP (1990) Inhibition of calcium carbonate and phosphate crystallization by peptides enriched in aspartic acid and phosphoserine. *In*: Surface Reactive Peptides and Polymers: Discovery and Commercialization. Sikes CS, Wheeler AP (eds) ACS Books, Washington, ch 5

Tanford C (1980) The hydrophobic effect: formation of micelles and biological membranes. John Wiley & Sons, New York

Teng H, Dove PM, Orme C, De Yoreo JJ (1998) Thermodynamics of calcite growth: Baseline for understanding biomineral formation. Science 282:724

Teng H, Dove PM, De Yoreo JJ (2000) Kinetics of calcite growth: Surface processes and relationships to macroscopic rate laws. Geochim Cosmochim Acta 64:2265

Travaille, AM, Donners JJM, Gerritsen, JW, Nico, NAJM, Nolte, RJM, van Kempen, H (2002) Aligned growth of calcite crystals on a self-assembled monolayer. Adv Mater 14:492-495

van Enckevort WJP, van der Berg ACJF (1998) Impurity blocking of crystal growth: a Monte Carlo study. J Cryst Growth 183:441-455

Vekilov PG, Feeling-Taylor AR, Yau S-T, Petsev DN (2002a) Solvent entropy contribution to the free energy of protein crystallization. Acta Crystallogr Section D 58:1611-1616

Vekilov PG, Feeling-Taylor AR, Petsev DN, Galkin O, Nagel RL, Hirsch RE (2002b) Intermolecular interactions nucleation and thermodynamics of crystallization of hemoglobin C. Biophys J 83:1147-1156

Vekilov PG, Chernov AA (2002) The physics of protein crystallization. *In*: Solid State Physics. Ehrenreich H, Spaepen F (eds) Academic Press, New York, p 1-147

Vekilov P (2003) Microscopic, mesoscopic, and macroscopic lengthscales in the kinetics of phase transformations with proteins. *In*: From Solid-Fluid Interfaces to Nanostructural Engineering. Vols. I and II. Liu XY, De Yoreo JJ (eds) Kluewer/Plenum Academic Press, New York

Voronkov, VV (1973) Dislocation mechanism of growth with a low kink density. Sov Phys Crystallogr 18:19-223

Voronkov VV, Rashkovich LN (1994) Step kinetics in the presence of mobile adsorbed impurity. J Cryst Growth 144:107-115

Wada K, Fujinuki T (1976) Biomineralization in bivalve mollusks with emphasis on the chemical composition of the extrapallial fluid. *In*: The Mechanisms of Mineralization in the Invertebrates and Plants. Watabe N, Wilbu KM (eds), Univ SC Press, Columbia, p 175-190

Walton AG (1969) Nucleation in liquids and solutions. *In*: Nucleation. Zettlemoyer AC (ed) Marcel Dekker, New York, p 225-307

Webb MA (1999) Cell-Mediated Crystallization of Calcium Oxalate in Plants. Plant Cell 11:751-761

Weiner S, Dove PM (2003) An overview of biomineralization processes and the problem of the vital effect. Rev Mineral Geochem 54:1-29

Wheeler AP, Low KC, Sikes CS (1990) $CaCO_3$ crystal-binding properties of peptides and their influence on crystal growth. *In:* Surface Reactive Peptides and Polymers: Discovery and Commercialization. Sikes CS, Wheeler AP (eds) ACS Books, Washington, ch. 6

Williams ED, Bartelt NC (1991) Thermodynamics of surface morphology. Science 251:393-400

Wilson DS, Dove PM, DeYoreo JJ (2002). Nanoscale effects of strontium on calcite growth: A baseline for understanding biomineralization in the absence of vital effects. Amer Geophys Union Fall Meeting Program, 228

Wong KK, Brisdon BJ, Heywood BR, Hodson GW, Mann S (1994) Polymer-mediated crystallisation of inorganic solids: Calcite nucleation on the surfaces of inorganic polymers. J Mater Chem 4:1387-1392

Yau S-T, Petsev DN, Thomas BR, Vekilov PG (2000a) Molecular-level thermodynamic and kinetic parameters for the self-assembly of apoferritin molecules into crystals. J Mol Biol 303: 667-678

Yau S-T, Thomas BR, Vekilov PG (2000b) Molecular mechanisms of crystallization and defect formation. Phys Rev Lett 85:353-356

Yau S-T, Vekilov PG (2000) Quasi-planar nucleus structure in apoferritin crystallisation. Nature 406:494-497

Yau S-T, Vekilov PG (2001) Direct observation of nucleus structure and nucleation pathways. J Am Chem Soc 123:1080-1089

4 Biologically Induced Mineralization by Bacteria

Richard B. Frankel
Department of Physics
California Polytechnic State University
San Luis Obispo, California 93407 U.S.A.

Dennis A. Bazylinski
Department of Biochemistry and Molecular Biology
Iowa State University
Ames, Iowa 50011 U.S.A.

INTRODUCTION

Bacteria are small, prokaryotic, microorganisms that are ubiquitous in surface and subsurface terrestrial and aquatic habitats. Prokaryotes comprise two Domains (Superkingdoms) in the biological taxonomic hierarchy, the Bacteria and the Archaea. They exhibit remarkable diversity both genetically and metabolically even within the same microenvironment and they are thought to play a major role in the deposition and weathering of minerals in the earth's crust. The synthesis of minerals by prokaryotes can be grouped into two canonical modes: 1) biologically induced mineralization (BIM) and 2) biologically controlled mineralization (BCM) (Lowenstam 1981; Lowenstam and Weiner 1989). In this chapter, we focus on biologically induced mineralization.

Minerals that form by biologically induced mineralization processes generally nucleate and grow extracellularly as a result of metabolic activity of the organism and subsequent chemical reactions involving metabolic byproducts. In many cases, the organisms secrete one or more metabolic products that react with ions or compounds in the environment resulting in the subsequent deposition of mineral particles. Thus, BIM is a presumably unintended and uncontrolled consequence of metabolic activities. The minerals that form are often characterized by poor crystallinity, broad particle-size distributions, and lack of specific crystal morphologies. In addition, the lack of control over mineral formation often results in poor mineral specificity and/or the inclusion of impurities in the mineral lattice. BIM is, in essence, equivalent to inorganic mineralization under the same environmental conditions and the minerals are therefore likely to have crystallochemical features that are generally indistinguishable from minerals produced by inorganic chemical reactions. In some cases, the metabolic products diffuse away and minerals form from solution. However, bacterial surfaces such as cell walls or polymeric materials (exopolymers) exuded by bacteria, including slimes, sheaths, or biofilms, and even dormant spores, can act as important sites for the adsorption of ions and mineral nucleation and growth (Beveridge 1989; Konhauser 1998; Banfield and Zhang 2001; Bäuerlein 2003).

BIM is especially significant for bacteria in anaerobic habitats including deep subsurface sites, or at oxic-anoxic interfaces. This is because under anaerobic conditions, many bacteria respire with sulfate and/or various metals including iron as terminal electron acceptors in electron transport. The metabolic products of these reductions, e.g., reduced metal ions and sulfide, are reactive and participate in subsequent mineral formation.

In BCM, minerals are usually deposited on or within organic matrices or vesicles within the cell, allowing the organism to exert a significant degree of control over the nucleation and growth of the minerals and thus over the composition, size, habit, and intracellular location of the minerals (Bazylinski and Frankel 2000a,b). These BCM mineral particles are structurally well-ordered with a narrow size distribution and species-specific, consistent, crystal habits. Because of these features, BCM processes are thought to be under metabolic and genetic control. Because intra-vesicular conditions (e.g., pH, Eh) are controlled by the organism, mineral formation is not as sensitive to external environmental parameters as in BIM. BCM by bacteria is discussed later in this volume (Bazylinski and Frankel 2003).

BIOLOGICALLY INDUCED MINERALIZATION ON ORGANIC SURFACES

Because of the high surface to volume ratio of bacteria, cell surfaces and the surfaces of exopolymers can be especially important in BIM processes. Negative charges on most cell and exopolymer surfaces can result in binding of cations by non-specific electrostatic interactions, effectively contributing to local supersaturation. Binding also helps stabilize the surfaces of nascent mineral particles, decreasing the free energy barrier for critical, crystal-nucleus formation. By this means, the rate of mineralization of amorphous to crystalline mineral particles can become several orders of magnitude faster than inorganic (i.e., without surface binding and nucleation) mineralization. In some cases this can result in a mineral layer that covers the cell.

Two surface BIM processes, known as passive and active, have been distinguished (Fortin and Beveridge 2000; Southam 2000). Passive mineralization refers to simple non-specific binding of cations and recruitment of solution anions, resulting in surface nucleation and growth of minerals. Active mineralization occurs by the direct redox transformation of surface-bound metal ions, or by the formation of cationic or anionic by-products of metabolic activities that form minerals on the bacterial surfaces.

Bacterial surface properties

Prokaryotes have various cell wall types whose chemistry determines the ionic charges present on the surface of the organism. In the Domain Bacteria, there are two general types of cell wall: gram-positive and gram-negative, the difference being the cell's reaction to a staining procedure used in light microscopy. The gram-positive cell wall is separated from the cytoplasm by a lipid/protein bilayer called the plasma or cell membrane and consists mainly of peptidoglycan (murein) that is rich in carboxylate groups that are responsible for the net negative charge of this structure (Beveridge and Murray 1976, 1980). Peptidoglycan forms a 15-25 nm thick sheet (Beveridge 1981) comprising multiple layers of repeating units of two sugar derivatives, N-acetylglucosamine and N-acetylmuramic acid, and a small group of amino acids. Peptidoglycan gives rigidity to the cell wall and its charged, multiple layers are mainly responsible for mineral formation (Beveridge and Murray 1976; Fortin et al. 1997; Fortin and Beveridge 2000). Additional components such as teichoic and/or teichuronic acids can be bound into peptidoglycan (Beveridge 1981). These polymers contain phosphoryl groups that further contribute to the net negative charge of the cell wall (Southam 2000).

The gram-negative cell wall is structurally more complex than, and differs from, the gram-positive type in that it has a thinner peptidoglycan layer (about 3 nm thick) and does not contain secondary polymers (Beveridge 1981). It is sandwiched between two lipid/protein bilayers, the outer and the plasma (or cell) membranes, within the space between the cell walls known as the periplasm. The outer membrane represents the cell's outermost layer. Unlike the plasma membrane, the outer membrane is not solely constructed of phospholipid and its outer face contains lipopolysaccharide (LPS) which is highly anionic. LPS consists of O-polysaccharide, the core polysaccharide, and lipid A. The O-sidechain can extend up to 40 nm away from the core polysaccharide which is attached to lipid A. Lipid A contains several strongly hydrophobic fatty acid chains that cement the LPS into the outer membrane bilayer. The core oligosaccharide and upper regions of lipid A are rich in phosphate groups that have an affinity for Mg^{2+} and Ca^{2+} (Ferris and Beveridge 1986b). The core has several keto-deoxyoctonate residues that provide available carboxylate groups while many O-sidechains also contain residues rich in carboxylate groups (Ferris and Beveridge 1986a). Phospholipid is mainly present in the inner face of the outer membrane. In gram-negative cells, it is the LPS that is the major factor in catalyzing mineral formation because of its high concentration of phosphate and carboxyl groups (Ferris and Beveridge 1984, 1986a).

Members of the Archaea also show gram-positive and gram-negative staining characteristics. However, the cell walls of the Archaea are very different chemically from the Bacteria and from each other (König 1988). Some gram-positive Archaea have cell walls composed of a layer of a peptidoglycan-like polymer, consisting of N-acetyltalosaminuronic acid and N-acetylglucosamine, called pseudomurein that overlies the plasma membrane. Others lack pseudomurein and have cell walls consisting of polysaccharide, glycoproteins, or protein. Some gram-negative Archaea lack a cell wall entirely but retain the plasma membrane. Thus, electrochemical charges present on the cell surfaces of the Archaea vary.

Other layers external to the bacterial cell wall that may be involved in mineral nucleation include S layers, capsules, slimes, and sheaths. S layers, very common in Archaea, are paracrystalline cell surface assemblages composed of protein or glycoprotein that self assemble and associate with the underlying wall through non-covalent interaction (Koval 1988). When S layers are present, they are the outermost layer of the cell facing the surrounding environment. S layers are acidic and possess a net negative charge thereby having an affinity for metal cations (Southam 2000). Capsules are dense, highly hydrated amorphous assemblages of polysaccharides or proteins that are chemically attached to the cell surface. They can be quite thick and extend up to 1 μm from the cell. Capsules are rich in carboxylate groups and may also contain a significant number of phosphate groups, both giving the structure a net negative charge. Because capsules are highly hydrated and cover the cell surface, there can be extensive interaction between the capsule and metal cations. In some cases, capsules are known to form in response to the presence of metal ions (Appanna and Preston 1987). Slime layers, a much more loosely packed version of the capsule, are similar to capsules chemically but are not attached to the cell so they can leave the cell entirely (Southam 2000). Sheaths are rigid hollow cylinders generally surrounding chains of cells or filamentous bacteria produced by a few species of prokaryotes (e.g, *Leptothrix*). In the Domain Bacteria, the sheath is a rigid homo- or heteropolymer of carbohydrate or carbohydrate and protein. In some organisms, the sheath is important in the nucleation of oxidized mineral precipitates and active biomineralization since it sometimes contains proteins that oxidize metals, e.g., as in the oxidation of manganese by *L. discophora* (Adams and Ghiorse 1986, 1987). In the Archaea, sheaths are composed of protein and are covalently linked to the cell wall.

Minerals known to be formed via BIM through passive surface-mediated mineralization include Fe, Mn, and other metal oxides, e.g., ferrihydrite ($5Fe_2O_3 \cdot 9H_2O$), hematite (α-Fe_2O_3), and goethite (α-FeOOH); metal sulfates, phosphates, and carbonates; phosphorite; Fe and Fe-Al silicates; and metal sulfides. Mineral formation results initially from the neutralization of chemically reactive sites on the cell, and proceeds via nucleation of additional metal ions with the initially sorbed metals (Southam 2000). Mineralization is most active at the sites of initial nucleation on the outer surface of the cell. Complete mineralization of the cell surface may eventually occur producing hollow minerals the size and shape of a bacterial cell (Southam 2000) (Fig. 1). It is interesting that nonliving cells may also form minerals in this way and, in one study, living cells of *Bacillus subtilis* bound less metal ions than nonliving cells (Urrutia et al. 1992). In this case, the membrane-induced proton motive force reduces the metal binding ability of the cell wall, most likely through competition of protons with metal ions for anionic wall sites.

There are a large number of examples of bacterial BIM resulting from active mineralization and the formation of reactive by-products. Some cyanobacteria precipitate a number of different minerals that result from the uptake of bicarbonate from solution and the release of hydroxyl anions. This causes an increase in the local pH of the cell. The S layer in some species (e.g., *Synechococcus* spp.) is the site of nucleation of gypsum ($CaSO_4 \cdot 2H_2O$) in weak light. However, during photosynthesis, an increase in pH at the S layer causes precipitation of calcite ($CaCO_3$) (Schultze-Lam et al. 1992, Fortin and Beveridge 2000). Cyanobacteria can also promote the precipitation of Fe and Mn oxides by increasing the pH and raising the O_2 concentration through oxygenic photosynthesis (Fortin and Beveridge 2000). The formation of iron sulfides by sulfate-reducing bacteria is also an excellent example of active mineralization from the formation of sulfide.

IRON AND MANGANESE MINERALIZATION PROCESSES

Biogenic iron and manganese minerals are particularly common products (Table 1) of BIM processes because of the relatively high concentrations of these elements in the earth's crust (4th and the 12th most abundant elements, respectively). Of these minerals, magnetite and maghemite are especially significant in geology because of their contribution to the magnetism of sediments. We will, therefore, emphasize BIM of iron minerals, especially magnetite. Our discussion is organized in terms of the major metabolic processes that cause deposition or dissolution of iron minerals, including metal oxidation and reduction, and sulfate oxidation and metal sulfide reduction.

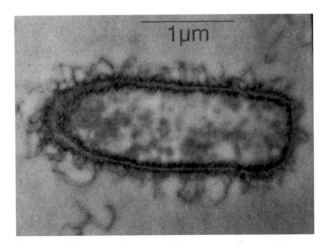

Figure 1. Unstained ultrathin section transmission electron micrograph of a "bacterial fossil" from a sulfate-reducing consortium. The cell has lysed but iron sulfide mineral encrustation has preserved the cell envelope. Figure kindly supplied by W. Stanley and G. Southam.

Table 1. Some biologically-induced iron and manganese minerals

Chemical Formula	Mineral Name
$Fe(OH)_3$ (approx.)	Ferric oxyhydroxide
$2Fe(OH)_3 \cdot Fe(OH)_2$ (approx.)	Green rust
α-$FeO(OH)$	Goethite
γ-$FEO(OH)$	Lepidocrocite
$5Fe_2O_3 \cdot 9H_2O$	Ferrihydrite
Fe_3O_4	Magnetite
γ-Fe_2O_3	Maghemite
$FeCO_3$	Siderite
$FePO_4 \cdot nH_2O$	Hydrous Ferric Phosphate
$Fe_3(PO_4)_2 \cdot 2H_2O$	Vivianite
FeS	Cubic FeS (Sphalerite-type)
FeS	Mackinawite (tetragonal FeS)
Fe_3S_4	Greigite
$Fe_{1-x}S$	Pyrrhotite
FeS_2	Pyrite
$KFe_3(SO_4)_2(OH)_6$	Jarosite
$Fe_8O_8SO_4(OH)_6$	Schwertmanite
$FeSO_4 \cdot 7H_2O$	Melanterite
$MnCO_3$	Rhodochrosite
$Mn_4O_7 \cdot H_2O$	Todorokite
$Na_4Mn_{14}O_{27} \cdot 9H_2O$	Birnessite

Adapted from Lowenstam and Weiner 1989.

Iron and manganese oxidation

Fe- and Mn-oxidizing bacteria are known to be responsible for the precipitation of oxides of both metals at acidic and neutral pH conditions. At low pH, where oxidized Fe(III) and Mn(IV) are soluble, active mineralization by organisms, such as the mesophilic, autotrophic Bacteria *Acidithiobacillus ferrooxidans* (formerly *Thiobacillus ferrooxidans*) (Kelly and Wood 2000) or *Leptospirillum* spp., that oxidize Fe(II), may be more important in iron oxyhydroxide precipitation (Fortin and Beveridge 2000; Southam 2000). The acidophiles are better known for their dissolution and bioleaching of minerals, particularly sulfide minerals such as pyrite, but are often involved in the nucleation and deposition of a secondary mineral, ferric oxyhydroxide, during Fe(II) oxidation (Fig. 2). Although mineral formation by BIM processes may not have been definitively demonstrated in every case, all Fe(II) oxidizers should be considered to have this potential. The known Fe(II)-oxidizing acidophiles are diverse and include: thermotolerant gram-positive species such as *Sulfobacillus* spp., *Acidimicrobium ferrooxidans*, and *Ferromicrobium acidophilus* (Blake and Johnson 2000); mesophilic Archaea such as *Ferroplasma* spp. (Edwards et al. 2000, 2001; Golyshina et al. 2000) that lack cell walls; and thermophilic Archaea such as *Sulfolobus* spp., *Acidianus brierleyi*, *Metallosphaera* spp., and *Sulfurococcus yellowstonensis* (Blake and Johnson 2000).

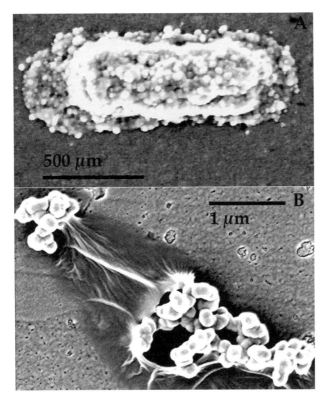

Figure 2. A) Scanning electron micrograph of a heavily mineralized cell of *Acidothiobacillus ferrooxidans* on a pyrite surface. The ferric oxyhydroxide deposits resulting from the oxidation of pyrite completely cover the cell. B) High resolution scanning electron micrograph of the ferric oxyhydroxide deposits on a cellular exopolymer. This figure was adapted from one kindly provided by K.J. Edwards.

At neutral pH, bacteria are thought to be more important in the passive formation of Fe(III) and Mn(IV) oxides although active mineralization of these oxides can also occur. There are several different physiological groups of bacteria that are known to oxidize Fe(II) at neutral pH, including both aerobes and anaerobes.

For the aerobic Fe(II) oxidizers to make a living at neutral pH, they must overcome several problems (Emerson 2000). First they must compete with inorganic oxidation of Fe(II) by O_2. Under aerobic conditions, the chemical oxidation of Fe(II) is relatively rapid. Acidophilic Fe(II)-oxidizing bacteria do not experience this problem because Fe(II) is very stable at low pH. A second problem is that the products of Fe(II) oxidation at neutral pH are insoluble Fe(III) oxyhydroxides. The cell must therefore oxidize Fe(II) at the exterior surface in order to prevent hydrolysis and precipitation of oxyhydroxides from occurring within the cell. Thus, cells must be able to transport electrons across the periplasm to the cell membrane where a chemi-osmotic potential is established. To solve the first problem, Fe(II) oxidizers grow under microaerobic conditions where the concentration of O_2 is low (e.g., oxic-anoxic interfaces), reducing the rate of inorganic iron oxidation. To solve the second problem, the Fe(II) oxidase, as well as the soluble electron transport components, are external to the cell membrane, as is apparently the case in *Acidithiobacillus ferrooxidans* (Rawlings and Kusano 1994).

Gallionella, originally described in the 1800s, was probably the first organism thought to be chemolithautotrophic based on Fe(II) oxidation (Hanert 2000a). When growing on Fe(II), each bean-shaped cell exudes a helically-twisted stalk composed mainly of ferric oxyhydroxides (Fig. 3). An organic matrix is present within the stalk (Hanert 2000a). Once formed, stalks appear to nucleate further mineralization and the iron mineral continues to accumulate on them (Hanert 2000a; Heldahl and Tumyr 1983). The stalks appear to be made of separate filaments; different species or strains synthesize different numbers of these filaments (Hanert 2000a). This is an interesting case of BIM in that there seems to be some control by the cell over the overall shape of the mineralized polymeric product and moreover, the product appears to be extruded from a specific site on the cell and is quite pure in composition. However, there is no obvious function to the structure. It is not essential for growth but Hallbeck and Petersen (1990) speculate that the stalk may represent a survival strategy. *Gallionella* appears to be a mesophilic chemolithoautotroph (Hallbeck and Petersen 1991) (it can also grow mixotrophically) and is phylogenetically associated with the β-subdivision of the Proteobacteria in the Domain Bacteria (Hallbeck et al. 1993).

Another group of Fe(II) oxidizers are also microaerophiles and grow at the oxic-anoxic interface of semi-solid O_2-gradient cultures (Emerson and Moyer 1997). These mesophilic organisms can use Fe(II) in iron sulfides or ferrous carbonate as electron donors and form Fe(III) oxides which are tightly associated with the cell wall of the bacteria. Although cells seem to encrust themselves with the metal oxides, they appear to be surrounded by a matrix where precipitation occurs. It is thought that this matrix may prevent them from being totally encased by the mineral. Phylogenetically, some of these organisms form a novel lineage within the *Xanthomonas* group in the γ-subdivision of the

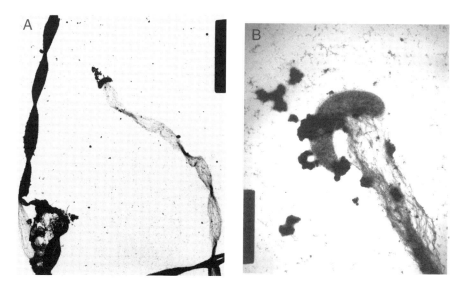

Figure 3. A) Transmission electron micrograph of unstained, whole cells of *Gallionella ferrugenia*. Contrast in electron density is primarily due to ferric oxyhydroxide mineral deposits. Newly synthesized stalk, like that attached to the cell near the upper center of the image, is composed of hair-like fibrils. Black bar along upper right edge is 5.7 μm. B) Higher magnification micrograph of a cell stained with ammonium molybdate, showing a newly synthesized stalk. Particles of ferric oxyhydroxide are attached to the cell surface and are beginning to deposit on the stalk. Black bar along lower left edge is 1.3 μm. Adapted from images kindly supplied by W. Ghiorse.

Proteobacteria (Emerson 2000). Phylotypes of these organisms have been identified from the Loihi Seamount near Hawaii (Moyer et al. 1995) where there is extensive, low-temperature, hydrothermal, venting and very large mats of hydrous Fe(III) oxides. These organisms appear to be very abundant at the site and a pure culture of a related organism has been obtained (Emerson and Moyer 2002). These organisms have also been associated with, and isolated from, the rhizosphere and Fe(III) hydroxide plaques on the roots of wetland plants (Emerson et al. 1999; Neubauer et al. 2002).

The anaerobic Fe(II) oxidizers that grow at or around neutral pH include several strains of phototrophic bacteria and some nitrate-respiring bacteria. Several freshwater strains of phototrophic bacteria are known to oxidize Fe(II) in iron sulfides, or in a mixture of ferrous carbonate and ferrous phosphate, to insoluble, rust-colored Fe(III) oxyhydroxides whose precise composition was not determined (Ehrenreich and Widdel 1994). These strains belong to the α- and γ-subgroups of the Proteobacteria. Two marine phototrophic strains of *Rhodovulum* (α-subgroup of Proteobacteria) growing on the same substrates produced the iron oxides ferrihydrite (~98%) and magnetite (trace amounts) (Straub et al. 1999). Both the freshwater and marine strains grow photoautotrophically and photoheterotrophically. The discovery of this novel type of microbial metabolism received much attention because it provided an alternative explanation for the development of the massive banded iron formations which formed in the absence of free dioxygen (Widdel et al. 1993; Ehrenreich and Widdel 1994).

An anaerobic group of Fe(II) oxidizers that uses nitrate as a terminal electron acceptor (Straub et al. 1996) includes a number of mesophilic strains belonging to the β- and γ-subgroups of the Proteobacteria (Buchholz-Cleven et al. 1997). All form rust-colored ferric oxyhydroxides from Fe(II) which probably contain considerable carbonate (Straub et al. 1996). A study using 16S rRNA-targeted probes designed from several strains showed that these organisms are quite widespread in diverse European sediments (Straub and Buchholz-Cleven 1998). This finding together with the fact that other known nitrate-reducing bacteria, including *Thiobacillus denitrificans* and *Pseudomonas stutzeri*, are also capable of Fe(II) oxidation suggests that this form of metabolism is widespread in anoxic habitats containing sufficient Fe(II) and nitrate (Emerson 2000). Chaudhuri et al. (2001) reported the isolation of *Dechlorosoma suillum* strain PS, a bacterium that is capable of oxidizing Fe(II) anaerobically with nitrate as the terminal electron acceptor. After the initiation of Fe(III) formation, the Fe(III), unreacted Fe(II), and carbonate in the medium were found to combine to form green rust which transformed into magnetite after prolonged incubation.

A hyperthermophilic member of the Archaea, *Ferroglobus placidus*, isolated from a shallow marine hydrothermal vent in Italy, is known to grow lithotrophically with Fe(II) as ferrous carbonate (Hafenbrandle et al. 1996). The optimum growth temperature of this organism is 85°C although the products of Fe(II) were not discussed. This organism can also reduce thiosulfate using hydrogen as the electron donor, and in the presence of Fe(II), produces iron sulfide minerals.

Some chemoheterotrophic bacteria also oxidize Fe(II). Two of the most well-described are the filamentous, sheathed bacteria *Sphaerotilus* and *Leptothrix*. Proteins in their sheaths catalyze the oxidation of Fe(II) and Mn(II) and nucleate the precipitation of Fe and Mn oxides, with which they are often encrusted. Members of the family Siderocapsaceae, which contains the genera *Siderocapsa*, *Naumanniella*, *Siderococcus*, and *Ochrobium*, seem to oxidize Fe(II) but the evidence is circumstantial in that most of the information about them is derived from environmental studies and enrichment cultures rather than studies with pure cultures (Hanert 2000b). In fact, it is seems questionable whether true strains of these genera actually exist (Emerson 2000).

Nonetheless they are widespread in aquatic environments and always associated with Fe(III) and Mn(IV) oxides. Many produce capsules which may be involved in mineralization (Hanert 2000b).

Several bacteria oxidize Mn(II) although none are known to grow lithotrophically with it (Emerson 2000). *Leptothrix discophora*, a mesophilic, sheathed bacterium mentioned earlier, oxidizes Mn(II) via a protein normally present in its sheath (Adams and Ghiorse 1986, 1987). Apparently, the protein is excreted by the cell and becomes associated with the sheath, resulting in the sorption of metal ions onto the sheath which eventually becomes encrusted with Mn oxides. Sheathless variants also secrete the protein; in the absence of the sheath, amorphous Mn(IV) oxides form as unattached particles in the growth medium. There are several theories concerning possible functions of the oxide crusts on *Leptothrix*; these include protection from protozoal grazing, attack from bacteriophages or UV radiation, detoxification of O_2 radicals, or sequestration of nutrients (Emerson 2000). The Mn-oxidizing protein from *Leptothrix*, MofA, has been identified and partially characterized as a multi-copper oxidase (Corstjens et al. 1997).

Two freshwater strains of the gram-negative, γ-Proteobacterium, *Pseudomonas putida*, are known to actively mineralize Mn oxide from Mn(II) (Brouwers et al. 1999). Cells deposit Mn oxide on their outer membranes and the oxidation is mediated by another multi-copper enzyme, CumA.

The dormant spores of several marine *Bacillus* species oxidize Mn(II) and become encrusted with amorphous Mn oxides (Rosson and Nealson 1982; Francis and Tebo 2002). This process also appears to be enzymatic: the oxidation of Mn(II) appears to be catalyzed by yet another multi-copper oxidase, MnxG (van Waasbergen et al. 1996).

Iron and manganese reduction

Dissimilatory metal-reducing bacteria are well recognized for their ability to utilize a number of diverse, oxidized, metal ions as terminal electron acceptors (Lovley 2000). Especially in the case of iron and manganese, this results in the dissolution of oxide minerals of these and any co-precipitated metals under anaerobic conditions (Lovley and Phillips 1986, 1988; Lovley 1991). Dissimilatory iron-reducing microorganisms respire with oxidized iron, Fe(III), usually in the form of amorphous Fe(III) oxyhydroxide (Lovley 1990, 1991) or crystalline iron oxides such as goethite, hematite, etc., under anaerobic conditions, and release reduced iron, Fe(II), into the environment. The Fe(II) can subsequently participate in adventitious interactions with anions resulting in the formation of various iron minerals. Iron-reducing bacteria are known to induce the precipitation of magnetite (Fe_3O_4), siderite ($FeCO_3$), and vivianite ($Fe_3(PO_4)_2 \cdot 8H_2O$), depending on the conditions and chemistry external to the cell (Moskowitz et al. 1989; Bazylinski and Frankel 2000a,b). For example, siderite was produced in cultures of *Geobacter metallireducens* along with magnetite when cells were grown in a bicarbonate buffering system (Lovley and Phillips 1988; Sparks et al. 1990), while vivianite, but neither magnetite nor siderite, was produced by the same organism with Fe(III) citrate as the terminal electron acceptor with a phosphate buffer (Lovley and Phillips 1988; Lovley 1990). Growing cells of the magnetotactic species, *Magnetospirillum magnetotacticum*, produced significant amounts of extracellular, needle-like crystals of vivianite (Fig. 4) while actively reducing Fe(III) in the form of Fe(III) oxyhydroxides (Blakemore and Blakemore 1990).

The biomineralization reactions described in the previous paragraph occur at neutral pH. There are a number of known chemolithoautotrophic and chemoheterotrophic, acidophilic, dissimilatory Fe(III)-reducing bacteria (Blake II and Johnson 2000) but little to nothing has been published as to any type of mineral formation at low pH.

Figure 4. Optical micrograph of extracellular crystals of vivianite, $Fe_3(PO_4)_2$, produced by cells of *Magnetospirillum magnetotacticum* in cultures containing high concentrations of Fe(III) buffered with phosphate. Under these conditions the cells reduce Fe(III) to Fe(II).

Interestingly, strictly anaerobic conditions do not seem to be required for these organisms to grow on Fe(III) although Fe(III) reduction may be most rapid under microaerobic conditions. The Archaean species *Sulfolobus adidocaldarius* reduces Fe(III) while growing heterotrophically on organic substrates. *Acidithiobacillus ferrooxidans* and *A. thiooxidans* oxidize reduced sulfur compounds coupling this reaction to the reduction of Fe(III). Cells of the α-Proteobacterium *Acidiphilium acidophilum* reduce Fe(III) with organic electron donors microaerobically. The Gram-positive, moderately-thermophilic Fe(II)-oxidizing, Bacteria *Sulfolobus* and *Acidimicrobium* also reduce Fe(III) and some are known to be capable of the reductive dissolution of Fe(III)-containing minerals (Bridge and Johnson 1998).

Many Fe(III)-reducing bacteria such as strains of *Shewanella* and *Geobacter* also reduce Mn(IV) to Mn(II). In this case, the organisms are well known for the dissolution of insoluble MnO_2; they reduce the Mn(IV) in MnO_2 to soluble to Mn(II). However, there are some instances of mineral formation during Mn(IV) reduction. Several metal-reducing strains of *Thermoanaerobacter* are known to form rhodochrosite ($MnCO_3$) during Mn(IV) reduction, uraninite (UO_2) during soluble U(VI) reduction, and gold metal during reduction of soluble Au(III) (Roh et al. 2002).

Biologically induced mineralization of magnetite

Fe(II) can react with excess, insoluble Fe(III) oxyhydroxide to form green rusts (mixed Fe(II) and Fe(III) oxyhydroxides) which can age to form magnetite. Magnetite particles, formed extracellularly by dissimilatory iron-reduction are typically irregular in shape and poorly crystallized (Moskowitz et al. 1989; Sparks et al. 1990) (Fig. 5). In addition, they have a relatively broad, lognormal, crystal size distribution with the mode in the superparamagnetic size range (< 35 nm) for magnetite. These crystal characteristics are typical of mineral particles produced by BIM or inorganic processes (Eberl et al. 1998).

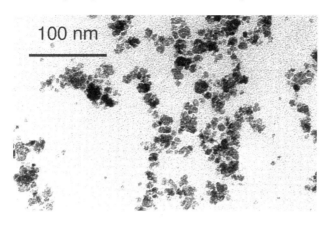

Figure 5. Magnetite crystals formed during reduction of ferric oxyhydroxide by the iron-reducing bacterium *Geobacter metallireducens*.

While many different species and physiological types of bacteria reduce Fe(III), not all conserve energy and grow from the reduction of this environmentally-abundant terminal electron acceptor (Myers and Nealson 1990) and form magnetite. *Geobacter metallireducens* and *Shewanella putrefaciens* are the most studied of this group and are phylogenetically associated with the δ- and γ-subdivisions, respectively, of the Proteobacteria (Myers and Nealson 1990; Lovley et al. 1993; Lonergan et al. 1996). *Shewanella* and *Geobacter* species are common in aquatic and sedimentary environments (DiChristina and DeLong 1993) and new species continue to be isolated (Caccavo et al. 1994; Rossello-Mora et al. 1994) suggesting that members of these genera may be the most environmentally-significant microbes involved in Fe(III) reduction and extracellular magnetite precipitation. BIM of magnetite has been demonstrated in cultures of *Shewanella*, *Geobacter* (Lovley et al. 1987; Lovley 1990), *Geothrix fermentans*, several thermophilic bacteria, including the Fe(III)-reducing bacterium strain TOR-39 (now known as a strain of the Gram-positive Bacterium *Thermoanaerobacter ethanolicus*) as well as other strains of the same genus (Liu et al. 1997; Zhang et al. 1998; Roh 2002), the Archaeon *Pyrobaculum islandicum* (Vargas et al. 1998) and the Bacterium *Thermotoga maritima* (Vargas et al. 1998). Magnetite is also formed in mixed cultures or consortia containing Fe(III) reducers (Bell et al. 1987; Liu et al. 1997; Zhang et al. 1997). It is likely that magnetite will be formed in a pure culture of any Fe(III)-reducing bacterium. Black, unidentified magnetic precipitates commonly observed in enrichment cultures or pure cultures of Fe(III)-reducing bacteria containing insoluble, amorphous, Fe(III) oxyhydroxide as the Fe(III) source (e.g., Greene et al. 1997; Slobodkin et al. 1997, 1999) probably consist primarily of magnetite. A halotolerant, facultatively-anaerobic, iron-reducing bacterium described by Rossello-Mora et al. (1994) most likely produces non-stoichiometric particles of magnetite with a composition intermediate between magnetite and maghemite (γ-Fe_2O_3) (Hanzlik et al. 1996).

The BIM magnetite particles produced by *Thermoanaerobacter ethanolicus* (strain TOR-39) have been well characterized (Zhang et al. 1998). Interestingly, like particles produced through BCM (Bazylinski 1995; Bazylinski and Frankel 2000a,b), the particles produced by *T. ethanolicus* have a size distribution that peaks in the single-magnetic-domain size range. The particles appear to be cuboctahedra with an average size of 56.2 ± 24.8 nm. *T. ethanolicus* is mildly thermophilic and growth and biomineralization experiments were performed at 65°C, raising the question of the role of temperature in size distribution of these crystals. Roh et al. (2001) later used this organism to produce metal-substituted magnetite crystals. Cobalt, chromium, and nickel were substituted into BIM magnetite crystals without changing the phase morphology. The incorporation of these metals into magnetite with the inverse spinel structure is of interest because of the unique magnetic, physical, and electrical properties of such crystals (Roh et al. 2001).

Cells of a magnetotactic species, *Magnetospirillum magnetotacticum*, have been shown to reduce Fe(III) in growing cultures and there is some evidence that iron reduction may be linked to energy conservation and growth in this bacterium (Guerin and Blakemore 1992). While extracellular BIM magnetite has never been observed in cultures of this organism, cells of *M. magnetotacticum* synthesize intracellular particles of magnetite (Frankel et al. 1979) via BCM (see Bazylinski and Frankel 2003).

Magnetite dissolution

In addition to magnetite mineralization, some iron-reducing bacteria are able to reduce ferric iron in magnetite—2 Fe(III) and 1 Fe(II) per formula unit—with release of Fe(II). *S. putrefaciens* was reported to reduce and grow on Fe(III) in magnetite (Kostka and Nealson 1995) whereas it appears *G. metallireducens* is unable to do so (Lovley and Phillips 1988). Dong et al. (2000) conducted reduction experiments in which *S. putrefaciens* strains CN32 and MR-1 respired with either biogenic or inorganic magnetite as electron acceptor and lactate as electron donor. In a medium buffered by bicarbonate (HCO_3^-), siderite ($FeCO_3$) precipitated, suggesting a dissolution-precipitation mechanism. Vivianite ($Fe_3(PO_4)_2$) precipitated in media with sufficient phosphate. The biogeochemical significance of this result is that some dissimilatory iron-reducing bacteria could utilize magnetite as an electron donor after the original pool of ferric iron, likely ferric oxyhydroxide, is exhausted. Thus it seems that some dissimilatory iron-reducing bacteria can both mineralize and dissolve magnetite under different Eh and pH conditions. Dong et al. (2000) note that magnetite is thermodynamically stable at pH 5–6.5 but is unstable at pH > 6.5. However, mineral formation that removes Fe(II) from solution tends to increase the pH range over which magnetite reduction is favorable. Thus BIM may function to shift thermodynamic equilibria in certain situations.

Sulfate reduction

Of all the metal sulfide minerals, iron sulfide mineralization is most often attributed to microbial activity (Southam 2000), more specifically to the activity of the dissimilatory sulfate-reducing bacteria. These ubiquitous, anaerobic, prokaryotes are a physiological group of microorganisms that are phylogenetically and morphologically very diverse and include species in the Domains Bacteria (δ-subdivision of Proteobacteria and Gram-positive group) and Archaea. Because all sulfate-reducing bacteria respire with sulfate under anaerobic conditions and release highly reactive sulfide ions, it is likely that all the species, regardless of phylogeny or classification, produce iron sulfide minerals through BIM under appropriate environmental conditions with excess, available, iron. Even a sulfate-reducing, magnetotactic bacterium, *Desulfovibrio magneticus* strain RS-1, is known to produce extracellular particles of iron sulfides through BIM while synthesizing intracellular crystals of magnetite via BCM (Sakaguchi et al. 1993). Sulfide ions react with the iron forming magnetic particles of greigite (Fe_3S_4) and pyrrhotite (Fe_7S_8) as well as a number of other non-magnetic iron sulfides including mackinawite (tetragonal FeS), pyrite (cubic FeS_2) and marcasite (orthorhombic FeS_2) (Freke and Tate 1961; Rickard 1969a,b). Mineral species formed in these bacterially-catalyzed reactions appear to be dependent on the pH and Eh of the growth medium, the incubation temperature, the presence of specific oxidizing and reducing agents, and the type of iron source in the growth medium. In addition, microorganisms clearly modify many of these parameters (e.g., pH, Eh) during growth. For example, cells of *Desulfovibrio desulfuricans* produced greigite when grown in the presence of ferrous salts but not when the iron source was goethite, FeO(OH) (Rickard 1969a).

Berner (1962, 1964, 1967, 1969) reported the chemical synthesis of a number of iron sulfide minerals, including marcasite, mackinawite, a magnetic, cubic iron sulfide of the spinel type (probably greigite), pyrrhotite, amorphous FeS, and even framboidal pyrite, a globular form of pyrite that was once thought to represent fossilized bacteria (Fabricus 1961; Love and Zimmerman 1961). Rickard (1969a,b) concluded that extracellular, biogenic iron sulfide minerals could not be distinguished from abiogenic (inorganic) minerals. However, in many cases, the iron sulfide minerals produced by the sulfate-reducing bacteria have not been systematically examined by high resolution electron microscopy. In addition, in many of early studies, the role of the cell in mineralization was not investigated.

More recent studies with sulfate-reducing bacteria show that mineralization proceeds initially by the immobilization of amorphous FeS on the cell surface (Fig. 6) through the ionic interaction of Fe^{2+} with anionic cell surface charges and biogenic H_2S (Fortin et al. 1994). Mineral transformations cause the production of other Fe sulfides, and eventually, pyrite (Fortin and Beveridge 2000; Southam 2000). Despite the results of Berner (1962, 1964, 1967, 1969), the bacterially-induced transformation of FeS to pyrite appears to be more efficient than that occurring under abiogenic conditions (Donald and Southam 1999).

Sulfide mineral oxidation

In addition to those bacteria that facilitate the mineralization of iron sulfides, there are bacteria that can oxidize iron sulfides such as pyrite (FeS_2) with molecular oxygen, with release of Fe(III) and sulfate (SO_4^{2-}) (Nordstrom and Southam 1997). This process is responsible for acid mine drainage and has also been put to use in enrichment and leaching of sulfide ores. The most studied organism is *Acidithiobacillus ferrooxidans*, an

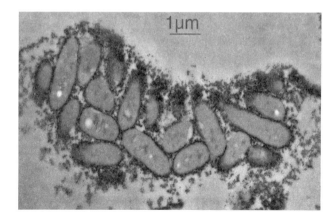

Figure 6. Unstained, ultrathin section transmission electron micrograph of a mineralized bacterial microcolony from a sulfate-reducing bacterial consortium grown with lactate in the presence of Fe(II). The cells in are encrusted with amorphous iron sulfides. Figure kindly supplied by W. Stanley and G. Southam.

acidophillic, autotrophic, bacterium. The oxidation process depends among other things on the properties of the pyrite, including grain size, crystallinity, defect structure, and trace metal impurities. Following oxidation, Fe(III) hydrolyses and initially precipitates as ferric oxyhydroxide. However, aging can result in a number of iron minerals including ferrihydrite and goethite, as well as iron-sulfate minerals jarosite ($KFe_3(SO_4)_2(OH)_6$) and schwertmanite ($Fe_8O_8SO_4(OH)_6$). Elemental sulfur is another possible reaction product. Other disulfide and monosulfide minerals can also be oxidized but result in substrate-mineral-specific products. Cells of *Acidithiobacillus* readily attach to the surfaces of sulfide minerals which maximizes the efficiency of the oxidation process. In general, microbe-mineral interactions in diverse environmental situations have become a major theme in biogeochemistry (Banfield and Hammers 1997; Fortin et al. 1997; Little et al. 1997; Edwards et al. 2001).

INTRACELLULAR BIOLOGICALLY INDUCED MINERALIZATION

Most of the examples of BIM discussed above involve extracellular deposition of minerals. However, there are several reports of intracellular deposition of minerals that seem to blur the line between BIM and BCM. For example, many bacteria have iron-storage proteins known as bacterioferritins (Chasteen and Harrison 1999). These are intracellular proteins comprising 24 identical subunits arranged in pairs that form a dodecahedral shell enclosing a 9 nm cavity. The cavity can accommodate up to 4000 iron atoms as an amorphous, ferric oxyhydroxy phosphate, with variable P/Fe ratio. The subunit pairs contain ferroxidase centers which catalyze the oxidation of ferrous iron and nucleation of the mineral in the cavity. While the organism provides an organic vesicle (the protein shell) for the deposition of the mineral, it apparently does not control the composition or crystallinity of the mineral. On the other hand, less crystallinity may allow greater access to the iron and perhaps phosphate stored as the mineral in the cavity. Addition of phosphate, a known glass former, may insure formation of an amorphous core mineral.

Intracellular iron-sulfide particles have been reported within cells of some sulfate-reducing bacteria, including *Desulfovibrio* and *Desulfotomaculum* species, when they were grown with relatively high concentrations of iron in the growth medium (Jones et al. 1976). The "particles" were randomly arranged in the cell and, based on electron diffraction, were not well-ordered crystals. They were also not separable by density gradient centrifugation. They are apparently not essential to the cell in that cells can be grown with much less iron where they do not form these structures.

Unidentified, presumably magnetic ("magnet-sensitive"), electron-dense particles were reported in cells of several purple photosynthetic bacteria including *Rhodospeudomonas palustris*, *R. rutilis* (both α-Proteobacteria), and *Ectothiorhodospira shaposhnikovii* (a γ-Proteobacterium) cultured in growth media containing relatively high concentrations of iron. The inclusions were spherical particles containing an electron-transparent core surrounded by an electron-dense matrix. The particles could be separated from lysed cells; X-ray microanalysis showed that the inclusions are Fe-rich but did not contain sulfur. The particles were arranged in a chain like magnetosomes (Bazylinski 1995) and possibly surrounded by a membranous structure (Vainshtein et al. 1997). Vainshtein et al. (2002) later showed that many other bacteria including non-photosynthetic members of both prokaryotic domains could be induced to form similar particles. Cells with the particles show a magnetic response but are not necessarily magnetotactic. The authors speculate that the particles function similarly to magnetosomes. This case of biomineralization appears to be almost intermediate between

BIM and BCM in that cells appear to control some features of these particles such as their arrangement in the cell.

Glasauer et al. (2001) reported unidentified iron oxide particles within the dissimilatory iron-reducing bacterium *Shewanella putrifaciens* grown in an H_2/Ar atmosphere with poorly-crystalline ferrihydrite (ferric oxyhydroxide) as electron acceptor. There is evidence from selected area electron diffraction that the intracellular iron oxide particles are magnetite or maghemite (γ-Fe_2O_3). Magnetite also formed outside the cell. Ona-Nguema et al (2002) found green rust with Fe(II)/Fe(III) ratio ~1 when *S. putrifaciens* was cultured under anaerobic conditions with formate as the electron donor and crystalline lepidochrocite (γ-FeOOH) as the electron acceptor. The green rust eventually remineralized as black magnetite/maghemite when the reaction culture medium was incubated at room temperature.

SIGNIFICANCE OF BIOLOGICALLY INDUCED MINERALIZATION

Biomineralization by prokaryotes is of great significance in scientific and commercial applications as well as having a major impact in microbiology, evolutionary biology, and geology. Bacterial metal sorption and precipitation can be important and useful in metal and radionuclide removal during the bioremediation of metal- and radionuclide-contaminated waters (Lovley 2000). The growth of Fe(II)- and Mn(II)-oxidizing bacteria that efficiently remove Fe and Mn ions from water by mineralization in wastewater treatment plants is promoted in France, thereby eliminating the problems of biofouling of pipelines by mineral deposits and of water discoloration (Mouchet 1992). This is a major problem in the use of groundwater sources.

Konhauser et al. (2002) have speculated that iron-oxidizing bacteria could have been responsible for the formation of the massive Precambrian banded iron formations (BIF) by BIM via oxidation of dissolved Fe(II) in the ancient ocean. Oxidation of Fe(II) to Fe(III) could have occurred by chemolithoautotrophy or by photosynthesis with Fe(II) as the electron donor. Based on the chemical analyses of BIF deposits dated to 2.5 Ga from Western Australia, they concluded that bacterial cell densities less than those found in modern Fe-rich environments would have been sufficient.

Bacterially-formed minerals, in one form or another, may be useful as biomarkers (indications of past life) when other remains of the cells or indications of the presence of the cells are no longer evident. In many situations only mineral encrustations that once encapsulated the cell are observed (Southam 2000). These biomarkers may be useful not only in determining when bacteria evolved on Earth but also as evidence of former life in extraterrestrial materials (Thomas-Keprta et al. 2000).

An interesting yardstick for the scale of bacterially-induced mineralization is provided by the wreck of the Titanic. When Robert Ballard found the Titanic in 1985, he noted rust-colored concretions hanging off the hull. The concretions have shapes similar to stalactites or icicles; hence Ballard called them "rusticles." Rusticles can be centimeters to meters in length and have a complex internal architecture with water channels the diameters of which are distributed over many orders of magnitude. Iron minerals comprise the major constituents of the rusticle, with ferric oxyhydroxides predominating on the outer surfaces and goethite on the inside. Associated with the rusticles are a microbial consortium of over twenty species that includes iron-oxidizing and sulfate-reducing bacteria (Wells and Mann 1997). This suggests that rusticles contain a number of micro-environments from oxic to anaerobic. SEM studies show heavily mineralized bacteria organized in chains (Fig. 7). From sequential observations over a

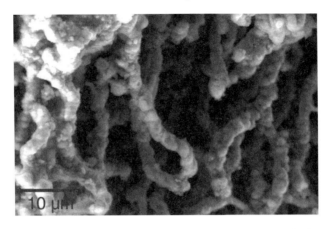

Figure 7. Scanning electron micrograph of "rusticles" recovered from the hull of the Titanic. The rusticles consist of ropes or chains of heavily mineralized bacteria. This figure was adapted from one kindly provided by H. Mann.

number of years, it has been estimated that the rusticle formation rate is about 1 ton per year over the ship. At this rate, the remaining lifetime of the hull must be measured in years, not centuries. This illustrates the fact that bacteria working over geologic time scales can effect enormous mineral transformations.

ACKNOWLEDGMENTS

We thank K.J. Edwards, W.C. Ghiorse, G. Southam and H. Mann for providing figures and T.J. Williams for reviewing the manuscript. DAB is grateful for support from National Science Foundation (NSF) Grant EAR-0311950 and National Aeronautics and Space Administration (NASA) Johnson Space Center Grant NAG 9-1115.

REFERENCES

Adams, LF, Ghiorse WC (1986) Physiology and ultrastructure of *Leptothrix discophora* SS-1. Arch Microbiol 145:126-135

Adams, LF, Ghiorse WC (1987) Characterization of extracellular Mn^{2+}-oxidizing activity and isolation of an Mn^{2+}-oxidizing protein from *Leptothrix discophora* SS-1. J Bacteriol 169:1279-1285

Appanna VD, Preston CM (1987) Manganese elicits the synthesis of a novel exopolysaccharide in an arctic *Rhizobium*. FEBS Lett 215:79-82

Bäuerlein E (2000) Biomineralization: From Biology to Biotechnology and Medical Application. Wiley-VCH, Weinheim, Germany

Bäuerlein E (2003) Biomineralization of unicellular organisms: An unusual membrane biochemistry for the production of inorganic nano- and microstructures. Angew Chem Int Ed 42:614-641

Banfield JF, Hammers RJ (1997) Processes at minerals and surfaces with relevance to microorganisms and prebiotic synthesis. Rev Mineral 35:81-122

Banfield JF, Zhang H (2001) Nanoparticles in the environment. Rev Mineral Geochem 44:1-58

Bazylinski DA (1995) Structure and function of the bacterial magnetosome. ASM News 61:337-343

Bazylinski DA, Frankel RB (2000a) Magnetic iron oxide and iron sulfide minerals within organisms. *In*: Biomineralization: From Biology to Biotechnology and Medical Application. Bäuerlein E (ed) Wiley-VCH, Weinheim, Germany, p 25-46

Bazylinski DA, Frankel RB (2000b) Biologically controlled mineralization of magnetic iron minerals by magnetotactic bacteria. *In:* Environmental Microbe-Mineral Interactions. Lovley DR (ed) ASM Press, Washington, DC, p 109-144

Bazylinski DA, Frankel RB (2003) Biologically controlled mineralization in prokaryotes. Rev Mineral Geochem 54:217-247
Bell PE, Mills AL, Herman JS (1987) Biogeochemical conditions favoring magnetite formation during anaerobic iron reduction. Appl Environ Microbiol 53:2610-2616
Berner RA (1962) Synthesis and description of tetragonal iron sulfide. Science 137:669
Berner, RA (1964) Iron sulfides formed from aqueous solution at low temperatures and atmospheric pressure. J Geol 72:293-306
Berner, RA (1967) Thermodynamic stability of sedimentary iron sulfides. Am J Sci 265:773-785
Berner RA (1969) The synthesis of framboidal pyrite. Econ Geol 64:383-393
Beveridge TJ (1981) Ultrastructure, chemistry, and function of the bacterial cell wall. Int Rev Cytol 72:229-317
Beveridge TJ (1989) Role of cellular design in bacterial metal accumulation and ineralization. Annu Rev Microbiol 43:147-171
Beveridge TJ, Murray RGE (1976) Uptake and retention of metals by cell walls of *Bacillus subtilis*. J Bacteriol 127:1502-1518
Beveridge TJ, Murray RGE (1980) Sites of metal deposition in the cell wall of Bacillus subtilis. J Bacteriol 141:876-887
Blake II R, Johnson DB (2000) Phylogenetic and biochemical diversity among acidophilic bacteria that respire on iron. *In:* Environmental Microbe-Mineral Interactions. Lovley DR (ed) ASM Press, Washington, DC, p 53-78
Blakemore RP, Blakemore NA (1990) Magnetotactic magnetogens. *In*: Iron Biominerals. Frankel RB, Blakemore RP (eds) Plenum Press, New York, p 51-67
Bridge TAM, Johnson DB (1998) Reduction of soluble iron and reductive dissolution of ferric iron-containing by moderately thermophilic iron-oxidizing bacteria. Appl Environ Microbiol 64:2181-2186
Brouwers G-J, de Vrind JPM, Corstjens PLAM, Cornelis P, Baysse C, de Vrind-de Jong EW (1999) *cumA*, a gene encoding a multi-copper oxidase, is involved in Mn^{2+} oxidation in *Pseudomonas putida* GB-1. Appl Environ Microbiol 65:1762-1768
Buchholz-Cleven BEE, Rattunde B, Straub KL (1997) Screening of genetic diversity of isolates of anaerobic Fe(II)-oxidizing bacteria using DGGE and whole-cell hybridization. Syst Appl Microbiol 20:301-309
Caccavo Jr F, Lonergan DJ, Lovley DR, Davis M, Stolz JF, McInerny MJ (1994) *Geobacter sulfurreducens* sp. nov., a hydrogen- and acetate-oxidizing dissimilatory metal reducing microorganism. Appl Environ Microbiol 60:3752-3759
Chasteen ND, Harrison PM 1999 Mineralization in ferritin: an efficient means of iron storage. J Struct Biol 126:182-194
Chaudhuri SK, Lack JG, Coates JD (2001) biogenic magnetite formation through anaerobic biooxidation of Fe(II). Appl Environ Microbiol 67:2844-2847
Corstjens PLAM, de Vrind JPM, Goosen T, de Vrind-de Jong EW (1997) Identification and molecular analysis of the *Leptothrix discophora* SS-1 mofA gene, a gene putatively encoding a manganese-oxidizing protein with copper domains. Geomicrobiol J 14:91-108
DiChristina TJ, DeLong EF (1993) Design and application of rRNA-targeted oligonucleotide probes for the dissimilatory iron- and manganese-reducing bacterium *Shewanella putrefaciens*. J Bacteriol 59:4152-4160
Donald R, Southam G (1999) Low temperature anaerobic bacterial diagenesis of ferrous monosulfide to pyrite. Geochim Cosmochim Acta 63:2019-2023
Dong H, Fredrickson JK, Kennedy DW, Zachara JM, Kukkadapu RK, Onsott TC (2000) Mineral transformations associated with the microbial reduction of magnetite. Chem Geol 169:299-318
Eberl DD, Drits VA, Srodon J (1998) Deducing growth mechanisms for minerals from the shapes of crystal size distributions. Am J Sci 298:499-533
Edwards KJ, Bond PL, Gihring TM, Banfield JF (2000) An Archaeal iron-oxidizing extreme acidophile important in acid mine drainage. Science 279:1796-1799
Edwards KJ, Hu B, Hamers RJ, Banfield JF (2001) A new look at microbial leaching patterns on sulfide minerals. FEMS Microbiol Ecol 34:197-206
Ehrenbach A, Widdel F (1994) Anaerobic oxidation of ferrous iron by purple bacteria, a new type of phototrophic metabolism. Appl Environ Microbiol 60:4517-4526
Emerson D (2000) Microbial oxidation of Fe(II) and Mn(II) at circumneutral pH. *In:* Environmental Microbe-Mineral Interactions. Lovley DR (ed) ASM Press, Washington, DC, p 109-144
Emerson D, Moyer CL (1997) Isolation and characterization of novel iron-oxidizing bacteria that grow at circumneutral pH. Appl Environ Microbiol 63:4784-4792

Emerson D, Moyer CL (2002) Neutrophilic Fe-oxidizing bacteria are abundant at the Loihi Seamount hydrothermal vents and play a major role in Fe oxide deposition. Appl Environ Microbiol 68:3085-3093

Emerson D, Weiss JV, Megonigal JP (1999) Iron-oxidizing bacteria are associated with ferric hydroxide precipitates (Fe-plaque) on the roots of wetland plants. Appl Environ Microbiol 65:2758-2761

Fabricus F (1961) Die Strukturen des "Rogenpyrits" (Kossener Schichten, Rat) als Betrag zum Problem der "Vererzten Bakterien". Geol Rundshau 51:647-657

Ferris FG, Beveridge TJ (1984) Binding of a paramagnetic cation to Escherichia coli K-12 outer membrane vesicles. FEMS Microbiol Lett 24:43-46

Ferris FG, Beveridge TJ (1986a) Site specificity of metallic ion binding in *Escherichia coli* K-12 lipopolysaccharide. Can J Microbiol 32:52-55

Ferris FG, Beveridge TJ (1986b) Physicochemical roles of soluble metal cations in the outer membrane of *Escherichia coli* K-12. Can J Microbiol 32:594-601

Fortin D, Beveridge TJ (2000) Mechanistic routes to biomineral surface development. *In*: Biomineralization: From Biology to Biotechnology and Medical Application. Bäuerlein E (ed) Wiley-VCH, Weinheim, Germany, p 7-24

Fortin D, Ferris FG, Beveridge TJ (1997) Surface-mediated mineral development by bacteria. Rev Mineral 35:161-180

Fortin D, Southam G, Beveridge TJ (1994) An examination of iron sulfide, iron-nickel sulfide and nickel sulfide precipitation by a *Desulfotomaculum* species: and its nickel resistance mechanisms. FEMS Microbiol Ecol 14:121-132

Francis CA, Tebo BM (2002) Enzymatic manganese(II) oxidation by metabolically dormant spores of diverse *Bacillus* species. Appl Environ Microbiol 68:874-80

Frankel RB, Bazylinski DA, Johnson M , Taylor B 1997 Magneto-aerotaxis in marine, coccoid bacteria. Biophys J 73:994-1000

Frankel RB, Blakemore RP, Wolfe RS (1979) Magnetite in freshwater magnetotactic bacteria. Science 203:1355-1356

Freke AM, Tate D (1961) The formation of magnetic iron sulphide by bacterial reduction of iron solutions. J Biochem Microbiol Technol Eng 3:29-39

Glasauer S, Langley S, Beveridge TJ (2001) Intracellular iron minerals in a dissimilatory iron-reducing bacterium. Science 295:117-119

Golyshina OV, Pivovarova TA, Karavaiko GI, Kondrateva TF, Moore ER, Abraham WR, Lunsdorf H, Timmis KN, Yakimov MM, Golyshin PN (2000) *Ferroplasma acidiphilum* gen. nov., sp. nov., an acidophilic, autotrophic, ferrous-iron-oxidizing, cell-wall-lacking, mesophilic member of the Ferroplasmaceae fam. nov., comprising a distinct lineage of the Archaea. Int J Syst Evol Microbiol 50:997-1006

Greene AC, Patel BKC, Sheehy AJ (1997) *Deferribacter thermophilus* gen. nov., sp. nov., a novel thermophilic manganese- and iron-reducing bacterium isolated from a petroleum reservoir. Int J Syst Bacteriol 47:505-509.

Guerin WF, Blakemore RP (1992) Redox cycling of iron supports growth and magnetite synthesis by *Aquaspirillum magnetotacticum*. Appl Environ Microbiol 58:1102-1109

Hafenbradl D, Keller M, Dirmeier R, Rachel R, Roβnagel S, Burggraf S, Huber H, Stetter KO (1996) Ferroglobus placidus gen nov., sp. nov., a novel hyperthermophilic archaeum that oxidizes Fe^{2+} at neutral pH under anoxic conditions. Arch Microbiol 166:308-314

Hallbeck L, Petersen K (1990) Culture parameters regulating stalk formation and growth rate of *Gallionella ferruginea*. J Gen Microbiol 136:1675-1680

Hallbeck L, Petersen K. (1991) Autotrophic and mixotrophic growth of *Gallionella ferruginea*. J Gen Microbiol 137:2657-2661

Hallbeck L, Ståhl F, Petersen K (1993) Phylogeny and phenotypic characterization of the stalk-forming and iron-oxidizing bacterium *Gallionella ferruginea*. J Gen Microbiol 139:1531-1535

Hanert HH (2000a) The Genus *Gallionella*. *In*: The Prokaryotes. Dworkin M et al.(eds) Springer-Verlag New York, Inc., New York (on the web at http://www.springer-ny.com/)

Hanert HH (2000b) The Genus *Siderocapsa* (and other iron- or manganese-oxidizing Eubacteria). *In*: The Prokaryotes. Dworkin M et al.(eds) Springer-Verlag New York, Inc., New York (on the web at http://www.springer-ny.com/)

Hanzlik MM, Petersen N, Keller R, Schmidbauer E (1996) Electron microscopy and ^{57}Fe Mössbauer spectra of 10 nm particles, intermediate in composition between Fe_3O_4–γ-Fe_2O_3, produced by bacteria. Geophys Res Lett 23:479-482

Harrison PM, Arosio P (1996) The ferritins: molecular properties, iron storage function and cellular regulation. Biochim Biophys Acta 1275:161-203

Heldal M, Tumyr O (1983) *Gallionella* from metalimnion in a eutrophic lake: morphology and X-ray energy-dispersive microanalysis of apical cells and stalks. Can J Microbiol 29:303-308

Jones HE, Trudinger PA, Chambers LA, Pyliotis NA (1976) Metal accumulation by bacteria with particular reference to dissimilatory sulphate-reducing bacteria. Z Allg Mikrobiol 16:425-435

Kelly, DP, Wood AP (2000) Reclassification of some species of *Thiobacillus* to the newly designated genera *Acidithiobacillus* gen. nov., *Halothiobacillus* gen. nov., and Thermithiobacillus gen. nov. Int J Syst Bacteriol 50:511-51

Konhauser KO (1998) Diversity of bacterial iron mineralization. Earth-Sci Rev 43:91-121

Konhauser KO, Hamade T, Raiswell R, Morris RC, Ferris FG, Southam G, Canfield DE (2002) Could bacteria have formed the Precambrian banded iron formations? Geology 30:1079-1082

König H (1988) Archaebacterial cell envelopes. Can J Microbiol 34:395-406

Kostka JE, Nealson KH (1995) Dissolution and reduction of magnetite by bacteria. Environ Sci Technol 29:2535-2540

Koval SF (1988) Paracrystalline surface arrays on bacteria. Can J Microbiol 34:407-414

Lam JS, Graham LL, Lightfoot J, Dasgupta T, Beveridge TJ (1992) Ultrastructural examination of lipopolysaccharides of *Pseudomonas aeruginosa* strains and their isogenic mutants by freeze-substitution. J Bacteriol 174:7159-7167

Little BJ, Wagner PS, Lewandowski Z (1997) Spatial relationships between bacteria and mineral surfaces. Rev Mineral 35:123-159

Liu SV, J Zhou, C Zhang, DR Cole, M Gajdarziska-Josifovska, TJ Phelps (1997) Thermophilic Fe(III)-reducing bacteria from the deep subsurface: the evolutionary implications. Science 277: 1106-1109

Lonergan DJ, HL Jenter, JD Coates, EJP Phillips, TM Schmidt, DR Lovley (1996) Phylogenetic analysis of dissimilatory Fe(III)-reducing bacteria. J Bacteriol 178:2402-2408

Love LG, DO Zimmerman (1961) Bedded pyrite and microorganisms from the Mount Isa Shale. Econ Geol 56:873-896

Lovley DR (1990) Magnetite formation during microbial dissimilatory iron reduction. *In*: Iron Biominerals. Frankel RB, Blakemore RP (eds) Plenum Press, New York, p 151-166

Lovley DR (1991) Dissimilatory Fe(III) and Mn(IV) reduction. Microbiol Rev 55:259-287

Lovley DR (ed) (2000) Environmental Microbe-Mineral Interactions. ASM Press, Washington, DC

Lovley DR, SJ Giovannoni, DC White, JE Champine, EJP Phillips, YA Gorby, S Goodwin (1993) *Geobacter metallireducens* gen. nov. sp. nov., a microorganism capable of coupling the complete oxidation of organic compounds to the reduction of iron and other metals. Arch Microbiol 159:336-344

Lovley DR, Phillips EJP (1986) Organic matter mineralization with the reduction of ferric iron in anaerobic sediments. Appl Environ Microbiol 51:683-689

Lovley DR, Phillips EJP (1988) Novel mode of microbial energy metabolism: organic carbon oxidation coupled to dissimilatory reduction of iron or manganese. Appl Environ Microbiol 54:1472-1480

Lovley DR, Stolz JF, Nord Jr. GL, Phillips EJP (1987) Anaerobic production of magnetite by a dissimilatory iron-reducing microorganism. Nature 330:252-254

Lowenstam HA (1981) Minerals formed by organisms. Science 211:1126-1131

Lowenstam HA, Weiner S (1989) On Biomineralization. Oxford University Press, New York

Moskowitz BM, Frankel RB, Bazylinski DA, Jannasch HW, Lovley DR (1989) A comparison of magnetite particles produced anaerobically by magnetotactic and dissimilatory iron-reducing bacteria. Geophys Res Lett 16:665-668

Mouchet P (1992) From conventional to biological removal of iron and manganese in France. J. Am Water Works Assoc 84:158-167

Moyer C, Dobbs FC, Karl DM (1995) Phylogentic diversity of the bacterial component from a microbial mat at an active, hydrothermal vent system, Loihi Seamount. Appl Environ Microbiol 61:1555-1562

Myers CR, Nealson KH (1990) Iron mineralization by bacteria: metabolic coupling of iron reduction to cell metabolism in *Alteromonas putrefaciens* MR-1. 131-149. *In*: Iron Biominerals. Frankel RB, Blakemore RP (eds) Plenum Press, New York, p 131-149

Neubauer SC, Emerson D, Megonigal JP (2002) Life at the energetic edge: kinetics of circumneutral iron oxidation by lithotrophic iron-oxidizing bacteria isolated from the wetland-plant rhizosphere. Appl Environ Microbiol 68:3988-3995

Nordstrom DK, Southam G (1997) Geomicrobiology of sulfide mineral oxidation. Rev Mineral 35:361-390

Ona-Nguema G, Abdelmoula M, Jorand F, Benali O, Gehin A, Block J-C, Genin, J-M R (2002) Microbial reduction of lepidochrocite γ-FeOOH by *Shewanella putrefaciens*; the formation of green rust. Hyp Interact 139/140: 231-237

Pósfai M, Buseck PR, Bazylinski DA, Frankel RB (1998) Iron sulfides from magnetotactic bacteria: structure, compositions, and phase transitions. Am Mineral 83:1469-1481

Rawlings DE, Kusano T (1994) Molecular genetics of *Thiobacillus ferrooxidans*. Microbiol Rev 58:39-55

Rickard DT (1969a) The microbiological formation of iron sulfides. Stockholm Contrib Geol 20:50-66
Rickard DT (1969b) The chemistry of iron sulfide formation at low temperatures. Stockholm Contrib Geol 20:67-95
Roh Y, Lauf RJ, McMillan AD, Zhang C, Rawn CJ, Bai J Phelps TJ (2001) Microbial synthesis and the characterization of metal-substituted magnetites. Solid State Commun 110:529-534
Roh Y, Liu SV, Li G, Huang H, Phelps TJ, Zhou J (2002) Isolation and characterization of metal-reducing *Thermoanaerobacter* strains from deep subsurface environments of the Piceance Basin, Colorado. Appl Environ Microbiol 68:6103-6020
Rossello-Mora RA, Caccavo Jr. F, Osterlehner K, Springer N, Spring S, Schüler D, Ludwig W, Amann R, Vannacanneyt M, Schleifer K-H (1994) Isolation and taxonomic characterization of a halotolerant, facultative anaerobic iron-reducing bacterium. Syst Appl Microbiol 17:569-573
Rosson RA, Nealson KH (1982) Manganese binding and oxidation by spores of a marine bacillus. J Bacteriol 174:575-585
Sakaguchi T, Burgess JG, Matsunaga T (1993) Magnetite formation by a sulphate-reducing bacterium. Nature 365:47-49
Schultze-Lam S, Harauz G, Beveridge TJ (1992) Participation of a cyanobacterial S layer in fine-grain mineral formation. J Bacteriol 174:7971-7981
Slobodkin AI, Jeanthon C, Haridon SL, Nazina T, Miroschnichenko M, Bonch-Osmolovskaya E (1999) Dissimilatory reduction of Fe(III) by thermophilic bacteria and archaea in deep subsurface petroleum reservoirs of western Siberia. Curr Microbiol 39:99-102
Slobodkin AI, Reysenbach A-L, Strutz N, Dreier M, Wiegel J (1997) *Thermoterrabacterium ferrireducens* gen. nov., sp. nov., a thermophilic anaerobic dissimilatory Fe(III)-reducing bacterium from a continental hot spring. Int J Syst Bacteriol 47:541-547
Southam G (2000) Bacterial surface-mediated mineral formation. In: Environmental Microbe-Mineral Interactions. Lovley DR (ed) ASM Press, Washington, DC, p 257-276
Sparks NHC, Mann S, Bazylinski DA, Lovley DR, Jannasch HW, Frankel RB (1990) Structure and morphology of magnetite anaerobically-produced by a marine magnetotactic bacterium and a dissimilatory iron-reducing bacterium. Earth Planet Sci Lett 98:14-22
Straub KL, Benz M, Schink B, Widdel F (1996) Anaerobic, nitrate-dependent microbial oxidation of ferrous iron. Appl Environ Microbiol 62:1458-1460
Straub KL, Buchholz-Cleven BEE (1998) Enumeration and detection of anaerobic ferrous iron-oxidizing, nitrate-reducing bacteria from diverse European sediments. 64:4846-4856
Straub KL, Rainey FA, Widdel F (1999) Isolation and characterization of marine phototrophic ferrous iron-oxidizing purple bacteria, *Rhodovulum iodosum* sp. nov. and *Rhodovulum robiginosum* sp. nov. Int J Syst Bacteriol 49:729-735
Thomas-Keprta KL, Bazylinski DA, Kirschvink JL, Clemett SJ, McKay DS, Wentworth SJ, Vali H, Gibson Jr. EK, Romanek CS (2000) Elongated prismatic magnetite (Fe_3O_4) crystals in ALH84001 carbonate globules: potential martian magnetofossils. Geochim Cosmochim Acta 64:4049-4081
Urrutia M, Kemper M, Doyle R, Beveridge TJ (1992) The membrane-induced proton motive force influences the metal binding activity of *Bacillus subtilis* cell walls. Appl Environ. Microbiol 58:3837-3844
Vainshtein M, Suzina N, Sorokin V (1997) A new type of magnet-sensitive inclusions in cells of photosynthetic bacteria. Syst. Appl Microbiol 20:182-186
Vainshtein M, Suzina N, Kudryashova E, Ariskina E (2002) New magnet-sensitive structures in bacterial and archaeal cells. Biol Cell 94:29-35
van Waasbergen LG, Hildebrand M, Tebo BM (1996) Identification and characterization of a gene cluster involved in manganese oxidation by spores of the marine *Bacillus* sp. strain SG-1. J Bacteriol 12:3517–3530
Vargas M, Kashefi K, Blunt-Harris EL, Lovley DR (1998) Microbiological evidence for Fe(III) reduction on early Earth. Nature 395:65-67
Wells W, Mann H (1997) Microbiology and formation of rusticles. Res Environ Biotechnol 1:271-281
Weiner S, Dove PM (2003) An overview of biomineralization processes and the problem of the vital effect. Rev Mineral Geochem 54:1-26
Widdel F, Schnell S, Heising S, Ehrenbach A, Assmus B, Schink B (1993) Ferrous iron oxidation by anoxygenic phototrophic bacteria. Nature 362:834-836
Zhang C, Liu S, Phelps TJ, Cole DR, Horita J, Fortier SM (1997) Physiochemical, mineralogical, and isotopic characterization of magnetite-rich iron oxides formed by thermophilic iron-reducing bacteria. Geochim Cosmochim Acta 61:4621-4632
Zhang C, Vali H, Romanek CS, Phelps TJ, Lu SV (1998) Formation of single domain magnetite by a thermophilic bacterium. Am Mineral 83:1409-1418

5 The Source of Ions for Biomineralization in Foraminifera and Their Implications for Paleoceanographic Proxies

Jonathan Erez

Institute of Earth Sciences
The Hebrew University of Jerusalem
Jerusalem 91904, Israel

INTRODUCTION

The global carbon cycle is strongly perturbed by fossil fuel burning leading to atmospheric CO_2 increase. Climatic warming followed by polar ice melting and global sea level rise are predicted due to the greenhouse effect of increasing CO_2 in the atmosphere (Houghton et al. 1995). The ocean plays a major role in neutralizing the excess CO_2 because the amount of inorganic carbon available for exchange with the atmosphere in the ocean is approximately 50–60 times larger than in the atmosphere (e.g., Siegenthaler and Sarmiento 1993). The bulk of the atmospheric CO_2 excess will eventually be neutralized by $CaCO_3$ dissolution in the deep marine environment. This is, however, is a relatively slow process that operates on the time scale of ocean circulation (1000 yrs) and is therefore causing an accumulation of CO_2 in the atmosphere. This phenomenon will result in potentially severe consequences to the well being of global ecological systems. Obviously the scientific attention of many biogeochemists is focused on processes controlling the response of the marine system to changes in atmospheric CO_2 concentrations (e.g., Archer and Maier-Reimer 1994; Broecker 1997; Sigman and Boyle 2000; Berger 2002).

Foraminifera, corals, and coccolithophores in the global carbon cycle

There are four dominant processes involved in neutralizing the excess atmospheric CO_2 in the ocean. These are: 1) gas exchange at the air-sea interface and reaction with the carbonate ion to form bicarbonate, 2) net primary productivity, 3) $CaCO_3$ production and dissolution and 4) ocean circulation. Most of these processes are biologically mediated and may have special importance in shallow tropical environments where the exchange with the atmosphere is more direct. In this review we address the mechanism of biomineralization in one of the major groups that precipitates $CaCO_3$ in the ocean—the foraminifera. As we will show, this group plays an important role in the oceanic carbon cycle as a potential atmospheric CO_2 modifier.

Marine photosynthesis and organic matter oxidation are closely linked to calcification and dissolution to control the pH of the oceans, and to a large extent, the atmospheric CO_2 levels. The general reactions are given by:

$$Ca^{2+} + 2HCO_3^- \rightarrow CaCO_3 + H_2O + CO_2 \rightarrow CaCO_3 + CH_2O + O_2 \quad (1a)$$
$$\text{\textit{calcification}} \qquad \text{\textit{photosynthesis}}$$

$$Ca^{2+} + 2HCO_3^- \leftarrow CaCO_3 + H_2O + CO_2 \leftarrow CaCO_3 + CH_2O + O_2 \quad (1b)$$
$$\text{\textit{dissolution}} \qquad \text{\textit{respiration}}$$

There is a clear spatial decoupling between the forward reactions, which are limited to the upper ocean (mainly in the photic zone), and the backward reactions, which occur primarily on the ocean floor. To be more precise, most of the organic matter oxidation occurs in surface water but $CaCO_3$ dissolution occurs mainly in the deeper water and on the ocean floor. The calcification process in today's ocean is almost entirely biogenic and in the absence of photosynthesis, it releases CO_2 to the atmosphere (Reaction 1a). Calcification occurs mainly in surface water, and gravitation removes the calcite shells rapidly to the sea floor. The CO_2 released will eventually be re-absorbed after dissolution of $CaCO_3$ on the sea floor and upwelling of the deep water to the surface. The time scale of this process, however, is on the order of 1000 years (Broecker and Peng 1982). If the calcification rates in the surface ocean were reduced, less CO_2 will be emitted to the atmosphere, the surface ocean alkalinity will increase, and more CO_2 will be absorbed from the atmosphere.

Reactions (1a,b) suggest that calcification and photosynthesis enhance each other, while $CaCO_3$ dissolution is closely linked to organic matter oxidation (i.e., respiration). It is therefore not surprising (although poorly understood) that the major calcifiers in the ocean; the algae, foraminifera, corals and even some mollusks, have a direct association with photosynthesis. Calcareous algae (mainly the open water unicellular coccolithophores) photosynthesize by their very nature as algae, while hermatypic corals and many planktonic and benthic foraminifera have photosynthetic symbiotic algae (Lee and Anderson 1991). Despite the fact that the ocean's surface waters are supersaturated with respect to calcite and aragonite (the two dominating $CaCO_3$ phases in the ocean), chemical precipitation is very rare in the open ocean due to kinetic barriers (Morse and MacKenzie 1990). Marine biogenic calcification and its connection to photosynthesis are therefore two of the most important processes in the global carbon cycle, but least understood scientifically. Much effort has been invested during the last years in studying calcification processes in coccolithophores (e.g., Westbroek et al. 1989; Holligan et al. 1993), however, the net effect of these organisms as calcifiers in the global carbon cycle may be reduced because they photosynthesize and calcify at roughly the same rate (Sikes and Wilbur 1982). Hermatypic corals form coral reefs, and these magnificent coastal ecosystems have important geochemical roles in marine calcification. However, quantitatively their role is much smaller than that of foraminifera as shown below.

Recently Schiebel (2002) estimated the contribution of planktonic foraminifera to the global carbon budget to be 0.36–0.88 Gt $CaCO_3$ yr^{-1}. This is roughly 32–80% of the $CaCO_3$ deposited in the deep ocean. The rate of Ca supply to the oceans by rivers is 1.3×10^{19} µmole yr^{-1} (Broecker and Peng 1982). To keep a constant Ca concentration in the ocean this quantity (which equals 0.16 Gt C yr^{-1} must be deposited as net $CaCO_3$ accumulation in oceanic sediments. Based upon calculations made by Smith (1978) and Milliman (1993), the shallow tropical shelves, including coral reefs, precipitate roughly 40 to 50% of the Ca supply as $CaCO_3$. This suggests a maximum net accumulation for pelagic $CaCO_3$ of 0.09 Gt C yr^{-1}. Even if we assume that foraminifera represent 50% of this amount (the rest would be coccolithophores) then their net accumulation rate for the sedimentary record should be 0.045 Gt C yr^{-1}. This is only 50% of the sedimentation flux, suggesting that at least 50% dissolves while on the sea floor. Some of these foraminifera dissolve in the sediments above the CCD, and part of them probably dissolve above the lysocline as well (Archer et al. 1989). The gross $CaCO_3$ production of foraminifera in the open ocean was estimated by Schiebel (2001) to be 3.24 Gt $CaCO_3$ yr^{-1}, which is roughly 5 times higher than the net accumulation rate. Roughly 65% of this amount must dissolve while settling in the deeper undersaturated water. The gross production is obviously a significant component of the present day global carbon cycle. It is roughly equal to 10% of the annual CO_2 accumulation in the atmosphere (Siegenthaler and Sarmiento 1993). Because of atmospheric CO_2 increase, the pH of the surface ocean is decreasing (Brewer et al. 1997). Based on our observations the rate of calcification in foraminifera is strongly related to the pH in their surrounding water (ter Kuile et al.

1989b), with lower rates at lower pH. If foraminiferal calcification is theoretically completely stopped and other processes are kept constant, a net atmospheric CO_2 sink of 0.4 Gt C yr^{-1} will be formed. The magnitude of this sink will not decline for at least several hundred years because the alkalinity/C_T ratio in the upper ocean will increase as calcification stops while the dissolution on the sea floor will continue. This is because the residence time of water in the upper ocean is roughly 100 years, while that of the deep sea where dissolution continues is ~1000 yr. It is reasonable to assume that within 100 years the surface alkalinity will become similar to the deep alkalinity (i.e., higher by 200 meq/l). Judging from estimates made by Boyle (1988) for the last glacial ocean, this may lower atmospheric CO_2 by 200 ppm. Obviously, these are only rough calculations and a detailed model is necessary to evaluate such a scenario. Nevertheless, a macro-scale reduction in foraminiferal calcification may produce a significant CO_2 sink. As will be discussed below, the coccolithophores and hermatypic corals also reduce their calcification rates when pH is reduced as a result of CO_2 increase (Kleypass et al. 1999; Riebessell et al. 2000). Hence, this sink may further increase.

The contribution of coral reefs and shallow carbonate shelves to global calcification is much lower. It was estimated by Smith (1978) to be roughly 0.07 Gt C yr^{-1}, i.e., 43% of the supply of Ca to the oceans. However, at least for several thousands of years, this is a net accumulation term. During the Pleistocene when sea level fluctuated roughly 120 m down and up between glacials and interglacials, the area occupied by coral reefs decreased and increased, respectively. This could have influenced the global cycle and may have been the cause for atmospheric CO_2 fluctuations, as suggested by Opdyke and Walker (1992). Out of this 0.07 Gt C yr^{-1} most of the area (90%) is calcifying at slower rates of roughly 1 kg $CaCO_3$ $m^{-2}yr^{-1}$, while the reef flats and fore reef areas are calcifying at a rate of 5 kg $CaCO_3$ $m^{-2}yr^{-1}$. In the low calcifying zones foraminifera are probably the major $CaCO_3$ producers. This is based on the reports of Muller (1974) for *Amphistegina madagaskariensis* and Smith and Wiebe (1977) for *Marginopora vertebralis* both calcifying at rates of 0.5 kg $CaCO_3$ $m^{-2}yr^{-1}$.

Importance of foraminifera and corals for paleoceanographic reconstructions

The significance of paleoceanographic studies stems from the wide recognition that the oceans, through their productivity and circulation patterns, play a major role in determining our climate. It is also well documented that large climatic changes occurred in the near geological past (Pleistocene). The deep sea record of stable isotopes ($\delta^{18}O$ and $\delta^{13}C$) and trace elements (mainly Cd, Ba and Mg) in foraminifera shells provided accurate information on the paleotemperatures, paleocirculation and paleochemistry of the oceans during the Pleistocene (e.g., Emiliani 1955, Shackleton and Pisias 1985; Broecker and Peng 1989, 1982; Boyle 1988,Wefer et al. 1999 and many others). This paleoceanographic information is essential for the tests and calibrations of global circulation models trying to predict the response of the atmosphere-ocean system to changes such as atmospheric CO_2 increase. In this regard, one of the most intriguing observations is the glacial-interglacial fluctuation in pCO_2 discovered in the polar ice cores (e.g., Barnola et al. 1987, Petit et al. 1999). The causes for these cycles and the role of the ocean in these changes may provide further testing of the models and thus help to predict the fate of the anthropogenic CO_2 input and its consequences (e.g., Archer and Maier-Reimer 1994).

An important achievement of the past two decades is the experimental evidence that foraminifera faithfully record the changes in seawater temperature and chemistry (e.g., Erez and Luz 1982, 1983; Delaney et al. 1985; Lea and Spero 1992, 1994; Russel et al. 1994; Lea et al. 1999). These studies have shown that planktonic and benthic foraminifera display constant distribution coefficients for various elements and thus validated the field calibration studies (e.g., Hester and Boyle 1982; Lea and Boyle 1989; Boyle 1992; Boyle et al. 1995; Rosenthal et al. 1997). New and existing information was provided by Spivack et al. (1993), Sanyal et al. (1995, 1996, 1997) on the paleo-pH of the oceans based on stable boron isotopes ($\delta^{11}B$). At this stage, however, this analysis requires large sample size (milligram quantities). Hence presently it cannot be applied to single species of benthic foraminifera. Spero et al. (1997) have shown that the CO_3^{2-} ion affects $\delta^{13}C$ and possibly $\delta^{18}O$. The possible effects of physical parameters (i.e., temperature and pressure) on the distribution coefficient of some trace and minor elements was shown by Rosenthal et al. (1997). Temperature effects on Mg distribution coefficient in foraminifera have recently been shown (Nurnberg et al. 1996; Rosenthal et al. 1997; Hastings et al. 1998; Rosenthal and Lohmann 2002). This has opened the possibility of separating the ice volume effect from the temperature effect hidden in the $\delta^{18}O$ record (Elderfield and Ganssen 2000; Lear et al. 2000).

Despite the wide use of foraminifera in paleoceanographic reconstructions, it is well recognized by all scientists involved in these studies that foraminifera are organisms that precipitate their shells by a complex physiological process. This process is biologically controlled (Lowenstam and Weiner 1989) and hence does not necessarily obey the chemical thermodynamics associated with inorganic precipitation of $CaCO_3$. These so-called vital effects are observed both in stable isotopes and also in trace element composition (e.g., Shackleton et al. 1973; Erez 1978; Erez and Honjo 1981; Boyle 1995; Spero et al. 1997; Weiner and Dove 2003). In part the symbiotic algae that are found in many species of planktonic and in shallow benthic foraminifera were often suggested as the source of this variability (Erez 1978). However studies on deep benthic species and other symbiont-barren species also demonstrate significant deviations from expected isotopic and chemical equilibrium fractionations, and these must be connected to the actual mechanism of shell formation (e.g., Erez 1993; Boyle 1995; Elderfield et al. 1996). As a result of these deviations the paleoceanographic work is limited to only those species that are close to equilibrium, or assumptions are made that the deviations are constant with time (see below).

Subjects that will not be included

This review will not address the following subjects:

1. Biomineralization in coccolithophores is of major importance for the pelagic $CaCO_3$ production. Young and Henriksen (2003) discuss coccolithophorid biomineralization in Chapter 7 of this volume.

2. Biomineralization in hermatypic corals will not be included in the review, although many inferences are made due to their importance in the global carbon cycle and as paleoceanographic indicators. When similar phenomena are found in corals and foraminifera, they will be discussed.

3. Various aspects of mineralogy and crystallography in foraminifera and corals. These subjects may be covered by other reviews in this volume and in the classical books of Lowenstam and Weiner (1989) and Simkiss and Wilbur (1989).

4. The structure, texture and skeletal architecture of foraminifera and the relations between structure and function. These subjects have been discussed at length in various papers and textbooks related mainly to their taxonomy (e.g., Loeblich and Tappan 1964, 1986, 1987). Much is known about these subjects on various levels from high resolution TEM and SEM and through light microscopy.

Geological aspects of rock-forming organisms

It should be pointed out that foraminifera have played a major role in the sedimentation history of the oceans and seas and have left impressive rock formations built almost entirely of their shells. Notable are the Mesozoic and Cenozoic chalks in Europe and the Middle Eastern deposits composed mainly of planktonic foraminifera and coccolithophores. Numulitic limestones of the Eocene are made almost entirely of very large benthic foraminifera (that may have had symbiotic algae), and fusulinids produced high accumulations of limestone in the Permian. Deposition of these formations must have modified the global carbon cycle of their times and are an important part of the sedimentary rock record.

BIOMINERALIZATION IN FORAMINIFERA

Introduction

Foraminifera are unicellular calcifying ameba, taxonomically part of the Protista (Loeblich and Tappan 1964, 1986, 1987). They are strictly marine organisms, although recently there have been some reports of fresh water non-calcified granoreticulate ameba that may be considered as foraminifera (Holzmann and Pawlowski 2002; Holzmann et al. 2003). Their evolution started in the mid-Paleozoic some 400 my ago, and their species diversity is very high. Over 40,000 species have been identified in the fossil record, while roughly 4,000 species are extant. These data suggest that the average rate of evolution per species is 1/10,000 yr. This high rate of evolution, the coexistence of many species, and the good preservation of their calcitic shells provide very detailed biostratigraphic division of the geological record based on fossil foraminifera (Loeblich and Tappan 1964, 1986, 1987). As a result, much attention has been given to their shell structure and texture and with the advent of SEM and TEM a large body of knowledge accumulated on these subjects (e.g., Towe and Cifelli 1967; Hansen and Reiss 1972). Biostratigraphic methods that are based upon foraminiferal zoning became the main tool for correlations and reconstruction of subsurface geology that were highly needed for oil and gas prospecting. This entire field has benefited greatly from the Deep Sea Drilling Project (DSDP) and its successor the Ocean Drilling Project (ODP). These projects provided continuous well-preserved records of oceanic sediments through the Cenozoic (e.g., Vincent and Berger 1981). In addition to obtaining well-preserved foraminifera, these projects greatly assisted in the transformation from biostratigraphy to absolute chronostratigraphic time scale (e.g., Berggren et al. 1995).

The process of biomineralization is a major modifier of the paleoceanographic proxies stored in their shells. More reliable and new information will be extracted from foraminiferal shells if their biomineralization process is well understood. In addition, the foraminifera are very large eukaryotic cells that can be cultured in the laboratory and provide a very informative experimental preparation to study biomineralization at the cellular level. As will be shown below, these organisms are particularly suitable for light and confocal microscopy. Finally, it should be mentioned that while calcification studies in foraminifera can teach us a great deal about the interpretation of their trace elements and isotopes, the latter may also teach us a most interesting lesson on biomineralization mechanisms in these organisms.

Recent work (e.g., Langer et al. 1993, Darling et al. 2000, 2003; Huber et al.1997; de Vargas et al. 2002; Pawlowski et al.2002) has shown that extant foraminiferal DNA produces a taxonomy that in general fits very well the classical one based on shell structure. Obviously new insights are obtained from these studies, but in general they confirm the strong genetic control on shell structure and possibly function.

Light microscopy, TEM and SEM of test structure and function

Detailed descriptions of foraminiferal morphology, composition and texture can be found in the micropaleontological literature and in several textbooks (e.g., Loeblich and Tappan 1987; Hedley and Adams 1978). With regard to their test (shell), it is possible to divide the foraminifera into four groups: the first two groups are with organic tests and with agglutinated tests (composed of particles collected by the organism and inserted in the test). The latter may have organic matter or $CaCO_3$ as their cementing agent and hence may be considered as performing biomineralization. The other two groups precipitate $CaCO_3$ shells and are known as the imperforate or porcelaneous (mainly Milliolids) and the perforate or calcitic radial. The perforate foraminifera are the ones that dominate today's oceans, hence most of this review will be dedicated to their biomineralization.

The perforate foraminifera test can be very simple with one or only few chambers, or it can be complex with many chambers arranged in various three-dimensional configurations. The most common arrangement is spherical coiling with planispiral or low trochospiral tests. The trochospiral test has an umbilical side, which contains the last whorl and the aperture, and the spiral side showing the spiral arrangement of earlier chambers. In all foraminifera the apertures of earlier chambers (called foramens, hence the name foraminifera) are connected, thus forming a continuous space where the cytoplasm is interconnected.

The imperforates (porcelaneous) are more primitive and evolved before the calcitic-radial perforate ones (probably in the Paleozoic). There is a major difference in the calcification mechanisms of these two groups. The porcelaneous species precipitate their needle shaped calcite crystals (usually high Mg calcite) within intracellular vacuoles (Angell 1980; Hemleben et al. 1986). The crystals are then deposited at the sites of chamber formation usually without any preferred orientation, within an organic matrix, which forms the shape of the chamber. The random orientation of the crystals blocks the light and hence the appearance of the test is opaque. Often the outer layer of crystals is deposited in a dense organized orientation, which forms a shiny light-reflecting veneer (op. cit.). Both characters contribute to the porcelaneous appearance of these tests. Most of the porcelaneous foraminifera are not perforated as opposed to the calcitic radial ones. Yet their shell is rather permeable to gases and solutions as can be deduced from the measurements of microelectrode gradients (Kohler-Rink and Kuhl 2000, 2001). Another observation is that radioactive exchange of ^{14}C and ^{45}Ca are higher in the porcelaneous foraminifera compared to the perforate ones (ter Kuile and Erez 1987). An unreported observation in our laboratory is that when some recent porcelaneous foraminifera are treated with NaOCl to remove the organic components, the shell partially disintegrates, and the $CaCO_3$ needle-shaped crystals are left behind.

The perforate foraminifera evolved in the Mesozoic and Cenozoic and are more advanced in many of their characteristics. Their shells are usually made of low Mg calcite (although some may approach 6–7 mole % Mg). Their crystals are oriented in a structure known as calcitic radial. This texture is formed by multilayered calcite platelets that have a radial appearance with their C-axes perpendicular to the shell surface (Hansen and Reiss 1972; Bellemo 1974). This group is also perforated with numerous microscopic pores that are found on most of the surface area of the shells. The pores can range from a few microns up to 10 μm in diameter and are sealed by an organic pore plug that prevents the intra-shell cytoplasm from flowing out of the shell (Hemleben et al. 1989). As in all foraminifera, the flow of cytoplasm is via the aperture or apertures through which pseudopods can propagate to perform the life function of these organisms including feeding, excretion, and of course chamber formation (see below). One major feature found in many perforate calcitic-radial foraminifera is their lamination (Fig. 1).

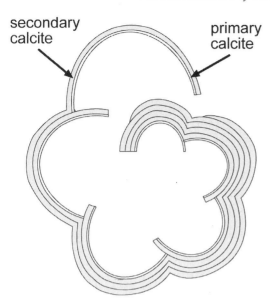

Figure 1. Lamination scheme in lamellar perforate foraminifera, after Reiss (1957). Each newly added chamber is composed of 2 layers of calcite—primary calcite, which outlines the new chamber and becomes the inner lamella and secondary calcite that covers the inner lamella and also the entire existing shell.

Lamination is obtained when these organisms cover their pre-existing shell with a new layer of calcite every time that the organism builds a new chamber. The shell is therefore composed of many layers of thick radial calcite and its thickness depends on the number of chambers per whorl (Reiss 1957). The bulk of the $CaCO_3$ shell is composed of this secondary lamination, although little is known about its calcification mechanism. The pores and the spines (when present) are not covered by the secondary lamination. Towards the end of their life cycle many planktonic species deposit several different types of $CaCO_3$ either in the form of thick crust or what was termed as gametogenic calcification. This $CaCO_3$ often shows different chemical and possibly isotopic compositions (Be 1980, 1982; Be et al. 1977). This is in part because it is deposited in deeper water where temperature, illumination and water chemistry are different from those where the rest of the skeleton was deposited, and in part because the mechanism of calcification may be different.

While the exact function of foraminiferal tests has not been investigated directly, there are many speculations (e.g., Hallock et al. 1991). Protection against predators is often mentioned, and, in this regard, small microbial or other microorganisms (ciliates) are probably not less important then the macro benthos (such as Echinoderms). In spinose planktonic foraminifera, the shell also serves as basis for the spines, which allow for the construction of a wide pseudopodial "spider web" that help to catch their prey (small copepods). This is important particularly in turbulent shallow water where they are found. The spines may also assist in dispersing the symbiotic algae in such a way as to provide CO_2 and to reduce oxygen toxicity by diffusion. In benthic foraminifera thick shells may assist in preventing buoyancy and in resisting abrasive wave action. It is obvious, however, that despite the sometimes massive structures of foraminiferal tests (i.e., the fossil Nummulites and the extant Rotalids), one major feature of all foraminiferal tests is a tendency to allow for cytoplasmic flow from the inside towards the environment. Pseudopodial networks are most important for food gathering, movement, shell building, respiration and extraction of waste. Hence many of the intricate and complex structures of foraminiferal tests are designed to allow fast and effective communication of cytoplasm between the outside environment and the intra-shell endoplasm (e.g., Reiss and Hottinger 1984, Hottinger et al. 1993).

Not much is known about the organic matrices and their role in the calcification process in foraminifera. King and Hare (1972) characterized the amino acids associated with tests of planktonic foraminifera and Weiner and Erez (1984) described the matrix of the benthic species *Heterostigina depressa*. The abundance of acidic amino acids and the presence of glycosaminoglycans have been documented in tests and organic lining of various foraminifera (reviewed by Langer 1992). Obviously an organic entity is involved in determining the shape of the newly formed chamber as will be described below. It may also function as a nucleating and as an inhibiting agent of mineralization.

Chamber formation in perforate foraminifera

Chamber formation has been described in many different foraminifera including imperforate and perforate, planktonic and benthic (e.g., Angell 1979; Be 1982; Spero 1988; Hemleben et al. 1986, 1989). These studies utilized primarily light microscopy, SEM and TEM observations. In perforate foraminifera the first step is delineation of a space that partially isolates the organism from its environment. This is done with the aid of the ectoplasmic pseudopods. The next step is the formation of the organic template in the shape of the newly formed chamber. This is achieved by formation of a cytoplasmic bulge that serves as a mold (called anlage) for the organic matrix, which will also serves as a template for nucleation. The third step is the precipitation of $CaCO_3$ on both sides of a thin organic layer (termed the Primary Organic Membrane or POM). POM is a somewhat misleading term because the same acronym often used in oceanography for particulate organic matter and also because it is clearly not a cell membrane in the strict sense of the word. Perhaps the term Primary Organic Sheet (POS; as suggested by L. Hottinger, pers. comm.) is better. During the entire process there is an intensive involvement of the granoreticulate pseudopodia, with strong cytoplasmic and vacuole streaming. TEM observations all show involvement of small electron-dense vesicles of various types. It should be emphasized here that the newly-formed chamber represents only a small fraction of the $CaCO_3$ deposited during a growth instar. This is because the secondary layering which covers the entire organism with a new layer of calcite probably represents ~90% of the newly deposited $CaCO_3$ (Fig. 1). Despite the importance of this secondary lamination, there are very few observations on this process (see below).

Rates of calcification: radiotracers, weight increase and microsensor studies

Rates of growth and calcification for planktonic foraminifera are reported by Caron et al. (1981), Erez (1983), Be et al. (1982), Anderson and Faber (1984) and are summarized in Hemleben et al. (1989). If well fed, planktonic foraminifera add a new chamber every day and grow at a rate of roughly 25% d^{-1}. Normal *G. sacculifer* reach the size of 600–700 µm and their weight is 30–40 µg each. For large symbiont bearing benthic foraminifera (both porcelaneous and calcitic radial) the rates are much lower, ranging from less than 0.5 to 4% d^{-1} (e.g., Erez 1978; Duguay and Taylor 1978; Duguay 1983; ter Kuile and Erez 1984). For a 1200 µm *A. lobifera* with shell weight of 1 mg this would be roughly 20 µg d^{-1}, so that the absolute amount is similar to that of the planktonic ones. Growth measurements are almost always calculated for the shells. There is a hidden assumption that the organic matter grows in parallel to the shell. For the benthic species *A. lobifera* and *A. hemprichii* the dry weight of the organic matter is 7 to 10% of the total weight.

One explanation often used in trying to explain vital effects in isotopes and trace elements is the variable rate of calcification (Boyle 1995; Elderfield et al. 1996). This should be reconsidered, as the actual rates of biogenic precipitation by foraminifera are probably several orders of magnitude lower then precipitation rates reported for inorganic calcites. The comparisons are very difficult because inorganic rates are reported in units of mass precipitated per unit surface area per unit of time. The surface area is a difficult parameter to estimate when dealing with biogenic carbonates. If we use BET specific surface areas as measured by Honjo and Erez (1978) for planktonic foraminifera, it is clear that the specific surface area is very high (in the range of 2–3 m^2 g^{-1}) as opposed to, for example, reagent grade calcite (0.6 m^2 g^{-1}). The rates of the inorganic experiments discussed by Morse and Bender (1990) were between 10^2–10^5 μmol m^{-2} hr^{-1}, and kinetic effects were observed mainly in the higher range of these values. The highest calcification rates reported for foraminifera are those for planktonic foraminifera (e.g., Erez 1983) reaching 25% day^{-1}. When normalized to the surface area the rate is ~40 μmol m^{-2} hr^{-1}, a factor of 2.5 lower than the lowest rates cited for the inorganic experiments. In shallow benthic foraminifera the rates are probably lower by an order of magnitude, and in deep benthic foraminifera, calcifying at temperatures below 5 °C, the rates must be even lower. Hence the suggestion made by many authors that calcification rate is an important factor in determining trace element distribution coefficients is probably incorrect. Figure 2 shows the results of several laboratory experiments with live foraminifera carried out to determine the distribution coefficient of Sr (D_{Sr}) in planktonic and benthic foraminifera (Cutani 1984). Both radioactive ^{89}Sr and stable Sr were used in these experiments at various Sr/Ca in the seawater. D_{Sr} showed considerable variability of 0.12 to 0.25 for the benthic *A. lobifera* and 0.05 to 0.15 for the planktonic *G. sacculifer*. Both within and between experiments the general trend was that D_{Sr} decreased with

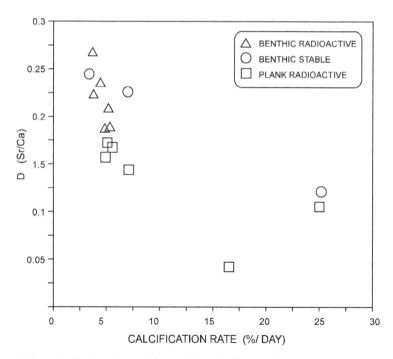

Figure 2. Change in D_{Sr} of benthic and planktonic foraminifera as a function of growth rate. Results of culture experiments using short term ^{89}Sr uptake (marked radioactive) and long term growth (marked stable). Note that D_{Sr} decreases with increasing growth rate, contrary to the expected from kinetic experiments.

increasing calcification rate. This is contrary to the expected, if the kinetics of $CaCO_3$ precipitation control the value of D, because at higher precipitation rates D should increase (Lorens 1981; Morse and Bender 1990). The explanation for the observed data may be that foraminifera with higher growth rates behave as an open system while those with lower growth rates are more closed systems. (see discussion below on distribution coefficients).

Microsensor studies of foraminifera, started by Jorgensen et al. (1985), who documented the intensive activity of the symbiotic algae in the planktonic foraminiferan *G. sacculifer*. They showed that a boundary layer of several hundred μm surrounds this spinose species and that the chemical conditions in this layer are significantly different from the surrounding seawater. Highest oxygen levels in the light were ~500 μM (a supersaturation of 200%) and the pH rose to 8.6 relative to the ambient levels of 8.2. In the dark the oxygen dropped to 120 μM while the pH dropped to 7.9. More recent studies (Rink et al. 1998) showed in the planktonic *Orbulina universa* a significant difference between gross and net photosynthesis (roughly a factor of 2), suggesting that in the light respiration is enhanced relative to dark respiration. Similar observations were also made in the benthic *A. lobifera* and *A. hemprichii* (Kohler-Rink and Kuhl 2000, 2001). In addition these authors also showed light and dark pH dynamics as well as $CO_{2(aq)}$ and Ca^{2+} gradients. The latter were typical concentrations of seawater, i.e., 10 ± 0.5 μM, with somewhat inconsistent fluctuations and gradients between light and dark. The calcification rates calculated from these gradients for *A. lobifera* were lower by a factor of 30 compared to the rates reported by ter Kuile and Erez (1984), suggesting that this technique still needs some improvement.

Photosynthesis and calcification

The process of light-enhanced calcification in symbiont bearing corals and foraminifera as well as in calcareous algae (including coccolithophores), is commonly explained by removal of CO_2 by photosynthesis which enhances calcification as clearly implied from Reaction (1) above. While possible for calcareous algae, this mechanism is doubtful for most of the symbiotic associations. Doubts about the validity of this mechanism come from several lines of evidence. Pearse and Muscatine (1971) have shown that the fastest calcifying parts of hermatypic corals are the tips of the branches, which are devoid of symbionts. The location of the symbiotic algae in corals is within the cells of the gastrodermis, well separated from the chalicoblastic ephithelium responsible for the calcification process. Also in the foraminifer *A. lobifera* there is a clear separation between the symbiotic algae, which calcify inside the shell, and the calcification process, which occurs in the extra-shell cytoplasmic space (see later discussion of Fig. 7). The most direct approach to this question comes from the radiotracer experiments of Erez (1983) and ter Kuile et al. (1989b) on planktonic and benthic foraminifera, respectively. Using DCMU—the compound 3(3,4-dichlorophenyl)-1,1-dimethylurea, a photosynthesis inhibitor that stops the electron flow in photosystem II—we showed that in foraminifera light-enhanced calcification could proceed when photosynthetic CO_2 uptake was completely inhibited (Fig. 3). The decoupling between these two processes is also clear from the fact that in planktonic foraminifera calcification is absolutely dependent on feeding (Be 1982; Hemleben et al. 1989). while photosynthesis can proceed at very high rates without calcification (Jorgensen et al. 1985). In the symbiont-bearing benthic foraminifera, it was shown that both calcification and photosynthesis could be inhibited without affecting each other (ter Kuile et al. 1989a,b). Furthermore, introduction of the enzyme carbonic anhydrase to the seawater where these foraminifera grew caused enhancement of photosynthesis, but reduction in calcification. In fact, this treatment caused a complete crash of the internal carbon pool that supplies calcification. These and

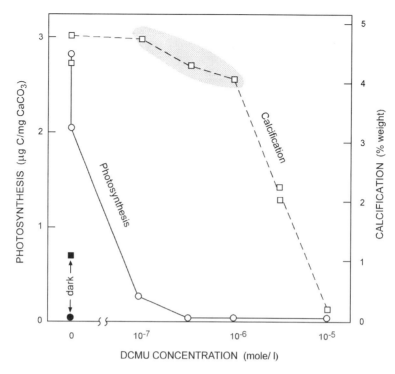

Figure 3. Calcification and photosynthesis experiments in the planktonic foraminiferan *G. sacculifer*. Light enhanced calcification is shown in the left side of the figure. Addition of the photosystem II inhibitor DCMU at high concentration (10^{-5}M), inhibits both processes. But at lower concentrations of DCMU light enhanced calcification can be observed (dotted area).

other observations suggested that the symbiotic algae might actually compete with their host for inorganic carbon (ter Kuile et al. 1989a). The competition for C_T may now be explained in view of the new observations on the mechanism of calcification in foraminifera as mediated by alkaline seawater vacuolization (see below). The conclusion from these experiments is that although there is light enhanced calcification in foraminifera, CO_2 removal and /or photosynthesis are not directly involved. Two alternative mechanisms may explain these observations: one is that DCMU does not inhibit photosystem I, which may proceed to produce ATP in a cyclic mode. Since ATP is needed for the calcification process (ter Kuile et al. 1989a), this may explain light enhanced calcification. Alternatively, the photosynthetic activity may provide a sink for protons, which combine with HCO_3^- to provide CO_2 for the symbiotic algae (McConnaughey 1994; McConnaughey and Whelan 1997). Recently, Al-Horani et al. (2003) showed the activity of a light-dependent Ca^{2+}-ATPase at the calcification space just below the chalicoblastic epithelium. This pump may exchange Ca^{2+} for $2H^+$ and thus enhance calcification and at the same time photosynthesis. It is certainly possible that foraminifera have a similar system to mutually enhance the two processes.

Internal carbon and calcium pools

Carbon pool. The studies of ter Kuile and Erez (1987, 1988) and ter Kuile et al. (1989 a,b), have demonstrated that the perforate foraminifera *A. lobifera H. depressa* and

Operculina ammonoides have a large internal carbon pool that serves for calcification, but not for photosynthesis (Fig. 4). On the other hand the imperforate species *A. hemprichii* and *Borelis schlumbergeri* do not show such a pool when similar techniques are employed (Fig. 5). The internal inorganic carbon pool was demonstrated using the radiotracer ^{14}C by three independent methods:

1. Lag time of roughly 12 hrs in the uptake of ^{14}C into the skeleton (Fig. 4).
2. Uptake of ^{14}C into the skeleton during the chase period in pulse-chase experiments (Fig. 4).
3. Direct measurement of the inorganic carbon pool by comparison of ^{14}C uptake in wet and dry individuals.

It turns out that the internal carbon pool varies in *A. lobifera* as a function of its age and physiological state (ter Kuile et al. 1989b). Biologically active uptake into the carbon pool was demonstrated (ter Kuile et al. 1989b), and it was also shown that metabolic carbon is entering the pool and eventually the skeleton. Uptake of carbon into the pool and the skeleton shows an enzyme kinetics type behavior (Michaelis-Menten) with respect to CO_3^{2-} ion. The total uptake shows a Hill-Wittingham type kinetics that is diffusion-limited at low concentrations and enzyme-limited at higher concentrations than seawater (ter Kuile et al. 1989b). These observations have direct implications for the carbon isotopic composition of these foraminifera and may affect trace elements that occupy the lattice position of the CO_3^{2-} ion in foraminiferal calcite. The existence of an internal carbon pool may influence the distribution coefficients of these elements in the same way that the Ca pool influences the cationic trace elements.

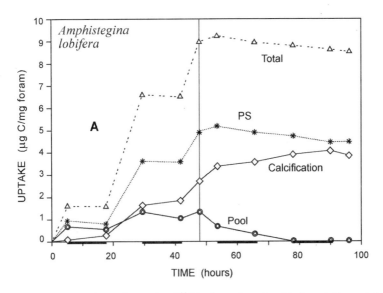

Figure 4. Pulse chase experiment for uptake of ^{14}C in the perforate benthic foraminiferan, *A. lobifera*. Dark hours are marked as a thick black line on the x-axis. The vertical line at 48 hrs marks the pulse period (seawater labeled with ^{14}C) and the chase period where incubation continues in regular seawater. In the perforate, uptake for photosynthesis and the inorganic carbon pool are without delay while calcification lags for the first day and only then begins at the normal light enhanced calcification. During the chase period the inorganic pool is depleted from its ^{14}C, which is all transferred to calcification. This continues for almost 2 days.

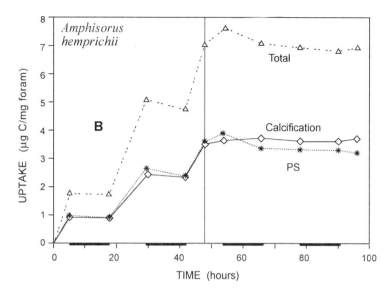

Figure 5. Pulse chase experiments for uptake of ^{14}C in the imperforate benthic foraminiferan, *A. hemprichii*. Graphics and experimental details as in Figure 4. Uptake for calcification and photosynthesis start without delay and both rates are similar suggesting 1:1 stoichiometry and self-enhancement of both processes according to Reaction (1) in text. No carbon pool is detected for this species.

While the existence of the inorganic carbon pool is clearly demonstrated, its nature and the concentration of carbon in this pool are still a mystery. The pool size in *A. lobifera* according to ter Kuile and Erez (1988) is roughly 1 µg C/mg foraminiferan. Given the diameter, weight and volume relations of ter Kuile and Erez (1984), the concentration of carbon in the pool is ~190 mM. This is a minimum estimate because this calculation assumes that the volume is all fluid. Even if we assume a very effective carbon concentration mechanism (CCM) in foraminifera similar to that known for algae (e.g., Raven and Johnston 1991), it is difficult to imagine concentration by a factor of 100 relative to seawater. If the diameter of the pseudopodial network of *A. lobifera* were increased roughly by a factor of four then the internal concentration will be around that of seawater. Alternatively the pool may not be in solution but in some solid unstable phase (see similar discussion on Ca below).

Calcium pool. An internal calcium pool, serving for calcification, was first observed by Anderson and Faber (1984) in the planktonic foraminiferan *G. Sacculifer*. They observed a gap (of 40%) between the actual $CaCO_3$ deposited and the incorporation of the radiotracer ^{45}Ca into the skeleton in short incubation of 24 hrs. This gap diminished after 48 and 72 hrs of incubation with the radiotracer. In a similar study conducted in our laboratory on *A. lobifera*, we also observed a large gap between the incorporation of ^{45}Ca into the skeletons of and the actual calcification (measured as $CaCO_3$ weight added) during the experiment (Table 1). The data showed that for small fast-growing specimens there was a good agreement between the two methods, while for the older and larger specimens the radiotracer underestimated the actual calcification by a factor up to 9. Obviously these data cannot be explained by isotopic discrimination against the radiotracer because the mass difference between ^{40}Ca and ^{45}Ca is only five mass units and such fractionations are not known in nature. The most likely explanation for these observations is that suggested by

Table 1. Calcification experiments with *A. lobifera* in which growth was measured simultaneously by weight and size increase and by ^{45}Ca uptake into the shell. For most experiments, a large discrepancy was observed in the calcification ratio (Wt/^{45}Ca) where the radiotracer is lagging behind the true calcification rate. The ratio of the two methods ranged from 8.6 in large, slow-growing individuals to 1.2 in small, fast-growing juveniles. The Ca pool size was calculated assuming that all the missing Ca in the radioactive assay is non-radioactive Ca from the internal pool.

Measured Size (mm)	Calculated Weight (mg)	Actual % $CaCO_3$ added	Calcification Ratio (Wt/^{45}Ca)	Pool Size (μmole/ind)	μmole/ind added
		10 days incubation			
1.49	0.975	14	8.6	1.207	1.365
0.94	0.246	18	4.8	0.351	0.443
0.57	0.0487	90	2.3	0.248	0.439
0.23	0.0126	190	1.2	0.040	0.240
		15 days incubation			
1.52	1.033	17	4.3	1.348	1.757
1.13	0.431	36	2.8	0.998	1.552
0.88	0.200	52	2.6	0.641	1.042
0.69	0.092	48	2.3	0.249	0.441
		15 days incubation, 10 days chase			
1.5	0.994	17	5.5	1.383	1.691
1.14	0.443	44	3.1	1.320	1.948
0.9	0.215	64	2.7	0.866	1.376
0.74	0.115	62	3.6	0.517	0.715
		15 days incubation, 20 days chase			
1.44	0.883	15	2.9	0.868	1.324
1.44	0.883	18	4.1	1.202	1.589

Anderson and Faber (1984), i.e., that the foraminifera have an internal calcium pool that is large enough to cause this discrepancy. During the initial period after the radiotracer is added to the water, the pool is not yet equilibrated with ^{45}Ca in the seawater. Only after equilibration (which is a function of the pool size and its exchange rate with the environment) will the ^{45}Ca calcification rate be equal to the true calcification rate.

The observation that the small, fast growing specimens have reached a ratio of one suggests that their pools have equilibrated with ^{45}Ca, and thus the radiotracer measures the correct calcification rate. In the experiments that had a "cold" chase period following the radioactive incubation, the ratio was lower than in equivalent experiments without chase, suggesting incorporation of ^{45}Ca into the skeleton from the internal pool. However, it is surprising that the gap between the two methods is so large and that during the chase the gap was not completely closed. Lea et al. (1995) have claimed that in their ^{48}Ca uptake experiments, they found no indication of a Ca pool in the planktonic species *Orbulina universa*. However, careful examination of their data shows a lag time for the Ca tracer incorporation on the order of that described by Anderson and Faber (1984).

To clarify this subject we conducted experiments with *A. lobifera*, where true calcification was estimated by comparison of initial weight and final weight and by changes in alkalinity during the course of the experiment (not shown). Again large discrepancy was observed between the methods. The first experiment, conducted in continuous light, demonstrated that during the first 24 hrs the total ^{45}Ca uptake is much lower than in the next days. The time needed to saturate the Ca-pool with ^{45}Ca in normal light/dark cycle was longer than 48 hrs. After the pool is filled up the uptake becomes linear with time. In these experiments we were able to measure directly the Ca-pool in the cytoplasm by the difference between the total ^{45}Ca uptake and the NaOCl-cleaned CaCO$_3$ uptake. The pool represents almost 40% of the total uptake in the first experiment and roughly 20% in the second experiment. In the second experiment after the 5th day the total ^{45}Ca uptake approached 75% of the true calcification rate. But again the discrepancy could not be fully bridged.

When the concentration of Ca in the pool is calculated using the total volume of the organisms (as shown above for the carbon pool), we obtain concentrations that are very high, ranging from 2 M to 20 M. These concentrations are unrealistically high unless they represent Ca sequestration in a solid phase. A similar calculation for Anderson and Faber (1984) experiment yields a concentration of ~80 mM, 8 times higher than seawater. It is possible that these Ca-pools may be connected to small polarizing granules that were recently observed in the endoplasm (Bentov and Erez 2001; Bentov et al. 2001; Erez et al. 2001, 2002) which may store Ca for the calcification process (see below). Alternatively there may be an amorphous CaCO$_3$ phase that is involved in the calcification process and has not been observed yet.

Incorporation of trace elements

The modern data on trace elements in foraminifera came from the study of Bender et al. (1975) and Graham et al. (1982) who used neutron activation for this measurement. An important step toward the use of trace elements was the original work of Boyle (1981) who designed a cleaning procedure to remove the ferro-manganese coating from sedimentary foraminifera and introduced Cd as a tracer for phosphate in the ocean. In a series of papers that followed, other tracers were introduced e.g., Ba, Li, F, U, V, δ^{11}B and recently Mg (e.g., Boyle 1988, 1992; Hester and Boyle 1982; Lea and Boyle 1989; Rosenthal and Boyle 1993; Russell et al.1994; Boyle et al. 1995; Nurnberg et al. 1996; Rosenthal et al. 1997; Hastings et al. 1996, 1998; Elderfield and Gansen 2000; Rosenthal and Lohmann 2002; Lear et al. 2002; Anand et al. 2003). As with all proxy development, problems always arise, but one comforting observation in many experimental and field studies is that for most of the trace elements that have been investigated (Mg, Sr, Cd, Ba, U, V, Li), there seems to be a constant partition coefficient (which may be temperature or pressure dependent). This means that when the Me/Ca ratio in seawater is increased the Me/Ca in the CaCO$_3$ also increases (e.g., Delaney et al. 1985; Lea and Boyle 1989; Lea and Spero 1992, 1994; Russell et al.1994; Hastings et al. 1996; Lea et al. 1999). This of course does not prove that these elements are incorporated into the CaCO$_3$ as solid solution substitution. A debate on this issue was raised with respect to Ba (Lea and Spero 1992; Pingitore 1993; Lea and Boyle 1993). Regardless of the lattice position of these trace elements these linear relations (op. cit.) form the basis for many paleoceanographic reconstructions. More recent observations showed that some of these trace elements, are incorporated into the skeleton of foraminifera in a differential manner. Different parts of the shell have different concentrations (Szafranek and Erez 1993; Erez et al. 1994; Nurnberg et al. 1996; Eggins et al 2003). Indeed certain parts of the calcitic shells are enriched in trace elements and others are depleted (Fig. 6). This will be discussed in relation to our new observations on the calcification mechanisms of foraminifera.

Inhomogeneous distribution in the test

Szafranek and Erez (1993) used Electron-Probe Micro Analysis (EPMA), EDS and WDS in a study that found inhomogeneous distribution of some minor elements in foraminiferal shells. Data are shown for *A. lobifera* for which we performed a profile across the knob in the vertical section (Fig. 6), but similar concentration changes were also found across the horizontal section. Both sections represent the growth history of the organism (i.e., the low numbers on the x-axis are younger stages). Profiles were made for the following elements: Ca, Mg, Sr, S, and Na. In this calcitic-radial species, Mg ranges from 1 to 4.5 mol % (average of 170 analyses is 2.3 mol % on 4 tests). Sr: 0.11–0.28 mol % (average value of 0.22 mol %), SO_4^{2-}: 0.013–0.57 mol % (Av. 0.1 mol %), Na: 0.78–1.35 mol % (average value of 1.04 mol %) In the porcelaneous *A. hemprichii* (AH), Mg concentrations ranges from 11% to 14.8 mol % (average value 13.6 mol % for 30 analyses from 3 tests), Sr: 0.14–0.19 mol % (average value of 0.16 mol %), SO_4^{2-}: 0.69–0.87 mol % (average value of 0.73 mol %). SO_4^{2-} was measured as S in the EPMA but was determined also gravimetrically as $BaSO_4$ showing bulk SO_4^{2-} concentrations of 0.7–0.8%. For both species, Mg shows negative correlation with Ca ($R^2>0.99$) suggesting simple Ca/Mg substitution in the carbonate phase. However, Mg distribution is also well correlated with SO_4^{2-} ($R^2 = 0.6–0.8$). The correlation is particularly tight in the lamellar part of *A. lobifera* where microprobe profiles revealed highest concentrations of Mg, SO_4^{2-}, and Na in the thin dark boundaries between the calcitic lamellae (Fig. 6). Perhaps the most important finding of this study is the systematic variations of Mg and S in the knob area of *A. lobifera*. These variations show low concentrations for these 2 elements within the calcitic lamellae (which represent roughly 90% of the calcite) and much higher concentration (factor of 5 for Mg) in the region of the dark layers that separate one lamella from the next (Fig. 6). The dark layers showed also higher concentrations of Na and Cl but not always for Sr. The source of this variability is not fully understood, but the most attractive explanation is that they represent alternations of primary and secondary calcite as will be discussed below. Duckworth (1977) showed Mg variability (but not in Sr) in *Globorotalia truncatolinoides* and attributed it to temperature at depth of calcification. Nurnberg et al. (1996) have shown variability in Mg in cultured planktonic foraminifera with high concentrations in the gametogenic calcite. Recently Eggins et al. (2003) have reported inhomogeneous distributions of trace elements in various planktonic species. Some of the variability was within the inner parts of the chamber wall, but most of the variability was because of high concentrations of all trace elements (except Sr) in the outer veneer of the foraminifera. It is possible that this may represent gametogenic calcite for the samples taken from sediments. One major concern in the use of Mg as a proxy for temperature is dissolution of the Mg rich phases (e.g., Brown and Elderfield 1996). Based on the data presented above, the tests of planktonic foraminifera may be described as thick secondary calcite (low in Mg, S and perhaps other trace elements) sandwiched between internal primary calcite (Fig. 1) and external gametogenic calcite

---

Figure 6 (on facing page). Inhomogeneity of trace elements in the test of the perforate foraminiferan *A. lobifera* measured by electron probe WDS. (A) Detailed profile of 400 data points with beam size of ~1μm showing systematic variations in Mg and S. (B) Detail of the last 100 data points of A. Note that each peak is composed of several data points and the good fit between Mg and S. (C) Detail of the knob area where the profile shown in A has been measured. (D) The original 2 profiles with normal beam size of ~10 μm in this backscatter image, reveal the dark (organic?) boundaries which show the high concentrations. (E) Detail of the area marked in rectangle in D. It is possible to see some variability in the texture of the calcite layers suggesting that the layers close to the dark region are primary calcite. (F) Vertical cross-section in *A. lobifera* showing the knob area, where the entire growth history of the foraminiferan can be found as alternations of primary and secondary calcite. The hatched area is where the profiles were measured.

Biomineralization of Foraminifera

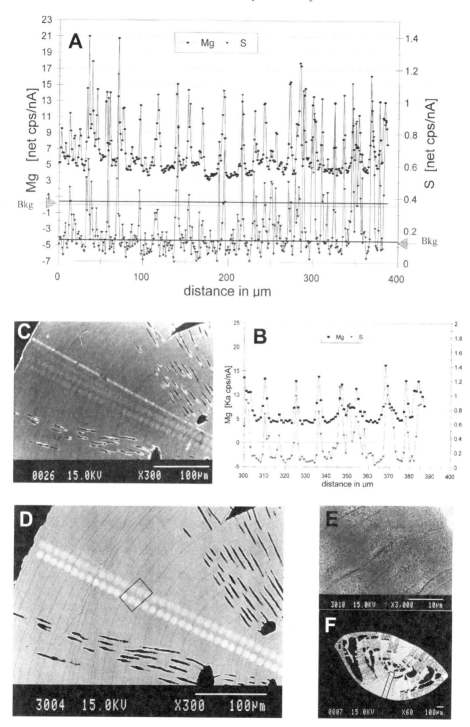

Figure 6. Caption on facing page.

which are both richer in trace elements. Since high Mg calcite is more soluble than low Mg calcite (Morse and MacKenzie 1990), these layers are more soluble and this may explain the observations of Brown and Elderfield (1996).

The possible implications of the calcium pool on the distribution coefficients (D) of trace metals may be demonstrated by our preliminary results on the distribution coefficients of Sr in cultured foraminifera shells (Fig. 2). These data show that the D_{Sr} in foraminifera is high at low calcification rates and becomes lower (approaching the inorganic D) at high calcification rates. In these experiments the low calcification rates are associated with the larger specimens, which have a larger calcium pool (Table 1) and hence may represent heterogeneous precipitation from this partially restricted calcium reservoir. In the extreme case of a totally restricted reservoir D will approach 1, while at the other extreme, when calcium pools do not exist and all the Ca is taken directly from seawater, D should approach the inorganic value (0.04). The average D_{Sr} and D_{Ba} in foraminifera are larger than the inorganic D by factors of 3.2 and 3.7, respectively, thus supporting the precipitation from a semi-closed reservoir. A further test for this hypothesis is the behavior of Cd with respect to the calcium pool in foraminifera. That is because the inorganic D_{Cd} in calcite is larger than 1 and may be between 7–15 (Boyle 1988). If the Ca pool is also active with respect to Cd distribution, we expect the D in foraminifera to be lower than the inorganic value, again approaching 1. It is significant to note that the average inorganic D_{Cd} is around 11, which is roughly 3.7 larger than D_{Cd} in foraminifera. This factor is rather similar to the factors by which D_{Sr} and D_{Ba} deviate from their inorganic D. An attempt to quantify the reservoir effect is presented in Elderfield et al. (1996) where we assumed a Rayleigh fractionation model to describe the distribution of Sr, Ba and Cd from a closed Ca reservoir. For these trace elements, 90% of the reservoir is utilized in order to obtain the foraminiferal distribution coefficients. If this model can be verified and elaborated it may be possible to correct the fossil record for these biogenic effects, using a tracer like Sr which should be rather conservative (over short periods) and influenced only by the degree of openness of the Ca pool. This may allow for a "normalization" procedure for the vital effects in order to obtain better paleoceanographic information from Cd and Ba. However, at the time that this model was suggested, the nature of the calcification reservoir was not known. The new observations described below may shed light on this subject.

New observations and the role of seawater vacuolization

An important advance in biomineralization research of foraminifera was achieved in our laboratory when during decalcification experiments using EDTA with live organisms, we obtained viable "naked" amoeboid cells. These shell-less foraminifera contain symbionts, remain alive and function for several weeks. They show intensive pseudopodial activity and precipitate skeletal $CaCO_3$ (Bentov 1997, Bentov et al. 2001, Bentov and Erez 2001, Erez and Bentov 1998, 2002, Erez et al 1994, 2001, 2002). The ameboids were first produced from specimens of *A. lobifera* and later from other species including planktonic foraminifera. These amoeboids allow direct polarized light microscopy of the endoplasm, an essential step in order to study the biomineralization process. Furthermore, these amoeboids allow the use of advanced fluorescence imaging microscopy to study the calcification process and also TEM studies that do not require the decalcification step. The second and perhaps more important preparation was discovered from the mother individuals which produced the amoeboids. These partially dissolved individuals when transferred to normal seawater continue to calcify and precipitate their newly formed chambers flat on the glass substrate. These individuals allow detailed light and fluorescent microscopy observations on calcification processes that were totally obscured before.

Many amoeboids freshly isolated from their mother foraminiferan contain small-mineralized light polarizing granules usually of 0.5–1.0 µm. These granules cycle with the endoplasm stream and show typical yellow birefringence. Within a few days, the density of the granules increases and small, actively growing microspherulites (1–5 µm) appear that display also elongated bone-like shapes. Light microscopy suggests that the polarizing granules are membrane-bound and may be produced in special vacuoles within the endoplasm. The birefringent granules were also observed in the cytoplasm of small non-calcified offspring of *A. lobifera* produced during asexual reproduction and in the cytoplasm of adult individuals that were crushed physically without any chemical treatment. These phenomena were also observed in the benthic species *Amphistegina lessoni* and *Hetetostigina depressa*, and the planktonic species *Globigerinoides sacculifer, Globigerinoides ruber* and *Orbulina universa*.

The next step of biomineralization in the amoeboids is a massive deposition of large $CaCO_3$ spherulites of 20–60 µm in diameter. The large spherulites are attached to glass below the amoeboid and therefore seem to be extracellular. In reality they are precipitated in the realm of the ectoplasm as displayed by recovering individuals (see below). The microspherulites (1–5 µm) are both extracellular and intracellular, and can sometimes be observed moving within the endoplasm. They are made of concentric $CaCO_3$ crystals as shown by the extinction cross. SEM shows that each unit is made of calcite platelets similar to the skeletal elements which compose the two-dimensional wall of a newly formed chamber in association with the organic matrix. Spherulite composition determined by EPMA changes with size: The small microspherulites (1–10 µm) are enriched in Mg (average of 18 and up to 25 mole %) and are very low in Sr (below detection when smaller then 10 µm). The larger spherulites (>20 µm) are low Mg calcite (~3%) and have roughly 1% Sr. It is important to note that in normally growing *A. lobifera,* the skeleton is composed of alternations of Mg rich and Mg poor layers (Szafranek and Erez 1993, as discussed above). An X-ray analysis of powdered *A. lobifera* reveals that indeed the skeleton is composed of two phases. The rare phase is Mg rich and is probably composed of the microspherulites while the common phase is low Mg calcite which forms the bulk of the calcitic radial test. The intracellular polarizing granules are highly soluble in HCl, NaOCl, and even distilled water. The granules are clearly bundles of needle-like crystals sometimes in a cross-like arrangement. The EPMA showed that they contain Mg, Ca, and P at an approximate ratio of 2:1:3, respectively. They are similar to granules described in *Crustacea* and in other unicellular organisms (Brown 1982; Taylor and Simkiss 1989). The fact that these granules are so soluble suggests that they may serve as a temporary cellular storage for Ca. As mentioned above these granules may be part of the internal Ca pool.

Observations on recovering individuals

The mother individual, while recovering from partial dissolution, instead of making normal chambers over the pre-existing skeleton, precipitates its newly-formed chambers flat on the glass substrate. Some of these individuals make several chambers and allow for detailed observations of the calcification process that were totally obscured before. Unlike the amoeboids, these individuals retain the ability to form the chambers including the primary matrix-mediated calcification, the lobes of each chamber, the pores and the precipitation of secondary calcite. The initial calcite deposition is on the organic template, marking the external outline of the newly formed chamber. In *A. lobifera* the external suture of the newly formed chamber appears as elongated lobes that separate the dense internal endoplasm from the extralocullar cytoplasm. It is possible that the polarizing granules (which are found only in the endoplasm) provide the Ca and possibly the organic matrix for this process, and hence the primary calcite is enriched in Mg. This being the case, the granules may exocytose their soluble mineral content just at the boundary between the dense endoplasm and the ectoplasm. It is also possible that primary calcite involves initial precipitation of amorphous $CaCO_3$, as has been demonstrated in other organisms (Raz et al. 2000, 2003). The next stage is the precipitation of the secondary calcite. This occurs in the realm of the extralocullar cytoplasm by a mechanism that is not clear yet. Initially a line of spherulites is precipitated, and later the entire space between the lobes and this perimeter is calcified. The biomineralization process is fast, and the mineral precipitated is low Mg calcite.

Using fluorescence imaging confocal microscopy we conducted pulse-chase experiments with the cell-impermeant probe, FITC-Dextran (Bentov et al. 2001; Bentov and Erez 2001; Erez et al. 2002). We demonstrated intensive seawater vacuolization forming large intracellular vacuoles that were labeled within minutes. Extralocullar cytoplasmic vacuoles, which surround the growing crystals, contained labeled seawater as well. With time, smaller labeled endoplasmic vacuoles appeared, probably budding out of the larger ones. After a few hours the dye was washed away from the seawater, but the crystals that continued to grow displayed strong additional fluorescence, which must have come from the internal labeled vacuoles. Similar results were obtained in experiments with Ca-binding dye Calcein. Both experiments suggest that seawater vacuoles that remain within the organisms serve as the main ion source (calcium and possibly carbonate) for the calcification process. The exact mode by which the FITC-Dextran labeled the growing calcite crystals is not clear, but it may be associated with the organic matrix. Initial experiments were conducted to measure the pH of the seawater in the vacuoles by ratio imaging of the pH indicator Snarf-Dextran to Ca green-Dextran. Preliminary qualitative analysis indicates that the pH of the seawater in the vacuoles increased significantly with time to values between 8.5 and possibly 9. Labeling the cells with Snarf AM (cell-permeant ester) shows that extralocullar cytoplasm pH, especially around the growing crystals, is higher than the endoplasmic pH (which is around 7.2), suggesting that the cell elevates the pH of the vacuoles. Photosynthesis by the symbiotic algae elevates the pH in the microenvironment (Jorgensen et al.1985, Rink et al. 1998) where these vesicles are produced. This may be another way in which symbionts may enhance calcification by the host. These vesicles are transported from the endoplasm by an pseudopodial network to the sites of secondary calcification where they provide their contents for the formation of the secondary low Mg calcite. The membrane probe FM 1–43 demonstrated that the ectoplasm forms a continuous thin layer that completely covers the growing crystals. This ectoplasmic layer probably isolates the crystals from the surrounding seawater, thus creating a delineated space, necessary for biologically controlled calcification (Figs. 7 and 8).

Summary and working hypothesis

The following working hypothesis for the calcification process in perforate foraminifera may be formulated: Calcium is concentrated in the endoplasm in a highly soluble, birefringent mineral phase composed of Ca, Mg, P possibly S. The granules are membrane-bound and may contain organic matrix or some of its components. The granules provide Ca for the first $CaCO_3$ crystals that precipitate over the newly formed organic matrix. At this stage the chamber consists of a two-dimensional primary wall made of Mg-rich calcitic microspherulites embedded within the organic matrix. The second stage of the calcification process involves massive deposition of a low-Mg calcite wall. This secondary calcite is made of layered crystal aggregates with their c-axes perpendicular to the test wall. In the amoeboids these units form large spherulites. In the intact foraminifera these units form the secondary lamination and are responsible for the bulk of the skeleton deposition. The biomineralization process forming secondary calcite

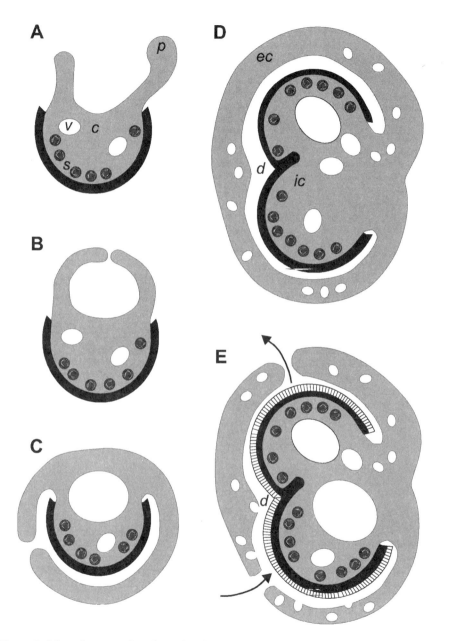

Figure 7. Schematic presentation of pseudopodia and vacuolization processes in perforate foraminifera. (A) The black cup represents the $CaCO_3$ shell which is exposed to the seawater. The letters $v,c,s,$ and p, represent vacuole, cytoplasm, symbionts and pseudopod, respectively. (B) The formation of a seawater vacuole. (C) Self-vacuolization of the organism separating the shell from the environment. (D) Building of a new chamber in a delimited space (d) with clear separation of intralocullar and extralocullar cytoplasm (ic and ec, respectively). The process of secondary lamination involves vacuolization of seawater. Arrows represent seawater exchange in the delimited space.

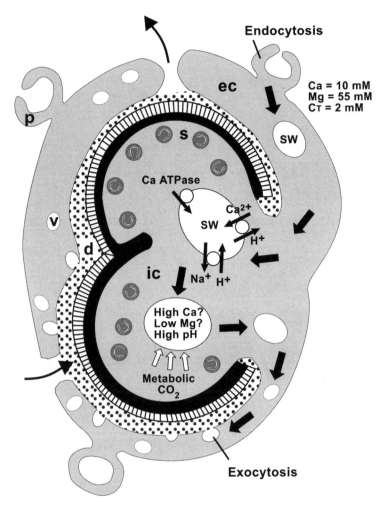

Figure 8. An overall scheme describing the secondary calcification process in perforate foraminifera. The symbols are: v – vacuole, p – pseudopodia, s – symbiotic algae, d – delimited biomineralization space, ic – intralocullar cytoplasm, ec – extralocullar cytoplasm, sw – seawater. Seawater vacuoles are formed by endocytosis, and during their pathway within the cell various pump and possibly channels operate to increase the pH and possibly Ca^{2+} and the C_T in these vacuoles. These modified seawater vacuoles are exocytosed into the delimited biomineralization space (marked with dense dots) where $CaCO_3$ is precipitated over the existing shell. It is possible that Mg concentration is lowered in the seawater vacuoles or alternatively an unknown process in the delimited space may sequester it in such a way that low Mg calcite can precipitate.

involves vacuolization of seawater and its modification within the cytoplasm perhaps to reduce the Mg:Ca ratio and to elevate the pH (Figs. 7 and 8). Diffusion of CO_2 into the basic vacuoles or direct CO_3^{2-} ion transport across the membrane may be responsible for the carbon pool that has been described above (ter Kuile et al. 1989a). The photosynthesis of symbiotic algae during daytime helps to elevate the pH in the boundary layer where the vacuoles are formed. However, inside the endoplasm, competition for CO_2 may occur between the symbionts and the alkaline vacuoles needed for calcification. Our EPMA work described earlier (Fig. 6) which demonstrated alternations of thin Mg and S rich layers with thick layers of low Mg and S may represent alternations of primary and secondary calcite, in good agreement with this scheme.

Relations between Ca^{2+} and CO_3^{2-} during $CaCO_3$ precipitation

If indeed seawater is the major component from which the minerals are formed there is a large discrepancy between the concentration of Ca^{2+} and CO_3^{2-} in seawater. Ca^{2+} concentration is between 10 and 11 mM while CO_3^{2-} is only 100–300 µM, a factor of ~50 lower then Ca^{2+}. Under these conditions the CO_3^{2-}, and not Ca^{2+}, is the limiting compound for the precipitation of $CaCO_3$. This situation is further accentuated because during $CaCO_3$ precipitation the alkalinity drops at a rate double that of the C_T causing a drop in the pH, which further lowers the CO_3^{2-}. Under these conditions the best strategy for the organism is to raise the pH. This will immediately increase the CO_3^{2-} concentration even if the C_T continues to drop. Furthermore, if the pH is higher than the bulk solution, $CO_{2(aq)}$ will diffuse into the basic compartment and provide C_T for the continued calcification process. This may also be the strategy preferred by corals, as indeed reported by Al Horani et al. (2003). Using a pH microelectrode they showed that the pH below the chalicoblastic epithelium is around 9. Under these conditions the CO_3^{2-} concentration is half of the C_T (which, if close to that of seawater, will be around 1 mM). In the case of the radial foraminifera the pH increase occurs in the seawater vacuoles as they mature (on the track from the endoplasm to the calcification space). These basic vacuoles (again pH around 9) may serve as a sink for $CO_{2(aq)}$ either from respiratory (metabolic) origin or from seawater. If the vacuoles are inside the endoplasm they may accumulate very high C_T concentrations because the pH in the cytosol is around 7.2 (based on normal marine cells and on preliminary microelectrode measurements on ameboids). These high C_T containing vacuoles may compete for $CO_{2(aq)}$ with the symbiotic algae as indeed observed by ter Kuile et al. (1989a). Furthermore, the internal carbon pool, which we discovered and described above (ter Kuile and Erez 1987, 1988; ter Kule et al. 1989), is most probably these basic seawater vacuoles. The pulse-chase experiments with ^{14}C (Fig. 4) showed that the internal carbon pool serves only for calcification and that inorganic carbon uptake into the pool occurs during daytime only. As showed above the concentration of C_T in the pool of *A. lobifera* is 190 mM (2 orders of magnitude higher than seawater concentration). However, the volume of the organism must include also the pseudopodial network that may extend and increase the volume of the organism considerably. For example if the thickness of the network is 400 µm then the C_T would be roughly 40 mM. C_T concentrations of 20–40 mM at pH 8.5 to 9 would provide enough CO_3^{2-} to match the amount of Ca^{2+} in the seawater vacuoles. This requirement is brought forward because our Rayleigh distillation model for distribution coefficients of Ba, Sr and Cd in foraminifera suggests that most of the Ca^{2+} in the calcification reservoir is precipitated (Elderfield et al. 1996). We suggest that these independent lines of reasoning support each other. In fact, in the old models for foraminifera (ter Kuile et al. 1989b) Ca^{2+} is pumped into the cells from seawater. If, however, modified seawater is the solution from which $CaCO_3$ precipitates, there is no need to concentrate Ca^{2+} from a strict stoichiometry point of view. However, in the case of foraminifera the presence of Mg may impose another role for Ca^{2+} concentration (see below).

Global CO_2 considerations

As discussed in the introduction, the contribution of foraminifera to the global carbon budget is significant. The possible negative feedback of CO_2 increase on the calcification of these organisms has already been pointed out (Barker et al. 2003). Changes in the CO_2 concentration between glacials and interglacials were used to calculate the carbonate ion concentrations and a response in shell weights where demonstrated (Barker and Elderfield 2002). In their Figure 1, Barker and Elderfield (2002) show an increase of 50% in the size-normalized weight of planktonic foraminifera when the CO_3^{2-} ion increases from 200 to 260 µM. Given all the complications involved in the biomineralization process it is surprising that foraminifera would respond to such a small change in CO_3^{2-}. Similar results were obtained in our experiments with benthic foraminifera (ter Kuile et al. 1989b). These data are shown in Figure 9 for both *A. lobifera* and *A. hemprichii*. A strong reduction in the calcification rate can be observed in *A. lobifera* as the pH is lowered from 9 to 7 (Fig. 9a). This trend is not linear; it stops above pH 9 and reverses at 9.5. When these data plotted against the CO_3^{2-} ion concentration it shows an increase of ~200% in the calcification over the range of CO_3^{2-} of 100 to 400 µM, quite similar to the response of the planktonic foraminifera (Barker and Elderfield 2002). It should be noted that the imperforate *A. hemprichii* shows an exponential increase in the rate of calcification over the range from 7.5 to 9.5 (Fig. 9b). In view of the discussion at the beginning of this section it is not surprising that foraminifera respond to the CO_3^{2-} concentration in their ambient seawater. If indeed seawater vacuolization is the starting point, then its carbonate chemistry sets the boundary conditions for the modifications that need to be made by the organism. Obviously more saturated seawater (with higher pH) would require less effort to elevate the pH, and hence calcification rates will increase. As a response to atmospheric CO_2 increase, it can be expected that planktonic foraminifera (as well as shallow benthic ones) will reduce their calcification rate, and that would help to reduce CO_2 in the atmosphere.

Concentration mechanism of ions inside seawater vacuoles and what prevents the precipitation of $CaCO_3$

If foraminifera utilize alkaline seawater vacuoles as their main calcification mechanism, a problem may rise as they increase the pH inside the vacuole. This is because very quickly the solution may become supersaturated and hence $CaCO_3$ will precipitate spontaneously. Our experience with Gulf of Eilat seawater (S = 40.7‰; C_T = 2060 µmol kg^{-1} and Total Alkalinity (A_T) = 2490 µEq kg^{-1}) showed that if the A_T is increased simply by adding NaOH, to roughly 3000 µEq kg^{-1}, $CaCO_3$ (most probably aragonite) is precipitated spontaneously (when the solution was equilibrating with the atmosphere). Under these conditions the pH is 8.7 and CO_3^{2-} is 650 µmol kg^{-1} and Ω for aragonite is 9.6. When the pH is increased to 9, the A_T is 3440 µEq kg^{-1}, and Ω for aragonite is 14.7. Clearly we can expect spontaneous precipitation of aragonite inside the vacuoles. This does not occur, as we have never observed inorganic precipitation inside the seawater vacuoles. A simple explanation is that as A_T and pH increase in the vacuoles, their C_T also increases rapidly. When C_T has increased to 2600 µmole kg^{-1}, the pH drops to 8.5, and Ω for aragonite is roughly 9. The process can continue if the organism keeps increasing the alkalinity and the C_T at a 1:1 ratio, and the pH will not change significantly. In addition to these considerations it is possible that the organism can prevent precipitation of $CaCO_3$ by providing organic or inorganic (such as PO_4^{3-}) inhibitors into the vacuoles. One important role of Mg may be to slow the kinetics of $CaCO_3$ precipitation inside the seawater vacuoles when Ω is high. Relatively high supersaturations may thus be achieved inside the vacuoles because Mg is not complexed or neutralized. Once the high pH vacuole is opening into the site of biomineralization Mg may be complexed in such a way that low Mg calcite will precipitate. Finally, it is worthwhile to consider the cost of elevating pH as a strategy for calcifying organisms (Zeebe and Sanyal 2002). Obviously the best strategy is to elevate the concentration of CO_3^{2-}. An increase of pH from 8 to 9 for normal seawater with C_T of 2mM, would increase the CO_3^{2-} concentration from 100 µM to 1000 µM. At the same time this would also reduce the H$^+$ concentration by a factor of 10. If the Na$^+$/H$^+$ exchanger achieves this change, it would cause an insignificant change in the Na$^+$ concentration of the cell.

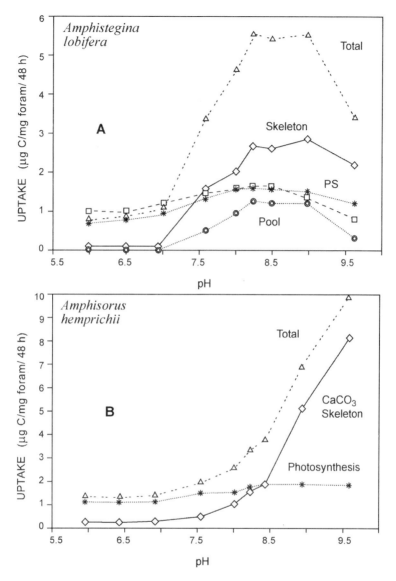

Figure 9. Calcification as function of pH in the benthic foraminifera (A) *A. lobifera* and (B) *A. hemprichii* based on ^{14}C uptake. In the pH range of 7 to 9 *A. lobifera* shows strong increase of calcification (skeleton uptake). In *A. hemprichii* the increase of calcification with pH is exponential.

THE ROLE OF Mg IN FORAMINIFERA

In foraminifera one additional factor that we should consider is the need to concentrate Ca^{2+} or alternatively remove Mg^{2+} from the seawater vacuoles in order to precipitate low Mg calcite. This is particularly true for the planktonic species, which have very low concentrations of Mg in their shells. It is difficult to conceive the nature of such a mechanism because Mg concentrations in seawater are 5 times higher then Ca. Hence this subject requires careful further investigation, particularly since Mg content in foraminifera shells is used now as a paleotemperature proxy (e.g., Lear et al. 2002). It should also be remembered that the primary calcite is rich in Mg (up to 20 mole %), while the secondary calcite is much lower in Mg. Different proportions of primary and secondary calcite may bias the paleotemperature estimates based on Mg content. Or alternatively the proportions between these to types of calcite is determined by temperature.

There are at least two possibilities to explain the exclusion of Mg from the shells. One is the removal of Mg from the seawater vacuoles by an active process (a pump). The other is discrimination against Mg during the formation of the mineral phase over and above the normal distribution coefficient. Raz et al. (2000) did note that Ca was enriched in the cores of spherulites composed of amorphous calcium carbonate, leaving the periphery with more Mg relative to Ca. There may be other as yet undiscovered processes, possibly involving the components of the organic matrix, that discriminate against Mg and in favor of Ca.

Another observation may shed light on the removal of Mg at the site of calcification. In inorganic and biogenic carbonates (including foraminifera) when Mg is higher, Sr is also higher (Carpenter and Lohmann 1992). This suggests that Mg may be included in the Rayleigh distillation model mentioned above for fractionation of divalent elements from a semi-closed reservoir. However, in order to extend the model to Mg we need to lower its activity in the solution by a factor of 10 for benthic foraminifera and 50 for planktonic ones. It is highly unlikely that an electrogenic pump will achieve this. A more probable solution is the involvement of some organic molecules that complex the Mg and do not let it poison the growth of low Mg calcite.

One difference between corals and foraminifera is that the corals precipitate aragonite, while foraminifera precipitate calcite. To precipitate aragonite from supersaturated seawater there is no need to remove Mg because aragonite is the kinetically preferred mineral. Corals precipitate their $CaCO_3$ without influence of pseudopods in the space formed between the chalicoblastic ectoderm and the skeleton. In the foraminifera, precipitation occurs under strong influence of pseudopodial network (i.e., bilipid membranes). Is it possible that these membranes or some of their components bind in some other way the Mg in the environment of crystal formation? This may neutralize the inhibition effects of Mg on kinetics of calcite formation. Such a mechanism may also explain why the foraminifera precipitate low Mg calcite (which is also more supersaturated!). Is it possible that this is the role of vesicles that are rich in fibrillar material (Angell 1967) from the electron dense bodies?

Finally it may be asked why most of the radial foraminifera (except of one family, the aragonitic Robertinacea) precipitate calcite. One possibility is that they evolved in periods where Mg:Ca ratio in the ocean was much lower than today. Such variations in ocean chemistry have indeed been reported (e.g., Hardie 1996, Stanley et al. 1999) suggesting that during the Phanerozoic there have been large variations in this ratio ranging from 1:1 (e.g., most of the Cretaceous and the Paleogene) to 5:1 as observed in the oceans today. These variations stem from changes in rates of sea floor spreading which, when high, also enhance hydrothermal circulation at the ocean ridges. During this process Mg is exchanged for Ca and this ratio decreases. It has been suggested that when Mg:Ca is high, the oceans are dominated by aragonite producers, while calcite producers are abundant during periods that have low Mg:Ca ratio. It is possible that when the modern radial foraminifera evolved (later Cretaceous and the Paleogene), the Mg:Ca was around 1:1, and the preferred $CaCO_3$ mineral was calcite. Calcite is advantageous because of its lower solubility product relative to aragonite. It can be speculated that when the Mg:Ca ratio started to increase, the foraminifera adapted by retaining the mineral calcite and found some ways to remove Mg from the site of biomineralization.

The carbon pathway and stable isotopes

As discussed above it is conceivable that bicarbonate ion enters the cell by an active transport mechanism (e.g., HCO_3^-/Cl^- exchanger). Otherwise $CO_{2(aq)}$ will diffuse out of the cell according to the pH gradient (7.2 in the cell and 8.2 in seawater) and this will diminish the C_T in the cytosol. The $\delta^{13}C$ of seawater HCO_3^- is roughly +1 ‰. When combined with protons inside the cell and in the absence of the enzyme carbonic anhydrase, the $CO_{2(aq)}$ that is formed may not fully fractionate with respect to the HCO_3^-; hence this $CO_{2(aq)}$ is probably enriched in ^{13}C. In the cell the symbionts fix $CO_{2(aq)}$ with some small fractionation resulting in $\delta^{13}C$ of roughly –12‰ , based upon several measurements that we have carried out on the cytoplasm of *A. lobifera* and *A. Hemprichii*. Similarly the $\delta^{13}C$ of symbionts in corals (e.g., Land et al. 1977) is also high, ranging from –11 to –15‰, a relatively suggesting low fractionation compared to normal marine phytoplankton with $\delta^{13}C$ ranging between –17 and –35‰. This ^{13}C enriched composition may in part be caused by their $CO_{2(aq)}$, which is enriched in ^{13}C as described above. Still, if the algae fractionate relative to the C_T in the cytosol, its $\delta^{13}C$ will increase. In the calcification zone however there are no symbionts. Seawater vacuoles are basic (pH = 9), and CO_2 may be attracted into these vacuoles either from the more acidic cytosol (although it is very thin) or from the activity of the mitochondria (respiration) that are highly abundant in the calcification region. Another source of carbon for the calcification is the original C_T of the seawater in the vacuoles. Using microsensors, Kohler-Rink and Kuhl (2001) showed that respiration rates in these foraminifera is enhanced during the light period. Similar light-enhanced respiration was also reported for hermatypic corals (Al Horani et al. 2002). This light-enhanced respiration must be fueled by photosynthesis of the symbionts, and it will provide isotopically light carbon to the calcification zone. This scheme may explain the observations and interpretation of Erez (1978), showing decrease in $\delta^{13}C$ of coral and foraminifera skeletons with increased photosynthesis.

The active transport mechanism of C_T from seawater into the cell may not be an exchanger but a co-transporter bringing in protons (which lower the pH) together with HCO_3^-. It may also be mediated by a Na^+/H^+ exchanger. Bringing in HCO_3^- together with protons is advantageous because it will help to increase the C_T, but not alkalinity. If the C_T concentration in the cytosol is 10–20 mM (as in normal cells) this internal C_T at pH of 7.2 serves for symbiont photosynthesis as well as sources of carbon into the basic vacuoles involved in calcification (Fig. 8). The symbionts may be separated spatially from the vacuoles (which may be concentrated at the calcifying regions).

SUMMARY AND CONCLUDING REMARKS

Foraminifera skeletons provide the most important source of paleoceanographic information and are a major part of the global oceanic carbon cycle. In addition to the traditional stable isotopes and trace elements, new proxies for paleoceanographic reconstructions are being introduced. Among them are various anions and stable isotopes of metals (Ca, Mg and possibly others). Despite the extensive use of foraminiferal proxies in paleoceanography, it is now widely recognized that in order to extract more reliable information, a better understanding of the biomineralization process of these organisms is needed.

The process of biomineralization in foraminifera is based on seawater vacuolization, which provides the basic component from which calcification starts. Calcification starts with a strong biological control over the site, shape and form of the shell. This is manifested by the formation of organic matrix serving both for nucleation of the first crystals and providing the overall shape of the newly formed chambers. The first crystals that precipitate over the organic matrix or within its complex three-dimensional web are quite different from the rest of the skeleton. These crystals have different shapes, chemical compositions and possibly also different isotopic compositions.

The bulk of the calcification process continues with secondary lamination involving layers of epitaxial crystal growth with tight biological control. The secondary calcification occurs in a delimited space created by the pseudopodia, and seawater vacuoles are the major source of the ions for the biomineralization process (Fig. 9). This is perhaps the main reason why foraminifera are so reliable in providing paleoceanographic reconstructions of seawater compositions. The seawater vacuoles are modified by the organism mainly by increase in the pH and thus also increase in the dissolved inorganic carbon. Precipitation of low Mg, calcite within the delimited space must overcome the kinetic barrier imposed by Mg but the mechanism for this process is unknown at this stage. The existence of a Ca reservoir must be evoked to explain the dynamics of Ca tracers during calcification. The nature of this reservoir, however, is not clear, and may involve a solid, perhaps amorphous, phase.

Vital effects in stable isotopes and in trace elements may be caused by several processes and their combinations (e.g., Weiner and Dove 2003). Modification of the seawater by the organism prior to the precipitation process is probably more important for trace elements such as Mg, which may be actively removed from the modified seawater vacuoles in foraminifera as suggested by Erez et al. (1994, 2001, 2002). It is also important for the effect of the carbonate ion on $\delta^{18}O$ and to some degree on $\delta^{13}C$ (Spero et al.1997; Zeebe et al. 1999). The increase in the CO_3^{2-} concentration near the calcification site is controlled by the organism (in foraminifera through the basic seawater vacuoles and in corals probably by Ca^{2+}-ATPase). This will cause a thermodynamic decrease in $\delta^{18}O$ as explained by Zeebe (1999) in foraminifera and by Adkins et al. (2003) for corals. The decrease in $\delta^{13}C$ was not explained thermodynamically and may be connected to kinetic effects (see paragraph on "Baertschi Effect" below).

Precipitation from a semi-closed reservoir will tend to bring the distribution coefficients of trace elements towards unity (or no fractionation). Good examples are trace elements in corals and foraminifera such as Sr, Ba, Cd, U and possibly others. In the case of carbon isotopes this will cause no fractionation, which may be observed in the organic matter of the symbiotic algae of corals and foraminifera as their photosynthesis is initiated from a rather closed reservoir. The result is usually enrichment in ^{13}C relative to what is expected from Rubisco.

Some of the so-called vital effects reflect our lack of knowledge of the exact environmental conditions where organisms precipitate their shells. For example some deep benthic foraminifera precipitate their skeletons in interstitial water within the sediments where they may encounter water with very different $\delta^{13}C$ than in the overlying water (e.g., McCorkle et al. 1997; Tachikawa and Elderfield 2002). The chemical conditions in these habitats are also quite different (e.g., pH, CO_3^{2-} concentrations which may affect both $\delta^{11}B$ and $\delta^{18}O$). Another classical example is the oxygen and carbon isotopes in shells of planktonic foraminifera (with or without symbionts). It has been shown quite convincingly that shells of these organisms represent the integral of their ontogenetic migration through the water column (e.g., Erez and Honjo 1981; Hemleben et al. 1989; Erez et al. 1991). Hence it is difficult to assign a specific depth to their growth. Deviations from "equilibrium" are possible if a wrong depth is assigned for their formation. In addition dissolution may modify this picture with a bias towards the thicker, isotopically heavier shells. Another well-described example is the seasonal growth of foraminifera (e.g., Anand et al. 2003). The life span of most foraminifera is probably shorter then a year, and for many it is in the range of only one to several weeks (Lee and Anderson 1991; Hemleben et al. 1989). They may, therefore, record the temperature of the period in which they calcified. If there is a temperature (or salinity and $\delta^{18}O$ of water) change between seasons, it will be difficult to reconstruct the exact conditions which different species record. This may explain some of the interspecies differences both in $\delta^{18}O$ and $\delta^{13}C$. Under this category we may also include the microenvironmental conditions created by the organism itself. These have been well demonstrated using microelectrodes and are present both in symbiont and non-symbiont bearing organisms (with rather complex and unexpected differences). Net respiration will decrease the pH and increase the C_T from metabolic sources. On the other hand, separation of calcification between night and day may bias this picture. Finally, the recent observations of Bentov and Erez in foraminifera (2001) and Al Horani et al. (2003) in corals show that the organism may modify the site of crystal formation by increasing the pH and perhaps by concentrating Ca^{2+} using Ca^{2+}-ATPase (Allemand et al. 1998; Al-Horani et al. 2003).

Some kinetic effects may also be apparent as suggested by McConnaughey (1989), especially in corals where, as mentioned above, some of the crystals may be formed almost inorganically. Fast precipitation will have similar effects as a closed reservoir with respect to trace elements, i.e., it will cause the distribution coefficients to shift towards unity. For stable isotopes, on the other hand, kinetic effects can either eliminate expected fractionation or bias the composition towards the lighter isotopes. For example, there is a −8 to −9‰ fractionation between $CO_{2(aq)}$ and HCO_3^-, elimination of which would produce isotopically heavy $CO_{2(aq)}$. On the other hand, fast diffusion of $CO_{2(aq)}$ into an alkaline solution may cause kinetic fractionations of up to −15‰ (termed as a "Baertschi Effect" by Lazar and Erez 1991, 1992). Since the calcification compartment in corals and foraminifera is alkaline (around pH 9), if $CO_{2(aq)}$ were to diffuse into this compartment, it may cause such a kinetic fractionation. Finally it difficult to ignore the real vital effects that stem from the different types of crystals precipitated by the organisms. In the case of corals and foraminifera it is clear that these different types of crystals have different chemistries, and hence if their proportions within the organisms change, they may deviate from expected behavior. Another clear example of a vital effect is seen in the oxygen and carbon isotope signatures of young specimens of planktonic foraminifera. Here, for reasons that are not yet clear, it has been shown that the thin, lightly-calcified skeleton of *G. sacculifer* and possibly other species (Berger et al. 1978; Erez and Luz 1983) may be out of equilibrium, and the general explanation of higher metabolism for their fractionation is not satisfactory.

The new era of high-resolution multicollector ICP-MS, when coupled with laser ablation techniques, opens new frontiers for the study of foraminifera as the ideal fossils for paleoceanographic reconstructions. The new advances in laser confocal microscopy and in electron microscopy with its micro-analytical capabilities will enhance our future understanding in biomineralization process of these unique organisms. This is just the beginning...

ACKNOWLEDGMENTS

I would like to thank my devoted graduate students, Shmuel Bentov, Jack Silverman, Kenny Schneider, Inbal Meidani, Mor Grinstein and Einat Segev for stimulating and productive discussions. Special thanks are to Elwira Halicz for her professional help, to S. Weiner and P. Dove for helpful comments and important encouragement, and to Karin Roman for the graphics. Many parts of this review were supported over the years by grants from the US-Israel BSF, the German BMBF RSP and the ISF. This review is dedicated to memory of our mentor and a pioneer in the field of foraminiferal research, Prof. Zeev Reiss.

REFERENCES

Adkins JF, Boyle EA, Curry WB, Lutringer A (2003) Stable isotopes in deep-sea corals and a new mechanism for "vital effects." Geochim Cosmochim Acta 67(6):1129-1143

Al-Horani FK, Al-Moghrabi SM, de Beer D (2002) Microsensor study of photosynthesis and calcification in the scleractinian coral *Galaxea fascicularis*: active internal carbon cycle. J Exp Mar Biol Ecol 4088:1-15

Al-Horani FK, Al-Moghrabi SM, de Beer D (2003) The mechanism of calcification and its relation to photosynthesis and respiration in the scleractinian coral *Galaxea fascicularis*. Mar Biol 142:419-426

Allemand D, Furla P, Bénazet-Tumbutté S (1998) Mechanisms of carbon acquisition for endosymbiont photosynthesis in Anthozoa. Can J Bot 76:925-941

Anand P, Elderfield H, Conte MH (2003) Calibration of Mg/Ca thermometry in planktonic foraminifera from a sediment trap time series. Paleoceanography 18(2):article no-1050

Anderson OR, Faber WW Jr (1984) An estimation of calcium carbonate deposition rate in a planktonic foraminifer *Globigerinoides sacculifer* using ^{45}Ca as a tracer: a recommended procedure for improved accuracy. J Foram Res 14:303-308

Angell RB (1967) The process of chamber formation in the foraminifer *Rosalina floridans*. J Protozool 14:566-574

Angell RB (1979) Calcification during chamber development in *Rosalina floridana*. J Foram Res 9(4):341-353

Angell RB (1980) Test morphogenesis (chamber formation) in the foraminifer *Spiroloculina hyaline* schulze. J Foram Res 10(2):89-101

Archer D, Emerson S, Reimers C (1989) Dissolution of calcite in deep-sea sediments: pH and O_2 microelectrode results. Geochim Cosmochim Acta 53:2831-2845

Archer D, Maier-Reimer E (1994) Effect of deep-sea sedimentary calcite preservation on atmospheric CO_2 concentration. Nature 367: 260-263

Barker S, Elderfield H (2002) Foraminiferal calcification response to glacial-interglacial changes in atmospheric CO_2. Science 297(5582):833-836

Barker S, Higgins JA, Elderfield H (2003) The future of the carbon cycle: review, calcification response, ballast and feedback on atmospheric CO_2. Phil Trans R Soc Lond A 361:1977-1999

Barnola JM, Raynaud D, Korotkevich YS, Lorius C (1987) Vostok ice core provides 160,000-year record of atmospheric CO_2. Nature 329:408-414

Be AWH (1980) Gametogenic calcification in a spinoso planktonic foraminifer *Globigerinoides sacculifer* (Brady). Mar Micropaleont 5:283-310

Be AWH (1982) Biology of planktonic foraminifera. *In*: Foraminifera: Notes for a Short Course. Studies in Geology. Vol. 6. Broadhead TW (ed) Univ Tenn, Knoxville, p 51-92

Be AWH, Hemleben C, Anderson OR, Spindler M, Hacunda J, Tunitivate-Choy S (1977) Laboratory and field observation of living planktonic foraminifera. Micropaleont 23(2):155-179

Be AWH, Spero HJ, Anderson OR (1982) Effects of symbiont elimination and reinfection on the life processes of the planktonic foraminifer *Globigerinoides sacculifer*. Mar Biol 70:73-86

Bellemo S (1974) Ultrastructures in recent radial and granular foraminifera. Bull Geol Inst Univ Uppsala New Series 4:117-122

Bender ML, Lorens RB, Williams DF (1975) Sodium, magnesium and strontium in the test of planktonic foraminifera. Micropaleont 21:448-459

Bentov S (1997) Biomineralization processes in the unicellular Foraminifera. MSc Thesis, The Hebrew University of Jerusalem. (in Hebrew with English abstract)

Bentov S, Erez J (2001) New observations on the calcification mechanisms of foraminifera and implications for trace elements distribution in their shells (abstract). 7th Int Conf Paleocanography, Sapporo, Japan, p 82

Bentov S, Erez J, Brownlee C (2001) Confocal microscope observations on the calcification processes of the foraminiferan *Amphistegina lobifera* (abstract). XI Int Congress of Protozool, Salzburg, Austria, p 40

Berger WH (2002) Climate history and the great geophysical experiment. *In:* Climate Development and the History of the North Atlantic Realm. Wefer G, Berger W, Behre KE, Jansen E (eds) Springer Verlag, p 1-16

Berger WH, Killingley JS, Vincent E (1978) Stable isotopes in deep sea carbonates: Box core ERDC-92, West Equatorial Pacific. Oceanologica Acta 1(2):203-216

Berggren WA, Kent DV, Swisher III CC, Aubrey M-P (1995) A revised Cenozoic geochronology and chronostratigraphy. *In:* Geochronology, timescales and global stratigraphic correlation. Berggren WA, Kent DV, Aubrey M-P, Haardengol J (eds) Soc Econ Paleontol Mineral Spec Publ 54:129-212

Boyle EA (1981) Cadmium, zinc, copper, and barium in foraminifera tests. Earth Planet Sci Lett 53(1):11-35

Boyle EA (1988) Cadmium: Chemical Tracer of Deepwater Paleoceanography. Paleoceanography 3(4):471-489

Boyle EA (1988) The role of vertical chemical fractionation in controlling Late Quaternary atmospheric carbon dioxide. J Geophys Res 93(C12):15701-15714

Boyle EA (1992) Cadmium and $\delta^{13}C$ paleochemical ocean distributions during the stage 2glacial maximum. Ann Rev Earth Planet Sci 20:245-287

Boyle EA (1995) Limits on benthic foraminiferal chemical analyses as precise measures of environmental properties. J Foram Res 25:4-13

Boyle EA, Labeyrie L, Duplessy JC (1995) Calcitic foraminiferal data confirmed by cadmium in aragonitic *Hoeglundina*—application to the last glacial maximum in the northern Indian-ocean. Paleoceanography 10(5):881-900

Brewer PG, Goyet C, Frienderich G (1997) Direct observation on the oceanic CO_2 increase revisited. Geophys Proc Nat Acad Sci USA 94:8308-8313

Broecker WS (1997) Thermohaline circulation, the Achilles Heel of our climate system: Will man-made CO2 upset the current balance? Science 278:1582-1588

Broecker WS, Peng T-H (1982) Tracers in the Sea. Eldigio Press, Palisades, New York

Broecker WS, Peng T-H (1989) The cause of the glacial to interglacial atmospheric CO_2 change: a polar alkalinity hypothesis. Global Biogeochem Cycles 3(3):215-239

Brown BE (1982) The form and function of metal containing "granules" in invertebrate tissues. Biol Rev 57:621-667

Brown SJ, Elderfield H (1996) Variations in Mg/Ca and Sr/Ca ratios of planktonic foraminifera caused by postdepositional dissolution: Evidence of shallow Mg-dependant dissolution. Paleoceanography 11(5):543-551.

Caron DA, Be AWH, Anderson OR (1981) Effects of variations in light intensity on life processes of the planktonic foraminifer *Globigerinoides sacculifer* in laboratory culture. J Mar Biol UK 62:435-451

Carpenter SJ, Lohmann K (1992) Sr/Ca ratios of modern marine calcite: Empirical indicators of ocean chemistry and precipitation rate. Geochim Cosmochim Acta 56:1837-1849

Cutani Y (1984) Sr/Ca ratios in marine carbonates from the Gulf of Eilat. MS Thesis, The Hebrew University of Jerusalem (in Hebrew)

Darling KF, Kucera M, Wade CM, von Langen P, Pak D (2003) Seasonal distribution of genetic types of planktonic foraminifer morphospecies in the Santa Barbara Channel and its paleoceanographic implications. Paleoceanography 18(2): article no-1032

Darling KF, Wade CM, Stewart IA, Kroon D, Dingle R, Brown AJL (2000) Molecular evidence for genetic mixing of Arctic and Antarctic subpolar populations of planktonic foraminifers. Nature 405(6782):43-47

Delaney ML, Be AWH, Boyle EA (1985) Li, Sr, Mg, and Na in foraminiferal calcite shells from laboratory culture, sediment traps, and sediment cores. Geochim Cosmochim Acta 49:1327-1341

Duckworth D (1977) Magnesium concentration in the test of the planktonic foraminifer *Globorotalia truncatolinoides*. J Foramin Res 7(4):304-312

Duguay LE (1983) Comparative laboratory and field studies on calcification and carbon fixation in foraminiferal algal associations. J Foramin Res 13(4):252-261

Duguay LE, Taylor DL (1978) Primary production and calcification by the soritid foraminifer *Archais angulatus* (Fitchel & Moll). J Protozool 25(3):356-361

Eggins S, De Deckker P, Marshall J (2003) Mg/Ca variation in planktonic foraminifera tests: Implications for reconstructing palaeo-seawater temperature and habitat migration. Earth Planet Sci Lett 212(3-4):291-306

Elderfield H., Bertram CJ, Erez J (1996) A biomineralization model for the incorporation of trace elements into foraminiferal calcium carbonate. Earth Planet Sci Lett 142:409-423

Elderfield H, Ganssen G (2000) Past temperature and $\delta^{18}O$ of surface ocean waters inferred from foraminiferal Mg/Ca ratios. Nature 405(6785):442-445

Emiliani C (1955) Pleistocene temperatures. J Geol 63:538-578

Erez J (1978) Vital effect on stable-isotope composition seen in foraminifera and coral skeletons. Nature 273:199-202

Erez J (1983) Calcification rates, photosynthesis and light in planktonic foraminifera. *In:* Biomineralization and Biological Metal Accumulation. Westbroek P, de Jong E (eds) D Reidel Publishing Company, p 307-312

Erez J (1993) Internal pools of carbon and calcium in foraminifera and their influence on stable isotope and trace element fractionation. EOS Trans, Am Geophys Union, p 183

Erez J, Almogi A, Abraham S (1991) On the life history of planktonic foraminifera: Lunar reproduction cycle in *Globigerinoides sacculifer* (Brady). Paleoceanography 6(3):295-306

Erez J, Bentov S (1998) The mechanism of calcification in perforate foraminifera (abstract). Foraminifera' 98 Int Symposium on Foraminifera, Monterrey, Mexico, p 34

Erez J, Bentov S (2002) Calcification processes in foraminifera and their paleoceanographic implications (abstract). Forams 2002 Int Symposium on Foraminifera, Perth, Australia, p 32-33

Erez J, Bentov S, Brownlee C (2001) A model for the calcification mechanisms in perforate foraminifera based on seawater vacuolization (abstract). XI Int Congress of Protozool, Salzburg, Austria, p 40

Erez J, Bentov S, Brownlee C, Raz M, Rinkevich B (2002) Biomineralization mechanisms in foraminifera and corals and their paleoceanographic implications. (abstract). Abstracts 12th Annual VM Goldschmidt Conference, Davos, Switzerland. Geochim Cosmochim Acta 66(15a):A216

Erez J, Bentov S, Tishler C, Szafranek D (1994) Intracellular calcium storage and the calcification mechanism of perforate foraminifera (abstract). PaleBios 16(2):30

Erez J, Honjo S (1981) Comparison of isotopic composition of planktonic foraminifera in plankton tows, sediment traps and sediments. Paleogeog Paleoclimat Paleoecol 33:129-156

Erez J, Luz B (1982) Temperature control of oxygen- isotope fractionation of cultured planktonic foraminifera. Nature 297(5863):220-222

Erez J, Luz B (1983) Experimental paleotemperature equation for planktonic foraminifera. Geochim Cosmochim Acta 47:1025-1031

Graham DW, Bender LM, Williams DF, Keigwin LD Jr (1982) Strontium-calcium ratios in Cenozoic planktonic foraminifera. Geochim Cosmochim Acta 46:1281-1292

Hallock P, Rottger R, Wetmore K (1991) Hypotheses on form and function in foraminifera. *In*: Biology of Foraminifera. Lee JJ, Anderson OR (eds) Academic Press p 41-72

Hansen HJ, Reiss Z (1972) Scanning electron microscopy of wall structures in some benthonic and planktonic foraminifera. Rev Esp Micropaleontol 4:169-179

Hardie LA (1996) Secular variation in seawater chemistry: an explanation for the coupled secular variation in the mineralogies of marine limestones and potash evaporates over the past 600 m.y. Geol 24(3):279-283

Hastings D, Emerson S, Erez J, Nelson KB (1996) Vanadium in foraminiferal calcite: Evaluation of a method to determine paleo-seawater vanadium concentrations. Geochim Cosmochim Acta 60(19):701-3715

Hastings DW, Russell AD, Emerson SR (1998) Foraminiferal magnesium in *Globigerinoides sacculifer* as a paleotemperature proxy. Paleoceanography 13(2):161–169

Hedley RH, Adams CG (eds) (1978) Foraminifera. Vol 3. Academic Press, London

Hemleben C, Anderson RO, Berthold W, Spindler M (1986) Chamber formation in Foraminifera – a brief overview. *In*: Biomineralization in Lower Plants and Animals. Leadbeater BS, Riding R (eds) The Systematics Association Spec Vol 30:237-249

Hemleben C, Spindler M, Anderson OR (1989) Modern Planktonic Foraminifera. Springer-Verlag, New York

Hester K, Boyle EA (1982) Water chemistry control of cadmium content in recent benthic foraminifera. Nature 298(5871):260-262

Holligan PM. Fernandez E, Aiken J, Balch WM, Boyd P, Burkill PH, Finch M, Groom SB, Malin G, Muller K, Purdie DA, Robinson C, Trees CC, Turner SM, Vanderwal P (1993) A biogeochemical study of the coccolithophore, *Emiliania huxleyi*, in the North Atlantic. Glob Biogeochem Cycles 7:879-900

Holzmann M, Habura A, Giles H, Bowser SS, Pawlowski J (2003) Freshwater foraminiferans revealed by analysis of environmental DNA samples. J Eukaryotic Microbiol 50(2):135-139

Holzmann M, Pawlowski J (2002) Freshwater foraminiferans from Lake Geneva: Past and present. J Foramin Res 32 (4):344-350

Honjo S, Erez J (1978) Dissolution rates of calcium carbonate in the deep ocean: an in-situ experiment in the North Atlantic Ocean. Earth Plan Sci Lett 40(2):287-300

Hottinger L, Halicz E, Reiss Z (1993) Recent foraminiferida from the Gulf of Aqaba, Red Sea. Ljubljana, Akademi ja Znanosti in Umetnosti Slovenia

Houghton JT, Maskell K, Bruce JP, Callander BT, Harris NB (eds)(1995) Climate Change 1994: Radiative Forcing of Climate Change and an Evaluation of the IS92 Emission Scenario. Cambridge Univ Press, Cambridge

Huber BT, Bijma J, Darling K (1997) Cryptic speciation in the living planktonic foraminifer *Globigerinella siphonifera* (d'Orbigny). Paleobiol 23(1):33-62

Jorgensen BB, Erez J, Revsbech NP, Cohen Y (1985) Symbiotic photosynthesis in a planktonic foraminiferan *Globigerinoides sacculifer* (Brady), studied with microelectrodes. Limnol Oceonog 30(6):1253-1267

King K, Hare PE (1972) Amino acid composition of the test as a taxonomic character of living and fossil planktonic foraminifera. Micropaleonontol 18:285-293

Kleypass JA, Buddemeier RW, Archer D, Gattuso JP, Langdon C, Opdyke BN (1999) Geochemical consequences of increased atmospheric carbon dioxide on coral reefs. Science 284:118-120

Kohler-Rink S, Kuhl M (2000) Microsensor studies of photosynthesis and respiration in large symbiotic foraminifera. I The physico-chemical microenvironment of Marginopora vertebralis, *Amphistegina lobifera* and *Amphisorus hemprichii*. Mar Biol 137:473-486

Kohler-Rink S, Kuhl M (2001) Microsensor studies of photosynthesis and respiration in large symbiont bearing foraminifera. *Amphistegina lobifera* and *Amphisorus hemprichii*. Ophelia 55(2):111-122

ter Kuile B, Erez J (1984) *In-situ* growth rate experiments on the symbiont bearing foraminifera *Amphistegina lobifera* and *Amphisorus hemrprichii*. J Foramin Res 14(4)262-276

ter Kuile B, Erez J (1987) Uptake of inorganic carbon and internal carbon cycling in symbiont-bearing benthonic foraminifera. Mar Biol 94:499-510

ter Kuile B, Erez J (1988) The size and function of the internal inorganic carbon pool of the foraminifer *Amphiseigina lobifera*. Mar Biol 99:481-487

ter Kuile B, Erez J, Padan E (1989a) Competition for inorganic carbon between photosynthesis and calcification in the symbiont-bearing foraminifer *Amphistegina lobifera*. Mar Biol 103:253-259

ter Kuile B, Erez J, Padan E (1989b) Mechanisms for the uptake of inorganic carbon by two species of symbiont-bearing foraminifera. Mar Biol 103:241-251

Land LS, Lang JC, Smith BN (1977) Preliminary observations on the carbon isotopic composition of some reef coral tissues and symbiotic zooxanthellae. Limnol Oceanog 20:283-287

Langer MR (1992) Biosynthesis of glycosaminoglycans in foraminifera: a review. Mar Micropaleont 19:245-255

Langer MR, Lipps JH, Piller WE (1993) Molecular paleobiology of protist: amplification and direct sequencing of foraminiferal DNA. Micropaleont 39:63-68

Lazar B, Erez J (1991) Extreme carbon isotope depletions in seawater-derived brines and their implication to the past geochemical carbon cycle. Geol 18:1191-1194

Lazar B, Erez J (1992) The geochemistry of marine derived brines: I. Variations in carbon isotopes, total CO_2 and alkalinity, and the role of microbial mats on their spatial distribution. Geochim Cosmochim Acta 56:335-345

Lea D, Boyle E (1989) Barium content of benthic foraminifera controlled by bottom-water composition. Nature 338:751-753

Lea DW, Boyle EA (1993) Reply to comment by N.E. Pingitore Jr. On "Barium in planktonic foraminifera." Geochim Cosmochim Acta 57:471-473

Lea DW, Martin PA, Chan DA, Spero HJ (1995) Calcium-uptake and calcification rate in the planktonic foraminifera *Orbulina-universa*. J Foramin Res 25(1):4-23

Lea DW, Mashiotta TA, Spero HJ (1999) Controls on magnesium and strontium uptake in planktonic foraminifera determined by live culturing. Geochim Cosmochim Acta 63(16):2369-2379

Lea DW, Spero HJ (1992) Experimental determination of barium uptake in shells of the planktonic foraminifera *Orbulina universa* at 22°C. Geochim Cosmochim Acta 56:2673-2680

Lea DW, Spero HJ (1994) Assessing the reliability of paleochemical tracers: Barium uptake in the shells of planktonic foraminifera. Paleoceanography 9(3):445-452

Lear CH, Elderfield H, Wilson PA (2000) Cenozoic deep-sea temperatures and global ice volumes from Mg/Ca in benthic foraminiferal calcite. Science 287(5451):269-272

Lear CH, Rosenthal Y, Slowey N (2002) Benthic foraminiferal Mg/Ca-paleothermometry: A revised core-top calibration. Geochim Cosmochim Acta 66(19):3375-3387

Lee JJ, Anderson OR (eds) (1991) Biology of Foraminifera. Academic Press, San Diego

Loeblich Ar, Tappan H (1964) Sacrcodina chiefly "Thecamoebians" and Foraminifera. *In*: Treatise of Invertebrate Paleontology, part Protista 2. Moore TC (ed), Geol Soc Am and Univ Kansas Press, p 1-900

Loeblich Ar, Tappan H (1986) Some new and revised genera and families of hyaline calcareous foraminiferida (Protozoa). Trans Am Microscop Soc 105(3):239-265

Loeblich Ar, Tappan H (1987) Foraminifera genera and their classification. New York, Van Nostrand Reinhold Company

Lorens RB (1981) Sr, Cd, Mn and Co distribution coefficients in calcite as a function of calcite precipitation rate. Geochim Cosmochim Acta 45:553-561

Lowenstam HA, Weiner S (1989) On Biomineralization. Oxford University Press
McConnaughey TA (1989) ^{13}C and ^{18}O isotopic disequilibrium in biological carbonates: II. In vitro simulation of kinetic isotope effects. Geochim Cosmochim Acta 53:163-171
McConnaughey TA (1994) Calcification, photosynthesis, and global carbon cycles. In: Past and Present Biomineralization Processes. Doumeng F (ed), Mus'ee Oc'eanographique Monaco, p 137-162
McConnaughey TA, Whelan JF (1997) Calcification generates protons for nutrient and bicarbonate uptake. Earth Sci Rev 42:95-117
McCorkle DC, Corliss BH, Farnham CA (1997) Vertical distributions and stable isotopic compositions of live (stained) benthic foraminifera from the North Carolina and California continental margins. Deep-Sea Res 44(6):983-1024
Milliman JD (1993) Production and accumulation of calcium carbonate in the ocean: budget of a nonsteady state. Global Biogeochem Cycles 7:927-957
Morse JW, Bender ML(1990) Partition coefficients in calcite: Examination of factors influencing the validity of experimental results and their application to natural systems. Chem Geol 82:265-277
Morse JW, MacKenzie FT (1990) Geochemistry of sedimentary carbonates. Developments in Sedimentology 48. Elsevier Science Publishers
Muller PH (1974) Sediment production and population biology of the benthic foraminifer *Amphistegina madagaskariensis*. Limnol Oceanog 19:802-809
Nurnberg D, Bijma J, Hemleben C (1996) Assessing the reliability of magnesium in foraminiferal calcite as a proxy for water mass temperature. Geochim Cosmochim Acta 60(5):803-814
Opdyke BN, Walker JCG (1992) Return of the coral reef hypothesis: Basin to shelf partitioning of $CaCO_3$ and its effect on atmospheric CO_2. Geol 10:733-736
Pawlowski J, Holzmann M, Berney C, Fahrni J, Cedhagen T, Bowser SS (2002) Phylogeny of allogromiid foraminifera inferred from SSU rRNA gene sequences. J Foram Res 32(4):334-343
Pearse VB, Muscatine L (1971) Role of symbiotic algae (Zooxanthellae) in coral calcification. Biol Bull 141:350-363
Petit JR, Jouzel J, Raynaud D, Barkov NI, Barnola JM, Basile I, Bender M, Chappellaz J, Davis M, Delaygue G, Delmotte M, Kotlyakov JM, Legrand M, Lipenkov VY, Lorius C, Pepin L, Ritz C, Saltzman E, Stievenard M (1999) Climate and atmospheric history of the past 420,000 years from the Vostok ice core, Antarctica. Nature 399:429-436
Pingitore NE Jr (1993) Comment. On "Barium in planktonic foraminifera" by Lea D.W. and Boyle E.A. Geochim Cosmochim Acta 57:469-470
Raven JA, Johnston AM (1991) Mechanisms of inorganic carbon acquisition in marine phytoplankton and their implications for the use of other resources. Limnol Oceanogr 36:1701-1714
Raz S, Hamilton P, Wilt F, Weiner S, Addadi L (2003) Proteins from sea urchin larval spicules mediate the transient formation of amorphous calcium carbonate on the way to calcite (in press)
Raz S, Weiner S, Addadi L (2000) Formation of high-magnesian calcites via an amorphous precursor phase: possible biological implications. Adv Mater 12(1):38-42
Reiss Z (1957) The Bilamellidea, nov. superfam. and remarks on Cretaceaous Globorotaliids. Contrib Cushman Found Foram Res VIII:127-145
Reiss Z, Hottinger L (1984) The Gulf of Aqaba. Ecological Micropaleontology. Billings WD, Golley F, Lange OLF, Olson JS, Remmert H (eds) Ecological Studies 50:48-66
Riebesell U, Zondervan I, Rost B, Tortell PD, Zeebe RE, Morel FMM (2000) Reduced calcification of marine plankton in response to increased atmospheric CO_2. Nature 407:364-367
Rink S, Kuhl M, Bijma J, Spero HJ (1998) Microprocessor studies of photosynthesis and respiration in the symbiotic foraminifer *Orbulina universa*. Mar Biol 131:583-595
Rosenthal Y, Boyle EA (1993) Factors controlling the fluoride content of planktonic foraminifera: an evaluation of its paleoceanographic applicability. Geochim Cosmochim Acta 57:335-346
Rosenthal Y, Boyle EA, Slowey N (1997) Temperature control on the incorporation of magnesium, strontium, fluorine, and cadmium into benthic foraminiferal shells from little Bahama bank: prospects for thermocline paleoceanography. Geochim Cosmochim Acta 61(17):3633-3643
Rosenthal Y, Lohmann GP (2002) Accurate estimation of sea surface temperatures using dissolution-corrected calibrations for Mg/Ca paleothermometry. Paleoceanography 17(3):article no-1044
Russell AD, Emerson S, Nelson KB, Erez J, Lea D (1994) Uranium in foraminiferal calcite as a record of seawater uranium concentrations. Geochim Cosmochim Acta 58(2):671-681
Sanyal A, Hemming NG, Hanson GN, Broecker WS (1995) Evidence for a higher pH in the glacial ocean from boron isotopes in foraminifera. Nature 373(6511):234-236
Sanyal S, Hemming NG, Broecker WS, Lea DW, Spero HJ, Hanson GN (1996) Oceanic pH control on the boron isotopic composition of foraminifera: evidence from culture experiments. Paleoceanography 11(5):513-517

Sanyal A, Hemming NG, Broecker WS, Hanson GN (1997) Changes in pH in the eastern equatorial pacific across stage 5-6 boundary based on boron isotopes in foraminifera. Global Biogeochem Cycles 11(1):125-133

Schiebel R (2002) Planktonic foraminiferal sedimentation and the marine calcite budget. Global Biogeochem Cycles 16(4):13-1–13-21

Shackleton NJ, Pisias NG (1985) Atmospheric carbon dioxide, orbital forcing, and climate. In: The carbon cycle and atmospheric CO_2: natural variations Archean to Present. Sundquist ET, Broecker WS (eds) Geophys. Monogr, Am Geophys Union, Washington, DC, p 303-317

Shackleton NJ, Wiseman JD, Buckley JD (1973) Non-equilibrium isotopic fractionation between sea-water and planktonic foraminiferal tests. Nature 242:177-179

Siegenthaler U, Sarmiento JL (1993) Atmospheric carbon dioxide and the ocean. Nature 365(9):119-125

Sigman DM, Boyle EA (2000) Glacial/interglacial variations in atmospheric carbon dioxide. Nature 407(6806):859-869

Sikes CS, Wilbur KM (1982) Functions of coccolith formation. Limnol Oceanog 27(1):18-26

Simkiss K, Wilbur KM (1989) Biomineralization: Cell Biology and Mineral Deposition. Academic Press

Smith SV (1978) Coral-reef area and the contributions of reefs to processes and resources of the world's oceans. Nature 273:225-226

Smith DF, Wiebe WJ (1977) Rates of carbon fixation, organiccarbon release and translocation in a reef-building foraminifer *Marginopora vertebralis*. Australian J Mar Freshwater Res 28:311-319

Spero HJ (1988) Ultrastructural examination of chamber morphogenesis and biomineralization in the planktonic foraminifer *Orbulina universa*. Mar Biol 99:9-20

Spero HJ, Bijma J, Lea DW, Bemis BE (1997) Effect of seawater carbonate concentration of foraminiferal carbon and oxygen isotopes. Nature 390(6659):497-500

Spivack AJ, You C-F, Smith HJ (1993) Foraminiferal boron isotope ratios as a proxy for surface ocean pH over the past 21 Myr. Nature 363:149-151

Stanley SM, Hardie LA, Blaustein MK (1999) Hypercalcification: paleontology links plate tectonics and geochemistry to sedimentology. GSA Today 9(2):1-7

Szafranek D, Erez J (1993) Chemistry of Mg, SO_4^{2-}, Sr, Na and Cl in live foraminifera shells (abstract). 7th Int Symp On Biomineralization, Monaco. Biomineralization 93:36

Tachikawa K, Elderfield H (2002) Microhabitat effects on Cd/Ca and $\delta^{13}C$ of benthic foraminifera. Earth Planet Sci Let 202(3-4):607-624

Taylor MG, Simkiss K (1989) Structural and analytical studies on metal ion-containing granules. In: Biomineralization, Chemical and Biochemical Perspectives. Mann S, Webb J, Williams RJP (eds) Springer Verlag, New York, p 427-460

Towe KM, Cifelli R (1967) Wall ultrastructure in the calcareous foraminifera: crystallographic aspects and a model for calcification. J Paleontol 41(3)

de Vargas C, Bonzon M, Rees NW, Pawlowski J, Zaninetti L (2002) A molecular approach to biodiversity and biogeography in the planktonic foraminifer *Globigerinella siphonifera* (d'Orbigny). Mar Micropaleontol 45(2):101-116

Vincent E, Berger WH (1981) Planktonic foraminifera and their use in paleoceanography. In: The Sea. Vol 7. The Oceanic Lithosphere. Emiliani C (ed) John Wiley, New York, p 1025-1119

Wefer G, Berger WH, Bijma J, Fischer G (1999) Clues to ocean history: a brief overview of proxies. In: Use of Proxies in Paleoceanography: Examples from the South Atlantic. Fisher G, Wefer G (eds) Springer-Verlag, Berlin and Heidelberg, p 1-68

Weiner S, Dove PM (2003) An overview of biomineralization processes and the problem of the vital effect. Rev Mineral Geochem 54:1-29

Weiner S, Erez J (1984) Organic matrix of the shell of the foraminifer, *Heterostegina depressa*. J Foramin Res 14:206-212

Young J, Henriksen K (2003) Mineralization within vesicles: the calcite of coccoliths. Rev Mineral Geochem 54:189-215

Westbroek P, Young JR, Linschooten K (1989) Coccolith production (biomineralisation) in the marine alga *Emiliania huxleyi*. J Protozool 36:368-373

Zeebe R (1999) An explanation of the effect of seawater carbonate ion concentration on foraminiferal oxygen isotopes. Geochim Cosmochim Acta 63(13/14):2001-2007

Zeebe RE, Bijma J, Wolf-Gladrow DA (1999) A diffusion-reaction model of carbon isotope fractionation in foraminifera. Mar Chem 64:199-227

Zeebe RE, Sanyal A (2002) Comparison of two potential strategies of planktonic foraminifera for house building: Mg^{2+} or H^+ removal? Geochim Cosmochim Acta 66(7):159-1169

6

Geochemical Perspectives on Coral Mineralization

Anne L. Cohen
Geology and Geophysics, MS#23
Woods Hole Oceanographic Institution
Woods Hole, Massachusetts 02543 U.S.A.

Ted A. McConnaughey
1304 Cedar Lane
Selah, Washington 98942 U.S.A.

INTRODUCTION

Corals open an exceptional window into many phenomena of geological, geochemical, climatic, and paleontological interest. From the Paleozoic to the present, corals provide some of the finest high-resolution archives of marine conditions. Corals are likewise exceptional for chronometric purposes, and even the terrestrial ^{14}C timescale has now been calibrated against coral $^{230}Th/^{234}U$. Corals also represent a testing ground for basic ideas about mineralogy and geochemistry. The shapes, sizes, and organization of skeletal crystals have been attributed to factors as diverse as mineral supersaturation levels and organic mediation of crystal growth. The coupling between calcification and photosynthesis in symbiotic corals is likewise attributed to everything from photosynthetic alkalinization of the water, to efforts by the coral to prevent photosynthetic alkalinization. Corals also leave a significant geochemical imprint on the oceans. Their aragonite skeletons accept about 10 times more strontium than does calcite, hence the proportion of marine aragonite precipitation affects the oceanic chemical balance. Biological carbonates represent the biosphere's largest carbon reservoir, hence calcareous organisms affect the ocean's pH, CO_2 content, and ultimately global temperatures through the greenhouse gas connection. Finally, corals present some geochemical puzzles for ecology and conservation. How do symbiotic corals obtain nutrients in some of the most nutrient deficient parts of the planet? Are global geochemical changes partially responsible for the widespread declines in coral reefs during recent decades? We will address many of these issues, but will concentrate on coral skeletal structure and calcification mechanism. These topics bear most directly on the biomineralization process and generally affect the choice of skeletal materials and analytical techniques used in geochemical investigations.

The coral reef is probably the planet's most spectacular biomineralization product. These grand and complex ecosystems build on the accumulated skeletal debris of countless generations of organisms, especially calcareous algae and symbiotic foraminifera and corals. The algae produce much of the reef mass and help to cement it together, while the corals build much of the erosion-resistant framework. Coral reefs dominate much of the world's tropical coastline and cover abut 15% of the seabed shallower than 30 m (Smith 1978).

Charles Darwin (1842) originally showed that reef corals grow almost exclusively in shallow waters. He nevertheless hypothesized that reef sediments might sometimes extend to great depths, having formed over millions of years near the sea surface, on the flanks of subsiding volcanoes. Deep drilling in the Marshall Islands ultimately confirmed Darwin's theory. Darwin didn't know why reef corals grew fastest in shallow water, but a century later Kawaguti and Sakumoto (1949) showed that they calcify fastest in the light, and Muscatine (1967) found that reef corals obtain much of their nutrition from symbiotic algae, called zooxanthellae. While these results tied up some of the loose ends left by Darwin, they revealed whole new problems worthy of geochemical investigation—like how and why reef corals couple calcification to photosynthesis, and how they flourish in some of the planet's most nutrient deficient regions.

Corals are clonal animals, sometimes consisting of thousands of small, genetically identical anemone-like polyps. The polyps remain connected and promote the common good, but can generally survive and continue to grow and reproduce on their own. Corals often reproduce largely by asexual colony expansion and fragmentation, and like other clonal organisms, sometimes enjoy phenomenal longevity. Coral colonies can live for several centuries, during which their continuous calcification creates layered skeletal archives that record past marine conditions.

Coral skeletons contain a wealth of environmental information. Ancient temperatures can often be inferred to a precision of better than 1°C from skeletal strontium, magnesium, uranium, and oxygen-18 levels, as well as skeletal densities and growth rates. River discharges, oceanic upwelling, and other hydrographic conditions leave their marks on various trace metals, ^{14}C, humic acids, and clay particles incorporated into the skeleton. Skeletal ^{13}C sometimes reflects cloudiness. Human activities affect such parameters as lead content. These environmental indicators become especially useful in the context of a layered skeleton with excellent chronological control. Radiometric dating tools have undergone explosive development in corals, and good chronometers are now available on most timescales. Many corals also contain recognizable annual bands, like tree rings, and many produce recognizable daily bands too. By counting the number of daily bands within annual bands of ancient corals, Wells (1963) was able to monitor the gradual slowing of the earth's rotation due to tidal friction.

Skeletal chemistry obviously depends on how corals extract materials from seawater and place them in their skeletons. Here we encounter an apparent contradiction. For the most part, the coral skeleton resembles the assemblage that might be expected from inorganic aragonite precipitation from seawater. Most trace elements and even small particles occur in the skeleton in proportions reflecting their abundance in seawater, and their tendencies to become incorporated either within or among the aragonite crystals. On the other hand, the isotopes of oxygen and carbon, which are widely used as environmental indicators, are not incorporated at the expected ratios. This dichotomy apparently reflects a fascinating difference in the ways that materials reach the calcification site—calcium, strontium and most trace components largely by fluid transport, but carbon and oxygen largely as carbon dioxide, which diffuses through the coral cells and reacts in the calcifying space, with interesting consequences for skeletal carbon and oxygen isotopic composition.

Corals also challenge some widespread notions about biomineralization and the importance of organic components in organizing and promoting mineral growth. Corals do incorporate organic materials into their skeletons, at relatively low levels—on the order of one percent. But corals don't shape or organize their skeletal crystals with anything approaching the care seen in molluscan nacre. Yet corals calcify a hundred times faster than inorganic calcification rates on the reef, and faster than most other animals, and thereby display a strong command over the biomineralization process. Biomineralization physiology has also tended to emphasize the importance of calcium transport to the skeleton, yet recent results show that calcium concentrations at the calcification site are only slightly above ambient. So how do corals create the high calcium carbonate supersaturations apparently needed for rapid crystal growth? This too has a logical explanation.

This paper reviews the current status of knowledge of the coral mineralization process, with emphasis on insights gained through geochemical measurements, particularly stable isotopes and trace elements. We start with a description of coral skeletal structure, then consider how corals precipitate an aragonite skeleton with its various impurities. We end with the question of why reef corals calcify so fast, and how calcification may relate to recent declines in coral health over much of the world (Pandolfi et al. 2003).

THE SCLERACTINIAN SKELETON:
MORPHOLOGY, MINERALOGY, GROWTH AND CHEMISTRY

Reef corals belong to the order Scleractinia, the "true" corals, all of which accrete hard exoskeletons (Fig. 1) and are distinguished from the soft corals (Octocorallia and Antipatheria) which permeate their tissue with supportive $CaCO_3$ spicules. Scleractinian corals are broadly divisible into two groups, the reef builders (the focus of this paper) and the non-reef builders. The majority of the reef-building scleractinia are colonial and hermatypic, meaning that they host symbiotic algae, or zooxanthellae, in the polyp tissue endoderm. Indeed, it is the intricate and interdependent relationship between the polyp and these single-celled dinoflagellates that enables the coral skeleton (and thus, the reef) to grow faster than it is eroded by wave action and boring organisms. However, this dependence on photosynthetic algae also means that the hermatypic corals are restricted in their geographic distribution to the shallow sunlit oceans of the tropics and subtropics, and are rarely found in waters where temperatures dip below 18°C for extended periods of time. These same restrictions do not apply to the second group of Scleractinia, the

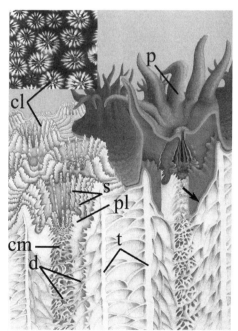

Figure 1. Surface view (inset) and T/S through coral skeleton showing detail of mesoscale skeletal architecture and location of polyp (p) within the corallite (cl). The corallite is a tube enclosed by a wall, the theca (t) which is intercepted by flattened plates, the septa (s) radiating out from the tube center. The paliform lobes (pl) are outgrowths of the septa. Extensions of the paliforms lobes meet in the center to form the columella (cm). The dissepiments (d) are thin horizontal sheets accreted at regular (monthly) intervals. The arrow at right points to the calcifying region, at the interface between tissue and skeleton. Figure is adapted from Veron (1986).

ahermatypes which do not have zooxanthellae and grow throughout the world's oceans to great depths. That ahermatypic corals also calcify, assembling complex aragonitic skeletons, indicates that photosynthesis is not a prerequisite for calcification. The basic building blocks of the coral skeleton are structurally similar in hermatypic and ahermatypic corals (Wainwright 1964) and corals with symbionts build skeleton both day and night (dark calcification) (Barnes 1985; Chalker et al. 1985) as do corals without symbionts (Jacques et al. 1980, 1983)

The polyp

The animal responsible for skeletal formation is the polyp, a double-walled sack of simple design. The innermost cell layer, the endoderm contains the zooxanthellae. A jelly-like cell-less connective layer, the mesoglea, separates the endoderm from the outer cell layer, the ectoderm. The ectoderm consists of two histologically distinct regions. That lying adjacent to the skeletal surface, the calicoblastic layer, is differentiated from the remainder of the ectoderm during larval fixation, preceding the first skeletal accretion, and is considered to be involved in some way in the calcification process. The degree of physical closeness of the calicoblastic ectoderm with the underlying skeleton is considered an indication of its level of involvement in skeletogenesis. A scenario in which calcification is most rapid where the tissue lies flush with the skeletal surface may indicate that crystal nucleation and growth is initiated, controlled and inhibited by tissue (Johnston 1980). On the other hand, a model which predicts rapid calcification within pockets created where the calicoblastic ectoderm is lifted away from the skeletal surface implies that the coral tissue plays a less direct role in crystal growth (Barnes 1970, 1972).

While the nature of the relationship between the calicoblastic ectodermal cells and the skeleton is at this point largely unknown, there is general agreement—amidst a confusing nomenclature—about the ultrastructural characteristics of the scleractinian skeleton. These are the basic morphological criteria upon which the scleractinian taxonomy is based. A coral colony is essentially a collection of the individual skeletons, or corallites, of its resident polyps. The corallite can be thought of as a tube, the theca, intercepted by radiating vertical partitions called septa and their attendant structures (Fig. 1). The base of the polyp sits on a thin horizontal sheet, the dissepiment. The surface of the corallite extends as the polyp pulls itself up the walls of its corallite tube, leaving the old dissepiment behind. Formation of a new dissepiment several millimeters higher up the tube essentially seals the living tissue from the now unoccupied skeleton below. Hence, only the top few millimeters of the skeleton of any massive coral colony is occupied by living tissue and only in the tissue layer does biomineralization take place. The skeleton beneath the tissue layer, which in massive colonies represents by far the greatest proportion of the skeletal mass, continues to be bathed in seawater and aragonite crystals continue to grow, albeit very slowly, within the porous spaces once occupied by gastrovascular canals and tissue. These abiotic crystals are distinguished from crystals grown in the presence of tissue by their morphology, growth rate and chemical composition (Enmar et al. 2000). This distinction indicates that biological and/or physiological factors are either directly (e.g., organic-matrix mediated) or indirectly (e.g., through modulation of calcifying fluid chemistry) involved in the calcification of coral skeletons.

The sclerodermites

The basic building blocks of all parts of all coral skeletons are the sclerodermites, consisting of fine aragonite crystals or fibers arranged in three dimensional fans about a calcification center (Fig. 2). The aragonite fibers, ~0.05–4 μm in diameter, are preferentially elongated in the c-axis direction. They grow as spherulites, grouped into

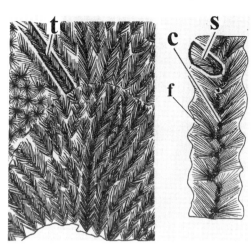

Figure 2. Detail of the scleractinian septal structure (s in Fig. 1). At right, aragonite fiber bundles (f) emerge from centers of calcifciation (c) and grouped into sclerodermites (s). Groups of sclerodermites growing upward together form the trabecula (t). This septum of *Galaxea* sp. is a palisade of trabeculae, shown at left. From Wells (1956).

fishscale-shaped bundles termed fascicles (Ogilvie 1896) or fasciculi (Constantz 1989). The diameters and morphologies of individual aragonite fibers are taxonomically distinct. A number of sclerodermites growing upwards together in the plane of the upfolded tissue develop into a vertical spine called a trabecula. Groups of trabeculae, united with or without intervening spaces (or pores) form the septa, the primary structures of the coral skeleton. Each trabecula depicted in Figure 2 terminates in a dentation at the growing tip of the septum. The dentation is made up of a delicate array of fine spikes splayed like fingers on a hand (Fig. 3). At the center of each spike is a vertical line of calcification centers (Fig. 4). Aragonite fibers in fasciculi grow out at low angles from each calcification center until they meet crystals growing out from the calcification centers of adjacent fingers, at which point mutual interference prevents further growth. Addition of material at the tip of each dentation lengthens the fingers and thus the trabecula. This is how the skeleton extends. Growth of aragonite fibers fills in spaces between the fingers which consolidate basally to form the hand and eventually the spine shown in Figure 2. The spine will continue to thicken for as long as the aragonite fibers are in contact with tissue (Barnes and Lough 1983). In porous skeletons such as *Porites*, fingers from adjacent dentations link laterally at regular intervals to form horizontal supporting rungs called synapticulae (Fig. 3).

Centers of calcification

Examination of thin-sections of coral skeleton in transmitted light indicates that the fasciculi emerge from dark "blobs" first recognized as calcification centers by Ogilvie (1896) (Fig. 4). Vertical lines of calcification centers in trabeculae form either discrete spots (e.g., *Porites* spp.), continuous lines (as in *Lophelia* spp.) or some combination thereof that defines the arrangement of the trabecular axes and thus, the septal structures upon which the classification of the Scleractinian suborders are broadly based (Wells 1956). Apart from their taxonomic usefulness, calcification centers are fundamental to any model for coral mineralization because they are traditionally considered to be nucleation sites for growth of the aragonite fibers (Bryan and Hill 1941; Gladfelter 1983; Constantz 1986, 1989; Constantz and Meike 1989; Le Tissier 1988; Cohen et al. 2001). Within the calcification centers are submicron-sized granular crystals (Constantz 1986, 1989; Cohen et al. 2001) (Fig. 5) bundled into discrete "nuclear packets" each 2–4 μm across. Examination of coral skeleton both in transmitted light and with SEM reveals the intimate relationship between centers of calcification and the fasciculi, the simplest

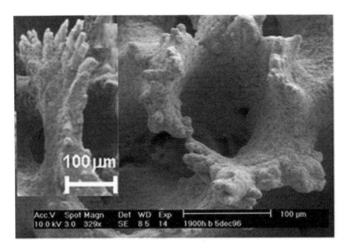

Figure 3. The tip of a growing septum is covered in dentations, each consisting of an array of fine spikes emerging like fingers from a hand (inset). Spikes from adjacent hands periodically link laterally to form synapticulae. Images courtesy of Dr. David J. Barnes (Australian Institute for Marine Science, pers. comm.).

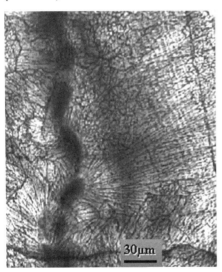

Figure 4. Petrographic thin-section of *Porites lutea* skeleton showing the vertical line of discrete centers of calcification and emergent bundles of aragonite fibers. In this specimen, the calcification centers are daily accretions. Note the fine growth bands within and perpendicular to the growth direction of the fiber bundles. During the day, osmotic pressure builds up between the skeleton and the calicoblastic ectoderm creating a space into which the crystals extend. The width of each band (2–3 µm) probably represents this daily daytime extension and is indicative of the size of the calcifying space. At night, the tissue lies flat against the skeletal surface inhibiting further growth hence the appearance of growth bands. Photo by Anne Cohen.

interpretation being that each granular "seed" crystal produces a spherulite of aragonite fibers and each center of calcification produces numerous fasciculi.

Recently, Cuif and Dauphin (1998) challenged the traditional interpretation of the sclerodermite. They propose that the calcification centers and the fasciculi are separate skeletal entities and the aragonite fibers nucleate on organic matrix sheets. Although it is difficult to reconcile the spherulitic growth morphology of the aragonite fibers with a polycyclic model of crystal growth, the test of this hypothesis lies in a better understanding of the nature and function (if any) of organic material within the skeleton. An alternative role for the organic matrix sheets is proposed later in this chapter.

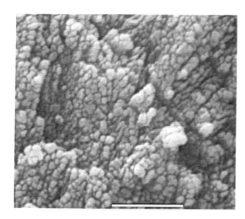

Figure 5. SEM of centers of calcification (regions of diverging fasciculate bundles) in *Porites lutea* reveal sub-micron size granular shaped crystals. Scale bar = 1 μm. Image by Anne Cohen.

The formation of submicron-sized granular shaped crystals and their aggregation into nuclear packets within centers of calcification is a subject that begs further investigation. The small grain size of the crystals may indicate intracellular mineralization, a process adopted by many unicellular mineralizing organisms (Constantz 1989; Lowenstam and Weiner 1989; Mann 2001). Densely packed vesicles in the apical membrane of the calicoblastic epithelium were first reported by Johnston (1980) and later confirmed by several workers (Isa and Yamazato 1981; Isa 1986, 1989; Le Tissier 1988, 1991; Clode and Marshall 2002). Johnston described movement of the vesicles across the apical membrane into the calcifying space, and proposed this as the mechanism whereby calcium ions and organic precursors were transported to the calcification site. This interpretation is not supported by the recent work of Clode and Marshall (2002) who observed large and small, oval-shaped, membrane-bound vesicles within the calicoblastic ectodermal cells but no evidence of vesicle transport across the apical membrane. They attribute Johnston's observations of vesicles entering the subectodermal space to artifacts of the preparation process rather than true structural features.

Nevertheless, it is probable that the *contents* of intracellular vesicles are transported across the apical membrane and exocytosed into the calcifying space. Indeed, this is a likely route for seawater entry. No mineralized structures have been detected within the vesicles. Therefore, it is unlikely that calcification or mineralization of the granular seed crystals occurs intracellularly, as suggested by Hayes and Goreau (1977). However, high Ca^{2+} concentrations and the presence of organic material within the vesicles is intriguing (Marshall and Wright 1993; Clode and Marshall 2002). Amorphous $CaCO_3$ (ACC) is one of several types of unstable, non-crystalline precursors prevalent in the early stages of many biomineralization systems, including plants and sea urchins (e.g., Beniash et al. 1997; see discussion of ACC in Weiner and Dove 2003). The amorphous $CaCO_3$ is enclosed in an impermeable organic sheath for stabilization during biomineralization. In the case of larval sea urchins, the organic sheath is a protein (Raz et al. 2000). It is feasible therefore, that the intracellular vesicles in the apical membrane of the calicoblastic ectoderm, with their organic contents, are sites of production and stabilization of amorphous $CaCO_3$, precursors of the granular seed crystals that occupy centers of calcification. The geometry of the nuclear packets indicates that they are incorporated into the skeleton in a non-rigid state (Constantz 1989). Cohen et al (2001) showed that each discrete 30-μm long calcification center in a skeletal spine of *Porites lutea* is accreted within a day. Indeed, daily growth bands in the vertical line of successive centers are visible in thin-sections of skeleton viewed in transmitted light.

Their evidence indicates that each group of nuclear packets making up the calcification center is added to the growing tip of the skeleton within 24 hours. However, whether entire packets of crystalline aggregates are precipitated at once or whether individual amorphous granules are exocytosed into the subskeletal space, transported within their protein sheaths to the site of calcification and added one by one to a growing nuclear packet remains an open question. It is tempting to consider that the high concentrations of organic material detected in centers of calcification (Cuif et al. 2003) may be remnants of the organic wrappings of thousands of tiny granular crystals (see discussion of granular crystals in foraminifera in Erez 2003).

In considering the possibility of an amorphous $CaCO_3$ precursor of the granular seed nuclei, identification of calcite in centers of calcification of *Mussa angulosa* by Constantz and Meike (1989) may have relevance because phase transformation would favor the lowest energy state, calcite, over aragonite. While the latter study did not replicate the finding of Houck et al. (1975) who reported up to 46% calcite in two *Porites* skeletons, it is not inconsistent with the report of Vandermeulen and Watabe (1973) of trace amounts of calcite in the larval plate of *Pocillopora damicornis*. However, the problem with having calcite granules at centers of calcification is the unlikelihood of them being seed nuclei for the growth of aragonite fibers. Furthermore, the combined results of recent independent investigations of the mineralogy and chemistry of centers of calcification in a range of Scleractinian species do not support this proposition. Cuif and Dauphin (1998), using Raman spectroscopy, found no evidence of calcite in either the calcification centers or the fibers of any of fifteen Scleractinian species, including *Mussa angulosa* (Fig. 6). They did find that calcification centers are preferentially invaded by endolithic algae which might explain observations of calcite in centers of calcification of *M. angulosa*. Bacterial membrane encrustation might have been the source of calcite in the larval plate, i.e., bacteria located on the substrate before larval fixation and metamorphosis (Jean-Pierre Cuif, personal communication 2003).

Sr/Ca geochemistry

Geochemical measurements also indicate that crystals in calcification centers are aragonitic (Cohen et al. 2001, 2004). Strontium, with an ionic radius 28% larger than that of Ca^{2+} prefers the open crystal structure of the orthorhombic aragonite to the hexagonal structure of calcite. The experimentally determined exchange co-efficient for Sr/Ca (K_d) in aragonite determined by Kinsman and Holland (1969) is >1 while that for calcite is ~0.08 (Lorens 1981; Tesoriero and Pankow 1996; Huang and Fairchild 2001). Therefore, the Sr/Ca ratio of a $CaCO_3$ crystal is a good indication of its mineralogy. Crystals within calcification centers of *Diploria labyrinthiformis* measured selectively by SIMS ion microprobe have Sr/Ca ratios as high as 9.7 mmol/mol Sr/Ca (Cohen et al. 2004) (Fig. 7). Given an average seawater Sr/Ca value of 8.56 mmol/mol (de Villiers et al. 1994), the exchange coefficient for Sr/Ca in crystals within centers of calcification ($K_d = 1.10$) is close to that for Sr/Ca in aragonite precipitated experimentally at 25°C ($K_d = 1.13$) (Kinsman and Holland 1969). These data support independent evidence for the aragonitic mineralogy of the calcification centers.

The Sr/Ca ratio of crystals in the calcification centers also provide information about the calcification process. The similarity amongst Sr/Ca ratios in inorganic aragonite crystals precipitated from seawater (Sr/Ca = 9.7 mmol/mol at 25°C; Kinsman and Holland 1969), aragonite precipitated abiotically within skeletal pore spaces evacuated by coral tissue (Sr/Ca = ~10 mmol/mol; Enmar et al. 2000) and the Sr/Ca ratio of crystals within the calcification centers indicates that crystals at centers of calcification are precipitated from a solution with a Sr/Ca ratio close to that of seawater.

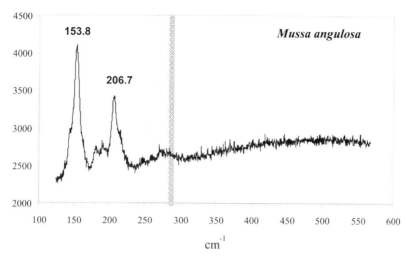

Figure 6. Raman spectrum of the calcification center of *M. angulosa* shows no evidence of calcite. The 282 cm^{-1} band is prominent in calcite. The 206 cm^{-1} band is specific for aragonite. The 154 cm^{-1} band is stronger than the 282 cm^{-1} band in Raman spectra of aragonite standards. Data communicated to us by J-P Cuif (2003), originally published in Cuif and Dauphin (1998).

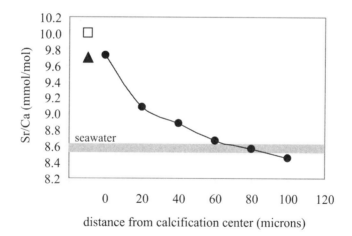

Figure 7. Ion microprobe analyses of Sr/Ca following the growth of a sclerodermite in the brain coral *Diploria labyrinthiformis* (solid circles). Sr/Ca ratio in the calcification center (at x = 0 microns) is close to Sr/Ca ratios of inorganic aragonite precipitated experimentally (triangle) and secondary aragonite infilling in evacuated skeleton (open square). Sr/Ca is depleted in the aragonite fibers and shows a progressive decline along the length of the fasciculus. This pattern is caused by changes in the Sr/Ca ratio of the calcifying solution. (Data from Cohen et al. 2004)

Strontium and calcium ions enter the coral's calcifying space by both passive and active transport (Ip and Krishnaveni 1991; Ferrier-Pages 2002). Passive entry occurs by way of seawater transported via invaginated vacuoles (see also Erez 2003), leaking or diffusing into the calcifying space (Kinsman 1969; Ip and Krishnaveni 1991). Active transcellular transport of both ions occurs enzymatically, via the Ca^{2+}-ATPase "pump."

The enzyme has a higher affinity for Ca^{2+} than for Sr^{2+} (Yu and Inesi 1995), fractionating the Sr/Ca ratio between fluid in the calicoblastic cells and fluid in the calcifying space. Because the pump is activated by exposure of the polyp to light (Al-Horani et al. 2003), active transport is likely the dominant pathway for Sr^{2+} and Ca^{2+} entry during daytime. Thus, skeleton accreted during the daytime is likely to be Sr-depleted. On the contrary, passive transport is likely to dominate at night (or in darkness) when the Ca^{2+}-ATPase pump is slow or inactive. At night, the Sr/Ca ratio of the calcifying fluid should be close to that of seawater and skeleton accreted at night will have a Sr/Ca ratio equivalent to an inorganic aragonite precipitated from seawater.

Figure 7 shows how the diurnal shift between passive transport-dominated and active transport-dominated results in an ontogenetic change in Sr/Ca ratio of the sclerodermite as it grows out from the nucleation site to fill the calcifying space. The Sr/Ca analyses made by SIMS ion microprobe start at the calcification center and follow the growth axis of the fasciculus up to the edge of the skeletal spine. The Sr/Ca content of the crystals is high in the calcification centers but low in the aragonite fibers. Along the length of the fibers, Sr/Ca shows a progressive decline as the crystals elongate away from the calcification center. The average Sr/Ca ratio of this aragonite fiber bundle, accreted during the summertime, is ~8.6 mmol/mol and comparable with values obtained from bulk coral skeletal samples analyzed by thermal ionization mass spectrometry (TIMS) (Alibert and McCulloch 1997). Because the aragonite fibers bundles contribute 99% of the skeletal mass of all corals, Sr/Ca ratios in bulk skeletal samples are depleted relative to inorganic aragonite precipitates (Weber 1973, Smith et al. 1979). Although kinetic effects are considered responsible for this depletion (de Villiers et al. 1995), the observed drop-off in Sr/Ca content of the fasciculus is unlikely to be the result of solution boundary layer-related processes (Rimstidt et al. 1998). The reason is that once Sr^{2+} and Ca^{2+} traverse the calicoblastic ectoderm and enter the micron-sized calcifying space, their rate of diffusion to the crystal growth surface through a solution with the viscosity of seawater must be extremely rapid. Therefore, a depletion of either ion at the solution-mineral interface relative to the bulk calcifying solution is unlikely and is not the cause of Sr depletion in the crystal.

A more probable scenario is that changes in the Sr/Ca ratio of the bulk calcifying solution, modulated by the daytime activity of the transport enzyme Ca^{2+}-ATPase, are responsible for the low (and declining) Sr/Ca content of the aragonite fiber bundles. When the pump is active, the proportion of Ca^{2+} entering the calcifying space is large compared with Sr^{2+}, causing a corresponding decline in the Sr/Ca ratio of both the calcifying fluid (Ferrier-Pages et al. 2003) and of course, the crystals growing into it. According to this model, diurnal, seasonal and interannual changes in the activity of the pump—which is linked to zooxanthellate photosynthesis and is sensitive to various factors including temperature, nutrient availability, cloudiness—will cause corresponding changes in both skeletal calcification rate and the Sr/Ca content of the aragonite crystals. The model predicts that, amongst corals in general, the skeletal Sr/Ca of rapid calcifiers will be lower than that of slow calcifiers, consistent with the observations of Weber (1973), de Villiers et al (1995), Cohen et al. (2002a) and others.

Cohen et al. (2002a) found that symbiotic colonies of *Astrangia poculata* incorporate progressively less Sr as they grow, especially during the summertime (Fig. 8a). In contrast, non-photosynthetic *Astrangia* colonies experiencing the same environmental conditions did not exhibit such a progressive downward drift in skeletal Sr/Ca, and their Sr/Ca ratios exhibited about the same temperature sensitivity as inorganically precipitated aragonite (Fig. 8b). Non-symbiotic *Astrangia* thus resembled the nighttime skeletal crystals from symbiotic colonies of *Porites* in this regard (Cohen et al. 2001).

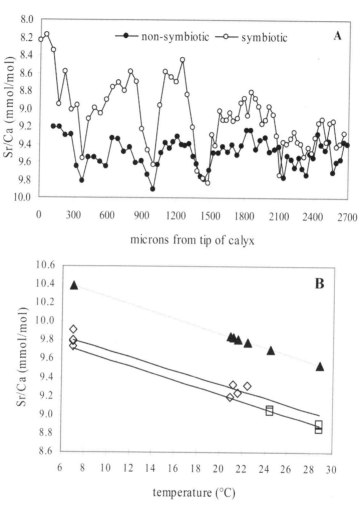

Figure 8. (a) Seasonally resolved Sr/Ca profiles from symbiotic (open circles) and asymbiotic (solid circles) skeleton of *Astrangia poculata* colonies over a four year period. Skeletal Sr/Ca in the first year of life is the same in both samples but the similarities decrease as the corallites mature. The divergence of Sr/Ca ratios is caused by extremely low Sr/Ca ratios in the symbiotic colony during the summer. This indicates that the underlying mechanism is operative in summer. The corals are exposed to identical environmental conditions and temperatures. Therefore this difference reflects the impact of the light-driven Ca^{2+}-ATPase pump, the activity of which is ramped up by the presence of symbionts the symbiotic coral skeleton. (b) The Sr/Ca-SST relationship in asymbiotic *Astrangia* skeleton ($\diamond$ = −0.036x + 10.065) is compared with nighttime skeleton of the tropical reef coral *Porites* ($\square$ = −0.038x + 9.9806) and inorganic aragonite precipitated at equilibrium ($\blacktriangle$ = −0.039x + 10.66). The slope of the regression equations, indicative of the temperature sensitivity of Sr/Ca uptake into the coral skeleton, are similar for all three precipitates (−0.036, −0.038 and −0.039 respectively). This agreement establishes temperature as the primary control of Sr/Ca in the asymbiotic and nighttime crystals (from Cohen et al. 2002).

Photosynthesis, or perhaps the rapid rate of calcification that often accompanies photosynthesis, can therefore have strong effects on skeletal Sr/Ca ratios. The high calcification rates of symbiotic *Astrangia* are reflected in the density of the skeleton which increases progressively as they grow (Cohen et al. 2002b). Assuming the temperature sensitivity of inorganic aragonite applies to all coral skeletons, temperature accounted for as little as 35% of the total Sr/Ca variation observed in symbiotic *Astrangia*. Such observations have obvious implications for the coral Sr/Ca thermometer, which is extensively used to estimate the temperature variability of ancient seas (e.g., Guilderson et al. 1994; McCulloch et al. 1994). The coral Sr/Ca thermometer is usually assumed to have the same temperature sensitivity in ancient corals as in modern analogs. Temperature can often be deduced with impressive precision from the Sr/Ca ratios of individually calibrated modern corals, but different corals nevertheless produce different temperature calibrations, especially when collected at different sites with different temperature regimes (citation). This introduces an uncertainty into ancient temperature reconstructions that is larger than the uncertainties deduced from individual coral calibrations. The *Astrangia* example also sheds doubt on the common assumption that "vital effects" remain constant within individual corals. Basic research into the behavior of geochemical proxies for sea surface temperature therefore seems necessary (Weiner and Dove 2003).

Diurnal cycle of calcification

The similarity between the Sr/Ca content and temperature-dependence of crystals within calcification centers, non-photosynthetic corals and inorganic aragonite precipitates indicates that accretion of calcification centers in symbiotic corals occurs in the absence of algal photosynthesis. Cohen et al (2001, 2002) proposed that these crystals are accreted at night when photosynthesis and the Ca^{2+}-ATPase pump are inactive, strontium enters the calcifying space largely by diffusion and as a result, the Sr/Ca ratio of the nighttime calcifying fluid is close to that of seawater.

There is substantial evidence for a diurnal cycle in the coral calcification and skeleton-building process during which the types of crystals deposited, their distribution about the skeletal surface and the overall rate of calcification changes between day and night (Chalker 1976; Barnes 1970; Barnes and Crossland 1980; Gladfelter 1983a; Constantz 1986; Le Tissier 1988; Marshall and Wright 1998). First, there is a distinct diurnal cycle in coral calcification rate, with rates 3–5 times higher recorded in daylight, correlated with a similar rhythm in the photosynthetic capacity of algal symbionts (Chalker 1976). Second, there are two different processes involved in skeletal accretion that are decoupled over the diurnal cycle: calcification which is most rapid in daylight and skeletal extension or actual upward growth which is most rapid at night (Barnes and Crossland 1980; Vago et al. 1997). The processes responsible for this apparent paradox were revealed by Gladfelter (1982, 1983a) who showed that the types of crystals accreted, their arrangement and their function in the skeleton-building process also follows a diurnal cycle. She used SEM to examine the growing tips of the skeleton of *Acropora cervicornis* collected at four intervals over the diel cycle. She observed that nighttime calcification by *A. cervicornis* resulted in the accretion of randomly-oriented fusiform-shaped crystals, several microns across, which formed an extensive yet flimsy skeletal framework. The framework was filled in by the rapid growth of needlelike crystals during the day. Gladfelters' model of a nighttime framework that causes the skeleton to extend and daytime infilling that causes the skeleton to bulk up and thicken has been substantiated by observations of diurnal growth of *Porites* skeletons (Barnes personal communication; Fig. 9). Gladfelter's interpretation implies that large fusiform-shaped crystals occupy the calcification centers of *A. cervicornis* and are the nucleation

Geochemical Perspectives on Coral Mineralization

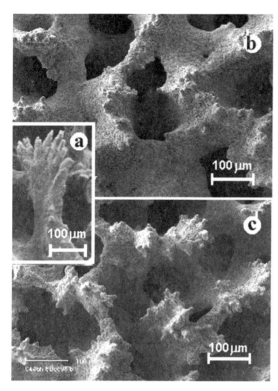

Figure 9. SEM of the tips of growing septa of *Porites lutea* over the diurnal cycle. Extension occurs by growth of fine spikes like fingers on a hand (a). At night, calcification results in elongation of the fingers by accretion of new calcification centers at the growing tips. Thus the septal surface appears spiky (c). During the day, the spaces between adjacent fingers is filled in by outward growth of aragonite fiber bundles. By the end of the day the surface of the septum has a smooth appearance. (b). Images courtesy of Dr David J. Barnes (Australian Institute for Marine Science, pers. comm)

sites for growth of the fasciculi. However, fusiform crystals have not been identified in any other scleractinian species examined thus far. Throughout the Scleractinia, the trabecula axes originally described by Ogilvie (1896) are occupied by submicron-sized granular shaped crystals (Constantz 1986; Cuif and Dauphin 1998; Cohen et al. 2001). Indeed, the distribution, origin and function of the fusiform crystals described by Gladfelter are intriguing and should be investigated further.

Constantz (1989) suggested that Gladfelters fusiform crystals are inorganically precipitated marine cements that nucleate at random on non-zooxanthellate portions of the corallum and are thus "non-biogenic" crystals precipitated at night in the absence of zooxanthellate activity. Indeed, the fusiform crystals in *A. cervicornis* bear a remarkable resemblance to inorganically-precipitated crystals in high magnesium calcite cements (Given and Wilkinson 1985). However, it is difficult to see how random precipitation would produce the organized framework of fusiform crystals that Gladfelter describes. Their architectural arrangement, albeit flimsy, indicates a process over which the coral exerts some degree of control.

Models of skeletogenesis: physicochemical

The morphology and arrangement of the aragonite fibers in the sclerodermites bear a remarkable resemblance to spherulitic crystal morphologies common to all inorganic crystalline systems (Fig. 10). The analogy between spherulites in rocks and the spherulitic morphology of coral crystals first discussed by Bryan and Hill (1941) is the basis of a physicochemical model of coral calcification that has dominated thinking about

Figure 10. (a) The fine, closed, fan spherulite morphology are the characteristic crystal arrangements in coral skeletons and are evident in the dissepiments of *Porites lutea* (Image by Anne Cohen). (b) Increasingly complex crystal forms identified in inorganic minerals represent increasingly large departures from equilibrium crystal growth. Crystal morphology changes systematically from (i) equant to (ii) tabular to (iii) dendritic to open (iv) coarse spherulitic to (v) closed fine spherulitic with increasing crystal growth rates.

coral calcification for several decades. Spherulites were first described in inorganic minerals as a radially disposed array of acicular (needle shaped) crystals in either a spherical or irregular, bow-tie or fan-shaped body (Cross 1891; Iddings 1891). The spherulitic morphology lies at the extreme end of a systematic progression of crystal shapes, from equant shaped to tabular to dendritic and finally spherulitic, that reveals information about their growth histories (Lofgren 1974, 1980; see Given and Wilkenson 1985 for treatment of inorganic marine calcite morphologies) (Fig. 10b).

Equant-shaped crystals in rock and mineral systems are considered to be formed under conditions close to equilibrium and thus characterize low rates of crystal growth. By contrast, spherulitic morphologies result when crystal growth is extremely rapid and growth is faster at the tips of the crystals than at their sides (Keith and Padden 1963). This model of spherulitic growth is based on systems in which crystal growth is driven by rapid and large changes in temperature. The coral system is essentially isothermal and the spherulitic morphology of coral crystals indicates that a rapid increase in the aragonite saturation state of the calcifying fluid must occur. How corals achieve extraordinarily high levels of aragonitic supersaturation is discussed later in this paper. Spherulitic crystal morphologies are found in other biogenic mineralized systems, including egg shells, and generally considered characteristic of precipitates from highly supersaturated solutions that form very rapidly (Lowenstam and Weiner 1989). The assertion by Adkins et al. (2003) that the spherulitic aragonite "bouquets" in the skeleton of a deep sea coral are characteristic of slow growth is therefore somewhat puzzling.

Despite the resemblance between crystals in corals and crystals in inorganic minerals, Bryan and Hill (1941) were hesitant to explain the fasciculate organization of the aragonite fibers as a purely physical system, a reluctance rooted in the observation that coral skeletons are highly intricate and complex structures at both microscopic and macroscopic scales. Thus, Bryan and Hill proposed the existence of an organic "gel" enveloping each fiber and penetrating deeply within the fibers to guide crystal growth. Several years later, a model proposed by Barnes (1970) showed that the morphology of the aragonite fibers and their organization into bundles are explicable entirely in terms of factors controlling abiotic crystal growth. Fundamental to this model is the idea that calcification occurs most rapidly in micron-sized spaces formed where the calicoblastic ectoderm lifts away from the skeletal surface. Barnes proposed that, given this limited space in which to grow, fast-growing crystals precipitated from a supersaturated solution will compete with each other. Crystals that happen to be oriented perpendicular to the calicoblastic ectoderm will extend most rapidly and occlude those growing horizontally or at low angles. The tendency for these crystals to diverge from the optimum axis of growth gives rise to three-dimensional fans (Fig. 10a). In this way, the fine aragonite needles grow as fan systems all over the basal plate, large fans outcompeting small ones for space until stable fan systems develop. These are the sclerodermites that grow upwards together to form the trabeculae.

Further compelling evidence for the predominance of physicochemical factors in the growth of aragonite fibers in coral skeletons is the correlation between fiber morphology and coral growth rate. Constantz (1986) observed very distinct and consistent differences in aragonite fiber morphology amongst the scleractinian taxa. In general, the narrowest fibers (~0.1 µm) are characteristic of the fastest growing genera, the Acroporidae. By contrast, the slow growing genera including the Favids have the widest fibers. Variations within the range of naturally-occurring spherulite morphologies in rocks and minerals can also be related to growth rate as the crystals of the spherulite become progressively finer and more tightly bunched together as growth rate increases (Lofgren 1974). Slower-growing fibers are larger and more widely spaced. Interestingly, the secondary aragonite crystals that grow within pore spaces of skeleton evacuated by coral tissue have the coarse, open morphology characteristic of slower-growing spherulites in rocks (Fig. 10b). Thus, the range of aragonite fiber morphologies found amongst the scleractinian taxa could be explained by basic theories of crystal growth in inorganic systems without the need for mediation by an organic macromolecular framework or matrix. These ideas form the basis for the physicochemical model of coral calcification.

Combining observations of diurnal changes in coral calcification, skeletal extension, crystal morphology and Sr/Ca ratio of the skeleton, we propose a model in which these aspects of coral mineralization can be explained in terms of the light-sensitive action of the Ca^{2+}-ATPase pump. The model is summarized in Figure 11(a-d) and shows how changes in the chemistry and pH of the calcifying fluid between night and day (Fig. 11a,c), and associated changes in the relationship between tissue ectoderm and skeletal surface (Fig. 11b,d) result in the observed cycle of calcification and extension rate, crystal morphology and chemistry. The dual role of the Ca^{2+}-ATPase enzyme in transporting cations into the calcifying space while removing protons, and the consequences for pH and aragonite saturation state is discussed in detail in the section on Calcification Mechanism (below).

Organic matrix models

Despite the remarkable similarities between the spherulitic structures found in rocks and those characteristic of coral skeleton, the role of organic material in coral skeletogenesis—either as an organic matrix framework or a seed for nucleation—remains a topic of debate, central to which is the fact that in almost all instances of biological mineralization, the mineral is associated with organic material (Watabe 1981). Indeed, some consider the presence of an organic matrix to be a prerequisite step for the formation and growth of most biominerals. While few would argue against some level of involvement of organic material in some part of the coral calcification process, it is both the level of control exerted over skeletogenesis and the type of control (promotional versus inhibitory) that is central to the debate. The classic organic matrix not only facilitates nucleation but also controls crystal mineralogy, orientation and growth. In these systems, oriented nucleation is considered to arise from specific molecular mechanisms that lower the activation energy of nucleation along a particular crystallographic direction (Mann 2001). Formation of mollusk shell nacre is a good example. In this case, the nuclei are crystallographically aligned with regard to the underlying organic matrix sheet. As a result, the plate-like aragonite crystals grow with their c-axes perpendicular to the organic surface. However, while the composition (proteins rich in aspartic and glutamic acids, acidic and sulfated polysaccharides—Crenshaw 1990;

Figure 11 (on facing page). Changes in crystal morphology, calcification rate, skeletal extension and Sr/Ca ratio of a *Porites* coral between night (a,b) and day (c,d) are explicable in terms of the light-sensitive action of the Ca^{2+}-ATPase pump. The skeletal surface depicted in (a) and (c) are three "fingers" shown in Figure 3. In (a) the Ca^{2+}-ATPase pump is turned off at night. As a result, pH within the calcifying space (CS) is ~8 (Al-Horani et al. 2003) and the aragonite saturation state is low (<10) (see Figs. 13 and 14). Low calcification and crystal growth rates cause low density aggregates of granular shaped, submicron aragonite crystals (CG) to precipitate on the old skeletal surface (SS). The main pathway for entry of strontium and calcium into the CS at night is in seawater transported via pericellular pathways (PC) and vacuoles (V) that form by invagination (IV) of the basal membrane (BM). The contents of the vacuoles are exocytosed (EX) through the apical membrane (AM) and into the CS (Clode and Marshall 2002). The Sr/Ca ratio of the calcifying fluid = seawater Sr/Ca (~8.6 mmol/mol)(de Villiers et al. 1994) and the Sr/Ca ratio of the precipitating crystals is the same as inorganic aragonite (~9-10 mmol/mol)(Kinsman and Holland 1969, Enmar et al. 2002). In (b) nighttime skeletal growth occurs mainly at the tips of the fingers (Barnes and Crossland 1980, Vago et al. 1997). The calicoblastic ectoderm is tight (TE) against the skeletal surface (SS) except at the tips where the tissue lifts away from the skeleton forming a small pocket (PO). The granular crystals are precipitated in bundles to form a new center of calcification (COC). Growth of aragonite fibers is inhibited at point of contact between tissue and skeleton forming a daily growth band (GB). In (c), The Ca^{2+}-ATPase pump is turned on in daylight. pH within the CS increases to ~9 (Al-Horani et al. 2003) and the aragonite saturation state increases (>100) (see Figs. 13 and 14). High calcification and crystal growth rates causes high densities of spherulites to grow from the granular surfaces of the new COC. Epitaxial crystal growth also continues along the entire skeletal surface lengthening the existing fasciculi until adjacent bundles meet and growth stops. The main pathway for entry of calcium into the CS is via the Ca^{2+}-ATPase pump, which may concentrate Ca^{2+} ions within the vacuoles (Marshall and Wright 1993) and/or transport Ca^{2+} ions directly across the apical membrane (AM). The relative transport of Sr^{2+} ions is low and the Sr/Ca ratio in the CS decreases relative to seawater (~7.9 mmol/mol). Assuming a constant K_d of 1.1, the Sr/Ca ratio of crystals precipitating from this fluid will drop to ~8.6 mmol/mol. In (d) daytime skeletal growth occurs mainly at the sides of the fingers which thicken and eventually consolidate to form a solid spine (Barnes 1970). Increased osmotic pressure in the CS pushed the ectoderm up off the skeletal surface creating a space (PO) into which the aragonite fibers grow. Growth continues until nighttime when the osmotic pressure within the CS decreases causing the tissue to sink back down onto the skeletal surface. MG = mesoglea, CE = calicoblastic epithelium, CS = calcifying space, CG = new crystal growth, SS = old skeletal surface.

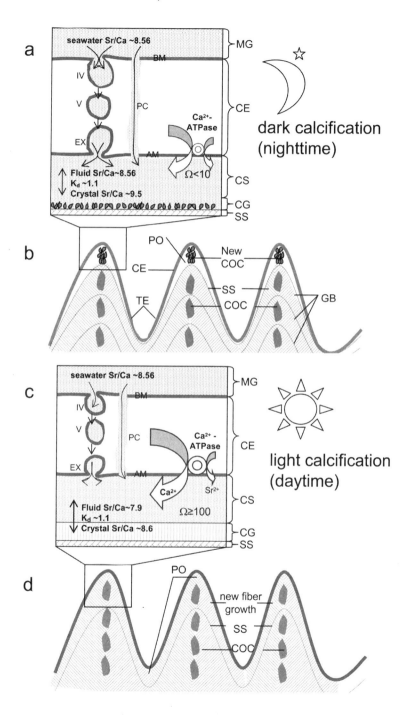

Figure 11. Caption on facing page.

Marxen et al. 1998) and role (Addadi and Weiner 1985; Weiner and Addadi 1991) of organic matrix in the formation of mollusk shells is relatively well characterized, no such well-defined organic structure has yet been isolated from coral skeleton. One argument is that the organic matrix is chemically unstable to preparative procedures for electron microscopy and is lost or displaced after decalcification (Vandermeulen 1975; Muscatine et al. 1997; Goldberg 2001). Another difficulty lies in identifying the origin of organic material extracted from coral skeleton because tissue, endolithic algae and skeleton are closely associated and difficult to separate. Organic material remaining after decalcification may be derived from the organism, dissolved organic substances leaked from cells or the remains of desmoid processes that anchor the tissue to the skeleton (Barnes and Chalker 1990). Sheets of tissue, compressed between indentations of the fasciculi in the undersurface of the calicoblastic epithelium may become trapped between growing crystals and incorporated into the skeleton (Barnes 1970) (Fig. 12a).

Nevertheless, compounds that in other biomineralizing systems constitute the primary components of organic matrices, that is, insoluble framework macromolecules and soluble acidic macromolecules, have been isolated from decalcified coral skeleton

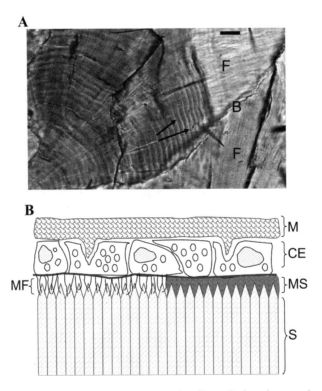

Figure 12. (A) Organic sheets trapped within aragonite fibers, F, in the septotheca of *Diploria labyrinthiformis* demarcate daily growth intervals (arrows) within the fasciculi but occur sporadically throughout sections of skeleton. A boundary line, B, forms where adjacent fasciculi meet and fiber growth stops. Scale Bar = 10 μm. From Cohen et al. (2004). (B) Schematic diagram of the structural relationships of cellular and skeletal components at the calcifying interface of *Galaxea fascicularis* as proposed by Clode and Marshall (2002). The likely role of the matrix sheath, MS, is to inhibit fiber growth. Organic sheets seen in (A) may be remnants of the matrix sheath. M = mesoglea, CE = calicoblastic epithelium, S = skeleton, MF = matrix fibrils. Figure adapted from Clode and Marshall (2002).

and interpreted as evidence of the existence of an organic matrix (Goreau 1959; Wainwright 1963; Young 1971; Mitterer 1978; Johnston 1980; Constantz and Weiner 1988; Allemand et al. 1998; Cuif et al. 1999; Dauphin 2001). The polysaccharide chitin, often a major component of the matrix framework in invertebrate skeletons, was identified in the skeleton of *Pocillopora damicornis* by Wainwright (1963). Glycoproteins, consisting of an acidic protein core with covalently linked polysaccharide side chains that often contain sulfate, have also been extracted from decalcified coral skeleton and thought to be intimately associated with, possibly occluded within, the mineral phase (Constantz and Weiner 1988). Although the precise role of acidic macromolecules in the mineralization process is not clear, the polysaccharide side chain identified as chondroitin sulfate (C-SO$_4$, also the main glycosaminoglycan in articular cartilage) is highly anionic and able to bind large numbers of Ca^{2+} ions (Constantz and Weiner 1988, Cuif and Dauphin 1998, Dauphin 2001).

The first organic matrix model, proposed by Goreau (1959) was of a muccopolysaccharide sheath as a template for crystallization. Johnston (1980) proposed a far more intricate meshwork of glycoproteins which envelopes each growing aragonite crystal. A recent model proposed by Cuif et al. (2003) supports the intimate involvement of sulfated organic compounds with both nucleation and growth of the aragonite fibers. XANES (X-ray absorption near edge structure spectroscopy) fluorescence revealed high concentrations of SO$_4$ within centers of calcification leading Cuif et al. (2003) to suggest that the aragonite fibers do not nucleate on crystals at centers of calcification but rather on sulfate-bearing organic compounds within them. Sulfate in the fibers was detected in lower concentrations than in calcification centers, in a banded pattern that corresponds with 2–4 μm wide growth bands that intercept the fasciculi in lines perpendicular to their growth direction (also discussed in Risk and Pearce 1992, Constantz 1989, Cohen et al. 2004). Based on these observations, Cuif et al. (2003) propose a polycyclic (as opposed to monocrystalline) model of crystal growth, involving step-by-step growth of aragonite fibers, each step initiated and guided by a sulfated organic matrix sheet. This model is not inconsistent with the suggestion by Constantz (1989) of a diurnal cycle in crystal growth within fasciculi of *Acropora cervicornis*, a new tuft of aragonite fibers emerging along the length of the fasciculus each day.

Alternatively, sheets of sulfated organic material at daily growth boundaries within the fasciculi could be inhibitory rather than promotional features, a way for the coral to prevent "runaway" crystal growth. Indeed, rapid crystal growth is relatively easily achieved once high levels of aragonite supersaturation are reached within the calcifying space. Slowing or stopping crystal growth to prevent impaling the overlying tissue might be a more difficult proposition. In this case, the Ca^{2+} binding properties of the organic material would be best served to reduce the activity of the cation and thus slow crystal growth at the edges of the fasciculi.

Although the origin, physical structure and function of the putative organic matrix in coral skeletons remains elusive, what is certain is that part of the ectoderm that lies apposed to the skeleton is an histologically distinct region. It is most likely to be associated with the calcification process, whether it be in a promotional or inhibitory way. An outstanding question is whether calcification occurs most rapidly within sub-epithelial spaces where the calicoblastic layer is lifted away from the skeleton (Wells 1969; Barnes 1970, 1972) or whether the calicoblastic layer remains closely applied to the skeleton and thus intimately involved with skeletal growth (Johnston 1980). If calcification occurs in a fluid-filled space beneath the uplifted tissue it is difficult to see how that tissue could physically initiate and modulate crystal growth. The role of the calicoblastic ectoderm in this instance would be to provide a compartment in which ions required for calcification could be concentrated, to prevent the ions from diffusing outward and to inhibit further crystal growth once the requisite length is obtained. Evidence is accumulating in support of the latter model. Clode and Marshall's (2002) recent work shows that calcification occurs in semi-isolated pockets between the skeleton and the calicoblastic ectoderm, possibly mediated in some way by a meshwork of organic fibrils observed therein. Their revised model of the calcifying interface shows the calicoblastic epithelium is tightly associated with the skeletal surface except where these small pockets form (Fig. 12b). The organic fibrils appear to entwine and penetrate the $CaCO_3$ crystals at the skeletal surface. The fibrils, also identified in fixed material of the deepwater coral *Mycetophyllia reesi* (Goldberg 2001), occur exclusively within these pockets although their function is undetermined.

The existence of a thin organic sheath adjacent to the apical border of the calicoblastic cells is intriguing. Its distribution, and in particular, its absence from sites where the fibrillar material is prevalent, may indicate that the role of the sheath is to control and restrict crystal growth thus preventing crystals from impaling the overlying cells (Clode and Marshall 2002). Marin et al. (1996) likewise reported that coral mucus and skeletal organic extracts inhibited $CaCO_3$ precipitation *in vitro*. Thus, such organic components may play critical roles in the coral calcification, but they may be analogous to the brakes on a car—essential for safe operation, but not responsible for movement.

CALCIFICATION MECHANISM

Calcium ATPase and CO_2 based calcification

Inorganic calcification from seawater provides a good reference point for viewing biological calcification. Aragonite can spontaneously precipitate from seawater when the $[Ca^{2+}][CO_3^{2-}]$ ion product exceeds the aragonite saturation product K_{AR}. This is equivalent to saying that the aragonite saturation state $\Omega = [Ca^{2+}][CO_3^{2-}]/K_{AR}$ is greater than 1, or that the aragonite supersaturation $(\Omega - 1) > 0$. Inorganic precipitation rates are usually related to $\Omega - 1$, for which Burton and Walter (1987) obtained a rate expression equivalent to $R = 12.1(\Omega - 1)^{1.7}$ microns/year at 25°C. An aragonite crystal might then grow 50 microns per year at a typical seawater pH of 8. Corals often precipitate aragonite a hundred times faster, implying that Ω reaches values of several hundred. This could theoretically be achieved by multiplying either the Ca^{2+} or CO_3^{2-} concentration by a similar factor.

Al-Horani et al. (2003) used micro-electrodes to measure Ca^{2+} and H^+ ion activities at various positions around the symbiotic coral *Galaxea* (Fig. 13). Beneath the calcifying epithelium, they observed small elevations of Ca^{2+} concentration which decreased in the dark and when the enzyme Ca^{2+} ATPase was inhibited. These results seem to confirm the long-standing inference that the coral pumps Ca^{2+} into the calcifying space using the enzyme Ca^{2+} ATPase (e.g., Kingsley and Watabe 1985). The elevations of Ca^{2+} concentration above seawater values were however far too small to significantly accelerate calcification.

Simultaneous with the Ca^{2+} increases, Al-Horani also observed pH increases, similar in character but less extreme than the synchronous pH and Ca^{2+} increases observed in calcareous algae by McConnaughey and Falk (1991). This alkalinization probably results from the "ping pong" catalytic cycle of Ca^{2+} ATPase, which expels Ca^{2+} from a cell, and then imports 2 protons (Niggli et al. 1982; Dixon and Haynes 1989) (Fig. 14). This proton removal from the calcification site converts $HCO_3^- \rightarrow CO_3^{2-}$ and thereby increases the aragonite saturation state Ω. By lowering the partial pressure of CO_2 in the calcifying space, proton removal also initiates a net CO_2 diffusion across the boundary

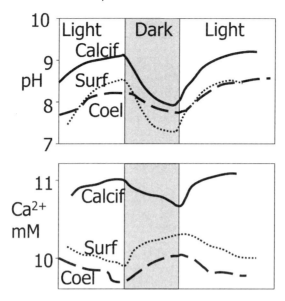

Figure 13. Ca^{2+} and pH in waters around the coral *Galaxea*, by Al-Horani et al. 2003. Composite drawing shows measurements over ~½ hr above the skeleton, in the coelenteron, and near the coral's outer surface.

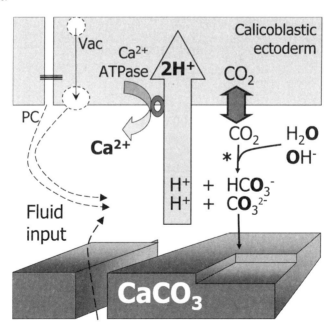

Figure 14. Physiological model for coral calcification. Ca^{2+}-ATPase adds Ca^{2+} and removes protons from calcifying fluid, raising its pH. CO_2 diffuses in and reacts with H_2O and OH^- to produce CO_3^{2-}. Much of this ion transport may actually take place across the membranes of vacuoles (vac) that transfer seawater through the cells of the basal epithelium. Seawater may also reach the calcifying space by diffusion through the porous skeleton and pericellular channels (PC) between the epithelial cells.

membrane, into the calcification space. This CO_2 reacts with H_2O and OH^- to produce more HCO_3^- and CO_3^{2-}. CO_3^{2-} theoretically accumulates until PCO_2 equilibrium across the membrane is re-established, at which time the concentrations of CO_3^{2-} on each side of the membrane are given by $[CO_3^{2-}] = [CO_2]K_1K_2/\{H^+\}^2$, where K_1 and K_2 are the first and second ionization constants of carbonic acid. Because $\{H^+\}$ is far less on the alkaline, calcifying side of the membrane however, CO_3^{2-} becomes concentrated on the calcifying side to the extent of $[CO_3^{2-}]_B/[CO_3^{2-}]_A = 10^{2(pH_B - pH_A)}$, where A and B refer to the acidic and basic sides of the membrane. A proton gradient of 1 pH unit can concentrate CO_3^{2-} 100-fold on a membrane's alkaline side, the calcifying side.

This calcification mechanism can produce very high aragonite supersaturations (Fig. 15a). A base case scenario (1×) assumes that the calcifying solution maintains CO_2

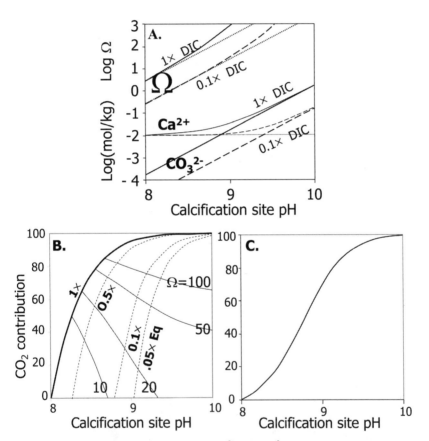

Figure 15. Chemistry of the calcification site. (a) Ca^{2+} and CO_3^{2-} ion concentrations, and aragonite saturation state Ω. (b) Fraction of skeletal carbon derived from CO_2. (c) Fraction of enzymatically added Ca. Calculations assume that the calcification site initially contains seawater at pH 8, with Ca^{2+} = 10.3 mMol/kg, PCO_2 = 360 ppm, total borate = 0.42 mMol/kg, and total alkalinity = 1.92 mEq/kg, where Alk = $OH^- - H^+ + HCO_3^- + 2CO_3^{2-} + B(OH)_4^-$. $Ca^{2+}/2H^+$ exchange adds 1 Ca^{2+} for each 2 units of alkalinity. The 1× scenario assumes that DIC in the calcification site maintains PCO_2 equilibrium with the environment at 360 ppm. The 0.1× (and other) scenarios assume proportionately lower DIC concentrations at the calcification site. Parts (b) and (c) subtract the initial DIC and Ca^{2+} content of seawater from the totals calculated for a particular pH to give the CO_2 contribution and pumped Ca^{2+} contributions to the skeleton. Calculations use CO_2 system equilibrium constants of Prieto and Millero (2002).

system equilibrium with an ambient PCO_2 of 360 ppm, and Ω reaches 300 by pH 9. The 0.1× scenario assumes only a tenth as much DIC, perhaps because coral photosynthesis reduces PCO_2, or because $CO_2 \rightarrow HCO_3^-$ reactions can't keep up with $CaCO_3$ precipitation. After all, the calcifying space is only about a micron thick, and its small fluid volume contains just a small amount of CO_2. Nevertheless, Ω still reaches 30 by pH 9. $CaCO_3$ supersaturation results mainly from CO_2 absorption by the alkaline fluid, not from Ca^{2+} accumulation. Ironically, the Ca^{2+} pump causes calcification not so much by pumping Ca^{2+} to the skeleton, but by pumping protons away. That indirectly concentrates CO_3^{2-} at the calcification site.

The calcifying membrane does not transport HCO_3^- or CO_3^{2-} in this scheme, and HCO_3^- or CO_3^{2-} transporters would actually allow these ions to leak out from the calcification site, where they are most concentrated. HCO_3^- transport has nevertheless been suggested, mainly because inhibitors of membrane anion exchange like DIDS and SITS slow coral calcification (Tambutté et al. 1996; Lucas and Knapp 1997). DIDS and SITS cause collateral damage to many enzymes including Ca^{2+}-ATPase however (Niggli et al. 1982), and that may account for the reduced calcification.

HCO_3^- transport to the skeleton has also been inferred from the importance of the enzyme carbonic anhydrase (CA) to calcification. Goreau's (1963) original idea was that CA might speed the reaction $2HCO_3^- \rightarrow CO_2 + H_2O + CO_3^{2-}$ within the calcifying space. The CO_3^{2-} would then precipitate, and CO_2 would diffuse back into the cells, taking the proton equivalents from calcification with it. Within the context of the CO_2 based calcification model however, CA facilitates CO_2 diffusion from the coelenteron, through the basal epithelium, and to the calcification site. In particular, CA counteracts CO_2 depletion at the membrane bounding the calcification site, ensuring an abundant supply of CO_2 for calcification.

Figure 15b estimates how much of skeletal carbon might derive from molecular CO_2. In the 1× scenario, CO_2 contributes more than half of skeletal carbon at pH>8.3, and about 95% of skeletal carbon at pH 9. Even when DIC levels are only a tenth of PCO_2 equilibrium, CO_2 still provides most of the skeletal carbon when alkalinization exceeds 0.5 pH units. Figure 14c suggests that most of the calcium could reach the skeleton by fluid routes, without enzymatic pumping. Thus Ca^{2+} and C apparently take different routes to the skeleton.

CO_2 BASED CALCIFICATION AND STABLE ISOTOPES

The kinetic model

The CO_2 based calcification model was originally formulated to explain why coral skeletons generally contain several ‰ less ^{18}O than "equilibrium" aragonite precipitated slowly from solution (Fig. 16). The reasoning was as follows (McConnaughey 1989b): The coral skeleton fails to equilibrate oxygen isotopes with water because rapid calcification buries CO_3^{2-} units within the crystal lattice before they can fully exchange oxygen isotopes with H_2O. Isotope exchange occurs by CO_3^{2-} escape from the crystal surface, oxygen isotope exchange between HCO_3^- and H_2O in solution, and CO_3^{2-} re-attachment to the crystal. The crucial ^{18}O exchange between HCO_3^- and H_2O appears to be the slowest step during crystal-solution isotopic equilibration, and can take hours at high pH. When $CaCO_3$ fails to isotopically equilibrate with H_2O, it follows that HCO_3^- also fails to isotopically equilibrate with H_2O. This implies that HCO_3^- doesn't come from seawater, where HCO_3^- has a long enough residence time for isotopic equilibration. So HCO_3^- in the calcification site probably derives instead from CO_2, which diffuses across the basal membrane. This

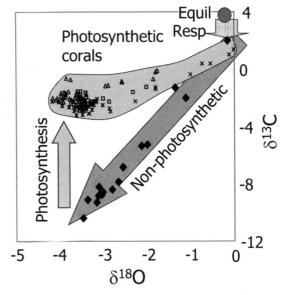

Figure 16. Sub-equilibrium ^{18}O and ^{13}C levels in photosynthetic and non-photosynthetic corals.

HCO_3^- must precipitate fairly rapidly to prevent oxygen isotopic equilibration with water, implying that $CO_2 \rightarrow HCO_3^-$ reactions have somewhat one-way character with respect to the slow process of isotopic equilibration. Kinetic rather than equilibrium isotope fractionations on $CO_2 \rightarrow HCO_3^-$ reactions therefore dominate.

$CO_2 \rightarrow HCO_3^-$ reactions produce DIC with low ^{18}O content partly because the H_2O and especially the OH^- with which the CO_2 reacts contain relatively little ^{18}O (Fig. 17). Kinetic discrimination against ^{18}O will further reduce the ^{18}O content on product HCO_3^-. Without complete isotopic equilibration, the skeleton will inherit some of this ^{18}O-depletion.

"Carbonate" scenarios

Spero et al. (1997) and Bijma et al. (1999) recently suggested that foraminiferal ^{18}O levels depend on the abundance of the isotopically light CO_3^{2-} ion in solution, as if foraminifera essentially precipitate ambient dissolved inorganic carbonates (DIC) (Fig. 18a). This casts doubt on both the "kinetic" isotope model and CO_2 based calcification, although the contradiction is less severe than it might seem. Spero's forams contained slightly less ^{18}O than ambient DIC, as if DIC isotopically equilibrated with water at a pH somewhat above ambient, or CO_2 contributed somewhat to the skeleton, as would be appropriate for mild alkalinization of the calcification site according to Figure 15b. A more serious problem for the "carbonate" explanation is that skeletal $\delta^{13}C$ varied even more than $\delta^{18}O$ (Fig. 18b), while the $\delta^{13}C$ ambient DIC remained constant. Hence some process within the forams caused their $\delta^{13}C$ variations. This same internal process probably controlled their $\delta^{18}O$, judging from the correlations between $\delta^{13}C$ and $\delta^{18}O$. Qualitatively similar but even more extreme isotopic patterns also showed up in corals that had probably never experienced much variation in ambient pH. For such reasons, corals and forams probably acquire their $\delta^{18}O$ ranges from internal processes, not by precipitating isotopically equilibrated ambient DIC.

Adkins et al. (2003) and Rollion-Bard et al. (2003) suggested that the low ^{18}O content of corals might result from ^{18}O equilibration between DIC and H_2O at the alkaline calcification site. Rollion-Bard also calculated pH at the calcification site from

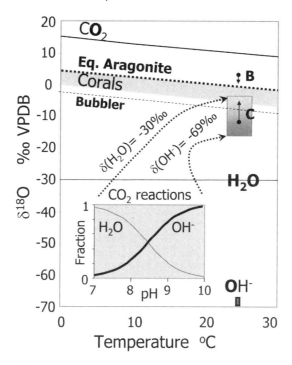

Figure 17. Calculated $\delta^{18}O$ of DIC produced by CO_2 reactions with H_2O and OH^- (at 25°C). Also shown are equilibrium aragonite, corals, "bubbler" simulation at pH < 8.5 by McConnaughey (1989b), and the equilibrium $\delta^{18}O$ of HCO_3^- (B) and CO_3^{2-} (C) at 25°C, based on analyses by McCrea (1950; triangles) and Usdowski and Hoefs (1993; dots). These $\delta^{18}O$ values apply to the named chemical species, not to CO_2 derived from that species for mass spectrometric analysis. Inset: Relative rates of CO_2 reactions with H_2O and OH^- in seawater at 25°C.

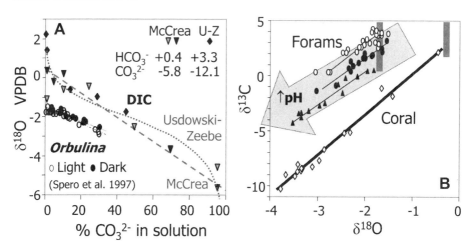

Figure 18. Isotopic composition of forams from pH experiments by Spero et al. 1997. (A) Foram $\delta^{18}O$ compared to DIC in equilibrium with water. (DIC equilibrium as modeled by McCrea 1950 and Usdowski and Hoefs 1993, as modified by Zeebe 1999). (B) $\delta^{18}O$ vs. $\delta^{13}C$ for Spero's forams, compared with a non-photosynthetic coral from Galapagos that presumably experienced only a limited pH range.

skeletal boron isotopic composition and obtained values from about 7.1 to 9.0 (Fig. 19). She considers the lower part of this range unlikely, and notes that an alternative δ^{11}B-pH calibration yields the more realistic pH range of 8.2–9.15. This higher pH range would also lower the projected δ^{18}O of DIC formed at the calcification site, due to increased importance of CO_2 hydroxylation, and extend the equilibration times needed for oxygen isotope exchange between DIC and H_2O.

These "carbonate" scenarios contain several weaknesses. DIC requires hours to equilibrate oxygen isotopes with H_2O at high pH, as shown in Figure 19, while DIC would probably not linger very long in the calcification site. Kinetic discrimination against the heavy isotopes during $CO_2 \rightarrow HCO_3^-$ reactions should also be considered. But most importantly, ^{18}O exchange between DIC and H_2O takes place through reactions that involve a CO_2 intermediate, i.e., $HCO_3^- \rightarrow CO_2 + OH^- \rightarrow HCO_3^-$. This CO_2 intermediate will quickly diffuse across the boundary membrane and escape from the calcification space, and not therefore equilibrate oxygen isotopes with H_2O at the elevated pH of the calcification site.

The exploration of "carbonate" effects has focused attention on the many remaining uncertainties about calcification physiology and isotopic geochemistry, and particularly on the role of pH. The scenarios of Adkins and Rollion-Bard are compatible with a Ca^{2+}-ATPase, CO_2 based calcification mechanism. Spero's forams probably lost ^{18}O as ambient pH increased partly because they kept their calcification sites slightly more alkaline than ambient waters, and as ambient pH increased, so did calcification site pH and the ratio of CO_2 hydroxylation to CO_2 hydration. Isotopic equilibration also slowed down. The skeleton therefore became isotopically lighter.

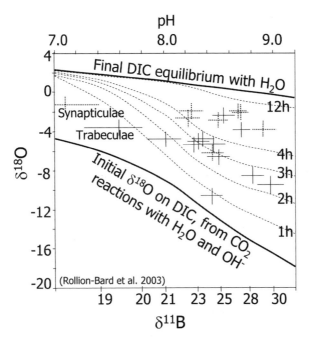

Figure 19. δ^{18}O and δ^{11}B in Porites skeleton. pH scale at top is calculated from δ^{11}B. Coral data is compared with a model for the δ^{18}O of DIC, whose initial composition is set by CO_2 reactions with H_2O and OH^-, but subsequently equilibrates with H_2O over several hours. (Rollion-Bard et al. 2003)

Respired CO_2 in the skeleton

Respired CO_2 is often considered a major source of skeletal carbon, based largely on a double isotope labeling technique pioneered by Goreau (1963). Using this technique, Goreau, Erez (1978) and Furla et al. (2000) suggested that corals build their skeletons mainly out of respired CO_2. Other observations minimize the skeletal contributions of respired CO_2 however. Spero and Lea (1986) fed ^{13}C labeled foods to forams and detected little in the skeletons. Griffin et al. (1989) and Adkins et al. (2003) showed that CO_2 probably contributed 5–10% of skeletal carbon in deep-sea corals (Fig. 20a). McConnaughey et al. (1997) compared the amount of CO_2 produced through respiration with the amount of environmental CO_2 flushed through an animal's body during the course of gas exchange with the environment, to obtain O_2. He concluded that aquatic invertebrates flush 10 times more CO_2 through their bodies than they produce by respiration, and therefore incorporate only about 10% respired CO_2 into their skeletons (Fig. 20b). Air breathers, on the other hand, dilute their respired CO_2 much less, because air contains 30 times less CO_2 relative to O_2.

^{13}C deficiencies, and $\delta^{18}O$ - $\delta^{13}C$ correlations in corals

CO_2 contains about 8‰ less ^{13}C than HCO_3^-, and CO_2 reactions with H_2O and OH^- discriminate against ^{13}C by about 7‰ and 27‰ respectively (Marlier and O'Leary 1984; Siegenthaler and Münnich 1981). DIC produced in the calcification site from CO_2 reactions can therefore be quite depleted in ^{13}C. These factors apparently cause most coral skeletons to contain less ^{13}C than aragonite precipitated in ^{13}C equilibrium with seawater DIC (Fig. 16). CO_2 exchange across the basal epithelium erodes this ^{13}C deficiency, bringing DIC in the calcification space back toward seawater $\delta^{13}C$ values. This CO_2 exchange carries C and O atoms together, causing simultaneous equilibration of both isotopes. Seawater input to the calcification site also adds DIC that is isotopically equilibrated with respect to both C and O isotopes. Non-photosynthetic corals therefore tend to display linear correlations between ^{18}O and ^{13}C, extending upward toward isotopic equilibrium for both isotopes.

Photosynthetic corals often contain about 10‰ more ^{13}C than non-photosynthetic corals, when one compares rapidly growing skeletal parts of the reef coral with materials from the non-photosynthetic coral showing similar degrees of ^{18}O disequilibrium

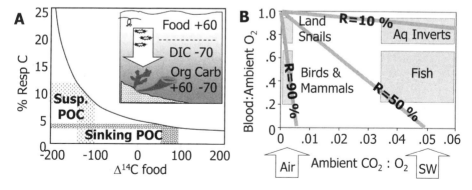

Figure 20. Skeletal incorporation respired CO_2. (a) Deep-sea corals of Griffin et al. (1989) and Adkins et al. (2003) illustrate how skeletal $CaCO_3$ resembles ambient DIC, not coral foods or tissues, in ^{14}C content. (b) McConnaughey et al.'s (1997) model for internal dilution of respired CO_2 by ambient CO_2, absorbed in the course of gas exchange with the environment.

(McConnaughey 1989a) (Fig. 16). This relatively strong photosynthetic ^{13}C enrichment probably occurs because reef corals calcify mainly in the light, so their skeletal isotopic compositions reflect conditions of maximal photosynthesis. Daytime photosynthesis is often several times faster than respiration, and corals also tend to close their mouths during the day, reducing DIC exchange with the environment and inviting strong ^{13}C enrichment of the internal carbon pool.

WHY DO REEF CORALS CALCIFY SO FAST?

Coral calcification clearly doesn't depend on photosynthesis, as reef corals calcify both day and night, as do non-photosynthetic corals. Reef corals calcify from tissues that don't contain symbionts, and the zooxanthellae live in tissues that don't calcify (Fig. 21a). Branching and foliose corals such as *Acropora* and *Agaricia* also calcify fastest at their branch tips, which contain lots of ATP but few symbiotic algae (Fang et al. 1983) (Fig.

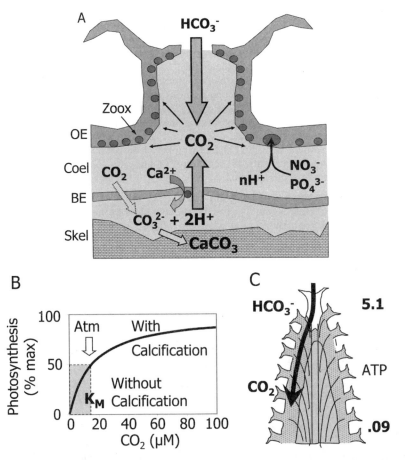

Figure 21. Proton Hypothesis. (a) Calcification mechanism and its connection to photosynthesis and nutrient uptake. (Abbr: <u>Coel</u>enteron, <u>Skel</u>eton, <u>Zoox</u>anthellae, OE, BE: oral & basal epithelia) (b) CO_2 dependence of photosynthesis. (c) Circulation in staghorn coral.

21c). Nevertheless, reef corals usually calcify fastest in the light (Kawaguti and Sakumoto 1948), accelerating calcification by a factor of 3 on average (Gattuso et al. 1999). The reasons for this "light-enhanced calcification" remain controversial despite a half century of intensive research. Does photosynthesis stimulate calcification by raising pH and increasing the $CaCO_3$ saturation state Ω? Do corals up-regulate calcification when the resulting proton flux would benefit their symbionts? We will examine these possibilities.

Photosynthesis does not cause rapid calcification

Photosynthesis increases CO_3^{2-} levels as suggested by the reaction $2HCO_3^- \rightarrow CH_2O + O_2 + CO_3^{2-}$. This increases the aragonite saturation state Ω, and could stimulate calcification. But is this important? Gattuso et al. (1999) estimated the average ratio of calcification to net photosynthesis (G/P_n) in reef corals is about 1.3. G/P_n ratios above 0.8 actually decrease Ω in seawater, so photosynthetic alkalinization of the water cannot produce G/P_n ratios of 1.3. Photosynthesis also drives down CO_2 concentrations, inhibiting itself and its ability to increase Ω. Healthy reefs seldom push CO_2 levels below 2 µM, or pH higher than 8.5 (Smith 1973; Frankignoulle et al. 1996), and Ω seldom exceeds 6. This pales by comparison to the high Ω values calculated earlier for the calcifying compartment (Fig. 15). Non-calcareous plants with high rates of photosynthesis can furthermore grow side by side with calcareous plants. Finally, increasing Ω by adding Ca^{2+} to seawater (beyond the 10 mM normally present) has little effect on coral calcification rates (Gattuso et al. 1998). All these examples point toward the conclusion that photosynthesis generally doesn't cause much calcification by elevating Ω.

Photosynthesis may however stimulate calcification by supplying food (Muscatine 1990) and oxygen (Rinkevich and Loya 1984) to the corals. Corals may also regulate calcification based partly on how efficiently the protons from calcification convert HCO_3^- to CO_2 (Fig. 21a). A model based on this idea produces Gattuso's average value for G/P_n = 1.3 at a typical reef pH value of 8 (McConnaughey et al. 2000). Below pH 8.5, G/P_n appeared to be relatively insensitive to pH and PCO_2, which could influence how reefs respond to glacial-interglacial and modern industrial CO_2 changes. This is a controversial topic, but the global CO_2 increases of recent decades often seem to have had little effect on coral skeletal growth (Lough and Barnes 1997; Bessat and Buiges 2001).

Calcification relieves CO_2 stress

High G/P_n ratios like Gatusso's average of 1.3 prevent most of the depletion of dissolved molecular CO_2 that would otherwise accompany photosynthesis. The photosynthetic kinetics shown in Figure 21b suggest that this could significantly stimulate coral photosynthesis. (The half saturation constant K_M for zooxanthellate photosynthesis appears to lie between about 6 and 60 µM CO_2; Legatt et al. 1999, 2002.) If zooxanthellae can fix more carbon when the coral calcifies rapidly, the coral has good reason to ramp up calcification during the day. The coral discharges the protons from calcification into its internal coelenteron cavity. Symbiotic zooxanthellae facing the coelenteron are well positioned to benefit from this proton flux (Fig. 21b). Al-Horani et al. (2003) observed only minor alkalinization and sometimes mild acidification of the coelenteron, even in illuminated corals (Fig. 13). This supports the idea that the protons from calcification keep coelenteron pH down and CO_2 levels up. Proton discharge into the coelenteron might also explain why branching and foliose corals such as Acropora (Fig. 21c) calcify fastest at their non-photosynthetic apical polyps, where Gladfelter (1983b) observed inward ciliary currents. Such currents could deliver acidified, CO_2 rich water from the branch tips to the highly photosynthetic tissues further down the branch.

McConnaughey and Whelan (1997) and McConnaughey et al. (2000) summarized evidence that calcification might stimulate photosynthesis. In contrast, Gattuso et al.

(2000) reported that coral photosynthesis continued unabated when ambient Ca^{2+} levels were reduced from >11 to <3 mM, and net calcification ceased. This disparity might have various explanations. For example, Gattuso's corals may have continued to pump protons from the skeleton to the coelenteron for the duration of the experiment, accumulating CO_3^{2-} and OH^- in the skeletal compartment. Or the corals may have exchanged coelenteron fluids with the ambient environment fast enough to avoid CO_2 depletion. But given the CO_2 dependence of photosynthesis and the ability of calcification to counteract CO_2 depletion, it is hard to see why calcification would not stimulate photosynthesis.

Calcification may stimulate nutrient uptake

Proton secretion by corals may also stimulate nutrient uptake, although some the best evidence comes unfortunately from land plants, where physiologists have done many of the crucial experiments. Land plants characteristically secrete protons from their roots when stressed for nutrients (Kochian 1991). Proton secretion usually concentrates somewhat above the root tips, where absorptive root hairs are often abundant. Where do the protons come from? Many plants precipitate large amounts of calcium oxalate (CaC_2O_4), and many also develop symbioses with soil fungi that precipitate calcium oxalate, especially under conditions of nutrient shortage. Calcium oxalate precipitation, like calcium carbonate precipitation, generates protons: $Ca^{2+} + H_2C_2O_4 \rightarrow CaC_2O_4 + 2H^+$. Plants growing in alkaline soils sometime precipitate calcite too, often filling the large central vacuoles of root cortical cells (Jaillard et al. 1991). This calcite precipitation occurs near the root hairs, where the root secretes protons and absorbs nutrients. Such calcification (both carbonates and oxalates) probably enables the plant to secrete more protons and obtain more nutrients.

Intensive research on land plants has shown that acid secretion leaches phosphate, iron, and ammonium from soil minerals and clays, and more interestingly, stimulates nutrient uptake. Proton secretion causes most of the membrane electrical potential (Kitsato 1968), which attracts cations like NH_4^+ and Fe^{2+} into the cell. Anions like NO_3^- and PO_4^{3-} are taken up in combination with various numbers of protons (Wollenweber 1997; Sakano 1990; Schachtman et al. 1998). McConnaughey et al. (2000) calculated that lowering pH at the cell surface by 1 pH unit (below what would prevail in the absence of calcification) theoretically improves nitrate uptake at least 10-fold and phosphate uptake even more, depending on the number of protons imported with each nutrient anion.

If corals calcify largely to help the zooxanthellae obtain nutrients, one might expect corals to calcify more (despite the metabolic expense) when the zooxanthellae need more nutrients. Experimental evidence to this effect surfaced a quarter of a century ago (Kinsey and Davies 1979), and subsequent experiments confirm that corals calcify more even though they photosynthesize less when phosphate, nitrate, ammonium, and iron are scarce (Marubini and Davies 1996; Marubini and Thake 1999; Ferrier-Pages et al. 2000, 2001) (Fig. 22).

The hypothesized connection between calcification and nutrient uptake may help to answer some the great conundrums of reef biology and geology: How do reefs flourish in some of the most nutrient deficient regions of the planet? Most corals feed on plankton, providing a source of nutrients beyond the reach of most plants. Corals also place their algae in a fixed location, often of high turbulence, enhancing their access to nutrients advected by in the ocean. But perhaps more importantly, corals generate protons through prolific calcification. The need for protons remains as long as the algae need nutrients. Corals therefore calcify year after year, long after they have built satisfactory skeletons. The same applies to calcareous algae. Over geologic time, this produces immense accumulations of limestone known as coral reefs. Coral reefs may occur in the

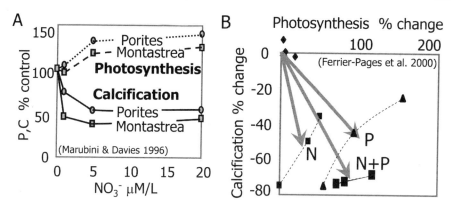

Figure 22. Nutrient scarcity reduces coral photosynthesis but stimulates calcification. (a) Nitrate (Marubini and Davies 1996) (b) Phosphate and ammonium (Ferrier-Pages et al. 2000)

tropics not so much because of high temperatures, but rather because of low nutrient levels and the greater need for calcification.

Industrial nitrogen fixation now rivals the natural process, and humans have enormously increased the rates of nutrient addition to coastal waters. "Nutrification" of the oceans opens up vast possibilities for plants that are less adept at scrounging the last remaining nutrients from oligotrophic waters. As a result, fleshy algae are increasingly invading coral reefs, often crowding out the calcareous algae and corals, and encouraging the proliferation of generalist consumers that sometimes eat the corals. The marked deterioration of coral reefs over the past few decades (e.g., Gardner et al. 2003) partially represents a phase shift from nutrient starvation to abundance.

Coral bleaching (i.e., temporary or permanent loss of zooxanthellae—the latter results in coral death) resembles a nutrient deficiency condition, particularly a condition called chlorosis in land plants. Bleaching tends to occur when the sea surface is particularly warm and highly stratified, and therefore receiving fewer nutrients from below the thermocline. Human activities especially fertilize the oceans with nitrogen, and the consequent proliferation of fleshy algae may further stress the iron or possibly the phosphate supply. The physiological sequence leading to bleaching remains to be positively identified, and could for example involve light-activated production of free radicals or increased metabolic expenses by the coral related to faster calcification. It is also possible that the dinoflagellate symbionts become annoying to the coral under conditions of nutrient stress. Phosphate deficiency induces greater saxitoxin production by some free-living dinoflagellates (Taroncher-Oldenburg et al. 1999), and if zooxanthellae also secrete more toxins in response to nutrient stress (perhaps to stimulate the coral to calcify faster), the coral might eventually tire of its demanding symbionts and toss them out.

ACKNOWLEDGMENTS

Anne Cohen is indebted to Drs. David J Barnes and Elizabeth Gladfelter for sharing ideas over many years about how coral make their skeletons; to Drs. Brent Constantz, Jean-Pierre Cuif and Gary Lofgren for clarifying aspects of their own research for this paper and to Drs. Glenn Gaetani, Nobuchimi Shimizu and Stanley Hart who periodically divert their attention from hard rocks to think about coral geochemistry and mineralogy (with unbounded enthusiasm). Part of this research was supported by NSF grants OCE-0241075 and OCE-0347328 to Anne Cohen. The Ocean Climate Institute and the Marine Policy Center at Woods Hole Oceanographic Institution are gratefully acknowledged for funding a meeting of the authors. This is WHOI contribution #11027

REFERENCES

Addadi L, Weiner S (1985) Interactions between acidic proteins and crystals: Stereochemical requirements in biomineralization. Proc Natl Acad Sci USA 92:4100-4114

Adkins JF, Boyle EA, Curry WB, Lutringer A (2003) Stable isotopes in deep-sea corals and a new mechanism for "vital effects." Geochim Cosmochim Acta 67:1129-1143

Al-Horani FA, Al-Moghrabi SM, de Beer D (2003) The mechanism of calcification and its relation to photosynthesis and respiration in the scleractinian coral *Galaxea fascicularis*. Mar Biol 142:419-426

Alibert C, McCulloch MT (1997) Strontium/calcium ratios in modern Porites corals from the Great Barrier Reef as a proxy for sea surface temperature: calibration of the thermometer and monitoring of ENSO. Paleoceanography 12:345-363

Allemand D, Tambutte E, Girard JP, Jaubert J (1998) Organic matrix synthesis in the scleractinian coral *Stylophora pistillata*: role in biomineralization and potential target of the organotin tributyltin. J Exp Biol 201:2001-2009

Barnes DJ (1970) Coral skeletons: An explanation of their growth and structure. Science 170:1305-1308

Barnes DJ (1972) The structure and formation of growth ridges on scleractinian coral skeletons. Proc R Soc London Ser B 182:331-350

Barnes DJ (1985) The effect of photosynthetic and respiratory inhibitors upon calcification in the staghorn coral *Acropora formosa*. Proc 5th Int Coral Reef Congr 6:161-165

Barnes DJ, Crossland CJ (1980) Diurnal and seasonal variations in the growth of a staghorn coral measured by time-lapse photography. Limnol Oceanography 25:1113-1117

Barnes DJ, Chalker BE (1990). Coral reef calcification and photosynthesis. *In*: Ecosystems of the World: Coral Reefs. Dubinsky Z (ed) Elsevier, Amsterdam Oxford New York Tokyo, p 109-131

Barnes DJ, Lough JM (1993) On the nature and cause of density banding in massive coral skeletons. J Exp Mar Biol Ecol 167:91-108

Beniash E, Aizenberg J, Addadi L, Weiner S (1997) Amorphous calcium carbonate transforms into calcite during sea urchin larval spicule growth. Proc R Soc Lond B 264:461-465

Bessat F, Buigues D (2001) Two centuries of variation in coral growth in a massive Porites colony from Moorea (French Polynesia): a response of ocean-atmosphere variability from south central Pacific. Palaeo Palaeo Palaeo 175:381-392

Bijma J, Spero HJ, Lea DW (1999) Reassessing foraminiferal stable isotope geochemistry: impact of the oceanic carbonate system (experimental results). *In*: Use of Proxies in Paleoceanography: Examples from the South Atlantic. Fisher G, Wefer G (eds) Springer, Berlin, p 489-512

Bryan WB, Hill D (1941) Spherulitic crystallization as a mechanism of skeletal growth in the hexacorals. Proc Roy Soc Queensland 52:78-91

Burton EA, Walter LM (1987) Relative precipitation rates of aragonite and Mg calcite from seawater: temperature or carbonate ion control? Geology 15:111-114

Chalker BE (1976) Calcium transport during skeletogenesis in hermatypic corals. Comp Biochem Physiol 54A:455-459

Chalker BE, Carr K, Gill E (1985) Measurement of primary production and calcification in situ on coral reefs using electrode techniques. Proc 5th Int Coral Reef Congr 6:167-172

Clode PL, Marshall AT (2002) Low temperature FESEM of the calcifying interface of a scleractinian coral. Tissue & Cell 34:187-198

Cohen AL, Layne GD, Hart SR, Lobel PS (2001) Kinetic control of skeletal Sr/Ca in a symbiotic coral: implications for the paleotemperature proxy. Paleoceanography 16(1):20-26

Cohen AL, Owens KE, Layne GD, Shimizu N (2002a) The effect of algal symbiosis on the accuracy of Sr/Ca paleotemperatures from coral. Science 296(5566):331-333

Cohen AL, Owens KE, Layne GD, Shimizu N (2002b) Sr/Ca in symbiotic corals is linked to light enhanced calcification. Geochim Cosmochim Acta 66(15A):A148-A148 Suppl.

Cohen AL, Smith SR, McCartney MS, van Etten J (2004) How brain corals record climate: an integration of skeletal structure, growth and chemistry. MEPS, in revision

Constantz BR (1986) Coral skeleton construction; a physiochemically dominated process. Palaios 1:52-157

Constantz BR (1989) Skeletal organisation in *Acropora*. *In*: Origin, Evolution and Modern Aspects of Biomineralization in Plants and Animals. Crick RE (ed) Plenum Press, New York, p 175-200

Constantz BR, Meike A (1989) Calcite centers of calcification in *Mussa angulosa* (Scleractinia). *In*: Origin, Evolution and Modern Aspects of Biomineralization in Plants and Animals. Crick RE (ed) Plenum Press, New York p 201-208

Constantz BR, Weiner S (1988) Acidic macromolecules associated with the mineral phase of scleractinian coral skeletons. J Exp Zool 248:253-258

Crenshaw MA (1990) Biomineralization mechanisms. *In*: Skeletal Biomineralization: Patterns, Processes and Evolutionary Trends, Vol. 1. Carter JG (ed) Van Nostrand Reinhold, New York, p1-9

Cross W (1891) Constitution and origin of spherulites in acid eruptive rocks. Wash Philosoph Soc Bull 11:411-444

Cuif JP, Dauphin Y (1998) Microstructural and physicochemical characterization of centers of calcification in septa of some recent Scleractinian corals. Pal Zeit 72:257-270

Cuif JP, Dauphin Y, Doucet J, Salome M, Susini J (2003) XANES mapping of organic sulfate in three scleractinian coral skeletons. Geochim Cosmochim Acta 67:75-83

Cuif JP, Dauphin Y, Freiwald A, Gautret P, Zibrowius H (1999) Biochemical markers of zooxanthellae symbiosis in soluble matrices of skeleton of 24 Scleractinian species. Comp Biochem Physiol 123(A):269-278

Darwin C (1842) The structure and distribution of coral reefs. Smith, Elder & Co., London

Dauphin Y (2000) Comparison of the soluble organic matrices of healthy and diseased shells of *Pinctada margartifera* L. and *Pecten maximus* L. (Mollusca, Bivalvia). J Invert Pathol 76:49-55

Dauphin Y (2001) Comparative studies of skeletal soluble matrices from some Scleractinian corals and molluscs. Int J Biol Macromolecules 28:293-304

de Villiers S, Nelson BK, Chivas AR (1995) Biological controls on coral Sr/Ca and ^{18}O reconstructions of sea surface temperatures. Science 269:1247-1249

de Villiers S, Shen GT, Nelson BK (1994) The Sr/Ca temperature relationship in coralline aragonite: influence of variability in (Sr/Ca)$_{seawater}$ and skeletal growth parameters. Geochim Cosmochim Acta 58:197-208

Dixon DA, Haynes DH (1989) Ca^{2+} pumping ATPase of cardiac sarcolemma is insensitive to membrane potential produced by K^+ and Cl^- gradients but requires a source of counter-transportable H^+. J Membr Biol 112:169-183

Enmar R, Lazar B, Bar-Matthews M, Sass E, Katz A, Stein M (2000) Diagenesis in live corals from the Gulf of Aqaba. I. The effect on paleo-oceanography tracers. Geochim Cosmochim Acta 64:3123-3132

Erez J (1978) Vital effect on stable-isotope composition seen in foraminifera and coral skeletons. Nature 273:199-202

Erez J (2003) The source of ions for biomineralization in foraminifera and their implications for paleoceanographic proxies. Rev Mineral Geochem 54:115-149

Fang L-S, Chen Y-W, Chen C-S (1983) Why does the white tip of stony coral grow so fast without zooxanthellae? Mar Biol 103:359-363

Ferrier-Pagès C, Boisson F, Allemand D, Tambutté E (2003) Kinetics of strontium uptake in the scleractinian coral *Stylophora pistillata*. MEPS 245:93-100

Ferrier-Pagès C, Gattuso JP, Dallot S, Jaubert J (2000) Effect of nutrient enrichment on growth and photosynthesis in the zooxanthellate coral *Stylophora pistillata*. Coral Reefs 19:103-113

Ferrier-Pagès C, Schoelzke V, Jaubert J, Muscatine L, Hoegh-Guldberg O (2001) Response of a scleractinian coral, *Stylophora pistillata*, to iron and nitrate enrichment. J Exp Mar Biol Ecol 259:249-261

Frankignoulle M, Gattuso JP, Biondo R, Bourge I, Copin-Montégut G, Pichon M (1996) Carbon fluxes in coral reefs. 2. Eulerian study of inorganic carbon dynamics and measurement of air-sea CO_2 exchanges. Mar Ecol Prog Ser 145:123-132

Furla P, Galgani I, Durand I, Allemand D (2000) Sources and mechanisms of inorganic carbon transport for coral calcification and photosynthesis. J Exp Biol 203:3445-3457

Gardner TA, Côté IM, Gill JA, Grant A, Watkinson AR (2003) Long-term region-wide declines in Caribbean corals. Science 301:958-960

Gattuso J-P, Allemand D, Frankignoulle M (1999) Photosynthesis and calcification at cellular, organismal, and community levels in coral reefs: A review of interactions and control by carbonate chemistry. Am Zool 39:160-183

Gattuso J-P, Frankignoulle M, Bourge I, Romaine S, Buddemeier RW (1998) Effect of calcium carbonate saturation of seawater on coral calcification. Global Planet Change 18:37-46

Gattuso JP, Reynaud-Vaganay S, Furla P, Romaine-Liourd S, Jaubert J, Bourge I, Frankignoulle M (2000). Calcification does not stimulate photosynthesis in the zooxanthellate scleractinian coral *Stylophora pistillata*. Limnol Oceanogr 45:246-250

Given RK, Wilkinson BH (1985) Kinetic control of morphology, composition and mineralogy of abiotic sedimentary carbonates. J Sed Pet 55:0109-0119

Gladfelter EH (1982) Skeletal development in *Acropora cervicornis*. 1. Patterns of calcium carbonate accretion in the axial corallite. Coral Reefs 1:45-51

Gladfelter EH (1983a) Skeletal development in *Acropora cervicornis*. Diel patterns of calcium carbonate accretion. Coral Reefs 2:91-100

Gladfelter EH (1983b) Circulation of fluids in the gastrovascular system of the reef coral *Acropora cervicornis*. Biol Bull 165:619-636

Goldberg WM (2001) Acid polysaccharides in the skeletal matrix and calicoblastic epithelium of the stony coral *Mycetophyllia reesi*. Tissue&Cell 33:376-387

Goreau TF (1959) The physiology of skeletal formation in corals I. A method for measuring the rate of calium deposition by corals under different conditions. Biol Bull 116:59-75

Goreau TF (1963) Calcium carbonate deposition by coralline algae and corals in relation to their roles as reef-builders. Ann NY Acad Sci 107:127-167

Griffin S, Griffin E, Druffel RM (1989) Sources of carbon to deep-sea corals. Radiocarbon 31:533-543

Guilderson TP, Fairbanks RG, Rubenstone JL (1994) Tropical temperature variations since 20,000 years ago: modulating interhemspheric climate change. Science 263:663-665

Hayes RL, Goreau NI (1977) Intracellular crystal-bearing vesicles in the epidermis of scleractinian corals, *Astrangia danae* (Agassiz) and *Porites porites* (Pallas). Biol Bull 152:26-40

Houck JE, Buddemeier RW, Chave KE (1975) Skeletal low magnesium calcite in living scleractinian corals. Science 189:997-999

Huang Y, Fairchild I (2001) Partitioning of Sr^{2+} and Mg^{2+} into calcite under karst-analogue experimental conditions. Geochim Cosmochim Acta 65:47-62

Iddings JP (1891) Spherulitic crystallization. Wash Philosoph Soc Bull 11:445-463

Ip YK, Krishnaveni P (1991) Incorporation of strontium ($^{90}Sr^{2+}$) into the skeleton of the hermatypic coral Galaxea fascicularis. J Exp Zool 258:273-276

Isa Y (1989) Calcium binding substance in the hermatypic coral *Acropora hebes* (Dana) *In*: Origin, Evolution and Modern Aspects of Biomineralization in Plants and Animals. Crick RE (Ed) Plenum Press, New York, p 167-174

Isa Y, Yamazato K (1981) The ultrastructure of calicoblast and related tissues in Acropora hebes (Dana) Proc 4th Int Coral Reef Symp 2:99-105

Isa Y (1986) An electron microscope study on the mineralization of the skeleton of the staghorn coral Acropora hebes. Mar Biol 93:91-101

Jacques TG, Marshall N, Pilson MEQ (1983) Experimental Ecology of the temperate scleractinian coral Astrangia danae II. Mar Biol 76:135-148

Jacques TG, Pilson MEQ (1980) Experimental Ecology of the temperate scleractinian coral Astrangia danae I. Mar Biol 60:167-178

Jaillard B, Guyon A, Maurin AF (1991) Structure and composition of calcified roots, and their identification in calcareous soils. Geoderma 50:197-210

Johnston IS (1980) The ultrastructure of skeletogenesis in hermatypic corals. Int Rev Cytol 67:171-214

Kawagitu S, Sakamuto D (1948) The effect of light on calcium deposition in corals. Bull Oceanogr Inst Taiwan 4:65-70

Kawaguti S, Sakumoto D (1949) The effect of light on the calcium deposition of corals. Bull Oceanogr Inst Taiwan 4:65-70

Keith HD, Padden FJ (1963) A phenomological theory of spherulitic crystallization. J Appl Physics 34:2409-2421

Kingsley RJ, Watabe N (1985) Ca-ATPase localization and inhibition in the gorgonian Leptogorgia virgulata (Lamarck) (Coelenterata: Gorgonacea). J Exp Mar Biol Ecol 93:157-167

Kinsey DW, Davies PJ (1979) Effects of elevated nitrogen and phosphorus on coral reef growth. Limnol Oceanogr 24:935-940

Kinsman DJJ (1969) Interpretation of the strontium concentration in carbonate minerals and rocks. J Sediment Petrol 39:486-508

Kinsman DJJ, Holland HD (1969) The co-precipitation of cations with $CaCO_3$-IV. The co-precipitation of Sr^{2+} with aragonite between 16°C and 96°C. Geochim Cosmochim Acta 33:1-17

Kitsato H (1968) The influence of H^+ on the membrane potential and ion fluxes of Nitella. J Gen Physiol 52:60-87

Kochian LV (1991) Mechanisms of micronutrient uptake and translocation in plants. *In*: Micronutrients in Agriculture. Mortvedt JJ (ed) Soil Science Society of America, Madison, p 229-296

Le Tissier M (1988) Diurnal patterns of skeletal formation in Pocillopora damicornis (Linneaus). Coral Reefs 7:81-88

Legatt W, Badger MR, Yellowlees D (1999) Evidence for an inorganic carbon-concentrating mechanism in the symbiotic dinoflagellate Symbiodinium sp. Plant Physiol 121:1247-1255

Leggat W, Marendy EM, Baillie B, Whitney SM, Ludwig M, Badger MR, Yellowlees D (2002) Dinoflagellate symbioses: strategies and adaptations for the acquisition and fixation of inorganic carbon. Funct Plant Biol 29:309-322

Lofgren G (1974) An experimental study of plagioclase crystal morphology: isothermal crystallization. Am J Science 274:243-273

Lofgren G (1980) Experimental studies on the dynamic crystallization of silicate melts. *In*: Physics of Magmatic Processes. Hargraves RB (ed) Princeton University Press, Princeton NJ, p 487-551

Lorens RM (1981) Sr, Cd, Mn and Co distribution coefficients in calcite as a function of calcite precipitation rate. Geochim Cosmochim Acta 45:3637-3655

Lough JM, Barnes DJ (1997) Several centuries of variation in skeletal extension, density and calcification in massive Porites colonies from the Great Barrier Reef: A proxy for seawater temperature and a background of variability against which to identify unnatural change. J Exp Mar Biol Ecol 211:29-67

Lowenstam HA, Weiner S (1989) On Biomineralization. Oxford University Press, New York p 207-251

Lucas JM, Knapp LW (1997) A physiological evaluation of carbon sources for calcification in the octocoral Leptogorgia virgulata (Lamarck). J Exp Biol 200:2653-2662

Mann S (2001) Biomineralization: principles and concepts in bioinorganic materials chemistry. Oxford University Press

Marin F, Smith M, Yeishin I, Muyzer G, Westbroek P (1996) Skeletal matrices, muci, and the origin of invertebrate calcification. Proc Natl Acad Sci USA 93:1554-1559

Marlier JF, O'Leary MH (1984) Carbon kinetic isotope effects on the hydration of carbon dioxide and the dehydration of bicarbonate ion. J Am Chem Soc 106:1054-1057

Marshall AT, Wright A (1998) Coral calcification: autoradiography of a scleractinian coral *Galaxea fascicularis* after incubation in ^{45}Ca and ^{14}C. Coral Reefs 17:37-47

Marshall AT, Wright OP (1993) Confocal laser scanning light microscopy of the extra-thecal epithelia of undecalcified scleractinian corals. Cell Tissue Res 272:533-543

Marubini F, Davies PS (1996) Nitrate increases zooxanthellae population density and reduces skeletogenesis in corals. Mar Biol 127:319-328

Marubini F, Thake B (1999) Bicarbonate addition promotes coral growth. Limnol Oceanogr 44:716-720

Marxen JC, Hammer M, Gehrke T, Becker W (1998) Carbohydrates of the organic shell matrix and the shell-forming tissue of the snail *Biomphalaria glabrata*. Biol Bull 194:231-240

McConnaughey TA (1989a) ^{13}C and ^{18}O isotopic disequilibrium in biological carbonates. I. Patterns. Geochim Cosmochim Acta 53:151-162

McConnaughey TA (1989b) ^{13}C and ^{18}O isotopic disequilibrium in biological carbonates. II. In vitro simulation of kinetic isotope effects. Geochim Cosmochim Acta 53:163-171

McConnaughey TA, Adey Wh, Small AM (2000) Community and environmental influences on reef coral calcification. Limnol Oceanogr 45:1667-1671

McConnaughey TA, Burdett JU, Whelan JF, Paull CK (1997) Carbon isotopes in biological carbonates: respiration and photosynthesis. Geochim Cosmochim Acta 61:611-622

McConnaughey TA, Falk RH (1991) Calcium-proton exchange during algal calcification. Biol Bull 180:185-195

McConnaughey TA, Whelan JF (1997) Calcification generates protons for nutrient and bicarbonate uptake. Earth-Science Rev 42:95-117

McCrea JM (1950) On the isotopic chemistry of carbonates and a paleotemperature scale. J Chem Phys 18:849-857

McCulloch MT, Tudhope AW, Esat, TM, Mortimer GE, Chappell J, Pillans B, Chivas, AR, Omura A. (1999) Coral record of equatorial sea-surface temperatures during the pen-ultimate deglaciation at Huon Peninsula. Science 283:202-204

Mitterer RM (1978) Amino acid composition and metal binding capability of the skeletal proteins of corals. Bull Mar Sci 28:173-180

Muscatine L (1967) Glycerol excretion by symbiotic algae from corals and Tridacna and its control by the host. Science 156:516-519

Muscatine L (1990) The role of symbiotic algae in carbon and energy flux in coral reefs. *In*: Coral Reefs. Dubinsky Z (ed) Elsevier Science Publishers BV, Amsterdam, p 75-87

Muscatine L, Tambutte E, Allemand, D (1997) Morphology of coral desmophytes, cells that anchor the calicoblastic epithelium to the skeleton. Coral Reefs 16:205-213

Niggli VE Sigel E Carafoli E (1982) The purified Ca^{2+} pump of the human erythrocyte membrane catalyzes an electroneutral $Ca^{2+}-H^+$ exchange in reconstituted liposomal systems. J Biol Chem 257:2350-2356

Ogilvie MM (1896) Systematic study of Madreporan corals. Phil Trans Roy Soc London 187:83-345

Pandolfi JM, Bradbury RH, Sala E, Hughes TP, Bjorndal KA, Cooke RG, McArdle, D, Loren McClenachan L, Newman MJH, Paredes G, Warner RR, Jackson JBC (2003) Global trajectories of the long-term decline of coral reef ecosystems. Science 301:955-958

Prieto FJM, Millero FJ (2002) The values of $pK_1 + pK_2$ for the dissociation of carbonic acid in seawater. Geochim Cosmochim Acta 66:2529-2540

Raz S, Weiner S, Addadi L (2000) Formation of high magnesium calcites via an amorphous precursor phase: possible biological implications. Adv Mater 12:38-42

Rimstidt JD, Balog A, Webb J (1998) Distribution of trace elements between carbonate minerals and aqueous solutions. Geochim Cosmochim Acta 62:1851-1863

Rinkevich B, Loya Y (1984) Does light enhance calcification in hermatypic corals? Mar Biol 80:1-6

Risk MJ, Pearce TH (1992) Interference imaging of daily growth bands in massive corals. Nature 358:572-573

Rollion-Bard C, Chaussidon M, France-Lanord C (2003) pH control on oxygen isotopic composition of symbiotic corals. Earth Planet Sci Lett (in press)

Sakano K (1990) Proton/phosphate stoichiometry in uptake of inorganic phosphate by cultured cells of *Catharanthus roseus* (L.) G Don Plant Physiol 93:479-483

Schachtman DP, Reid RJ, Ayling SM (1998) Phosphorus uptake by plants: form soil to cell. Plant Physiol 116:447-453

Siegenthaler U, Münnich KO (1981) $^{13}C/^{12}C$ fractionation during CO_2 transfer from air to sea. *In*: Carbon Cycle Modelling. Bolin B. (ed) Wiley, New York, p 249-257

Smith SV (1973) Carbon dioxide dynamics: a record of organic carbon production, respiration, and calcification in the Eniwetok reef flat community. Limnol Oceanogr 18:106-120

Smith SV (1978). Coral reef area and contributions to processes and resources of the world's oceans. Nature 273:225-226

Smith SV, Buddemeier RW, Redalje RC, Houck JE (1979) Strontium-Calcium thermometry in coral skeletons. Science 204:404-407

Spero HJ, Bijma J, Lea DW, Bemis BE (1997) Effect of seawater carbonate concentration on foraminiferal carbon and oxygen isotopes. Nature 370:497-500

Spero HJ, Lea D (1986) Experimental determination of stable isotope variability in Globigerina bulloides: implications for paleoceanographic reconstructions. Marine Micropaleontol 28:231-246

Tambutté É, Allemand D, Mueller E, Jaubert J (1996) A compartmental approach to the mechanism of calcification in hermatypic corals. J Exp Biol 199:1029-1041

Taroncher-Oldenburg G, Kulis DM, Anderson DM (1999) Coupling of saxitoxin biosynthesis to the G1 phase of the cell cycle in the dinoflagellate Alexandrium fundyense: temperature and nutrient effects. Nat Toxins 7:207-219.

Tesoriero AJ, Pankow JF (1996) Solid solution partitioning of Sr^{2+}, Ba^{2+} and Cd^{2+} into calcite. Geochim Cosmochim Acta 60:1053-1063

Usdowski E, Hoefs J (1993) Oxygen isotope exchange between carbonic acid, bicarbonate, carbonate, and water: a re-examination of the data of McCrea (1950) and an expression for the overall partitioning of oxygen isotopes between the carbonate species and water. Geochim Cosmochim Acta 57:3815-3818

Usdowski E, Michaelis J, Böttcher ME, Hoefs J (1991) Factors for the oxygen isotope equilibrium fractionation between aqueous and gaseous CO_2, carbonic acid, bicarbonate, carbonate, and water (19°C). Zeitschrift für Physikalische Chemie 170:S237-S249

Vago R, Gill E, Collingwood JC (1997) Laser measurements of coral growth. Nature 386:30-31

Vandermeulen JH (1975) Studies on Reef Corals III. Fine structural changes of calicoblast cells in *Pocillopora damicornis* during settling and calcification. Mar Biol 31:69-77

Vandermeulen JH, Watabe N (1973) Studies on reef corals I. Skeletal formation of *Pocillopora damicornis*. Mar Biol 23:45-57

Veron JEN (1986) Corals of Australia and the Indo-Pacific. University of Hawaii Press

Wainwright S (1963) Skeletal organisation in *Pocillopora damicornis* Q J Microsc Sci 104:169-183

Wainwright S (1964) Studies of the mineral phase of coral skeleton. Exp Cell Res 34:213-230

Watabe N (1981) Crystal growth of calcium carbonate in the invertebrates. *In*: Progress in crystal growth and characterization, Vol. 4(1-2). Ramplin B (ed) Pergamon Press, Oxford, p 99-147

Weber JN (1973) Incorporation of strontium into reef coral skeletal carbonate. Geochim Cosmochim Acta 37:2173-2190

Weiner S, Addadi L (1991) Acidic macromolecules of mineralized tissues. The controllers of crystal formation. Trends Biol Sci 16:252-256

Weiner S, Dove PM (2003) An overview of biomineralization processes and the problem of vital effect. Rev Mineral Geochem 54:1-29

Wells JW (1956) Scleractinia. *In*: Treatise on Invertebrate Paleontology. Part F Coelenterata. Moore RC (ed) Geological Society of America and University of Kansas Press, p F328-F479

Wells JW (1963) Coral growth and geochronometry. Nature 197:948-50

Wells JW (1969) The formation of dissepiments in zooantherian corals. *In*: Stratigraphy and Paleontology. KSW Campbell (ed) Australian National University press, Canberra, p17-26

Wollenweber (1997) A sensitive computer-controlled pH-stat system allows the study of net H^+ fluxes related to nitrogen uptake of intact plants in situ. Plant Cell Environ 20:400-408

Wright OP, Marshall AT (1991) Calcium transport across the isolated oral epithelium of scleractinian corals. Coral Reefs 10:37-40

Young SD (1971) Organic material from scleractinian coral skeletons 1. Variation in composition between several species. Comp Physiol Biochem 40B:113-120

Yu X, Inesi G (1995) Variable stichiometric efficiency of Ca^{2+} and Sr^{2+} transport by the sarcoplasmic reticulum ATPase. J Biol Chem 270(9):4361-4367

Zeebe R (1999) An explanation of the effect of seawater carbonate concentration on foraminiferal oxygen isotopes. Geochim Cosmochim Acta 63:2001-2007

7 Biomineralization Within Vesicles: The Calcite of Coccoliths

Jeremy R. Young
Palaeontology Department
The Natural History Museum
Cromwell Road, London SW7 5BD, United Kingdom

Karen Henriksen
NanoGeoScience, Geological Institute
University of Copenhagen
Øster Voldgade 10
Dk-1350 Copenhagen K, Denmark

INTRODUCTION

Coccolithophores are a group of unicellular plant plankton, which produce exoskeletons of minute calcite plates, called coccoliths. Despite their small size (1–10 μm across), coccoliths are remarkably elaborate biomineral structures characterized by precise control of both nucleation and growth of calcite by the organic system. They are also of considerable environmental and applied geological importance. Coccolithophores occur in enormous abundances, forming a significant proportion of total marine primary production and carbon fixation, especially in open ocean environments. Their biogeochemical impact is magnified by export of coccoliths to the ocean floor, where coccoliths are the largest single component of deep-sea sediments, forming vast accumulations of calcareous oozes and chalks, including the Late Cretaceous chalks of Northwestern Europe. Moreover, coccoliths are extensively employed by geologists as marker fossils to determine the ages of sediments, especially from drill cores. More recently, satellite imaging of immense coccolithophore blooms has led to extensive multi-disciplinary research into their biology and impact on the marine ecosystem and global carbon cycle (e.g., Westbroek et al. 1993; Paasche 2002; Thierstein and Young in press).

Biological affinities

Coccolithophores belong to the algal division (or phylum) Haptophyta. Other haptophytes are non-calcifying and include the well-known genera *Phaeocystis*, *Prymnesium*, *Pavlova* and *Chrysochromulina*. They are all characterized by possessing golden-brown chloroplasts, two smooth flagella and a third flagellum-like organelle, the haptonema. The haptonema shows distinctive coiling behavior and has a quite different microtubular sub-structure from flagella. Molecular genetic data have shown that the haptophytes are a group discrete from other algal protists that probably originated during the Precambrian (>600 Ma) protist radiation and acquired chloroplasts subsequently (possibly in the Late Palaeozoic, ca. 300–400 Ma) as a result of secondary endosymbiosis (Medlin et al. 1997).

Molecular genetic phylogenies of the haptophytes are still being developed, but all results to date (Edvardsen et al. 2000; Fujiwara et al. 2001; Saez et al. in press) show that the coccolithophores constitute a discrete clade within the haptophytes, thus indicating that calcification evolved only once within the phylum. The molecular information also indicates that they diverged relatively recently, long after the differentiation of haptophytes and probably within the Mesozoic. This is concordant with the geological record, which shows coccoliths first occurring in the Late Triassic (Bown 1998a,b) ca. 220 Ma. Calcification in coccolithophores occurs primarily in Golgi-derived vesicles. Since Golgi bodies are a primitive feature of eukaryotes, this makes an endosymbiotic origin of calcification extremely unlikely. So coccolith calcification has probably evolved *de novo* relatively recently and entirely independently from biomineralization in other groups. It may thus prove a useful outlier group for testing hypotheses of fundamental similarities in biomineralization mechanisms among groups that adopted biomineralization in the Cambrian radiation (Kirschvink and Hagadorn 2000).

Coccolith biomineralization has been rather extensively researched; reviews include de Vrind-de Jong et al. (1986, 1994), Westbroek et al. (1989), Pienaar (1994), de Vrind-de Jong and de Vrind (1997), Young et al. (1999), and Marsh (2000). This interest reflects both the importance of coccolithophores and the fact that, as rapidly reproducing protists, they are amenable to laboratory culture and study. At least some of the pioneers of research in this field also anticipated that, as unicellular organisms, the coccolithophores would show simpler biomineralization mechanisms than higher organisms (Westbroek pers comm.). However, coccolith biomineralization has proven to be under strong cellular control, lying at the biologically controlled end of the spectrum of biomineralization mechanisms. This probably reflects the fact that mineralization occurs within intracellular vesicles. This complex biomineralization process is still only partially characterized. The majority of research on coccolith biomineralization has come from detailed laboratory study of a very limited range of taxa, including cytological and biochemical aspects. These findings are briefly reviewed here, but the main focus is on morphological, crystallographic and structural aspects of coccoliths from across the spectrum of coccolith diversity. This approach is adopted to constrain mechanistic hypotheses and identify targets for future research.

Life cycles

One of the most distinctive aspects of coccolithophore biomineralization is that two very different biomineralization modes typically occur within the life cycle of single species. This is unusual and makes for interesting biomineralization research possibilities, but has been rather neglected, since the life cycle details have only recently been well established. Protist life cycles are often difficult to study, as they are heavily dependent upon fortuitous observations of laboratory cultures. Haptophytes definitely follow this pattern, as our knowledge of life cycles has long been confined to a few rather disparate case studies. The most notable example is the coccolithophore *Coccolithus pelagicus*, which was shown by Parke and Adams (1960) to have a two phase life cycle, with both phases producing coccoliths, but of very different types. Fresnel (1986) and Billard (1994) synthesized these observations and hypothesized that a common pattern could be observed, based on a haplo-diploid life cycle. Subsequently, a steady accumulation of new data (e.g., Green, et al. 1996; Cros et al. 2000; Geisen et al. 2002; Noel et al. in press) supported this inference, although without direct testing of the key hypothesis that coccolith type was directly related to ploidy level. More recently, life cycle transitions of this type have occurred in laboratory cultures of three additional species from families spanning the evolutionary diversity of coccolithophores. This allowed critical testing of the ploidy level hypothesis via flow cytometric measurement of DNA content (Houdan et al. in press).

Following this work, a basic model of coccolithophore life cycles can now be given (Fig. 1). As is typical for eukaryotes, the fundamental aspect is alternation of diploid and haploid phases; diploid cells have two copies of each chromosome, and so possess 2N chromosomes, while haploid cells have only one copy, and so have N chromosomes. Transition from the diploid to haploid phase occurs through meiosis, which involves cell

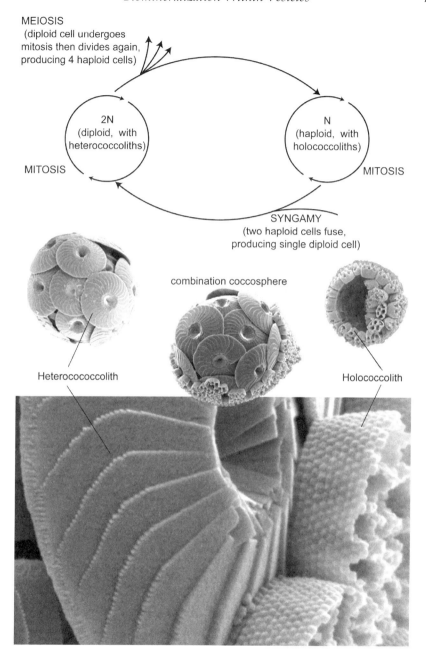

Figure 1. Coccolithophore life cycles. Diagram at top summarizes the basic life cycle of coccolithophores. Both diploid and haploid phases are chloroplast-bearing and can reproduce indefinitely by asexual binary fission. Both phases can calcify producing heterococcoliths and holococcoliths, respectively, as illustrated here by coccospheres of the two phases of *Calcidiscus leptoporus* ssp. *quadriperforatus*. Combination coccospheres with both coccolith types occur very rarely recording the life cycle transition, as shown by the centre SEM and enlarged view.

division, while transition from the haploid to the diploid phase occurs via syngamy, with fusion of two cells. In the life-cycles of most organisms one phase is dominant, but in haptophytes the two phases are equally well-developed, and both of them can undergo asexual binary fission (mitosis) apparently indefinitely, until some environmental trigger prompts transition to the alternate life-cycle phase. Thus each species of coccolithophore can occur in both haploid and diploid phases often with very different characteristics. In particular, it is now clear that the diploid and haploid phases have consistently different coccoliths, which must be a result of consistently different biomineralization modes (Billard 1994; Young et al. 1999, Cros et al. 2000; Geisen et al. 2002; Cros and Fortuño 2002; Billard and Inouye in press; Houdan et al. in press).

The diploid phase is characterized by *heterococcoliths*, which show radial symmetry and are formed of a limited number (typically <100) of calcite crystals with complex strongly modified shapes. The haploid phase, by contrast, is characterized by *holococcoliths* which lack the strong radial symmetry of heterococcoliths and are formed of numerous (several hundred to a few thousand) minute (ca. 0.1 µm), equant calcite crystallites. The holococcoliths and heterococcoliths are so different that in almost all species the separate phases were originally described and named as separate species. For example the holococcolith in Figure 1 was originally regarded as a separate species, *Syracolithus quadriperforatus,* but is now known to the haploid phase of the heterococcolith species illustrated, *Calcidiscus leptoporus*. (NB This example is actually slightly more complex taxonomically, since two closely related species or subspecies occur with very similar heterococcoliths but significantly different holococcoliths—see Geisen et al. 2002; Saez et al. 2003). Evidence linking these life cycle phases has come both from culture studies (e.g., Houdan et al. in press; Noel et al. in press) and from observation among plankton samples of rare combination coccospheres (Fig. 1). These are formed during transition between the life-cycle phases and so bear coccoliths of both types. For example, the specimen illustrated in Figure 1 consists of an inner coccosphere of heterococcoliths partially surrounded by holococcoliths. Because coccoliths are added to the coccosphere from the inner surface this is interpreted as a diploid cell encased in a new coating of heterococcoliths, forming beneath the old coating of holococcoliths inherited from the two haploid cells that produced it during syngamy (Geisen et al. 2002).

Life cycle variants do occur within coccolithophores; for example the well-studied genera *Pleurochrysis* and *Emiliania* calcify only in the diploid phases, producing heterococcoliths, and are non-calcifying in the haploid phases. However, enough examples have now been studied for us to predict that all heterococcoliths are formed on diploid cells and all holococcoliths on haploid cells. So coccolithophores present an unusual example of biomineralization in which two separate, but presumably not entirely independent, biomineralization mechanisms occur within single species. This provides a key perspective for this review.

HETEROCOCCOLITH BIOMINERALIZATION

Cytological aspects

Observations of coccolith formation are possible using transmission electron microscope (TEM) analysis of microtome sections through cells. One of the most well-studied coccolithophores is *Pleurochrysis carterae,* a coastal species that is very readily maintained in culture; investigations of coccolith development in *P. carterae* include Manton and Leedale (1969) Outka and Williams (1971), van der Wal et al. (1983), Marsh (1994, 1999). The sequence of coccolith development was worked out in particular detail by van der Wal et al. (1983), as summarized in Figure 2. Growth occurs in vesicles, or

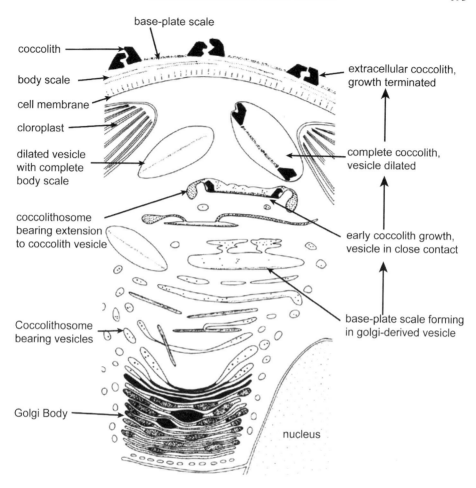

Figure 2. Coccolithogenesis in *Pleurochrysis carterae*. Schematic section, based on TEM images, through a coccolithophore showing coccolith formation within Golgi-derived vesicles. Reproduced with permission from van der Wal et al. 1983.

cisternae, derived from the Golgi body. The process commences with formation of an organic scale. These scales have a distinctive microfibrillar ultrastructure that is characteristic of scales in the haptophytes, and homologous non-calcified scales occur in most non-coccolithophorid haptophytes (e.g., Billard 1994). In some cases, the scales are exocytosed without further development, to form an inner layer of body scales. In others, the vesicle develops a rather complex form, with extensions containing dense particles termed coccolithosomes. Nucleation of calcite then occurs around the periphery of the base-plate scale followed by crystal growth upward and outward to form the complete coccolith. During coccolith growth, the vesicle gradually expands, remaining in rather close contact with the developing coccolith. After completion of the coccolith, the vesicle dilates and at this stage, a dense organic coating is visible around and between the coccolith crystals. Exocytosis then occurs, apparently by fusion of the vesicle membrane and cell membrane. In this species the coccolithosomes appear to play a key role in calcification and have been shown to be complexes of acidic polysaccharides with calcium ions (Marsh 1994). It is thought that they function as calcium vectors during biomineralization and that the polysaccharide phase forms the crystal coatings.

Detailed cytological observations on heterococcolith formation have also been made on *Emiliania huxleyi* (e.g., Wilbur and Watabe 1963; Klaveness 1976; van der Wal et al. 1983b, van Emburg et al 1986), while for a few other species there are isolated useful studies, including notably: Manton and Leedale (1969) – *Coccolithus pelagicus*; Inouye and Pienaar (1984) – *Umbilicosphaera sibogae*; Inouye and Pienaar (1988) – *Syracosphaera pulchra*. There is significant diversity in these observations, for instance coccolithosomes are only known in *Pleurochrysis*, and in *E. huxleyi* the base-plate scale is a very nebulous structure. Also in *E. huxleyi* calcification occurs in the "reticular body," an organelle that is separated from the Golgi body, but is thought to be derived from it. Nonetheless, a clear overall pattern is similar in all cases, with calcification occurring in Golgi-derived vesicles and commencing with nucleation of a proto-coccolith ring of simple crystals around the rim of a precursor base-plate scale, followed by growth of these crystals in various directions to form complex crystal units.

Biochemical aspects: organic molecules involved in coccolith biomineralization

The control by organisms on mineral expression is to a large extent attributed to the influence of organic molecules on nucleation and growth. Organic matrix frameworks can provide binding sites for the components of a mineral, selectively nucleating specific crystallographic faces, and organic carrier molecules can ensure supersaturation of a phase within mineralizing compartments. On a larger scale, shape regulation of such compartments can direct the morphology of biominerals forming within them (Mann 1993). In the case of coccoliths, the primary organic molecules involved are complex acidic polysaccharides (Fichtinger-Schepman et al. 1981; Marsh et al. 1992; Marsh 1994; Ozaki et al. 2001; Marsh et al. 2002).

The coccolith-associated polysaccharides (CAPs) of two species have been isolated and analyzed. In *Emiliania huxleyi*, one polysaccharide has been located, having a backbone of mannose with many complex sidechains of galacturonic acid as well as ester-bound sulphate groups (Fichtinger-Schepman et al. 1981). In the case of *Pleurochrysis carterae*, the situation is more complex, with 3 types of polysaccharide present (PS1, PS2, PS3; Marsh et al. 2002). PS1 and PS2 bind calcium and form coccolithosomes, with PS2 probably playing an important role in the nucleation of the proto-coccolith ring, as evidenced by mutant *P. carterae* cells deficient in this polysaccharide showing very little calcification (Marsh and Dickinson 1997). During coccolith growth, PS3 is located between the crystals and the vesicle, and its function is believed to be shape regulatory, because *P. carterae* cells not expressing PS3 produce a proto-coccolith ring that does not develop further (Marsh et al. 2002). It is likely that similar polysaccharides are active in other species of coccolithophore, instrumental in producing their elaborately complex coccoliths.

The active role inferred from this work for polysaccharides is unusual, since in most biomineralizing systems proteins are the key functional molecules. Proteins are also attractive targets for research since they are synthesized by RNA, allowing a range of genomic techniques to be applied to protein studies. However, despite extensive research, there has been only limited success to date in isolating and characterizing proteins associated with coccolith formation (Corstjens et al. 1998). This apparent anomaly is less surprising from an ecological perspective. Coccolithophores are photosynthetic algae, and a prime constraint on algal abundance in most oceanic environments is availability of nutrient elements for biosynthesis, especially N and P (e.g., Brand 1994). Sequestration of nutrient elements in proteins associated with biomineralization reduces their availability for synthesis of DNA and physiologically essential proteins and so is maladaptive in a nutrient-limited autotroph. Hence it is likely that proteins associated with coccolith formation have enzymatic functions, such as controlling production of polysaccharides, rather than having a direct role in regulating crystal growth, thus minimizing the total amount of protein used for biomineralization.

Morphological observations

An unusual aspect of coccoliths as biomineral structures is that they are routinely studied at the level of the biomineral fabric. In most other mineralized structures, for instance bivalves or vertebrates, or even foraminifera, the routine level of study is gross morphology and the biomineral fabric is only investigated in special higher resolution studies. With coccoliths, by contrast, the component crystals are directly visible and form the basis of classification and identification. This obviously applies to routine observations by scanning electron microscopy (SEM), but also less obviously to observation by light microscopy (LM). In normal illumination only the basic form of coccoliths can be seen; because coccoliths are formed of calcite they can be better studied in cross-polarized light. Coccoliths are typically about one micron thick, and the very high birefringence of calcite means that they show first-order polarization colors. The radial fabric of the coccoliths causes them to show bright extinction crosses (e.g., Young 1992). This is the routine means of study of fossil coccoliths (Bown and Young 1998), although it is less commonly used by biologists studying plankton. As a result of these light microscopy observations, it has long been appreciated that coccoliths consist in part of crystals with sub-radial orientations, forming bright pseudo-extinction crosses in polarized light but also in part of crystals with sub-vertical calcite c-axis orientations that appear dark in cross-polarized light in plan view (e.g., Kamptner 1954; Prins 1969; Romein 1979).

A key advance in knowledge of coccolith formation came from combining these light microscope observations with study of coccoliths at different growth stages, either from culture samples or low-diversity natural assemblages (Young and Bown 1991; Young et al. 1992). Representative structures are shown in Figures 3-6 and are described briefly below, then discussed in terms of nucleation and growth regulation. Terminology: *Biscutum, Coccolithus* and *Emiliania* all have *placolith* type coccoliths; these have a basic shape similar to that of a cable reel consisting of a central *tube* which connects a lower *proximal shield* and upper *distal shield* and encloses the *central area*. The placolith morphology allows the coccoliths to interlock closely on the coccosphere, forming a robust structure. This morphology has evolved repeatedly, but many other coccolith morphologies also occur. The structures shown have been worked out primarily from SEM and LM observations. Further details of the structures are given by Young et al. (1992, in press), Young (1992), and Henriksen et al. (in press b). SEM observations of multiple specimens, including broken ones and specimens at different growth stages, allow the shapes and interconnections of coccolith elements to be worked out. LM observations are the primary source of information on broad crystallographic orientation of the crystal units, although this has been supplemented by use of selected area electron diffraction (SAED), (e.g., Davis 1995; Marsh 1999), atomic force microscopy (Henriksen et al. in press a,b) and crystal face morphology (e.g., Black 1963; Young et al. in press).

Biscutum provides a straightforward example of a V/R structure. The distal (upper) shield and outer tube cycle is formed of V-units, which appear dark in cross-polarized light. The proximal shield, inner tube and central area elements all interconnect and are formed from R-units, with sub-radial calcite c-axes. The two unit types can be seen to alternate with each other in a ring on the proximal face of the coccolith and this is interpreted to be the proto-coccolith ring locus, i.e., the location where nucleation occurred.

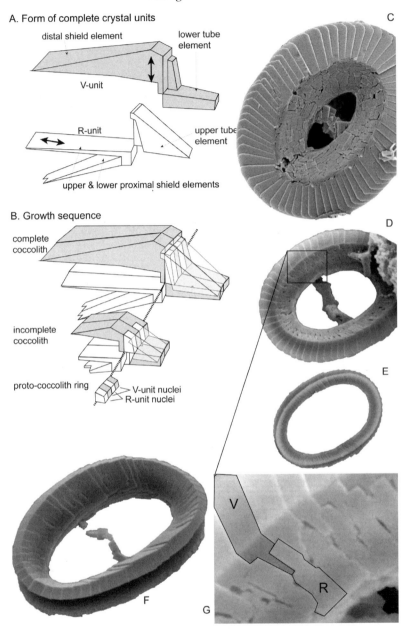

Figure 3. *Coccolithus pelagicus* growth and structure. (A) Schematic diagrams showing the form of the component crystal units. Double-headed arrow symbols indicate orientation of calcite *c*-axis. (B) Schematic diagrams of two segments at three growth stages. (C–E) SEMs of isolated coccoliths in distal view at different growth stages. *N.B.* The earliest growth stage illustrated here is a later growth stage than the simple proto-coccolith ring illustrated opposite. (F) SEM photomicrograph of another intermediate growth stage specimen, also showing a very early stage of bar formation as a belt of simple nuclei. (G) Detail of intermediate growth stage specimen D showing the intergrowth of V and R units, with one pair of units outlined (compare with drawing of incomplete coccolith).

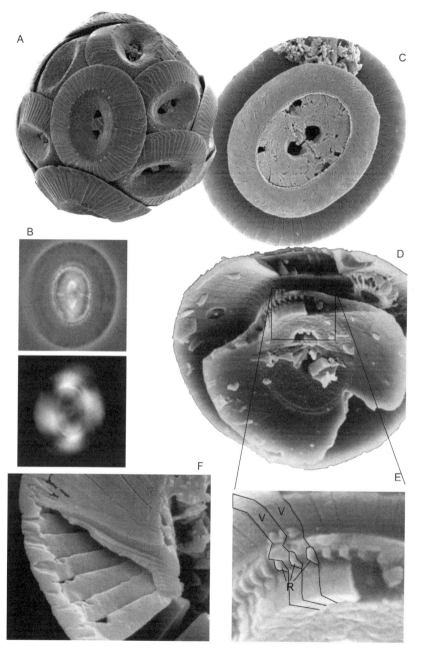

Figure 4. *Coccolithus pelagicus* structure details. (A) Complete coccosphere. (B) Light micrographs of isolated coccoliths in phase contrast (top) and cross-polarized light (bottom), only the proximal shield is birefringent in this orientation and so the specimen appears smaller than it does in phase contrast. (C) Complete coccolith in proximal view. (D) Oblique proximal view of broken coccolith. (E) Detail of D, with proto-coccolith ring locus visible. (F) Detail of proximal shield from a broken and slightly etched specimen showing the two layers in this structure (compare with crystal unit diagrams).

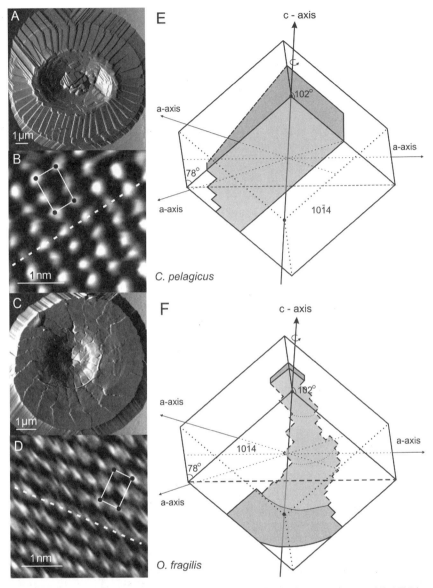

Figure 5. *Coccolithus pelagicus* and *Oolithotus fragilis* atomic force microscopy images. (A) AFM image of a complete *C. pelagicus* coccolith. Compare with Figure 4C, oblique stripes extending beyond the coccolith edge are tip artifacts. (B) Ultra high resolution AFM image from a large distal shield element, showing atomic pattern and demonstrating that rhombic faces are developed. A surface unit cell is shown, dashed line gives direction of atomic rows. (C–D) AFM images of complete *O. fragilis* coccolith and high resolution image from outer edge of coccolith showing atomic pattern and demonstrating that rhombic faces are developed. (E) Interpretation of relationship of faces developed on *C. pelagicus* distal shield elements (grey) to calcite rhomb. *N.B.* The element represented corresponds to those at bottom left of the specimen in A. (F) Interpretation of relationship of faces developed on *O. fragilis* distal shield elements (grey) to calcite rhomb; *N.B.* The element represented corresponds to those at bottom centre of the specimen in A, the pale grey portion of the element is a stepped surface. (B) and (D) have been corrected for distortion following the method in Henriksen et al 2002.

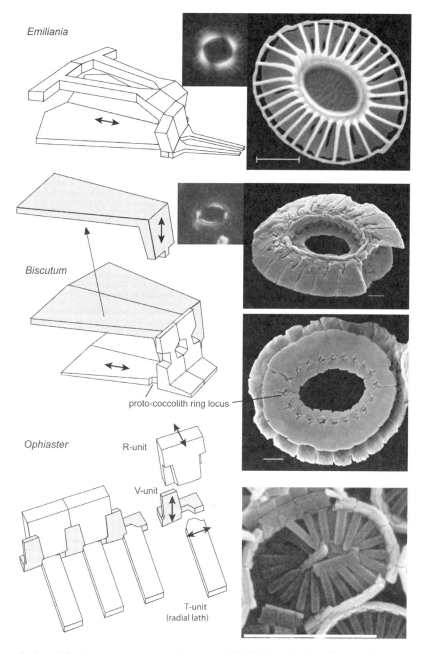

Figure 6. Coccolith structure in representative genera. (A) *Emiliania huxleyi*, diagram of crystal units; cross-polarized light micrograph of a coccolith (*N.B.* this micrograph is of *Pseudoemiliania lacunosa* a closely related fossil species with larger coccoliths); SEM of isolated coccolith in distal view. (B) *Biscutum magnum*, a fossil species from the Late Cretaceous: diagram of crystal units, cross-polarized light micrograph, and SEMs of distal and proximal surfaces. (C) *Ophiaster hydroideus*, an extant species from a plankton sample: diagram of crystal units and SEM of partially disintegrated coccolith. All scale bars one micrometer, double headed arrow symbols indicate orientation of calcite *c*-axis.

The *Coccolithus* structure is similar, but more complex—the V-unit again forms the distal shield but also extends onto the proximal surface as a lower tube element. The R-unit forms the proximal shield, but extends on the distal surface as an upper tube element (Fig. 3A-B). As a result, the two crystal unit types are intergrown and the proto-coccolith ring is embedded within the coccolith structure and only visible on broken specimens (Fig. 4D-E). Additional complexity is seen on the proximal shield with development of separate upper and lower elements growing at different angles. The upper unit is approximately radially directed (and elongate approximately parallel to the *c*-axis), while the lower element is oblique to this (Fig. 4F). Finally on complete coccoliths, the central area elements on both the proximal and distal surfaces appear to be composed of a mass of minute elements, but interconnections between them are often visible, and at intermediate growth phases the morphology is simpler; hence these are evidently a product of less regular growth at a late phase.

The *Emiliania* structure (Westbroek et al. 1989; Young 1989; Didymus et al. 1994) is unusual in being composed essentially of R-units, with the V-units confined to the proto-coccolith ring (Fig. 6A). As a result, the individual R-units have particularly complex form, being constructed of proximal shield, distal shield, inner tube and central area elements, each of which represents a separate growth direction from the nucleation site.

Ophiaster has a simple rim of alternating V- and R-units but the central area is floored by separate elements. These elements interdigitate with the rim units, but are crystallographically separate and have sub-tangential *c*-axis orientations. This structure with three separate crystal unit types has been described in detail for *Syracosphaera pulchra* (Young et al. in press).

Nucleation

The most remarkable common feature of the heterococcoliths is the development of the rim from a proto-coccolith ring of simple calcite crystals with alternating sub-vertical and sub-radial *c*-axis orientations—the V/R structure of Young et al. (1992). This is shown by each of the illustrated examples and has been demonstrated in numerous others (e.g., Young and Bown 1997; Marsh 1999; Young et al. 1999). In some cases, it has only been possible to demonstrate that the rim consists of two alternating crystal unit types, not to prove that these have different *c*-axis orientations. For instance, in the Papposphaeraceae the rim is clearly formed from two alternating crystal unit types (e.g., Manton and Oates 1975), but because the entire coccoliths are usually only 1-2 microns long and specimens are rare, it has not been possible for crystallographic orientations to be determined. However, it seems reasonable to predict that these fit the basic V/R model. This means that the first stage of biomineralization is nucleation of a ring of calcite nuclei with alternating orientations. This is a distinctive feature of heterococcoliths with no obvious parallels in any other biomineralization systems and has demonstrably remained constant through more than 200 million years of coccolith evolution (Young et al. 1992, 1999). Moreover, nucleation controls total calcite orientation, i.e., not simply *c*-axis orientation, but also *a*-axis orientation. This is evidenced by the fact that crystal faces on adjacent elements are sub-parallel (e.g., Fig. 1, 3C–G, 5A). It has also been demonstrated using SAED study of *Emiliania huxleyi* coccoliths (Davis et al. 1995) and AFM study of *Coccolithus pelagicus* and *Oolithotus fragilis* (Henriksen et al. in press a).

A final distinctive, nucleation-related feature is that all heterococcoliths show chirality. This is seen for instance in element orientations and cross-polarized light extinction crosses. The lower proximal shield elements of *Coccolithus* (Figs. 3A, 4F) are a good example. These grow at an oblique angle and always show clockwise offset from the radial direction in proximal view. This chirality is related to departures of the nucleation orientations from simple orthogonal radial and vertical directions, i.e., the nucleation is chiral (Didymus et al 1994; Young et al. 1999).

Thus, heterococcolith rim nucleation is characterized by a suite of features: alternation of *c*-axis orientations, control of *a*-axis orientations, stability through evolutionary history and chirality. This very precise nucleation control can most easily be explained in terms of control from a macromolecular template substrate. The principal alternative would be random nucleation followed by selection of favorably oriented nuclei, but this should only lead to selection of *c*-axis directions, as is seen in many other biomineral systems, not total crystal orientation, which is much more rare. Similarly, chirality is a predictable result of nucleation on a biochemical substrate since virtually all organic macromolecules are chiral and occur in natural systems as only one enantiomorph. Therefore, a plausible model for the precise nucleation control around the margin of the organic baseplate is that a belt of macromolecular substrate occurs in this position with arrays of potential binding sites for calcium or carbonate ions which promotes nucleation of oriented calcite crystallites. Young et al. (1992, 1999), and Marsh (1999) speculated on possible geometries of such a template that could result in the V/R pattern of evolutionarily stable alternation of nucleation directions.

This hypothesis is tempting, but it has not been substantiated by direct evidence. Furthermore, although all heterococcoliths appear to have rims conforming to the V/R model, additional, separately-nucleated crystals that do not fit this model often occur in the central area of heterococcoliths. *Ophiaster* illustrates one rather common pattern, seen in numerous coccoliths produced by the families Syracosphaeraceae and Rhabdosphaeraceae (Young et al. in press), an apparent pattern of V/R/T alternation. This would require a more complex template but could still conform to the general model. Such coccoliths, however, also often have additional elements separate from the rim. There are a few such elements visible in *Ophiaster* (Fig. 6C). A more extreme and very beautiful example is provided by *Rhabdosphaera* (Fig. 7). Here the rim, with inferred V/R structure, is very narrow and encloses a large mass of lamellar crystal units and a central spine of intergrown crystals (Kleijne 1992). The spine structure is remarkably elegant, consisting of five interwound spirals of crystals (Fig. 7C). The component crystals of the spine show regular rhombohedral faces with perfectly regular alignment spiraling up the structure. Evidently, nucleation here is not confined to the narrow belt around the baseplate but occurs across the base and up the spine, still with regular crystal alignment and consistent chirality. If a template is involved, it is unclear where it is located and it lacks the V/R alternation.

Growth

Regulation of calcite growth by coccolithophores is almost as remarkable as the regulation of nucleation. The crystal units sketched in Figures 3 and 6 are each single crystals of calcite, and their shapes are radically different from that of inorganic calcite. However, such crystals do not exist in isolation, but are parts of the complete coccolith, and much of the morphology is a product of interaction between adjacent crystals. Further, it is clear from the cytological observations that there is no precursor matrix for the coccolith, but rather that the coccolith grows in an expanding vesicle. The final structure of the completed coccoliths to a large extent an emergent result of inorganic growth of the crystals defined by the nucleation stage, within a space defined by the expanding vesicle.

Because inorganic growth directions are utilized, the precise orientation of the calcite crystal lattice in coccolith elements is important for developing morphology. Using atomic force microscopy, we have investigated *C. pelagicus* and *O. fragilis* coccoliths from

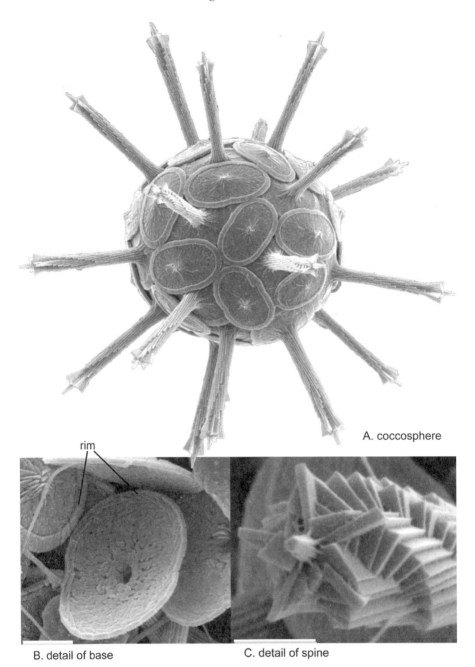

Figure 7. *Rhabdosphaera clavigera*. (A) Coccosphere. (B) Detail of base, showing the narrow rim, with V/R structure, surrounding a wide central area of separately nucleated lamellar elements, these have rhombic faces on their distal side and truncated surfaces on the proximal side. (C) A spine tip showing the spiral structure of spine formed from consistently aligned crystal units with rhombic faces. *N.B.* The pentaradiate apical structure is consistently developed. Scale bars = 1 μm.

micrometer to atomic scale and with these observations have established the precise relationships between element morphology and crystal lattice (Fig. 5; Henriksen et al. in press a). High resolution AFM of the flat surfaces of the elements (Fig. 5B and D) reveal an atomic pattern identical to that known from rhombic {10$\bar{1}$4} faces of inorganic calcite (Stipp et al. 1994; Stipp 1999). It shows the characteristic surface unit cell, atomic rows defining the orientation of the *a*-axis vector, and a pairing of these rows that constrains the crystallographic orientation fully (Henriksen et al. in press a). *C. pelagicus* has {10$\bar{1}$4} faces on the whole distal shield surface and has element edges defined by rhombic cleavage (Fig. 5E), with individual elements stuck together by imbrication. By contrast, in *O. fragilis* the crystal lattice is rotated so that acute crystallographic corners make the complex element edges. This enables elements to interlock without overlap, allowing for very thin and delicate coccoliths (Fig. 5F). The outer part of the distal shield consists of rhombic faces that are broken by a series of steps forming the slope towards the middle, probably induced by constraints on growth space by the vesicle. Thus, in both species the rhombic calcite motif is much in evidence, but detailed variations in orientation allow them to construct coccoliths with very different properties.

An example of the importance of the interplay of rhombic growth directions with vesicle geometry is provided by comparing coccoliths of the closely related species *Calcidiscus leptoporus*, *Umbilicosphaera foliosa* and *Umbilicosphaera sibogae* (Fig. 8). These belong to the family Calcidiscaceae and have their distal shields entirely formed from V-units (Young 1998; Young et al. in press). *Calcidiscus leptoporus* (Fig. 8A) shows flat, overlapping elements with curved radial edges. The surfaces are almost certainly rhombic calcite faces, as can be seen from overgrown specimens, from examination of face relationships in the central area, and by analogy with the related species *Coccolithus pelagicus* and *Oolithotus fragilis*. *Umbilicosphaera sibogae* (Fig. 8B) is similar, but has a wider central opening and radial edges showing a pronounced kink. *Umbilicosphaera foliosa* (Fig. 8C) also has flat, probably rhombic faces, but has two cycles in the shield. The inner cycle is imbricated clockwise, the outer is imbricated counter-clockwise, with both cycles showing crystallographic edges. Superficially, this is a very complex crystal structure. However, when a large number of specimens is investigated at high resolution, paying close attention to overgrown and malformed examples, it becomes evident that both shield cycles are made up of the same crystal units. This is illustrated in Figure 9. The crystal units actually have a rather simple basic shape, as indicated by the dotted lines on Figure 9B, which also reflect the shape of the elements on the proximal surface. This simplicity is in marked contrast to the apparent complexity on the distal surface, caused by counter-clockwise growth on the surface of the crystal in the outer part of the coccolith and clockwise growth in the inner part. The difference in growth direction is in fact a predictable result of the plane of the crystal surface intersecting obliquely with a conical surface (Fig. 9C). This is not seen in *C. leptoporus* since the elements are less obliquely oriented, and there are fewer of them. Figure 8D shows a specimen of *U. foliosa* with rather irregular growth that has the complete range of element types, from simple elements like those typical of *C. leptoporus,* through kinked elements like those of *U. sibogae,* to the offset bicyclic elements typical of *U. foliosa*. The fact that all three can occur on one coccolith as a result of malformation proves that no major changes in mechanism are necessary to form one rather than the other. Further, this suggests that the range of distal shield element morphologies seen in the three species arise as a product of interaction of crystal orientation and coccolith shape rather than through any direct control of crystal growth. This example is thought-provoking because it shows that in biomineralization, a trivial change in process can result in a large morphological effect.

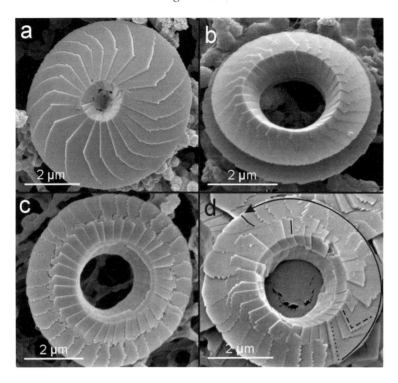

Figure 8. Growth pattern variations in *Calcidiscus* and *Umbilicosphaera*. Coccoliths of three closely related species, in distal view. In each of these species the entire distal shield is formed of V-units, and the visible surface are rhombic faces. (a) *Calcidiscus leptoporus*. (b) *Umbilicosphaera sibogae*. (c) well-formed *Umbilicosphaera foliosa* showing complex suture pattern and apparent division into two cycles of elements. (d) malformed *U. foliosa* showing a range of element morphologies from those typical of *C. leptoporus* to those typical of *U. foliosa*.

The surfaces shown in Figures 8 and 9 are all characterized by growth across the surface resulting in formation of inorganic rhombic crystal faces. These are common, especially on distal surfaces and on central area structures (e.g., Rhabdosphaera, Fig. 7). Another class of surfaces also occurs, characterized by absence of growth on the surface and development of non-crystal surfaces (Henriksen et al. in press b; Young et al. in press). The distal shield of *Coccolithus pelagicus* shows examples of both surface types; growth occurs incrementally on the upper surface, which develops as a rhombic face, while no growth occurs on the lower surface, which develops as a trailing surface behind the growth front. Some coccoliths, e.g., *Emiliania huxleyi*, appear to be almost entirely characterized by this surface type, with growth confined to narrow fronts at the tip of the element. With this type of growth it seems likely that crystal growth is blocked by organic coatings, most likely of coccolith-associated polysaccharides (Young et al. 1999). Such coatings are visible on TEM sections after staining for polysaccharides (e.g., Outka and Williams 1971; Marsh 1994; Klaveness 1976; Young et al. 1999). We have been able to demonstrate through AFM dissolution studies that these coatings do protect the completed crystals (Henriksen et al. in press b). It is reasonable to infer that a coating of polysaccharide that can prevent dissolution of an element would also serve to inhibit crystal growth. Therefore, the difference between the two calcite surface types seen in coccoliths—rhombic calcite

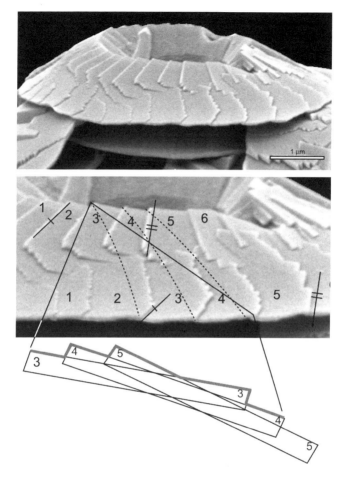

Figure 9. Complex element morphology from simple crystal growth. Side view of an *Umbilicosphaera foliosa* coccolith (top), detail of distal shield (middle), and interpretative diagram (bottom). Numbering on the crystal units indicates the degree of offset between the inner and outer cycles. The continuity of the crystal units between the cycles is directly visible on units 1 and 2 and is further indicated by development of parallel edges on the inner and outer cycles (lines on leading and trailing edges of units 2 and 5). Dotted lines indicate summary trajectory of units 3 and 4; this is approximately the pattern seen on the proximal surface of the distal shield. Diagram below: oblique section across the coccolith surface illustrating how the differing senses of apparent imbrication in the inner and outer cycles can be produced by a single set of crystal faces.

faces and non-crystalline faces—may be a function of selective blocking by polysaccharides. Nonetheless, is not clear why these two crystal surface types occur or how vesicles are able to block growth on some surfaces, but not others.

HOLOCOCCOLITHS

While heterococcoliths have been rather extensively studied, the alternative biomineral form, holococcoliths (Fig. 10), formed in the haploid life cycle stage, has received less attention and proven more enigmatic.

Figure 10. Holococcolith form. (A–C) *Syracosphaera anthos* holococcoliths (older name *Periphyllophora mirabilis*). (A) complete coccosphere; (B) detail of coccolith; (C) acid etched specimen (reproduced with permission from Halldal and Markali (1955), showing organic coatings around the coccoliths. (D) *Corisphaera* sp. partially collapsed coccolith with similar wall fabric to that of *S. anthos*, showing that it is formed of rhombohedral crystallites. (E) *Calyptrolithophora papillifera* showing perforate hexagonal crystallite arrangement. Again the constituent crystallites are rhombohedral. (F) *Coccolithus pelagicus* holococcoliths (older name *Crystallolithus hyalinus*), showing construction from rhombohedral crystallites in rhombohedral assembly.

Cytological observations

Very few holococcolithophore cultures have been maintained and only two species have been studied in cytological section. *Coccolithus pelagicus* was studied by Manton and Leedale (1963, 1969) and by Rowson et al. (1986) while *Calyptrosphaera sphaeroidea* was studied by Klaveness (1973). As with heterococcoliths these studies showed that the holococcoliths are underlain by organic base-plate scales and these scales could be seen developing in Golgi vesicles. However, despite numerous observations, no intracellular calcification could be seen, nor have intracellular holococcoliths been seen in light microscopy studies. Hence, it has been inferred that calcification occurs outside the cell membrane, after exocytosis of the base-plate scale. This poses obvious problems for understanding how calcification is regulated. A partial solution is provided by observations that, in motile species, exocytosis of coccoliths and scales occurs at the flagellar pole, and that, at least in the studied species, a delicate "skin" envelopes the coccosphere. Therefore, even if calcification does occur outside the cell membrane, it is likely to occur in a privileged and highly regulated environment. Alternatively, it is possible that calcification occurs just below the cell membrane but is a rapid process immediately preceding exocytosis and so has not been captured in cytological sections. Since holococcoliths in *statu nascendi* have not been observed, there is no evidence from cytology on the pattern of growth in holococcoliths.

Biochemical observations

There have been no studies on the biochemistry of holococcolith biomineralization. This reflects the need for biochemical work to focus on a few case studies and the very small number of holococcolith cultures available for study. In addition, the two most frequently cultured coccolithophore species, *Emiliania huxleyi* and *Pleurochrysis carterae* do not produce holococcoliths. The species have a similar haplo-diploid life-cycle to other coccolithophores, but the haploid phase cells do not calcify.

Morphological observations

The primary research carried out on holococcoliths is detailed study of morphology in order to establish their taxonomy (e.g., Kleijne 1991, Cros et al. 2000, Cros and Fortuño 2002). This has resulted in a rich archive of detailed morphological data. The primary data source is scanning electron microscopy; supplementary information comes from cross-polarized light microscopy, which has provided data on crystallographic orientation (Crudeli and Young 2003). Finally, invaluable observations of organic coatings were made by early Norwegian researchers, notably Halldal and Markali (1955), who decalcified plankton samples using hydrochloric acid. This technique left the organic coatings as ghost specimens. A representative image is included as Figure 10C. From this work several relevant observations can be made.

1. Holococcolith shape is highly regulated. Separate species have unique and very consistent morphologies. In the case of *Syracosphaera anthos* holococcoliths, (Fig. 10A–C) consistent features in addition to the gross shape, include the presence of a hole at the base of the leaf-like distal process and a double-layered structure to both the process and the main tube wall. In comparison to heterococcoliths, there is at least as much regulation of total coccolith shape.

2. The entire structure is made up of minute equant crystallites. This is obvious as a surface texture, but also applies where holococcoliths form thick basal structures. In such cases, the fabric consists of numerous layers of separate crystallites, rather than elongated crystallites. Moreover, observations of disintegrating holococcoliths (e.g., Fig. 10D) suggest that crystallites are predominantly and

perhaps invariably, euhedral rhombohedra. The crystallites are often referred to as hexagonal, but this is more accurately a description of the arrangement of the crystallites.

3. The crystallites do not interlock or intergrow. This is very nicely shown by the acid etched specimens (Fig. 10C), but again observations on numerous other specimens show it is a general pattern.
4. The crystallites are each enveloped in an organic coating. The best evidence for this comes from the acid etched specimens. Presumably, this is similar to the polysaccharide coatings of heterococcoliths. In SEM the holococcolith crystallites often have rough surfaces rather than smooth crystal faces; this may be due to these organic coatings.
5. The crystal faces of the crystallites are clearly aligned across large areas of the holococcoliths; thus the crystallographic orientation of the crystallites is completely controlled.
6. Although all holococcolith crystallites are broadly similar, there are consistent differences in crystallite size and surface morphology, both between species and occasionally even over different parts of a single holococcolith. This is shown to a limited extent by the *S. anthos* holococcoliths, where there is a field of larger, less regular crystallites at the base of the tube. Note also that the crystallites of *C. pelagicus* holococcoliths (Fig. 10F) are significantly larger (~0.2 µm across) than those of the other figured species (which are ~0.1 µm across).
7. Crystallite arrangement patterns are clearly controlled by the biomineral system, varying between species and between parts of single holococcoliths. The most common arrangement is a hexagonal fabric (Figs. 10A–D), in which rhombohedral crystallites are arranged edge to edge, with their *c*-axes directed perpendicular to the surface. This may be modified into a hexagonal mesh fabric, in which one in four crystallites is missing, giving a sieve-like appearance (Fig. 10E). Other less regular variations on this basic pattern occur. Rhombohedral array fabrics (Fig. 10F) are distinctively different; the rhombohedra are arranged with the *c*-axes oblique to the surface, and the crystallites are aligned face to face.

All these features tend to emphasize the basic observation that holococcolith biomineralization is a remarkably highly regulated process with the final morphology a product of biologically controlled mineralization, even if it occurs outside the cell membrane. There is, however, still essentially no model for how the growth regulation occurs. Crystallite growth does not show the remarkable sculpting of heterococcolith crystal units, but the regular termination of growth such that each crystallite is of regular size and shape is striking. This could be a product of biologically mediated growth; calcite rhombohedra of uniform size can be produced *in vitro* with surface inhibitors (Henriksen unpubl. data).

Control of crystallite orientation is difficult to understand, since the dispersed nucleation necessary to produce all the crystallites and the layered structure in some species makes it difficult to envisage how or where an organic template precursor could occur. An alternative that should be considered is that the crystallites are initially randomly oriented and are aligned by the coccolith shaping system. It is even conceivable that the crystallites are not formed *in situ* but are delivered preformed to the growing holococcolith.

It is frustrating that biomineralization in holococcoliths is so poorly understood and that it has not been possible to make any direct observations of the growth process, not least since this appears to be a rather flexible biomineralization mechanism capable of construction of elaborate structures from simple biomineral components. Therefore, it may prove a more useful model system than heterococcolith biomineralization for potential applications in nanofabrication.

THE EVOLUTION OF COCCOLITHOPHORE BIOMINERALIZATION MECHANISMS

As explained earlier, the haplo-diploid life cycles with holococcolith-heterococcolith alternation are common in coccolithophores and indeed occur in at least six of the 13 described families. These families span the phylogenetic diversity of coccolithophorids, as inferred from both the fossil record and molecular genetics (Young et al. 1999; Houdan et al. in press). Hence, it seems that holococcolith-heterococcolith differentiation must be a primitive evolutionary feature, with an origin near in time to that of calcification in coccolithophores as a whole. This is also supported by the fact that fossil holococcoliths, although much rarer than fossil heterococcoliths, have a record which extends back nearly as far through geological time. Definite holococcoliths are known back to the Early Jurassic ca. 190 Ma while definite heterococcoliths occur back to Late Triassic, ca. 220 Ma (Bown 1993, 1998a,b). Given that both heterococcoliths and holococcoliths are elaborate biomineral structures with numerous unique features, they must each have evolved only once. Hence, it is likely that biomineralization evolved first in one life cycle phase, then transferred to the other, although, for some reason, only part of the process could be transferred.

The occurrence of the heterococcoliths in the fossil record before that of holococcoliths provides weak evidence that they evolved first. This might be dismissed as a preservational artifact but it is supported by two other lines of evidence. First, all of the published molecular genetic analyses (Edvardsen et al. 2000; Fujiwara et al. 2001, Saez et al. in press) find the basal coccolithophore clade is comprised of the genera *Emiliania* and *Gephyrocapsa* (Family Noelaerhabdaceae). These genera only calcify in the diploid phase, so their basal position in the tree suggests that the absence of calcification in the haploid phase may be a primitive character. This cannot be taken as definite evidence however, since calcification is also absent in the haploid phase of *Pleurochrysis* and *Hymenomonas,* but these have a much higher position within the coccolithophore clade and therefore must have lost holococcolith biomineralization secondarily. Second, two coccolithophore life cycles are now known in which a heterococcolith phase alternates with a nannolith-producing phase (Fig. 11), i.e., a phase that produces calcareous structures which do not show the characteristic features of either holococcoliths or heterococcoliths (Young et al. 1999). These examples are (1) *Ceratolithus,* in which a heterococcolith-producing phase alternates with a phase producing large horseshoe-shaped nannoliths (Alcober and Jordan 1997; Young et al. 1998; Cros et al. 2000; Sprengel and Young 2000). (2) *Alisphaera,* in which a heterococcolith-producing phase (Fig. 11E) alternates with a phase producing minute, cup-shaped, aragonitic nannoliths (Fig. 11D–F), as evidenced by the occurrence of spectacular combination coccospheres (Cros and Fortuño 2002). These nannoliths, formerly regarded as a discrete genus, "*Polycrater*", show some similarities to normal heterococcoliths but are unique in being formed of aragonite (Manton and Oates 1980) rather than calcite. They also lack the typical rim features of heterococcoliths; consequently Young et al. (1999) highlighted their anomalous structure and considered them likely candidates for additional biomineralization modes within the coccolithophores. This prediction is strongly supported by the combination coccosphere evidence. Neither *Ceratolithus* nor *Alisphaera* have yet been isolated into culture so there

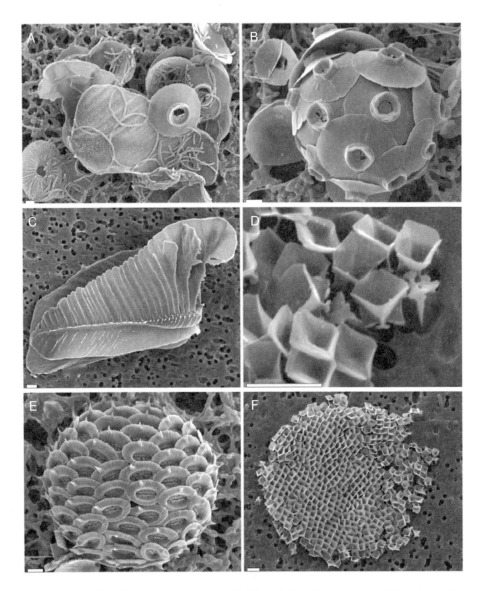

Figure 11. Nannoliths. Two examples of species with life cycles involving heterococcoliths and nannoliths rather than heterococcoliths and holococcoliths. (A–C) *Ceratolithus cristatus:* (A) collapsed combination coccosphere with heterococcoliths and ceratolith nannolith; (B) coccosphere of heterococcoliths; (C) isolated ceratolith. (D–F) *Alisphaera*: (D) isolated nannoliths, minute cup-shaped aragonitic liths (also known as *Polycrater galapagensis*); (E) coccosphere of the heterococcolith phase; (F) coccosphere of the nannolith ("*Polycrater*") phase.

are no direct observations on ploidy level (relative genome size) in the two phases. We hypothesize that, as in all studied cases to date, the heterococcolith-producing phase will prove to be diploid. The nannoliths produce by these two genera are very different from each other as well as from either heterococcoliths or holococcoliths so they are very unlikely to be directly related. Instead, we predict they will prove to be two additional examples of transfer of calcification from the diploid to the haploid phase with necessary evolution of a new mechanism for calcification resulting each time in distinctively different biomineral structures.

Summary

These inferences can be summarized in a schematic evolutionary tree (Fig. 12). Heterococcolith biomineralization occurs consistently in the diploid phase and so is likely to have been the primitive calcification mode. Subsequently calcification has apparently been transferred to the haploid phase on at least three occasions with a different biomineralization mode evolving in each case. Given this, a comparison of features between the biomineralization modes may be of value.

Crystallographic orientation—template control or a self-organizing system?

Precise control of crystallographic orientation, including placement and spacing of crystal nuclei and both c- and a-axis orientation is a pervasive feature of coccolith biomineralization shown equally well by heterococcolith rims, heterococcolith central area structures, nannoliths and holococcoliths. In general, this is not a common feature of

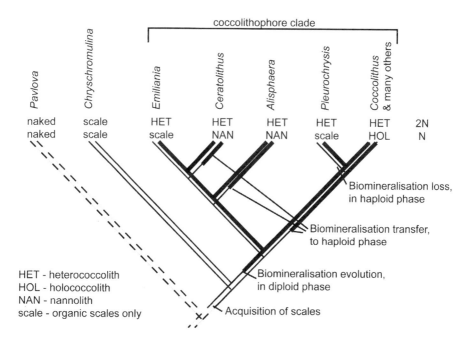

Figure 12. Evolution of biomineralization modes. Phylogenetic tree showing inferred pattern of acquisition of biomineralization modes in haptophytes. The double lines represent the diploid (upper) and haploid (lower) life cycle phases, thick lines indicate biomineralization within these phases. The position of *Ceratolithus* and *Alisphaera* within the tree is speculative, the rest of the tree topology corresponds to that from molecular genetic data (Edvardsen et al. 2000; Saez et al. in press).

biomineralization systems, hence it is reasonable to infer that the nucleation control mechanism should be similar for all coccolith cases. For the V/R alternation of nuclei in the heterococcolith rim, template control has been a tempting theory, particularly since the nucleation is confined to single, well-defined belts with the organic base-plate scale available as a substrate for the nucleation template (Young et al. 1992; Marsh et al. 1999). However, with holococcoliths and heterococcolith central area structures, crystallographic orientation control is equally precise, but nucleation is much more dispersed. Hence, a self-organizing system seems more likely than nucleation on a pre-designed template. Given this insight, *in vitro* experiments on the effects of coccolith-associated macromolecules become more important.

Crystal growth regulation

The styles of growth regulation seen in heterococcoliths, holococcoliths and nannoliths are rather variable. Heterococcolith rim crystals show selected growth in particular directions, complex development of rhombic faces to fill non-rhombohedral space and blocked growth to produce non-crystalline surfaces. Heterococcolith central area growth tends to show only rhombic face development, but with elaborate interlocking crystal growth. In holococcoliths, growth is confined to rhombohedra but blocked at a finite size. So there is great variation in the details of precise regulation within a basic motif.

Genomic approaches

The parallel approaches of biochemical and morphological research have been applied rather extensively to coccolithophore calcification and are arguably at a relatively mature state of knowledge, although new insights are emerging from atomic force microscopy (Henriksen et al. in press a,b) and from study of polysaccharide effects in mutant strains (Marsh and Dickinson 1997; Marsh 2000). Probably the most promising approach for the future will be genomic research aimed at identifying the suite of genes responsible for biomineralization and their functions. In this general field, the coccolithophores may prove to be ideal test cases for a number of reasons. First, as protests, they have relatively small genomes. Second, they are readily grown in culture and can be experimented upon without ethical dilemmas. Third, the considerable interest in coccolithophores from other perspectives means that genomic projects on them are already underway. Finally, the fact that different biomineralization modes occur within two phases of a single organism provides a natural experimental system for comparing the biomineralization-related genes expressed in the two life stages.

ACKNOWLEDGMENTS

We are grateful to many colleagues for scientific collaboration over an extended period, especially Paul Bown, Markus Geisen, Steven Mann, Ian Probert, Blair Steel, Susan Stipp, Mary Marsh and Peter Westbroek. Research funding has been provided by the UK Natural Environment Research Council, The European Union (through the CODENET FP4 TMR network project) and the Danish Research Council.

REFERENCES

Alcober J, Jordan RW (1997) An interesting association between *Neosphaera coccolithomorpha* and *Ceratolithus cristatus* (Haptophyta). Eur J Phycol 32:91-93

Billard C (1994) Life cycles. *In:* The Haptophyte Algae. Green JC, Leadbeater BSC (eds) Clarendon Press, Oxford, p 167-186

Billard C, Inouye I (in press) What's new in coccolithophore biology? *In:* Coccolithophores - From molecular processes to global impact. Thierstein HR, Young JR (eds) Springer, Heidelberg

Black M (1963) The fine structure of the mineral parts of coccolithophorids. Proceedings of the Royal Society of London 174:41-46
Bown PR (1993) New holococcoliths from the Toarcian-Aalenian (Jurassic) of northern Germany. Senckenbergiana lethaea 73:407-419
Bown PR (1998a) Calcareous Nannofossil Biostratigraphy. Chapman & Hall, London
Bown PR (1998b) Triassic. *In:* Calcareous Nannofossil Biostratigraphy. Bown PR (ed) Chapman & Hall, London, p 29-33
Bown PR, Young JR (1998) Techniques. *In:* Calacareous Nannofossil Biostratigraphy. Bown PR (ed) Chapman & Hall, London, p 16-28
Brand LE (1994) Physiological ecology of marine coccolithophores. *In:* Coccolithophores. Winter A, Siesser WG (eds) Cambridge University Press, Cambridge, p 39-49
Corstjens PLAM, Kooij Avd, Linschooten C, Brouwers G-J, Westbroek P, de Vrind-de Jong EW (1998) GPA, a Clcium-binding protein in the coccolithophorid Emiliania huxleyi (Prymnesiophyceae). J Phycol 34:622-630
Cros L, Fortuño J-M (2002) Atlas of northwestern Mediterranean coccolithophores. Scientia Marina 66:1-186
Cros L, Kleijne A, Zeltner A, Billard C, Young JR (2000) New examples of holococcolith-heterococcolith combination coccospheres and their implications for coccolithophorid biology. Mar Micropaleontol 39:1-34
Crudeli D, Young JR (2003) SEM-LM study of holococcoliths preserved in Eastern Mediterranean sediments (Holocene/Late Pleistocene). J Nannoplankton Res 25:39-50
Davis SA, Young JR, Mann S (1995) Electron microscopy studies of shield elements of *Emiliania huxleyi* Coccoliths. Botanica Mar 38:493-497
de Vrind-de Jong EW, Borman AH, Thierry R, Westbroek P, Grüter M, Kamerling JP (1986) Calcification in the coccolithophorids *Emiliania huxleyi* and *Pleurochrysis carterae* II. Biochemical aspects. Syst Assoc Spec Vol 30, p 205-217
de Vrind-de Jong EW, de Vrind JPM (1997) Algal deposition of carbonates and silicates. Rev Mineral 35:267-307
de Vrind-de Jong EW, van Emburg P, de Vrind JPM (1994) Mechanisms of calcification: *Emiliania huxleyi* as a model system. *In:* The Haptophyte Algae. Green JC, Leadbeater BSC (eds) Syst Assoc Spec Vol 51, p 149-166
Didymus JM, Young JR, Mann S (1994) Construction and morphogenesis of the chiral ultrastructure of coccoliths from marine alga *Emiliania huxleyi*. Proc R Soc Ser B 258:237-245
Edvardsen B, Eikrem W, Green JC, Andersen RA, Yeo Moon-van der Staay S, Medlin LK (2000) Phylogenetic reconstructions of the Haptophyta inferred from 18S ribosomal DNA sequences and available morphological data. Phycologia 39:19-35
Fichtinger-Schepman AMJ, Kamerling JP, Versluis C, Vligenthart JFG (1981) Structural studies of the methylated, acidic polysaccharide associated with coccoliths of *Emiliania huxleyi* (Lohmann) Kamptner. Carbohydrate Research 93:105-123
Fresnel J (1986) Nouvelles Observations sur une Coccolithacée rare: *Cruciplacolithus neohelis* (McIntyre et Bé) Reinhardt (Prymnesiophyceae). Protistologica 22:193-204
Fujiwara S, Tsuzuki M, Kawachi M, Minaka N, Inouye I (2001) Molecular phylogeny of the haptophyta based on the *rbc*L gene and ssequence variation in the spacer region of the RUBISCO operon. J Phycol 37:121-129
Geisen M, Billard C, Broerse ATC, Cros L, Probert I, Young JR (2002) Life-cycle associations involving pairs of holococcolithophorid species: intraspecific variation or cryptic speciation? Eur J Phycol 37:531-550
Green JC, Course PA, Tarran GA (1996) The life-cycle of *Emiliania huxleyi*: A brief review and a study of relative ploidy levels analysed by flow cytometry. J Mar Syst 9:33-44
Halldal P, Markali J (1955) Electron microscope studies on coccolithophorids from the Norwegian Sea, the Gulf Stream and the Mediterranean. Avh Norske VidenskAkad Oslo 1:1-30
Henriksen K, Stipp SLS, Young JR, Bown PR (in press-a) Tailoring calcite: Nanoscale AFM of coccolith biocrystals. Am Mineral 88:
Henriksen K, Young JR, Bown PR, Stipp SLS (in press-b) Coccolith biomineralization studied with atomic force microscopy. Palaeontology 47:
Henriksen K, Stipp SLS (2002) Image distortion in scanning probe microscopy. Am Mineral 87:5-17
Houdan A, Billard C, Marie D, Not F, Sáez AG, Young JR, Probert I (in press) Flow cytometric analysis of relative ploidy levels in holococcolithophore-heterococcolithophore (Haptophyta) life cycles. Systematics and Biodiversity

Inouye I, Pienaar RN (1984) New observations on the coccolithophorid *Umbilicosphaera sibogae* var. *foliosa* (Prymnesiophyceae) with reference to cell covering, cell structure and flagellar apparatus. Br Phycol J 19:357-369

Inouye I, Pienaar RN (1988) Light and electron microscope observations of the type species of *Syracosphaera, S. pulchra* (Prymnesiophyceae). Br Phycol J 23:205-217

Kamptner E (1954) Untersuchungen über den Feinbau der Coccolithen. Anz öst Akad Wiss Mathematisch-Naturwissenschaftliche Klasse 87:152-158

Kirschvink JL, Hagadorn JW (2000) A grand unified theory of biomineralization. *In:* Biomineralization: from Biology to Biotechnology and Medical Application. Bäuerlein E (ed) Wiley-VCH, Weinheim, p 139-149

Klaveness D (1973) The microanatomy of *Calyptrosphaera sphaeroidea*, with some supplementary observations on the motile stage of *Coccolithus pelagicus*. Norw J Bot 20:151-162

Klaveness D (1976) *Emiliania huxleyi* (Lohmann) Hay & Mohler. III. Mineral deposition and the origin of the matrix during coccolith formation. Protistologica 12:214-224

Kleijne A (1991) Holococcolithophorids from the Indian Ocean, Red Sea, Mediterranean Sea and North Atlantic Ocean. Mar Micropaleontol 17:1-76

Kleijne A (1992) Extant Rhabdosphaeraceae (coccolithophorids, class Prymnesiophyceae) from the Indian Ocean, Red Sea, Mediterranean Sea and North Atlantic Ocean. Scr Geol 100:1-63

Mann S (1993) Molecular tectonics in biomineralization and biomimetic materials chemistry. Nature 365:499-505

Manton I, Leedale GF (1963) Observations on the microanatomy of *Crystallolithus hyalinus* Gaarder and Markali. Arch Mikrobiol 47:115-136

Manton I, Leedale GF (1969) Observations on the microanatomy of *Coccolithus pelagicus* and *Cricosphaera carterae*, with special reference to the origin and nature of coccoliths and scales. J Mar Biol Ass U K 49:1-16

Manton I, Oates K (1975) Fine-structural observations on *Papposphaera* Tangen from the Southern Hemisphere and on *Pappomonas* gen. nov. from South Africa and Greenland. Br Phycol J 10:93-109

Manton I, Oates K (1980) *Polycrater galapagensis* gen. et sp. nov., a putative coccolithophorid from the Galapagos Islands with an unusual aragonitic periplast. Br Phycol J 15:95-103

Marsh ME (1994) Polyanion-mediated mineralization—assembly and reorganization of acidic polysaccharides in the Golgi system of a coccolithophorid alga during mineral deposition. Protoplasma 177:108-122

Marsh ME (1999) Coccolith crystals of *Pleurochrysis carterae*: Crystallographic faces, organization and development. Protoplasma 207:54-66

Marsh ME (2000) Polyanions in the $CaCO_3$ mineralization of coccolithophores. *In:* Biomineralization: from Biology to Biotechnology and Medical Application. Bäuerlein E (ed) Wiley-VCH, Weinheim, p 251-268

Marsh ME, Chang D, King GC (1992) Isolation and characterization of a novel acidic polysaccharide containing tartrate and glyoxylate residues from the mineralized scales of a unicellular coccolithophorid alga *Pleurochrysis carterae*. J Biol Chem 267:20507-20512

Marsh ME, Dickinson DP (1997) Polyanion mediated mineralization—mineralization in coccolithophore (*Pleurochrysis carterae*) variants which do not express PS2, the most abundant and acidic mineral-associated polyanion in wild-type cells. Protoplasma 199:9-17

Marsh ME, Ridall AL, Azadi P, Duke PJ (2002) Glacturonomannan and Golgi-derived membrane linked to growth and shaping of biogenic calcite. J Struct Biol 139:39-45

Medlin LK, Kooistra WHCF, Potter D, Saunders JB, Andersen RA (1997) Phylogenetic relationships of the "golden algae" (haptophytes, heterokont chromphytes) and their plastids. Pl Syst Evol Suppl 11:187-219

Noël M-H, Kawachi M, Inouye I (in press) Induced dimorphic life cycle of a coccolithophorid, *Calyptrosphaera sphaeroidea*. J Phycol

Outka DE, Williams DC (1971) Sequential coccolith morphogenesis in *Hymenomonas carterae*. J Protozool 18:285-297

Ozaki N, Sakuda S, Nagasawa H (2001) Isolation and some characterization of an acidic polysaccharide with anti-calcification activity from coccoliths of a marine alga, *Pleurochrysis carterae*. Biosci Biotechnol Biochem 65:2330-2333

Paasche E (2002) A review of the coccolithophorid *Emiliania huxleyi* (Prymnesiophyceae), with particular reference to growth, coccolith formation, and calcification-photosynthesis interactions. Phycologia 40:503-529

Parke M, Adams I (1960) The motile (*Crystallolithus hyalinus* Gaarder & Markali) and non-motile phases in the life history of *Coccolithus pelagicus* (Wallich) Schiller. J Mar Biol Ass U K 39:263-274

Pienaar RN (1994) Ultrastructure and calcification of coccolithophores. *In:* Coccolithophore. Winter A, Siesser WG (eds) Cambridge University Press, Cambridge, p 13-37
Prins B (1969) Evolution and stratigraphy of cocolithinids from the Lower and Middle Lias. *In:* First International Conference Planktonic Microfossils, Vol. 2. Bronnimann P, Renz HH (eds) Geneva, E. J. Brill, Leiden, p 547-558
Romein AJT (1979) Lineages in Early Paleogene calcareous nannoplankton. Utrecht Micropaleontol Bull 22:1-231
Rowson JD, Leadbeater BSC, Green JC (1986) Calcium carbonate deposition in the motile (*Crystallolithus*) phase of *Coccolithus pelagicus* (Prymnesiophyceae). Br Phycol J 21:359-370
Sáez AG, Probert I, Geisen M, Quinn P, Young JR, Medlin LK (2003) Pseudo-cryptic speciation in coccolithophores. Proc Natn Acad Sci USA 100:7163-7168
Sáez AG, Probert I, Young JR, Medlin LK (in press) A review of the phylogeny of the Haptophyta. *In:* Coccolithophores - From Molecular Processes to Global Impact. Thierstein HR, Young JR (eds) Springer, Heidelberg
Sprengel C, Young JR (2000) First direct documentation of associations of *Ceratolithus cristatus* ceratoliths, hoop-coccoliths and *Neosphaera coccolithomorpha* planoliths. Mar Micropaleontol 39:39-41
Stipp SLS, Eggelston CM, Nielsen BS (1994) Calcite surface structure observed at microtopographic and molecular scales with atomic force microscopy (AFM). Geochim Cosmochin Acta 58:3023-3033
Stipp SLS (1999) Toward a conceptual model of the calcite surface: Hydration, hydrolysis and surface potential. Geochim Cosmochim Acta 63:3121-3131
Thierstein HR, Young JR (in press) Coccolithophores - From molecular processes to global impact, Springer, Heidelberg
van Emburg PR, de Jong EW, Daems W Th (1986) Immunochemical localization of a polysaccharide from biomineral structures (coccoliths) of *Emiliania huxleyi.* J Ultrastruct Molecular Struct Res 94:246-259
van der Wal P, de Jong EW, Westbroek P, de Bruijn WC, Mulder-Stapel AA (1983) Polysaccharide localization, coccolith formation, and Golgi dynamics in the coccolithophorid *Hymenomonas carterae.* J Ultrastr Res 85:139-158
Westbroek P, Brown CW, van Bleijswijk J, Brownlee C, Brummer G-JA, Conte M, Egge JK, Fernandez E, Jordan RW, Knappersbusch M, Stefels J, Veldhuis MJW, van der Wal P, Young JR (1993) A model system approach to biological climate forcing. The example of *Emiliania huxleyi.* Global Planetary Change 8:27-46
Westbroek P, Young JR, Linschooten K (1989) Coccolith production (Biomineralization) in the marine alga *Emiliania huxleyi.* J Protozool 36:368-373
Wilbur KM, Watabe N (1963) Experimental studies on calcification in molluscs and the alga *Coccolithus huxleyi.* Ann N Y Acad Sci 109:82-112
Young JR (1989) Observations on heterococcolith rim structure and its relationship to developmental processes. *In:* Nannofossils and Their Applications. Crux JA, Heck SEV (eds) Ellis Horwood, Chichester, p 1-20
Young JR (1992) The description and analysis of coccolith structure. *In:* Nannoplankton Research. Hamrsmid B, Young JR (eds) ZPZ, Knihovnicha, p 35-71
Young JR (1998) Neogene. *In:* Calcareous Nannofossil Biostratigraphy. Bown PR (ed) Chapman & Hall, London, p 225-265
Young JR, Bown PR (1991) An ontogenetic sequence of coccoliths from the Late Jurassic Kimmeridge Clay of England. Palaeontology 34:843-850
Young JR, Bown PR (1997) Higher classification of calcareous nannofossils. J Nannoplankton Res 19:15-20
Young JR, Davis SA, Bown PR, Mann S (1999) Coccolith ultrastructure and biomineralization. J Struct Biol 126:195-215
Young JR, Didymus JM, Bown PR, Prins B, Mann S (1992) Crystal assembly and phylogenetic evolution in heterococcoliths. Nature 356:516-518
Young JR, Henriksen K, Probert I (in press) Structure and morphogenesis of the coccoliths of the CODENET species. *In:* Coccolithophores—From Molecular Processes to Global Impact. Thierstein HR, Young JR (eds) Springer, Heidelberg
Young JR, Jordan RW, Cros L (1998) Notes on nannoplankton systematics and life-cycles—*Ceratolithus cristatus, Neosphaera coccolithomorpha* and *Umbilicosphaera sibogae.* J Nannoplankton Res 20:89-99

8   Biologically Controlled Mineralization in Prokaryotes

Dennis A. Bazylinski
Department of Biochemistry, Biophysics, and Molecular Biology
Iowa State University
Ames, Iowa 50011 U.S.A.

Richard B. Frankel
Department of Physics
California Polytechnic State University
San Luis Obispo, California 93407 U.S.A.

INTRODUCTION

As stated in an Chapter 4, prokaryotes of both Domains or Superkingdoms, the Bacteria and the Archaea, mediate the formation of a large number of diverse minerals. They are known to do this either through biologically induced mineralization (BIM) (Lowenstam 1981) (the subject of Chapter 4 in this volume; Frankel and Bazylinski 2003) or biologically controlled mineralization (BCM). The latter has also been referred to as organic matrix-mediated mineralization (Lowenstam 1981) and boundary-organized biomineralization (Mann 1986). There are several important differences between BIM and BCM that will be discussed in detail in this chapter, most notably those dealing with aspects of the mineral crystals and the biomineralization process (see also Weiner and Dove 2003). However, there is another significant difference between the two modes of biomineralization and this is the aspect of functionality. Generally, in BIM there is no function to the biomineralized particles except, perhaps, as a solid substrate for attachment in the case of bacteria or as a form of protection against certain environmental conditions or attack from predators. However, it is easier to recognize the primary function of bone or shells produced by molluscs; two well-characterized examples of BCM by higher organisms. Functionality should always be examined when dealing with examples of biomineralization particularly in situations where the mineral product displays some qualities of both BIM and BCM.

In BCM, the organism exerts a great degree of crystallochemical control over the nucleation and growth of the mineral particles. For the most part, the minerals are directly synthesized at a specific location within or on the cell and only under certain conditions. The mineral particles produced by bacteria in BCM are characterized as well-ordered crystals with narrow size distributions, and specific, consistent particle morphologies. Because of these features, BCM processes are likely to be under specific chemical/biochemical and genetic control. In the microbial world, the most characterized example of BCM is magnetosome formation by the magnetotactic bacteria, a group of microorganisms in which BCM-produced magnetic crystals appear to have a relatively specific function. There are some examples of biomineralization that appear to be intermediate between BIM and BCM; these are covered in Chapter 4 (Frankel and Bazylinski 2003) since the mineralization product usually has more features in common with BIM. This chapter is, in effect, a review of the magnetotactic bacteria. However, it should be remembered that the production of magnetic minerals by the magnetotactic bacteria is used in this volume as an example of BCM. The microbiological, geological, chemical, biochemical, and physical approaches used to study BCM in the magnetotactic bacteria are universal and will certainly be utilized again in other examples of BCM in bacteria discovered in the future.

THE MAGNETOTACTIC BACTERIA

Classification and general features

The magnetotactic bacteria are a heterogeneous group of prokaryotes that passively align and actively swim along the Earth's geomagnetic field lines (Blakemore 1975, 1982). This group is morphologically, metabolically, and phylogenetically diverse and cellular morphotypes that have been observed include coccoid (roughly spherical or ovoid), rod-shaped, vibrioid (curved), spirilloid (helical) and even multicellular forms (Blakemore et al. 1982; Farina et al. 1983; Rogers et al. 1990; Bazylinski 1995). The term "magnetotactic bacteria" therefore has no taxonomic significance and should be interpreted as a collection of diverse bacteria that possess the apparently widely distributed trait of magnetotaxis (Bazylinski 1995).

Despite the diversity of these magnetotactic bacteria, they have several important features in common: 1) all that have been described are gram-negative members of the Domain Bacteria (it is possible that some members of the Domain Archaea produce magnetosomes but none have been reported); 2) they are all motile, generally by flagella (although this does not preclude the possibility of the existence of non-motile bacteria that synthesize magnetosomes which, by definition, would be magnetic but not magnetotactic); 3) all exhibit a negative tactile and/or growth response to atmospheric concentrations of oxygen; 4) all strains in pure culture have a respiratory form of metabolism (i.e., none are known to ferment substrates); and 5) they all possess a number of magnetosomes, the signature feature of the group (Bazylinski 1995; Bazylinski and Moskowitz 1997). The bacterial magnetosome is defined as an intracellular single-magnetic-domain crystal of a magnetic iron mineral enveloped by a membrane (Balkwill et al. 1980). The membrane is intracellular and may be connected to the cytoplasmic membrane. There is some evidence that the magnetosome membrane comprises a vesicle in which the magnetosome subsequently nucleates and grows (Gorby et al. 1988).

There are two general types of magnetotactic bacteria based upon the minerals they biomineralize: there are the iron-oxide types that mineralize crystals of magnetite (Fe_3O_4) and the iron-sulfide types that mineralize crystals of greigite (Fe_3S_4) (Bazylinski and Frankel 2000b). The iron-oxide type are obligate microaerophiles, facultative anaerobes (that are microaerophilic when growing with O_2), or obligate anaerobes while the iron-sulfide type appear to be obligate anaerobes.

Ecology of magnetotactic bacteria

Magnetotactic bacteria are ubiquitous in aquatic environments containing water with pH values close to neutrality and are not thermal, strongly polluted, or well-oxygenated (Moench and Konetzka 1978; Blakemore 1982). They are cosmopolitan in distribution (Blakemore 1982) and because magnetotactic bacterial cells are easy to observe and separate from mud and water by exploiting their magnetic behavior using simple laboratory magnets (Moench and Konetzka 1978), there are frequent reports of their occurrence in various freshwater and marine locations (e.g., Matitashvili et al. 1992; Iida and Akai 1996).

On a local basis, magnetotactic bacteria are generally present in the highest numbers at the microaerobic oxic-anoxic transition zone (OATZ) and just below it in the anaerobic zone (Bazylinski 1995). In many freshwater habitats, the OATZ is located at the sediment-water interface or just below it. However, in some brackish-to-marine systems, the OATZ is found, or is seasonally located, in the water column. The Pettaquamscutt Estuary (Narragansett Bay, RI, USA) (Bazylinski et al. 1995) and Salt Pond (Woods Hole, MA, USA) (Wakeham et al. 1984, 1987) are typical examples of this situation. Hydrogen sulfide, produced by sulfate-reducing bacteria in the anaerobic zone and sediment, diffuses upward while oxygen diffuses downward from the surface resulting in a double, vertical, chemical concentration gradient with a co-existing redox gradient. Strong pycnoclines and other physical factors, probably including the microorganisms themselves, stabilize the vertical chemical gradients and the resulting OATZ.

Several morphotypes of iron-oxide and iron-sulfide magnetotactic bacteria are found at sites like the Pettaquamscutt Estuary and Salt Pond. The magnetite-producing magnetotactic bacteria are generally present at the OATZ proper and appear to behave as microaerophiles. Experiments with pure cultures support this observation. Two strains of magnetotactic bacteria have been isolated from the Pettaquamscutt and are now in pure culture. One is a vibrio designated strain MV-2 (Meldrum et al. 1993b; Dean and Bazylinski 1999) and the other is a coccus designated strain MC-1 (DeLong et al. 1993; Meldrum et al. 1993a; Frankel et al. 1997; Dean and Bazylinski 1999). Both strains grow as microaerophiles although strain MV-2 can also grow anaerobically with nitrous oxide (N_2O) as a terminal electron acceptor. Other cultured magnetotactic bacterial strains, including spirilla (e.g., Blakemore et al. 1979; Schleifer et al. 1991; Schüler et al. 1999) and rods (Sakaguchi et al. 1993a), are microaerophiles or anaerobes or both. The greigite-producers appear to be strict anaerobes positioned in the more sulfidic waters just below the OATZ where oxygen is barely or not detectable (Bazylinski et al. 1990, 1992; Bazylinski 1995). To date, no greigite-producing magnetotactic bacterium has been grown in pure culture.

Phylogeny of the magnetotactic bacteria

Because magnetite particles resembling those of magnetotactic bacteria in ancient and recent sediments have been interpreted as magnetofossils (discussed later in the chapter), it is important to understand the phylogeny of the magnetotactic bacteria. Phylogenetic analysis, based on the sequences of 16S ribosomal RNA genes of many cultured and uncultured magnetotactic bacteria, initially showed that the magnetite-producing strains are associated with the α-subdivision of the Proteobacteria (Burgess et al. 1993; DeLong et al. 1993; Eden et al. 1991; Schleifer et al. 1991; Spring and Schleifer 1995; Spring et al. 1992, 1994, 1998), a vast assemblage of Gram-negative prokaryotes in the Domain Bacteria (Woese 1987), while an uncultured multicellular, greigite-producing bacterium was found to be associated with the sulfate-reducing bacteria in the δ-subdivision of the Proteobacteria (DeLong et al. 1993). Since the different subdivisions of the Proteobacteria are considered to be coherent, distinct evolutionary lines of descent (Woese 1987; Zavarzin et al. 1991), DeLong et al. (1993) proposed that the evolutionary origin of magnetotaxis was polyphyletic and that magnetotaxis based on iron oxide production evolved separately from that based on iron sulfide production. However, recent studies have shown that not all magnetotactic bacteria with magnetite-containing magnetosomes are associated with the α-subgroup of the Proteobacteria. The cultured, sulfate-reducing, magnetotactic bacterium, *Desulfovibrio magneticus* strain RS-1 (Sakaguchi et al. 1993a), has magnetite containing magnetosomes, yet is a member of the δ-subgroup of the Proteobacteria while an uncultured magnetotactic bacterium also with magnetite containing magnetosomes, *Magnetobacterium bavaricum* (Spring et al. 1993), is phylogenetically placed in the Domain Bacteria in the newly formed Nitrospira group and not with the Proteobacteria (Spring and Bazylinski 2000). These results suggest that magnetotaxis as a trait may have evolved several times and, moreover, may indicate that there is more than one biochemical-chemical pathway for the biomineralization of magnetic minerals by the magnetotactic bacteria. Alternatively, lateral transfer of a group or groups of genes responsible for magnetosome synthesis between diverse microorganisms might also explain these findings.

Biogeochemical significance of the magnetotactic bacteria

This section focuses on aspects of the known physiology of the magnetotactic bacteria and what they do in natural environments. Physiological experiments in general, have shown they have great potential in playing significant roles in the biogeochemical cycling of several key elements in natural aquatic habitats, more specifically at the oxic-anoxic interface and below. These include nitrogen, sulfur, and carbon as well as iron. Iron is not specifically mentioned here since it is discussed throughout this paper.

Several species of magnetotactic bacteria are known to facilitate important transformations of nitrogen compounds. All species tested thus far show acetylene-reducing activity, an indication they are capable of nitrogen-fixation (Bazylinski and Blakemore 1983b; Bazylinski et al. 2000; Bazylinski and Frankel 2000b). *Magnetospirillum magnetotacticum* is capable of denitrification, an agriculturally-important process, in which the organism respires with the fixed nitrogen oxides, nitrate and/or nitrite, as terminal electron acceptors converting them to the nitrogenous gases, nitrous oxide and/or dinitrogen (Bazylinski and Blakemore 1983a). *M. magneticum* strain AMB-1 also respires with nitrate (Matsunaga and Tsujimura 1993) but the products of nitrate respiration do not seem to have been reported. The marine vibrio, strain MV-1, is capable of the last step of denitrification, the reduction of nitrous oxide to dinitrogen (Bazylinski et al. 1988).

Many uncultured cells of magnetotactic bacteria collected from natural sulfidic environments contain elemental sulfur globules (e.g., Cox et al. 2002) indicating they oxidize reduced sulfur compounds. Several marine strains including MV-1 and MV-2, the coccus MC-1, and the marine spirillum MV-4, have been shown to be capable of lithotrophic growth on reduced sulfur compounds in the laboratory (Bazylinski and Frankel 2000b). *Desulfovibrio magneticus* strain RS-1 is a sulfate-reducing bacterium that grows anaerobically, respiring with sulfate and producing sulfide (Sakaguchi et al. 1993a). It is noteworthy that this organism produces iron-sulfide minerals via BIM when grown on sulfate as the terminal electron acceptor (Sakaguchi et al. 1993a). Magnetotactic bacteria are thus involved both in the oxidative and reductive parts of the sulfur cycle.

When strains MV-1, MV-2, MC-1, and MV-1 are grown lithotrophically on reduced sulfur compounds, they are also able to fix carbon dioxide as their major carbon source and are therefore capable of autotrophic growth. Thus these organisms are chemolithoautotrophs and can be considered as primary producers based on chemosynthesis (not photosynthesis). Strains MV-1 and MV-2 use the Calvin-Benson-Bassham pathway for autotrophy as do plants (Bazylinski et al. 2000b). Autotrophic pathways have not been determined in the other strains.

THE MAGNETOSOME

Composition of the magnetosome mineral phase

Magnetotactic bacteria generally biomineralize either iron-oxide or iron-sulfide magnetosomes. The iron-oxide type contains crystals of magnetite (Fe_3O_4) (e.g., Frankel et al. 1979) whereas the iron-sulfide type contains crystals of greigite (Fe_3S_4) (Mann et al. 1990; Heywood et al. 1990). Several other iron-sulfide minerals have been identified in iron-sulfide magnetosomes, including mackinawite (tetragonal FeS) and a cubic FeS, which are thought to be precursors to greigite (Pósfai et al. 1998a,b). Magnetic, monoclinic, pyrrhotite (Fe_7S_8) (Farina et al. 1990) and non-magnetic iron pyrite (FeS_2) (Mann et al. 1990) were also identified in a many-celled magnetotactic prokaryote that produces iron-sulfide magnetosomes but likely represent misinterpretations of electron diffraction patterns. An unusual rod-shaped bacterium, collected from the OATZ in the Pettaquamscutt Estuary, produces both iron-oxide and iron-sulfide magnetosomes (Bazylinski et al. 1993b, 1995), but has not been isolated and grown in pure culture. Figure 1 shows magnetite and greigite crystals co-aligned in a double chain within a cell of this organism.

The mineral composition of the magnetosome appears to be under strict chemical control since cells of several cultured magnetotactic bacteria continue to synthesize magnetite, not greigite, even when hydrogen sulfide is present in the growth medium (Meldrum et al. 1993a,b). Moreover, magnetite crystals in magnetosomes are of high chemical purity (Bazylinski 1995; Bazylinski and Frankel 2000b) and reports of impurities such as other metal ions within the particles are rare. Gorby (1989) showed that iron could not be replaced by other transition metal ions, including titanium, chromium, cobalt, copper, nickel, mercury, and lead, in the magnetite crystals of *Magnetospirillum magnetotacticum* when cells were grown in the presence of these ions. Towe and Moench (1981) reported trace amounts of titanium in the magnetite particles of an uncultured freshwater magnetotactic coccus collected from a wastewater treatment pond. Significant amounts of copper have been found in the greigite particles of a many-celled, magnetotactic microorganism (Bazylinski et al. 1993a). The concentration of copper was extremely variable and ranged from about 0.1 to 10 atomic % relative to iron. The copper appeared to be mostly concentrated on the surface of the particles and was only present in those organisms collected from a salt marsh in Morro Bay (CA, USA) and not in those collected from other sites. The presence of copper in the magnetosomes did not appear to affect their function since the organisms were still magnetotactic. Interestingly, magnetosome crystals in greigite-producing, rod-shaped magnetotactic bacteria collected from the same site did not contain copper. However, copper was present in the greigite-particles of rod-shaped magnetotactic bacteria collected from other sites (Pósfai et al. 1998b). These observations might indicate that the mineral phase of the magnetosomes in

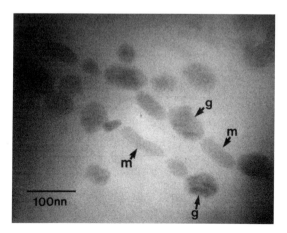

Figure 1. Brightfield scanning-transmission electron micrograph of tooth-shaped magnetite (m at arrows) and pleomorphic greigite (g at arrows) magnetosome crystals co-organized within the same chains in an uncultured magnetotactic bacterium collected from the oxic-anoxic transition zone of the Pettaquamscutt Estuary, RI.

these organisms is more susceptible to chemical and redox conditions in the external environment, and/or that the magnetosomes function in transition metal detoxification. Proteins are also not present within magnetite magnetosome crystals (Arakaki et al. 2003).

Size of the magnetosome mineral phase

There is important physical significance to the size of the magnetosome mineral phase which is reflected in its physical and magnetic properties. Mature magnetosome mineral crystals occur in a very narrow size range, from about 35 to 120 nm (Bazylinski et al. 1994; Frankel and Bazylinski 1994; Moskowitz 1995; Bazylinski and Moskowitz 1997; Frankel et al. 1998). Magnetite and greigite particles in this range are stable single-magnetic-domains (Butler and Banerjee 1975; Diaz-Ricci and Kirchvink 1992). Smaller particles would be superparamagnetic at ambient temperature and would not have stable, remnant magnetizations. Cells initially produce this type of particle which eventually grow into permanent, single-magnetic-domain size crystals (Bazylinski and Frankel 2000a). Larger particles would tend to form multiple domains, reducing the remnant magnetization. Thus, by producing single-magnetic-domain particles, cells have maximized the remnant magnetization of the magnetosome mineral phase (Bazylinski and Frankel 2000b). Magnetic crystals from magnetotactic bacteria have even been used in magnetic domain analyses (Futschik et al. 1989).

Morphology of the magnetosome mineral phase

The particle morphology of magnetite varies but is consistent within cells of a single bacterial species or strain (Bazylinski et al. 1994). Minor variations have been noted to occur in crystals of some species grown under different conditions (Meldrum at al. 1993a; Bazylinski and Frankel 2000a). It is not known whether this applies to greigite crystals in magnetotactic bacteria since there are no greigite-producing strains in pure culture. However, there seems to be more morphological variation in greigite crystals within cells collected from natural environments, i.e., several particle morphologies have been observed within a single cell (Pósfai et al. 1998b). Newly formed superparamagnetic crystals tend to have rounded edges and crystal faces which become more defined as the particles mature and increase in size to the single-magnetic-domain size range (Bazylinski and Frankel 2000a). Three general projected shapes of mature magnetite and greigite particles have been observed in magnetotactic bacteria using transmission electron microscopy (TEM) (Bazylinski and Frankel 2000a,b). They include: 1) roughly cuboidal (Balkwill et al. 1980; Mann et al. 1984a); 2) parallelepipedal or elongated pseudo-prismatic (quasi-rectangular in the horizontal plane of projection) (e.g., Mann et al. 1984b; Towe and Moench 1981; Meldrum et al. 1993a,b); and 3) tooth-, bullet- or arrowhead-shaped (anisotropic) (Mann et al. 1987a,b; Thornhill et al. 1994). Figure 2 shows examples of cubo-octahedral and elongated-prismatic magnetite crystals from various magnetotactic bacteria. A bacterium containing multiple chains of bullet-shaped magnetite crystals is shown in Figure 3.

High resolution transmission electron microscopy, selected area electron diffraction, and electron tomography studies have revealed that the magnetite particles within magnetotactic bacteria are of high structural perfection and have been used to determine their idealized morphologies (e.g., Mann et al. 1984a,b, 1987a,b; Meldrum et al. 1993a,b; Thomas-Keprta et al. 2000, 2001; Clemett et al. 2002). Fe_3O_4 and Fe_3S_4 have face-centered spinel crystal structures (Fd3m space group) (Palache et al. 1944). Macroscopic crystals of Fe_3O_4 often display habits of the octahedral {111} form, more rarely dodecahedral {110}, or cubic {100} forms (bracketed numbers represent specific crystal forms) (Palache et al. 1944). The habits of the cuboidal crystals are cuboctahedra, composed of {100} + {111} forms (Mann et al. 1994), with equal development of the six

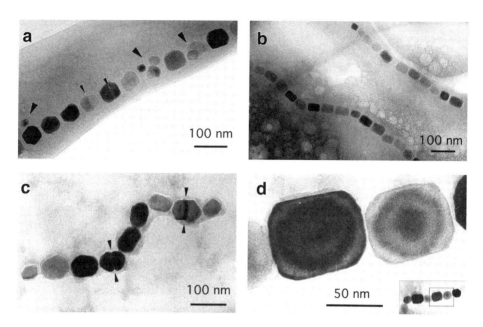

Figure 2. Transmission electron micrographs of magnetite magnetosomes in chains within cells of various magnetotactic bacteria. (a) Equidimensional cubo-octahedral crystals in *Magnetospirillum magnetotacticum*, small arrows denote twinned crystals and large arrows indicate smaller superparamagnetic crystals. (b-d) Elongated crystals with quasi-rectangular projected shapes: (b) two chains of crystals from the marine vibrio strain MV-1; (c) chain of crystals from the marine spirillum strain MV-4 with arrows denoting twinned crystals; (d) high magnification electron micrograph of crystals from the marine coccus strain MC-1, crystals are part of the chain shown in the inset.

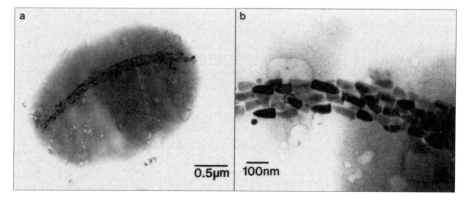

Figure 3. Transmission electron micrograph of an unstained, uncultured magnetotactic bacterium (a) collected from the oxic-anoxic transition zone of the Pettaquamscutt Estuary, RI and a high magnification electron micrograph (b) of the multiple chains of bullet-shaped anisotropic magnetite crystals within the cell. (Figure adapted from Devouard et al. 1998).

symmetry related faces of the {100} form and of the eight symmetry related faces of the {111} form. The habits of the non-equidimensional crystals in some strains can be described as combinations of {100}, {111} and {110} forms (Devouard et al. 1998). In these cases, the six, eight and twelve symmetry related faces of the respective forms that constitute the habits do not develop equally (Mann and Frankel 1989; Devouard et al. 1998). Examples of these crystal types are illustrated in Figure 4.

Whereas the cubooctahedral form of magnetite occurs in inorganically-formed magnetite crystals (Palache et al. 1944), the elongated, pseudo-prismatic structures, corresponding to the unequal development of some symmetry related faces, might occur either because of anisotropy in the growth environment (for example, concentration and/or temperature gradients) or the growth sites (Mann and Frankel 1989). Anisotropy could derive from an anisotropic flux of ions through the magnetosome membrane surrounding the crystal, or from anisotropic interactions of the magnetosome membrane with the growing crystal (Mann and Frankel 1989). In these cases, the growth process could break the symmetry of the faces of each form. These elongated crystals are so unusual that their presence in recent and ancient sediments and in the Martian meteorite ALH84001 has led to them being referred to as magnetofossils (Chang and Kirschvink 1989) and being used as evidence for the past presence of magnetotactic bacteria in aquatic habitats and sediments (Chang and Kirschvink 1989; Stolz et al. 1986, 1990) and life on ancient Mars (McKay et al. 1996; Thomas-Keprta et al. 2000, 2001, 2002) (this is discussed at length later in this chapter).

The most anisotropic crystal habits are those of the tooth-, bullet- or arrowhead-shaped magnetite crystals (Fig. 3b). The synthesis of these crystal habits appears to be

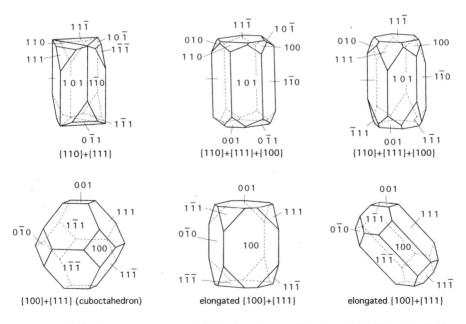

Figure 4. Idealized magnetosome crystal habits based on combinations of {100} (cube), {110} (dodecahedron), and {111} (octahedron) forms. All the habits except the equidimensional cubo-octahedron (lower left) have elongated projected shapes which could result from anisotropy in the crystal growth site. (Figure adapted from Devouard et al. 1998).

more complex than that of the other habits. They have been examined by high resolution TEM in one uncultured organism (Mann et al. 1987a,b) and their idealized morphology suggests that crystal growth of these particles occurs in two stages. The nascent crystals are cuboctahedra which subsequently elongate along a [112] axis to form a pseudo-octahedral prism with alternating (110) and (100) faces capped by (111) faces. Tooth-shaped greigite crystals have also been observed (Pósfai et al. 1998b).

ARRANGEMENT AND EFFECT OF MAGNETOSOMES WITHIN THE CELL

In most magnetotactic bacteria, the magnetosomes are arranged in one or more chains (Bazylinski et al. 1995; Bazylisnki and Moskowitz 1997) (see Figs. 1, 2, 3). Magnetic interactions between the magnetosome particles in a chain cause their magnetic dipole moments to orient parallel to each other along the chain length. In this chain motif, the total magnetic dipole moment of the cell is simply the sum of the permanent dipole moments of the individual, single-magnetic-domain, magnetosome particles. Magnetic measurements (Penninga et al. 1995), magnetic force microscopy (Proksch et al. 1995; Suzuki et al. 1998), and electron holography (Dunin-Borkowski et al. 1998, 2001) have confirmed this idea and show that the chain of magnetosomes in a magnetotactic bacterium functions like a single magnetic dipole and causes the cell to behave similarly. Electron holography of a magnetotactic bacterium is shown in Figure 5. Therefore, the

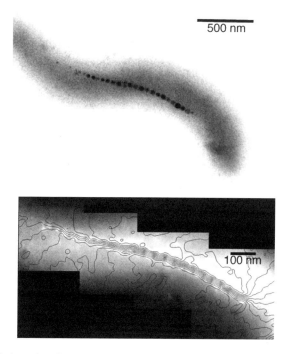

Figure 5. Electron holography of a magnetotactic bacterium showing magnetic field lines associated with the magnetosomes. (*top*) transmission electron micrograph of an unstained cell of *Magnetospirillum magnetotacticum* showing chain of magnetite magnetosomes. (*bottom*) Magnetic field lines generated from the magnetic contribution to the holographic phase overlaid onto the positions of the magnetosomes. Note that most of the magnetic field lines run parallel to the magnetosome chain showing that that the chain acts as a single magnetic dipole. (Figure adapted from Dunin-Borkowski et al. 1998).

cell has maximized its magnetic dipole moment by arranging the magnetosomes in chains. The magnetic dipole moment of the cell is generally large enough so that its interaction with the Earth's geomagnetic field overcomes the thermal forces tending to randomize the cell's orientation in its aqueous surroundings (Frankel 1984). In many magnetotactic bacteria, the newly-formed crystals appear to be at the end of magnetosome chains within the cell (Bazylinski and Frankel 2000a) suggesting that the magnetosome chain increases in size by the precipitation of new magnetosomes at the ends of the chain following cell division. However, there are some magnetotactic bacteria that do not show this pattern and have large gaps between magnetosomes where new magnetosomes could be formed (Bazylinski et al. 1995). Multiple chains of particles appear to be more common in those bacteria that produce greigite or tooth-, bullet-, and arrowhead-shaped magnetite crystals (Figs. 1, 3, 6).

There are some magnetotactic bacteria that do not have their magnetosomes arranged in chains, instead producing a clump at one end of the cell or clumps within partial chains. These include some magnetite-producing cocci (Moench 1988; Cox et al. 2002), some greigite-producing, rod-shaped bacteria (Heywood et al. 1990), and the greigite-producing, many-celled, magnetotactic prokaryote (Mann et al. 1990; Pósfai et al. 1998a,b). Nonetheless, even these organisms clearly show that they have a net magnetic dipole moment.

A rod-shaped magnetotactic bacterium, collected from the OATZ from the Pettaquamscutt Estuary (discussed earlier), was found to contain arrowhead-shaped

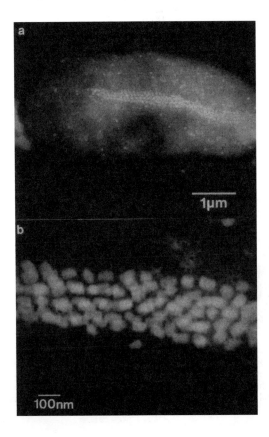

Figure 6. Scanning transmission electron micrograph of: (a) an uncultured magnetotactic bacterium that produces multiple parallel chains of greigite; (b) high magnification image of a portion of the multiple chains showing that this organism synthesizes elongated greigite crystals with a quasi-rectangular projected shape.

crystals of magnetite and rectangular prismatic crystals of greigite co-organized within the same chains of magnetosomes (Fig. 1) (Bazylinski et al. 1993b, 1995). The magnetite and greigite crystals in this organism occur with different, mineral-specific morphologies and sizes and are positioned with their long axes oriented along the chain direction (Fig. 1). Both particle morphologies have been found in organisms with single mineral component chains (Mann et al. 1987a,b; Heywood et al. 1990) suggesting that the magnetosome membranes surrounding the magnetite and greigite particles contain different nucleation templates and that there are differences in magnetosome vesicle biosynthesis. Thus, it is likely that two separate sets of genes control the biomineralization of magnetite and greigite in this organism (Bazylinski et al. 1995).

The magnetic dipole moment resulting from the presence of magnetosomes causes the cell to passively align along geomagnetic field lines while it swims (magnetotaxis). Cells are neither attracted nor pulled toward either geomagnetic pole. Dead cells, like living cells, align along geomagnetic field lines but do not move along them. Living cells behave like very small, self-propelled, magnetic compass needles (Frankel 1984).

Function and physics of magnetotaxis

Magnetotactic bacteria were originally thought to have one of two magnetic polarities, north- or south-seeking, depending on the magnetic orientation of the cell's magnetic dipole with respect to their direction of motion (Blakemore et al. 1980). The vertical component of the inclined geomagnetic field seems to select for a predominant polarity in each hemisphere by favoring those cells whose polarity causes them to migrate down towards the microaerobic sediments and away from potentially high, toxic concentrations of oxygen in surface waters. This hypothesis appears to be partially true: north-seeking magnetotactic bacteria predominate in the Northern hemisphere while south-seeking cells predominate in the Southern hemisphere (Blakemore et al. 1980). At the Equator, where the vertical component of the geomagnetic field is zero, both polarities co-exist in approximately equal numbers (Frankel et al. 1981). However, the discovery of stable populations of magnetotactic bacteria existing at specific depths in the water columns of chemically-stratified aquatic systems at higher latitudes (Bazylinski et al. 1995) and the observation that virtually all magnetotactic bacteria in pure culture form microaerophilic bands of cells below the meniscus of the growth medium (Frankel et al. 1997) are not consistent with this model of magnetotaxis. For example, according to this model, persistent, north-seeking, magnetotactic bacteria in the Northern hemisphere should always be found in the sediments or at the bottom of culture tubes.

Most free-swimming bacteria including magnetotactic species propel themselves forward in their aqueous surroundings by rotating their helical flagella (Silverman and Simon 1974). Unlike cells of *Escherichia coli* and several other chemotactic bacteria, magnetotactic bacteria do not display "run and tumble" motility (Bazylinski et al. 1995). Because of their magnetic dipole moment, magnetotactic bacteria align and migrate along the local magnetic field B. Some species, especially the magnetotactic spirilla (e.g., *Magnetospirillum magnetotacticum*) swim parallel or antiparallel to B and form aerotactic bands (Blakemore et al. 1979; Spormann and Wolfe 1984) at a preferred oxygen concentration [O_2]. In a homogeneous medium, approximately equal numbers of cells swim in either direction along B. Cells use the magnetic field as an axis for migration, with aerotaxis determining the direction of migration along the axis. This behavior has thus been termed axial magneto-aerotaxis (Frankel et al. 1997). The distinction between north-seeking and south-seeking does not apply to axial magneto-aerotactic bacteria. Most flagellated microaerophilic bacteria form aerotactic bands at a preferred or optimal [O_2] where the proton motive force is maximal (Zhulin et al. 1996), using a temporal sensory mechanism (Segall et al. 1996) that samples the local environment as they swim and compares the present [O_2] with that in the recent past (Taylor 1983). The change in [O_2] with time determines the sense of flagellar rotation (Manson 1992). The behavior of individual cells of *M. magnetotacticum* in aerotactic bands in thin capillaries is consistent with the temporal sensory mechanism (Frankel et al. 1997). Thus, cells in the band which are moving away from the optimal [O_2], to either higher or lower [O_2], eventually reverse their swimming direction and return to the band.

Unlike the magnetotactic spirilla, the bilophotrichously-flagellated (possessing two bundles of flagella on one side of the cell), magnetotactic cocci (Blakemore 1975; Frankel et al. 1997; Moench 1988; Moench and Konetzka 1978) and some other magnetotactic bacteria, swim persistently in a preferred direction relative to B when viewed microscopically in wet mounts (Blakemore 1975; Frankel et al. 1997). The persistent swimming direction of populations of magnetotactic cocci led to the discovery of magnetotaxis in bacteria (Blakemore et al. 1975). However, magnetotactic cocci in oxygen gradients, like cells of *Magnetospirillum magnetotacticum*, can also swim in both directions along B without turning around (Frankel et al. 1997). The cocci, like cells of *M. magnetotacticum*, also form microaerophilic, aerotactic bands of cells seeking a preferred [O_2] along the concentration gradient (Frankel et al. 1997). However, while the aerotactic behavior of *M. magnetotacticum* is consistent with the temporal sensory mechanism, the aerotactic behavior of the cocci is not. Instead their behavior is consistent with a two-state aerotactic sensory model in which the [O_2] determines the sense of the flagellar rotation and hence the swimming direction relative to B. Cells at [O_2] higher than optimum swim persistently in one direction relative to B until they reach a low [O_2] threshold at which they reverse flagellar rotation, and hence, swimming direction relative to B. They continue swimming until they reach a high [O_2] threshold at which they reverse again. In wet mounts, the [O_2] is above optimal, and the cells swim persistently in one direction relative to B. This model, termed polar magneto-aerotaxis (Frankel et al. 1997), accounts for the ability of the magnetotactic cocci to migrate to and maintain position at the preferred [O_2] at the OATZ in chemically-stratified, semi-anaerobic basins. Frankel et al. (1998) have developed an assay using chemical gradients in thin capillaries that distinguishes between axial and polar magneto-aerotaxis.

For both aerotactic mechanisms, migration along magnetic field lines reduces a three-dimensional search problem to a one-dimensional search problem. Magnetotaxis is advantageous to motile microorganisms in vertical concentration gradients because it increases the efficiency of finding and maintaining an optimal position in vertical concentration gradients, in this case, a vertical oxygen gradient. It is possible that there are other forms of magnetically-assisted chemotaxis to molecules or ions other than oxygen, such as sulfide, or magnetically-assisted redox- or phototaxis in bacteria that inhabit the anaerobic zone (e.g., greigite-producers) in chemically-stratified waters and sediments.

The function of cellular magnetotaxis described above is a consequence of the cell possessing magnetosomes. There is some conflicting evidence for the role of magnetosomes in magnetotaxis: many obligately microaerophilic bacteria find and maintain an optimal position at the OATZ without the aid of magnetosomes and cultured magnetotactic bacteria form microaerophilic bands of cells in the absence of a magnetic field. It therefore seems likely that iron uptake and magnetosome production is somehow also linked to the physiology of the cell and to other cellular functions as yet unknown.

STUDYING BCM IN BACTERIA: SYNTHESIS OF THE BACTERIAL MAGNETOSOME

Microbiologists try to answer questions regarding prokaryotes using approaches that are not necessarily typical of other types of biologists or geologists because of the uniqueness of the organisms they work with. Thus, it is worthwhile reviewing progress in the elucidation of the chemical/biochemical and molecular basis of the BCM process involved in magnetosome synthesis. Most studies are directed at the role of the magnetosome membrane for a number of reasons: 1) the magnetosome membrane vesicle seems to be a universal structural feature of magnetotactic bacteria and has been found in virtually all cultured and some non-cultured strains (Figs. 7 and 8); 2) it contains some proteins not present in the cytoplasmic and outer membranes; and 3) it is in close physical proximity to the magnetosome crystal. Other approaches include: examining environmental and physiological conditions that favor or support magnetosome synthesis; searching for chemical/biochemical links between physiological functions and BCM of the magnetosome; and using genetics to generate and study non-magnetotactic mutants that do not make magnetosomes, comparing their biochemical and physiological characteristics with wild-type strains. Finally, several genomes of magnetotactic bacterial species are available for examination and thus, genomics can be used in these studies. Ultimately the goals are to identify the genes and proteins required for magnetosome synthesis, to determine how these genes are regulated, and to determine the roles of specific genes and proteins in magnetosome synthesis. To achieve all these goals, working genetic systems in several bacterial strains are going to be necessary. There have been a number of serious problems in establishing genetic systems in the magnetotactic bacteria although the situation is rapidly improving. In this section, we review the techniques and progress in studying the molecular and biochemical basis of magnetosome formation.

Synthesis of the bacterial magnetosome is a complex process that involves a number of discrete steps including: 1) iron uptake by the cell; 2) magnetosome vesicle formation: 3) iron transport into the magnetosome vesicle; and 4) controlled magnetite or greigite biomineralization within the magnetosome vesicle; but not necessarily occurring in this order.

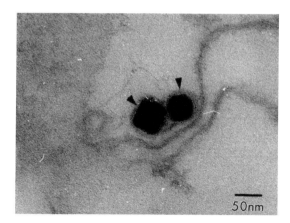

Figure 7. Transmission electron micrograph of a thin-section of several magnetite magnetosomes within a lysing cell of the marine coccus strain MC-1. Arrows denote the electron-dense magnetosome membrane surrounding each crystal. Note that magnetosome membrane is adjacent to cytoplasmic membrane.

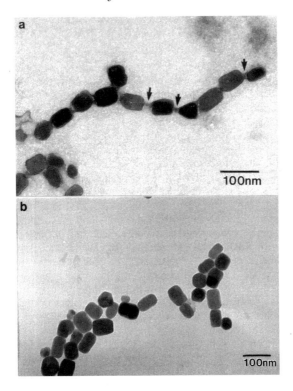

Figure 8. Transmission electron micrograph of: (a) purified magnetite magnetosomes released from cells of strain MV-1 negatively stained with 0.5% uranyl acetate; "halo" around crystals represents magnetosome membrane while material at arrows might indicate additional membranes holding chains together; (b) the same magnetosomes after treatment with 1% sodium deodecyl sulfate, a detergent that removes lipids.

Magnetite magnetosomes

Iron uptake in magnetotactic bacteria. Magnetotactic bacteria consist of up to 3% iron on a dry weight basis (Blakemore 1982), several orders of magnitude higher than non-magnetotactic species. However, there is no evidence that they possess unique iron uptake systems. Fe^{2+}, which is soluble up to 100 mM at neutral pH (Neilands 1984), is generally thought to be taken up by bacteria by non-specific means. In contrast, Fe^{3+} is so insoluble that most microbes produce and rely on iron chelators, called siderophores, which bind and solubilize Fe^{3+} for uptake. Siderophores are defined as low molecular weight (<1 kDa) specific ligands usually produced by bacterial cells under iron-limited conditions (Guerinot 1994). Their synthesis is repressed under high iron conditions. There are a number of studies directed at iron uptake in the magnetotactic bacteria.

Cells of *Magnetospirillum magnetotacticum* were thought to take up iron via a non-specific transport system (Frankel et al. 1983). Although iron is supplied as Fe^{3+} quinate to the cells, the growth medium also contains chemical reducing agents (e.g., ascorbic acid) potent enough to reduce Fe^{3+} to Fe^{2+}. Thus, both reduced and oxidized iron are present in the growth medium and it is not known which form is taken up by cells. However, cells of *M. magnetotacticum* were reported to produce a hydroxamate siderophore when grown under high ($\geq$ 20 µM) but not under low iron conditions ($\leq$ 5 µM), the siderophore production pattern here being the reverse of what is normally observed (Paoletti and Blakemore 1986).

There appear to be at least two iron uptake systems in *Magnetospirillum gryphiswaldense* (Schüler and Bäuerlein 1996). The iron in magnetite was mostly taken up as Fe^{3+} in an energy-dependent process. Fe^{2+} was also taken up by cells but by a slow, diffusion-like process. Fe^{3+} uptake followed Michaelis-Menten kinetics with a K_m of 3 µM and a V_{max} of 0.86 nmol min^{-1} (mg dry cell weight)$^{-1}$ suggesting that Fe^{3+} uptake by cells of *M. gryphiswaldense* is a low affinity but high velocity transport system. Spent culture fluid appeared to enhance iron uptake although no evidence for the presence of a siderophore was found.

Nakamura et al. (1993) found evidence for the involvement of a periplasmic binding protein called sfuC in the transport of iron by cells of *Magnetospirillum magneticum* strain AMB-1. In this study, siderophores were not detected in spent growth medium. More recently, this species was reported to produce both hydroxamate and phenolate siderophores (Calugay et al. 2003). Like cells of *M. magnetotacticum*, those of *M. magneticum* strain AMB-1 produce siderophores under growth conditions that would be considered to be at least iron-sufficient, if not iron-rich, for most non-magnetotactic bacteria. This unusual pattern of siderophore production might be due to the fact that iron is taken up rapidly and converted to inert magnetite that apparently cannot be used by cells. Therefore levels of iron available for growth would likely decrease relatively quickly causing the cells to experience iron-limiting conditions, stimulating siderophore production.

Recently, we have found that cells of a marine magnetotactic vibrio, strain MV-1, also synthesize a siderophore that appears to be a hydroxamate type (Dubbels et al. submitted for publication).The iron concentration pattern of siderophore production is similar to the *Magnetospirillum* species. We have also found biochemical and molecular evidence for the presence of a copper-dependent, high affinity Fe uptake system in strain MV-1 similar to that found in the yeast *Saccharomyces cerevisiae* (Van Ho et al. 2002).

Magnetosome vesicle formation. The magnetosome membrane in several *Magnetospirillum* species is made up of a lipid bilayer about 3-4 nm thick (Gorby et al. 1988) consisting of phospholipids, fatty acids, some unique proteins and some similar to those in the cytoplasmic membrane. These similarities between the magnetosome and the cytoplasmic membranes suggests that the magnetosome membrane vesicle originates from the cytoplasmic membrane. This may be the reason that magnetosomes in virtually all magnetotactic bacteria appear to be anchored to the cytoplasmic membrane as demonstrated by electron microscopy and electron tomography. However, no direct, unequivocal, evidence for the contiguity of these two membranes has ever been shown. Nonetheless, current thought is that the magnetosome membrane vesicle is a result of the invagination and pinching off of the cytoplasmic membrane. It is not clear whether the vesicle is produced prior to magnetite nucleation and precipitation or whether magnetite nucleation occurs in the periplasm and invagination of the cytoplasmic membrane occurs around the developing crystal. There is some evidence for the former as seemingly empty and partially-filled magnetosome vesicles have been observed in iron-starved cells of *M. magnetotacticum* (Gorby et al. 1988) and in strain MV-1.

Small GTPases, such as Sar1p, are essential in the budding reaction in the production of membrane vesicles and vesicle trafficking in eukaryotes (Kirchhausen 2000). Okamura et al. (2001) identified a 16 kDa protein, Mms16, that shows GTPase activity, in the magnetosome membrane vesicle of *Magnetospirillum magneticum* strain AMB-1 where it was the most abundant of 5 proteins present in the vesicle. Cells grown in the presence of AlF_4^-, a GTPase inhibitor, showed less overall magnetism and produced fewer magnetosomes, suggesting that GTPase activity is required for magnetosome synthesis and involved in vesicle formation.

Iron transport into the magnetosome membrane vesicle. Electron microscopy from early on has shown magnetite crystals in various stages of maturity and that the crystals increase in size within magnetosome vesicles. Thus, regardless of when the magnetosome membrane vesicle is formed, additional iron must be transported through the magnetosome vesicle for the crystal to grow. It is not known which redox forms of iron are transported into the magnetosome vesicle in most magnetotactic bacteria but there is evidence that Fe^{2+} is transported into vesicles of *Magnetospirillum magneticum* strain AMB-1 (Nakamura et al. 1995a).

The *magA* gene was identified through transposon mutagenesis that encodes for a protein with significant homology to the cation efflux proteins, KefC, a K^+ translocating protein in *Escherichia coli* and NapA, a putative Na^+/H^+ antiporter from *Enterococcus hirae* (Nakamura et al. 1995a,b). The MagA protein is present in both the cytoplasmic and magnetosome membranes of *Magnetospirillum magneticum* strain AMB-1. Inverted membrane vesicles prepared from *E. coli* cells that expressed *MagA* transported Fe^{2+} in an energy-dependent manner into the vesicle suggesting that MagA functions as a H^+/Fe^{2+} antiporter in *M. magneticum* strain AMB-1. However, *magA* was expressed to a much greater degree when wild-type *M. magneticum* strain AMB-1 cells were grown under iron-limited conditions than under iron-sufficient conditions where they produce more magnetosomes. Moreover, cells of the non-magnetotactic Tn5 transposon mutant over-expressed *magA* under iron-limited conditions although they did not make magnetosomes. Although MagA appears to be involved in iron transport, it alone is not responsible for magnetosome synthesis. Genes sharing significant homology with *magA* are present in other magnetotactic bacteria including *M. magnetotacticum* and the unnamed magnetotactic coccus, strain MC-1 (Grünberg et al. 2001).

Controlled Fe_3O_4 biomineralization within the magnetosome vesicle. Frankel et al. (1983) proposed a model in which Fe^{3+} is taken up by the cell, reduced to Fe^{2+}, and transported to the magnetosome membrane vesicle. It is then presumably reoxidized in the magnetosome vesicle to form hydrous Fe^{3+} oxides similar to the mineral ferrihydrite. One-third of the Fe^{3+} ions in ferrihydrite are reduced and with further dehydration, magnetite is produced. Contrary to this study, cells of *M. gryphiswaldense* take up Fe^{3+} and rapidly use it in the formation of magnetite without any apparent delay (Schüler and Bäuerlein 1998), suggesting that significant accumulation of a precursor to magnetite inside the cell does not occur, at least under conditions that appeared to be optimal for magnetite production by that organism.

The size and morphology of the magnetosome mineral crystal is thought to be controlled by the magnetosome membrane although how it does this is unclear. Specific proteins are perhaps distributed asymmetrically in the magnetosome membrane facilitating crystal growth in certain crystallographic directions but retarding it in others. In addition, it is possible that the magnetosome membrane vesicle places physical constraints on the growing crystal thereby limiting the size of the mineral crystal to the single-magnetic-domain size range. Arakaki et al. (2003) partially characterized a number of magnetosome membrane proteins that were tightly bound to the magnetite crystals in *Magnetospirillum magneticum* strain AMB-1. These proteins included Mms5, Mms6, Mms7, and Mms13. Mms6 was overexpressed in *Escherichia coli,* purified, and found to bind iron. More importantly, magnetite crystals, formed chemically in the presence of Mms6, had a size range of about 20 to 30 nm and a cuboidal morphology like those produced by intact cells. Those produced in the absence of Mms6 were non-homogeneous in shape and ranged in size from 1 to 100 nm. All of these Mms proteins contain hydrophobic N-terminal and hydrophilic C-terminal regions, the latter rich in carboxyl and hydroxyl groups thought to be involved in the binding of iron. All also have the common amino acid sequence LGLGLGLGAWGPXXLGXXGXAGA. Mms7 and Mms13 show homology to the MamD and MamC proteins (described below), respectively.

There are a number of other magnetosome membrane proteins and/or genes that encode for magnetosome membrane proteins that have been identified and partially characterized but whose role in magnetosome formation has not been elucidated. Okuda et al. (1996) identified three magnetosome membrane proteins with molecular weights of 12, 22 and 28 kDa in *Magnetospirillum magnetotacticum*. The amino acid sequence of the 22 kDa protein (called MAM22) showed significant homology with proteins in the tetratricopeptide repeat protein family that include mitochondrial and peroxisomal protein import receptors. Okuda et al. (1996) proposed that this protein functions as a receptor interacting with associated cytoplasmic proteins. The MAM22 gene was expressed in *Escherichia coli* and the resultant protein partially characterized (Okuda and Fukumori 2001). A structural model of MAM22 was proposed that contains five tetratricopeptide repeats and a putative sixth repeat. A similar protein, called MamA, and/or genes encoding for similar proteins have been found in *M. gryphiswaldense* and strain MC-1 (Grünberg et al. 2001).

Grünberg et al. (2001) identified a number of genes encoding for Mam proteins in *Magnetospirillum gryphiswaldense* that were assigned to two different genomic regions. The proteins exhibited the following homologies: MamA to tetratricopeptide repeat proteins; MamB to cation diffusion facilitators; and MamE to HtrA-like serine proteases. The sequences of MamC and MamD showed no homology to existing proteins except some of the Mms proteins described earlier. There were also other putative genes in these genomic regions that were conserved that may encode for other magnetosome membrane proteins. Similar gene clusters containing homologues to *MamA* and *MamB* were also found in *M. magnetotacticum* and strain MC-1 that contained additional genes that showed no homology with known genes or proteins in established databases. However, it was not shown whether the homologous genes in strain MC-1 actually encode for magnetosome membrane proteins in this organism although we recently confirmed that *MamC* encodes for a magnetosome membrane protein in MC-1.

Matsunaga et al. (2000) identified three major magnetosome membrane proteins of molecular masses of approximately 24.8, 35.6, and 66.2 kDa in *Magnetospirillum magneticum* strain AMB-1. One of these, *mpsA*, was cloned and sequenced. MpsA was found to have homology with alpha subunits of acetyl-CoA carboxylases and the CoA-binding motif. The function of this protein is presently unknown. A series of non-magnetotactic mutants of *M. magneticum* strain AMB-1 were generated by mini-Tn5 transposon mutagenesis (Wahyudi et al. 2001). One of these, designated strain NMA21, was recently isolated and characterized (Wahyudi et al. 2003) and found to contain a disrupted gene that encodes for a protein with significant homology to a tungsten-containing aldehyde ferredoxin oxidoreductase from the thermophile *Pyrococcus furiosis*. The protein was cytoplasmic and synthesized under microaerobic conditions. Cells of NMA21 did not produce magnetosomes and rates of iron uptake and growth of the mutant were lower than those of the wild-type.

Environmental and physiological conditions that appear to support magnetosome synthesis

Several factors have been found to influence BCM of magnetite by magnetotactic bacteria, the most important being the concentration of oxygen and the presence and reduction of nitrogen oxides. In addition, because iron is so redox reactive and functions as an efficient electron carrier, and because both redox forms of iron are present in magnetite, much attention has been devoted to the role of enzymatic iron oxidation and reduction in magnetite synthesis.

Effect of oxygen concentration. Blakemore et al. (1985) first reported that microaerobic conditions and some molecular O_2 are required for magnetite production by *Magnetospirillum magnetotacticum*. Cells of this organism grew in sealed, unshaken, culture vessels with 0.1 to 21% oxygen in the headspace. Maximum magnetite production occurred at an oxygen concentration of about 1% in the headspace while concentrations of oxygen greater than 5% were inhibitory. However, isotope fractionation experiments show the oxygen in magnetite is derived from water, not molecular O_2 (Mandernack et al. 1999). The role of molecular O_2 in magnetite synthesis is therefore unknown although it has been shown to affect the synthesis of specific proteins not necessarily involved in magnetite synthesis. For example, Sakaguchi et al. (1993b) showed the presence of O_2 in nitrate-grown cultures repressed the synthesis of a 140 kDa membrane protein in *M. magnetotacticum* and Short and Blakemore (1989) showed that increasing the oxygen tension from 1 to 10% of saturation in cultures caused cells to express increased activity of a manganese-type superoxide dismutase relative to that of an iron-type. Schüler and Bäuerlein (1998) showed that magnetite formation in *Magnetospirillum gryphiswaldense* is induced in non-magnetotactic cells by a low threshold oxygen concentration of about 2 to 7 µM at 30°C.

Other bacteria synthesize magnetite under microaerobic conditions but molecular O_2 is not required. Cells of *M. magneticum* strain AMB-1 synthesize magnetite microaerobically or anaerobically using nitrate as the terminal electron acceptor (Matsunaga et al. 1991, Matsunaga and Tsujimura 1993). Cells of the marine, magnetotactic vibrio, strain MV-1, synthesize magnetite microaerobically in semi-solid agar oxygen gradient cultures and anaerobically under 1 atm of N_2O which it uses as a terminal electron acceptor in respiration (Bazylinski et al. 1988).

Reduction of nitrogen oxides as teminal electron acceptors. The addition of nitrate to the growth medium of *Magnetospirillum magnetotacticum* as an additional terminal electron acceptor to oxygen appears to stimulate magnetite production in this organism (Blakemore et al. 1985). *M. magnetotacticum* is a microaerophilic denitrifier, converting nitrate to nitrous oxide (N_2O) and dinitrogen, but cannot grow under strict anaerobic conditions with nitrate (Bazylinski and Blakemore 1983). Attempting to understand the relationship between nitrate reduction, oxygen utilization, and magnetite synthesis in *M. magnetotacticum*, Fukumori and co-workers examined electron transport and cytochromes in this organism. Tamegai et al. (1993) reported a novel "cytochrome a_1-like" hemoprotein that was present in greater amounts in magnetic cells than non-magnetic cells. They found no evidence for a cytochrome a_1 once reported to be one of the terminal oxidases, or an *o*-type cytochrome, in *M. magnetotacticum* (O'Brien et al. 1987). A new *ccb*-type cytochrome *c* oxidase (Tamegai and Fukumori 1994), a cytochrome *c*-550 homologous to cytochrome c_2 in some photosynthetic bacteria (Yoshimatsu et al. 1995), and a cytochrome cd_1-type nitrite reductase (Yamazaki et al. 1995) were identified and purified from *M. magnetotacticum*. The latter protein may be important since it shows a novel Fe^{2+}:nitrite oxidoreductase activity which may be linked to the oxidation of Fe^{2+} in the cell and possibly to magnetite synthesis. Recently, a soluble periplasmic nitrate reductase was purified from *M. magnetotacticum* (Taoka et al. 2003). The enzyme is comprised of 86 and 17 kDa subunits and contains molybdenum, non-heme iron and heme *c*. Molydenum starvation of cells resulted in cell-free extracts with little periplasmic nitrate reductase activity but the magnetosome fraction still had almost half the iron present in the same fraction of cells grown with molybdenum. These results suggest that nitrate reduction in this organism is not essential for magnetite synthesis.

As mentioned previously, other magnetotactic bacteria including *Magnetospirillum magneticum* and strain MV-1 biomineralize magnetite when growing and reducing nitrate and nitrous oxide, respectively (Bazylinski et al. 1988; Matsunaga et al. 1991; Matsunaga and Tsujimura 1993). However, other than the fact that strain MV-1 appears to produce more magnetosomes when grown anaerobically than microaerobically, there have been no studies attempting to link anaerobic respiration with nitrogen oxides and magnetite synthesis in these organisms.

Iron oxidation and reduction. Several species of magnetotactic bacteria reduce or oxidize iron either as intact whole cells or as cell-free extracts or both. Cells of *Magnetospirillum magnetotacticum* reduce Fe^{3+} (Guerin and Blakemore 1992) and translocate protons when Fe^{3+} is introduced to them anaerobically (Short and Blakemore 1986) suggesting that cells conserve energy during the reduction of Fe^{3+}. Guerin and Blakemore (1992) reported anaerobic, Fe^{3+}-dependent growth of *M. magnetotacticum* in the absence of nitrate although growth yields on Fe^{3+} were low. Cells grown anaerobically with poorly-ordered (amorphous) Fe^{3+} oxides as the terminal electron acceptor were extremely magnetic and produced nearly double the amount of magnetosomes compared to nitrate-grown cells with 1% O_2 in the headspace (Guerin and Blakemore 1992). However, cells grew very slowly under these conditions and growth yields were poor compared to those on nitrate and/or oxygen. They further showed that iron oxidation may also be linked to aerobic respiratory processes, energy conservation, and magnetite synthesis in this bacterium. Fe^{3+} reductase activity has also been demonstrated in cell-free extracts of *M. magnetotacticum* (Paoletti and Blakemore 1988). Noguchi et al. (1999) purified a Fe^{3+} reductase from *M. magnetotacticum*. The enzyme appears to be loosely bound on the cytoplasmic face of the cytoplasmic membrane, has an apparent molecular weight of 36 kDa, and requires reduced nicotinamide adenine dinucleotide and flavin mononucleotide as an electron donor and cofactor, respectively. Enzyme activity was inhibited by zinc which also reduced the number of magnetosomes produced by cells when included in the growth medium as $ZnSO_4$.

Cell-free extracts of strain MV-1 show Fe^{3+} reductase activity and the activity appears to be cytoplasmic (Dubbels et al. submitted).

Greigite magnetosomes

Although all freshwater magnetotactic bacteria have been found to synthesize magnetite magnetosomes, many marine, estuarine, and salt marsh species produce iron sulfide-type magnetosomes which primarily consists of the magnetic iron sulfide, greigite (Mann et al. 1990; Heywood et al. 1990; Pósfai et al. 1998a,b). Reports of iron pyrite (Mann et al. 1990) and pyrrhotite (Farina et al. 1990) have not been confirmed and may represent misidentifications of additional iron sulfide species occasionally observed with greigite in cells (Pósfai et al. 1998a,b). Because of the lack of pure cultures of greigite-producing magnetotactic bacteria, little is known about the synthesis of greigite magnetosomes.

The iron sulfide-type magnetosomes examined thus far contain either particles of greigite or a mixture of greigite and transient, non-magnetic, iron sulfide phases that likely represent mineral precursors to greigite (Pósfai et al. 1998a,b). These phases include mackinawite and likely a sphalerite-type, cubic FeS. Based on transmission electron microscopic observations, electron diffraction, and known iron sulfide chemistry (Berner 1967, 1970), the reaction scheme for greigite formation in the magnetotactic bacteria appears to be (Pósfai et al. 1998a,b):

$$\text{cubic FeS} \rightarrow \text{mackinawite (tetragonal FeS)} \rightarrow \text{greigite (Fe}_3\text{S}_4)$$

The *de novo* synthesis of non-magnetic crystalline iron sulfide precursors to greigite aligned along the magnetosome chain indicates that chain formation and magnetosome orientation in the chain does not necessarily involve magnetic interactions. Under the strongly reducing, sulfidic conditions at neutral pH in which the greigite-producing magnetotactic bacteria are found (Bazylinski et al. 1990, 1992), greigite particles would be expected to transform into pyrite (Berner 1967, 1970). It is not known if this could occur during the lifetime of the cell or whether cells somehow prevent this transformation.

It is presently impossible to determine the conditions favorable for greigite production in magnetotactic bacteria without pure cultures. However, again given the anaerobic, sulfidic conditions where they are generally found (Bazylinski et al. 1990, 1992), it is likely that BCM of greigite by magnetotactic bacteria occurs only in the absence of oxygen.

Genetic systems in the magnetotactic bacteria

In many ways progress in the elucidation of the chemical and biochemical pathways involved in magnetite magnetosome synthesis, particularly in determining the function of specific proteins has been limited by the general absence of a workable genetic system in the magnetotactic bacteria. Problems in establishing genetic systems in the magnetotactic bacteria have included: 1) the lack of a significant number of magnetotactic bacterial strains; 2) their fastidiousness and general microaerophilic nature that require elaborate techniques for growth; 3) they are difficult to grow on the surface of agar plates to screen for mutants; and 4) there has been a lack of effective methods of DNA transfer in these microorganisms. In the last five years, however, this situation has changed drastically for the better.

Waleh (1988) and co-workers found that some genes from *Magnetospirillum magnetotacticum* can be functionally expressed in *Escherichia coli* and that the transcriptional and translational elements of the two microorganisms are compatible (a necessity for a functional genetic system). The *recA* gene from *M. magnetotacticum* was cloned, sequnced and characterized (Berson et al. 1989, 1990). In addition, a 2 kb DNA fragment from *M. magnetotacticum* that complemented the aroD (biosynthetic dehydoquinase) gene function in *Escherichia coli* and *Salmonella typhimurium* was also cloned and characterized (Berson et al. 1991). AroD mutants of these strains cannot take up iron from the growth medium. When the 2 kb DNA fragment from *M. magnetotacticum* was introduced into these mutants, the ability of the mutants to take up iron from the growth medium was restored. Although introduction of the cloned fragment restored iron uptake in siderophore-less, iron uptake-deficient mutants of *E. coli*, it did not mediate siderophore biosynthesis.

If cells of a magnetotactic bacterial strain can be induced to form colonies on agar plates, the selection of non-magnetotactic mutants that do not produce magnetosomes has proven to be a relatively easy task. Techniques for growing several magnetotactic bacteria strains including *Magnetospirillum magneticum* strain AMB-1 (Matsunaga et al. 1991), *M. magnetotacticum* (Matsunaga et al. 1991; Schultheiss and Schüler 2003), *M.*

gryphiswaldense (Schultheiss and Schüler 2003), and strain MV-1 (Dubbels et al. submitted) on the surfaces of agar plates have now been developed. However, when cells are grown aerobically, the oxygen concentration of the incubation atmosphere must be decreased to 0.5 to 2% depending on the strain. Cells of strain MV-1 form colonies not just microaerobically but also anaerobically under 1 atm of N_2O (Dubbels et al. submitted). Generally, cells that produce magnetosomes form dark-colored, even black, colonies while mutants that do not produce magnetosomes form lighter-colored colonies, usually white to pink-colored. For example, non-magnetic mutants of *M. magneticum* strain AMB-1, obtained by the introduction of transposon Tn5 by the conjugal transfer of plasmid pSUP1021, were easily detected using this screening technique (Matsunaga et al. 1992). Using these Tn5-derived non-magnetic mutants, Nakamura et al. (1995a) found that at least three regions of the *M. magneticum* strain AMB-1 chromosome were required for the successful synthesis of magnetosomes. One of these regions, consisting of 2975 base pairs, contained two putative open reading frames, one of which was *magA*, which was discussed above.

Magnetospirillum magneticum strain MGT-1 was recently reported to contain a 3.7 kb cryptic plasmid referred to as pMGT (Okamura et al. 2003). Recombinant plasmids were constructed by ligating pUC19 (pUMG) or a kanamycin resistance cassette (pMGTkm) into a *Bam*HI site of pMGT that were capable of replicating in both *Magnetospirillum* spp. and *Escherichia coli*. These plasmids were introduced into cells using a newly developed electroporation procedure. However, the authors found that cells containing magnetosomes were preferentially killed during electroporation and they therefore had to use aerobically-grown, non-magnetotactic cells.

Schultheiss and Schüler (2003) recently developed a genetic system in *Magnetospirillum gryphiswaldense*. By adding activated charcoal, dithiothreitol, and elevated concentrations of iron compounds known to decompose inhibitory, toxic, oxygen radicals produced during respiration in the growth medium, colony formation on agar surfaces by this strain at greater than a 90% plating efficiency was achieved. Cells even formed colonies (white) on agar plates incubated under fully aerobic conditions although cells from these colonies were non-magnetotactic. Protocols were also developed for the introduction of foreign DNA into cells by electroporation and high-frequency conjugation. A number of broad-host range vectors of the IncQ, IncP, and pBBR1 groups containing antibiotic resistance markers were shown to be capable of replicating in *M. gryphiswaldense*.

Genomics and BCM in magnetotactic bacteria

Genome arrangements and sizes of several different magnetotactic bacteria were determined by pulsed-field gel electrophoresis as a prelude to genomic studies involving magnetotactic bacteria. Two marine vibrios, strains MV-1 and MV-2, have genomes consisting of a singular circular chromosome of about 3.7 and 3.6 Mb, respectively (Dean and Bazylinski 1999). The marine coccus, strain MC-1, also has a single circular chromosome but is larger, about 4.5 Mb (Dean and Bazylinski 1999). No evidence for the presence of extrachromosomal DNA such as plasmids in any of these strains was found. The genome of the freshwater species, *Magnetospirillum magnetotacticum*, consists of a single, circular chromosome of about 4.3 Mb (Bertani et al. 2001).

The partially-sequenced draft genomes of two magnetotactic bacteria, *Magnetospirillum magnetotacticum* and strain MC-1, are available for examination at the Joint Genome Institute website at *http://www.jgi.doe.gov/JGI_microbial/html/index.html*.

SIGNIFICANCE OF BCM

Magnetofossils and biomarkers

A major problem for understanding the origin of life and the evolutionary origin and phylogeny of prokaryotes is the general lack of microbial fossils. Moreover, much of the material thought to be fossilized microbes is subject to alternate interpretations. For example, microbial fossils supposedly representing cyanobacterial species from some of the oldest rocks on Earth, 3.5 billion year old cherts from western Australia (Schopf and Packer 1987; Schopf 1993), have recently been under intense scrutiny (Dalton 2002). Years after the original microbial fossil interpretation of the structures in these rocks had become "textbook orthodoxy" (Dalton 2002), Brasier et al. (2002) reexamined the structures and offered an alternative explanation for their formation, i.e., the structures are secondary artifacts formed from amorphous graphite within multiple generations of metalliferous hydrothermal vein chert and volcanic glass, or as Dalton (2002) puts it more simply, they represent "carbonaceous blobs, probably formed by the action of scalding water on minerals." The point here illustrates that there is a significant need for unequivocal evidence for the past presence of microbes. If BCM-produced minerals are unusual enough (i.e., cannot be formed by geological or chemical processes alone) and persist over long periods of geologic time, they might prove be excellent evidence of the past presence of certain microbes. This is thought by some to be the case with certain types of magnetite crystals.

Based on a number of studies (e.g., Sparks et al. 1990; Bazylinski et al. 1994; Devouard et al. 1998) magnetite crystals produced via BCM by the magnetotactic bacteria have the following characteristics: a) high chemical and structural purity; b) high structural perfection; c) consistent habits within a given species or strain, most commonly equidimensional cubooctahedra or non-equidimensional, pseudo-hexagonal prisms with (110) side faces and truncated (111) end caps, elongated along the [111] axis perpendicular to the endcaps; d) a certain fraction (~10%) of twined crystals characterized by rotations of 180 degrees around [111] axis with a common (111) contact plane; e) a consistent width to length ratio; and f) an asymmetric crystal-size distribution with a sharp cutoff for larger sizes within the single magnetic domain size range (Devouard et al. 1998). In addition, they have novel magnetic properties (Moskowitz et al. 1989, 1993).

After magnetotactic cells die and lyse, their magnetosome crystals are released into the surrounding environment where they may settle into sediments and persist for some time. Nanometer-sized magnetite grains have been recovered from a number of soils (Maher and Taylor 1988; Maher 1990), freshwater sediments (Peck and King 1996), and modern and ancient deep sea sediments (Stolz et al. 1986, 1990; Petersen et al. 1986; Chang et al. 1989; Chang and Kirschvink 1989; Akai et al. 1991). In some cases, they were identified as biogenic by their shape and size similarity to crystals in magnetotactic bacteria and were thus referred to as "magnetofossils" (Kirschvink and Chang 1984; Chang et al. 1989; Chang and Kirschvink 1989). Regarding ancient sediments, putative fossil magnetotactic bacterial magnetite crystals have been found in stromatolitic and black cherts from the Precambrian some dating as far back as 2 billion years (Chang and Kirschvink 1989; Chang et al. 1989). If these crystals are actually magnetofossils from magnetotactic bacteria, they represent some of the oldest bacterial fossils on Earth. Many of the crystals found in these rocks were partially degraded and showed indistinct edges (compared to freshly-isolated magnetite crystals from magnetotactic bacteria) indicating that the particles underwent partial oxidation (Chang et al. 1987, 1989). The crystals were used as supporting evidence that free O_2 had begun to accumulate in the atmosphere before 2000 Ma ago, that the present level of the earth's magnetic field strength had appeared by 2000

Ma ago, and as an implication that matrix-mediated biomineralization (BCM) appeared on Earth as early as 2000 Ma ago (Chang et al. 1989).Because the cubooctahedral crystal habit is the equidimensional form of magnetite, it is commonly found in inorganically-produced magnetites (Palche et al. 1944). However, the elongated, non-equidimensional, pseudo-hexagonal prismatic forms of magnetite are very unusual and those found in sediments and soils may represent true fossils of magnetotactic bacteria.

McKay et al. (1996) included the presence of ultrafine-grained magnetite, pyrrhotite, and greigite in the rims of carbonate inclusions in the Martian meteorite ALH84001 as one of the lines of evidence for life on ancient Mars. This meteorite, which is estmated to be approximately 4.5 billion years old (Nyquist et al. 2001; Thomas-Keprta et al. 2002), contains magnetite crystals that are cuboid, teardrop, and irregular in shape with sizes ranging from about 10 to 100 nm (McKay et al. 1996; Thomas-Keprta et al. 2000). About 25% of the magnetite crystals present have an unusual morphology referred to as a truncated hexa-octahedral, essentially the same morphology as that found in the marine magnetotactic vibrio, strain MV-1 (Thomas Keprta et al. 2001).

The magnetite crystals have been examined to a great degree using a number of different techniques. The intracellular magnetite crystals produced by strain MV-1 display six distinctive properties that allow them to be distinguished from any known population of inorganically-produced magnetites (Thomas-Keprta et al. 2000, 2001, 2002). These six properties are: (i) narrow size range (a non-log-normal size distribution with the mean centered in the single-magnetic-domain size range); (ii) restricted width-to-length ratios (iii) high chemical purity (crystals are essentially stoichiometric magnetite); (iv) few crystallographic defects (crystals are defect-free with the exception of occasional twinning perpendicular to the [111] axis of elongation; (v) crystal morphology with unusual truncated hexa-octahedral geometry consisting of a combination of the three crystallographic forms for the $\frac{4}{m}3\frac{2}{m}$ point group, the cube {100}, the octahedron {111}, and the rhombic dodecahedron {110}, with only 6 of the 12 possible {110} faces being expressed, namely, those that satisfy the relationship {1$\underline{1}$0} • [111] = 0, with elongation defined to be along the [111] ↔ [$\bar{1}\bar{1}\bar{1}$] axis [i.e., ($\bar{1}$10), ($1\bar{1}$0), ($10\bar{1}$), ($\bar{1}$01), ($01\bar{1}$), ($0\bar{1}$1)] (the term truncated hexa-octahedron should not be confused with hexoctahedron, which is another crystallographic form in the isometric system with 48 equivalent faces in the $\frac{4}{m}3\frac{2}{m}$ point group denoted by the symbol {321}); and (vi) elongation along only one of the possible four threefold rotation axes of a regular octahedron. These six properties have been referred to as the magnetite assay for biogenicity and used as a set of standards for the determination of whether magnetite crystals have a biogenic origin or not (Thomas-Keprta et al. 2002).

The 25% of the magnetite crystals in ALH84001 that resemble those of strain MV-1 appear to fit the above criteria (Thomas-Keprta et al. 2001, 2002). However, some disagree that that these criteria are robust enough to distinguish between biogenically- and abiogenically-produced magnetite crystals (Buseck et al. 2001) and so the debate continues. An abiogenic hypothesis is based on the low temperature precipitation of carbonates (Golden et al. 2000) and thermal decomposition of iron-bearing carbonate to produce magnetite (Golden et al. 2001, 2002, 2003; Barber and Scott 2002) with the implication that in ALH84001 such an event occurred through impact shock heating. Evidence for this process comes from the observation that in addition to magnetite, nano-dimensional periclase (MgO) crystals are also associated with the carbonate globules in ALH84001, particularly the Mg-rich carbonate (Barber and Scott 2002). Both magnetite and periclase crystals are frequently associated with voids in the carbonate, suggesting a mineralization process in which CO_2 is released. Some faceted magnetite and periclase crystals in carbonate are crystallographically oriented with respect to the carbonate crystal lattice (Bradley et al. 1998; Barber and Scott 2002). This could be evidence that these magnetites formed abiogenically *in situ*. Golden et al. (2001, 2002, 2003) demonstrated that thermal decomposition of pure siderite ($FeCO_3$) above 450°C results in the formation of magnetite crystals with a size-range and projected shapes similar to those found in ALH84001. Some of these magnetite crystals are elongated along [111], as are the magnetite crystals in a number of magnetotactic bacteria, although there are some differences in the relative sizes of facets of the low index forms {100}, {110} and {111} (Golden et al. 2002, 2003). However, in MV-1 there are variations in the relative sizes of the corner faces from crystal to crystal, even in the same chain.

Although statistical analyses of the sizes and shapes of of fine-grained magnetite and greigite crystals (Devouard et al. 1998; Pósfai et al. 2001) as well as the characteristics listed above might prove to be useful criteria for distinguishing between biogenic and non-biogenic magnetite and greigite, it seems additional criteria are needed to distinguish between biogenic and non-biogenic nanophase magnetic iron minerals. This is particularly important in the case of magnetite produced via BIM by dissimilatory iron-reducing bacteria since these crystals morphologically resemble those produced abiogenically (Sparks et al. 1990). In addition, Zhang et al. (1998) showed that magnetite crystals formed via BIM by thermophilic iron-reducing bacteria had a size distribution that peaked in the single-magnetic-domain size range as does the BCM magnetite crystals produced by the magnetotactic bacteria. Mandernack et al. (1999) found a temperature dependent fractionation of oxygen isotopes in magnetite produced by *Magnetospirillum magnetotacticum* and strain MV-1 that closely matched that for extracellular magnetite produced by a bacterial consortium containing thermophilic iron-reducing bacteria (Zhang et al. 1997). No detectable fractionation of iron isotopes was observed in the magnetite. In contrast, Beard et al. (1999) found enrichment of Fe^{54} compared to Fe^{56} in the soluble Fe^{2+} produced by a dissimilatory iron-reducing bacterium, *Shewanella algae*, growing with the iron oxide mineral ferrihydrite. How this isotopic fractionation is reflected in magnetite formed by this organism and other iron-reducing bacteria remains to be seen, but could ultimately provide a means for distinguishing BIM magnetite from abiotically-produced magnetite.

An important question that arises from the discussion presented above is whether the magnetotactic bacteria are or can be considered as an ancient group of prokaryotes. We presently do not know how long magnetotactic bacteria have been on Earth or even how widely the magnetotactic phenotype is currently distributed among contemporary prokaryotes. Phylogenetic analysis (discussed earlier in this chapter), based on 16S rRNA gene sequences of numerous magnetite-producing magnetotactic bacteria, show that these organisms are associated with the α- and δ-subgroups of the Proteobacteria with one exception that belongs in Nitrospira group in the domain Bacteria. Only one greigite-producing magnetotactic bacterium has been analyzed to date, and this organism is also associated with the Proteobacteria, in the δ-subgroup (DeLong et al. 1993). Neither the Proteobacteria nor the Nitrospira group are deeply branching lineages in the Domain Bacteria (Woese 1987) and thus these groups are not generally considered to be ancient groups of procaryotes. Estimating the age at which different bacterial groups evolved is currently impossible or at best incredibly difficult. Moreover, in addition, phylogenetic analyses suggest that magnetotaxis may have evolved several times in the past (DeLong et al. 1993; Spring and Bazylinski 2000).

Magnetic sensitivity in other organisms

Magnetic sensitivity has been reported in other microbes including single-celled algae (Torres de Araujo et al. 1986) and various types of protists (Bazylinski et al. 2000) and a relatively large number of higher organisms (Wiltschko and Wiltschko 1995). In addition, magnetite crystals with morphologies similar to those produced by the magnetotactic bacteria have been found in single-celled algae (Torres de Araujo et al. 1986), in protists (Bazylinski et al. 2000), in the ethmoid tissues of salmon (Mann et al. 1988), in trout (Walker et al. 1997), and in the human brain (Kirschvink et al. 1992, Kobayashi and Kirschvink 1995). The fact that many higher creatures biomineralize single-magnetic-domain magnetite crystals suggests the intriguing idea that that all these organisms share the same or similar set of genes responsible for magnetite biomineralization (Kirschvink 2000) that would likely have originated in the magnetotactic bacteria. Elucidating how magnetotactic bacteria use BCM in synthesizing crystals of iron oxides and sulfides might have an significant scientific impact far beyond the studies of microbiology and geoscience.

ACKNOWLEDGMENTS

We thank B. Devouard and R. Dunin-Borkowski for use of figures; our many collaborators and students; and T. J. Williams and P. M. Dove for critically reviewing this manuscript. DAB is grateful for support from National Science Foundation (NSF) Grant EAR-0311950 and National Aeronautics and Space Administration (NASA) Johnson Space Center Grant NAG 9-1115.

REFERENCES

Akai J, Takaharu S, Okusa S (1991) TEM study on biogenic magnetite in deep-sea sediments from the Japan Sea and the western Pacific Ocean. J Electron Microsc 40:110-117

Arakaki A, Webb J, Matsunaga T (2003) A novel protein tightly bound to bacterial magnetite particles in *Magnetospirillum magneticum* strain AMB-1. J Biol Chem 278:8745-8750

Balkwill DL, Maratea D, Blakemore RP (1980) Ultrastructure of a magnetic spirillum. J Bacteriol 141:1399-1408

Barber DJ, Scott ERD (2002) Origin of supposedly biogenic magnetite in the Martian meteorite Allan Hills 84001. Proc Natl Acad Sci USA 99:6556-6561

Bazylinski DA (1995) Structure and function of the bacterial magnetosome. ASM News 61:337-343

Bazylinski DA, Blakemore RP (1983a) Denitrification and assimilatory nitrate reduction in *Aquaspirillum magnetotacticum*. Appl Environ Microbiol 46:1118-1124

Bazylinski DA, Blakemore RP (1983b) Nitrogen fixation (acetylene reduction) in *Aquaspirillum magnetotacticum*. Curr Microbiol 9:305-308

Bazylinski DA, Dean AJ, Schüler D, Phillips EJP, Lovley DR (2000) N_2-dependent growth and nitrogenase activity in the metal-metabolizing bacteria, *Geobacter* and *Magnetospirillum* species. Environ Microbiol 2:266-273

Bazylinski DA, Frankel RB (1992) Production of iron sulfide minerals by magnetotactic bacteria from sulfidic environments. *In*: Biomineralization Processes of Iron and Manganese: Modern and Ancient Environments. Skinner HCW, Fitzpatrick RW (eds) Catena, Cremlingen, Germany, p 147-159

Bazylinski DA, Frankel RB (2000a) Magnetic iron oxide and iron sulfide minerals within organisms. *In*: Biomineralization: From Biology to Biotechnology and Medical Application. Bäuerlein E (ed) Wiley-VCH, Weinheim, Germany, p 25-46

Bazylinski DA, Frankel RB (2000b) Biologically controlled mineralization of magnetic iron minerals by magnetotactic bacteria. *In*: Environmental Microbe-Mineral Interactions. Lovley DR (ed) ASM Press, Washington, DC, p 109-144

Bazylinski DA, Frankel RB, Garratt-Reed AJ, Mann S (1990) Biomineralization of iron sulfides in magnetotactic bacteria from sulfidic environments. *In*: Iron Biominerals. Frankel RB, Blakemore RP (eds) Plenum Press, New York, p 239-255

Bazylinski DA, Frankel RB, Jannasch HW (1988) Anaerobic production of magnetite by a marine magnetotactic bacterium. Nature 334:518-519

Bazylinski DA, Frankel RB, Heywood BR, Mann S, King JW, Donaghay PL, Hanson AK (1995) Controlled biomineralization of magnetite (Fe_3O_4) and greigite (Fe_3S_4) in a magnetotactic bacterium. Appl Environ Microbiol 61:3232-3239

Bazylinski DA, Garratt-Reed AJ, Abedi A, Frankel RB (1993a) Copper association with iron sulfide magnetosomes in a magnetotactic bacterium. Arch Microbiol 160:35-42

Bazylinski DA, Garratt-Reed AJ, Frankel RB (1994) Electron microscopic studies of magnetosomes in magnetotactic bacteria. Microsc Res Tech 27:389-401

Bazylinski DA, Heywood BR, Mann S, Frankel RB (1993b) Fe_3O_4 and Fe_3S_4 in a bacterium. Nature 366:218

Bazylinski DA, Moskowitz BM (1997) Microbial biomineralization of magnetic iron minerals: microbiology, magnetism and environmental significance. Rev Mineral 35:181-223

Bazylinski DA, Schlezinger DR, Howes BH, Frankel RB, Epstein SS (2000) Occurrence and distribution of diverse populations of magnetic protists in a chemically-stratified coastal salt pond. Chem Geol 169:319-328

Beard BL, Johnson CM, Cox L, Sun H, Nealson KH, Aguilar C (1999) Iron isotope biosignatures. Science 285:1889-1892

Berner RA (1967) Thermodynamic stability of sedimentary iron sulfides. Am J Sci 265:773-785

Berner RA (1970) Sedimentary pyrite formation. Am J Sci 268:1-23

Berson AE, Hudson DV, Waleh, NS (1991) Cloning of a sequence of *Aquaspirillum magnetotacticum* that complements the *aroD* gene of *Escherichia coli*. Mol Microbiol 5:2261-2264

Berson AE, Peters MR, Waleh NS (1989) Cloning and characterization of the *recA* gene of *Aquaspirillum magnetotacticum*. Arch Microbiol 152:567-571

Berson AE, Peters MR, Waleh NS (1990) Nucleotide sequence of *recA* gene of *Aquaspirillum magnetotacticum*. Nucl Acids Res 18:675

Blakemore RP (1975) Magnetotactic bacteria. Science 190:377-379

Blakemore RP (1982) Magnetotactic bacteria. Annu Rev Microbiol 36:217-238

Blakemore RP, Frankel RB, Kalmijn AJ (1980) South-seeking magnetotactic bacteria in the southern hemisphere. Nature 236:384-385

Blakemore RP, Maratea D, Wolfe RS (1979) Isolation and pure culture of a freshwater magnetic spirillum in chemically defined medium. J Bacteriol 140:720-729

Blakemore RP, Short KA, Bazylinski DA, Rosenblatt C, Frankel RB (1985) Microaerobic conditions are required for magnetite formation within *Aquaspirillum magnetotacticum*. Geomicrobiol J 4:53-71

Bradley JP, McSween Jr HY, Harvey RP (1998) Epitaxial growth of nanophase magnetite in Martian meteorite ALH 84001: Implications for biogenic mineralization. Meteorit Planet Sci 33:765-773

Brasier MD, Green OR, Jephcoat AP, Kleppe AK, Van Kranendonk MJ, Lindsay JF, Steele A, Grassineau NV (2002) Questioning the evidence for Earth's oldest fossils. Nature 416: 76-81

Burgess JG, Kanaguchi R, Sakaguchi T, Thornhill RH, Matsunaga T (1993) Evolutionary relationships among *Magnetospirillum* strains inferred from phylogenetic analysis of 16S rRNA sequences. J Bacteriol 175:6689-6694

Buseck PR, Dunin-Borkowski RE, Devouard B, Frankel RB, McCartney MR, Midgley PA, Pósfai M, Weyland M (2001) Magnetite morphology and life on Mars. Proc Natl Acad Sci USA 98:13490-13495

Butler RF, Banerjee SK (1975) Theoretical single-domain grain size range in magnetite and titanomagnetite. J Geophys Res 80:4049-4058

Calugay RJ, Miyashita H, Okamura Y, Matsunaga T (2003) Siderophore production by the magnetic bacterium *Magnetospirillum magneticum* AMB-1. FEMS Microbiol Lett 218:371-375

Chang S-BR, Kirschvink JL (1989) Magnetofossils, the magnetization of sediments, and the evolution of magnetite biomineralization. Annu Rev Earth Planet Sci 17:169-195

Chang S-BR, Stolz JF, Awramik SM, Kirschvink JL (1989) Biogenic magnetite in stromatolites. II. Occurrence in ancient sedimentary environments. Precamb Res 43:305-315

Chang S-BR, Stolz JF, Kirschvink JL (1987) Biogenic magnetite as a primary remanence carrier in limestone. Phys Earth Planet Inter 46:289-303

Clemett SJ, Thomas-Keprta KL, Shimmin J, Morphew M, McIntosh JR, Bazylinski DA, Kirschvink JL, McKay DS, Wentworth SJ, Vali H, Gibson Jr. EK, Romanek CS (2002) Crystal morphology of MV-1 magnetite. Am Mineral 87:1727-1730

Cox BL, Popa R, Bazylinski DA, Lanoil B, Douglas S, Belz A, Engler DL, Nealson KH (2002) Organization and elemental analysis of P-, S-, and Fe-rich inclusions in a population of freshwater magnetococci. Gemicrobiol J 19:387-406

Dalton R (2002) Squaring up over ancient life. Nature 417:782-784

Dean AJ, Bazylinski DA (1999) Genome analysis of several magnetotactic bacterial strains using pulsed-field gel electrophoresis. Curr Microbiol 39:219-225

DeLong EF, Frankel RB, Bazylinski DA (1993) Multiple evolutionary origins of magnetotaxis in bacteria. Science 259:803-806

Devouard B, Pósfai M, Hua X, Bazylinski DA, Frankel RB, Buseck PR (1998) Magnetite from magnetotactic bacteria: size distributions and twinning. Am Mineral 83:1387-1398

Diaz-Ricci JC, Kirschvink JL (1992) Magnetic domain state and coercivity predictions for biogenic greigite (Fe_3S_4): a comparison of theory with magnetosome observations. J Geophys Res 97:17309-17315

Dubbels BL, DiSpirito AA, Morton JD, Semrau JD, Bazylinski DA (submitted) Evidence for a copper-dependent iron transport system in the marine, magnetotactic bacterium strain MV-1. Microbiology Submitted for publication

Dunin-Borkowski RE, McCartney MR, Frankel RB, Bazylinski DA, Pósfai M, Buseck PR (1998) Magnetic microstructure of magnetotactic bacteria by electron holography. Science 282:1868-1870

Dunin-Borkowski RE, McCartney MR, Pósfai M, Frankel RB, Bazylinski DA, Buseck PR (2001) Off axis electron holography of magnetotactic bacteria: magnetic microstructure of strains MV-1 and MS-1. Eur J Mineral 13:671-684

Eden PA, Schmidt TM, Blakemore RP, Pace NR (1991) Phylogenetic analysis of *Aquaspirillum magnetotacticum* using polymerase chain reaction-amplified 16S rRNA-specific DNA. Int J Syst Bacteriol 41:324-325

Farina M, Esquivel DMS, Lins de Barros HGP (1990) Magnetic iron-sulphur crystals from a magnetotactic microorganism. Nature 343:256-258

Farina M, Lins de Barros HGP, Esquivel DMS, Danon J (1983) Ultrastructure of a magnetotactic bacterium. Biol Cell 48:85-88

Frankel RB (1984) Magnetic guidance of organisms. Annu Rev Biophys Bioeng 13:85-103

Frankel RB, Bazylinski DA (1994) Magnetotaxis and magnetic particles in bacteria. Hyperfine Interactions 90:135-142

Frankel RB, Bazylinski DA (2003) Biologically induced mineralization by bacteria. Rev Mineral Geochem 54:95-114

Frankel RB, Bazylinski DA, Johnson M, Taylor BL (1997) Magneto-aerotaxis in marine, coccoid bacteria. Biophys J 73:994-1000

Frankel RB, Bazylinski DA, Schüler D (1998) Biomineralization of magnetic iron minerals in bacteria. Supramol Sci 5:383-390

Frankel RB, Blakemore RP, Torres de Araujo FF, Esquivel DMS, Danon J (1981) Magnetotactic bacteria at the geomagnetic equator. Science 212:1269-1270

Frankel RB, Blakemore RP, Wolfe RS (1979) Magnetite in freshwater magnetotactic bacteria. Science 203:1355-1356

Frankel RB, Papaefthymiou GC, Blakemore RP, O'Brien W (1983) Fe_3O_4 precipitation in magnetotactic bacteria. Biochim Biophys Acta 763:147-159

Frankel RB, Zhang J-P, Bazylinski DA (1998) Single magnetic domains in magnetotactic bacteria. J Geophys Res 103:30601-30604

Futschik K, Pfützner H, Doblander A, Schönhuber P, Dobeneck T, Petersen N, Vali H (1989) Why not use magnetotactic bacteria for domain analyses? Phys Scr 40:518-521

Golden DC, Ming DW, Schwandt CS, Lauer Jr HV, Socki RA (2001) A simple inorganic process for formation of carbonates, magnetite, and sulfides in Martian meteorite ALH84001. Amer Mineral 8:370-375

Golden DC, Ming DW, Lauer Jr HV, Schwandt CS, Morris RV, Lofgren GE, McKay GA (2002) Inorganic formation of "truncated hexa-octahedral" magnetite: implications for inorganic processes in Martian meteorite ALH84001. Lunar Planet Sci Conf 33, abstract 1839, on CD-ROM

Golden DC, Ming DW, Morris RV, Brearley AJ, Lauer Jr HV, Treiman A, Zolensky E, Schwandt CS, , Lofgren GE, McKay GA (2003) Morphological evidence for an exclusively inorganic origin for magnetite in Martian meteorite ALH84001. Lunar Planet Sci Conf 34, abstr 1970, on CD-ROM

Golden DC, Ming DW, Schwandt CS, Morris RV, Yang SV, Lofgren GE (2000) An experimental study on kinetically-driven precipitation of calcium-magnesium-iron carbonates from solution: Implications for the low-temperature formation of carbonates in Martian meteorite Allan Hills 84001. Meteorit Planet Sci 35:457-465

Gorby YA (1989) Regulation of magnetosome biogenesis by oxygen and nitrogen. PhD dissertation. University of New Hampshire, Durham, New Hampshire

Gorby YA, Beveridge TJ, Blakemore RP (1988) Characterization of the bacterial magnetosome membrane. J Bacteriol 170:834-841

Grünberg K, Wawer C, Tebo BM, Schüler D (2001) A large gene cluster encoding several magnetosoem preoteins is conserved in different species of magnetotactic bacteria. Appl Environ Microbiol 67:4573-4582

Guerin WF, Blakemore RP (1992) Redox cycling of iron supports growth and magnetite synthesis by *Aquaspirillum magnetotacticum*. Appl Environ Microbiol 58:1102-1109

Guerinot ML (1994) Microbial iron transport. Annu Rev Microbiol 48:743-772

Heywood BR, Bazylinski DA, Garratt-Reed AJ, Mann S, Frankel RB (1990) Controlled biosynthesis of greigite (Fe_3S_4) in magnetotactic bacteria. Naturwiss 77:536-538
Iida A, Akai J (1996) Crystalline sulfur inclusions in magnetotactic bacteria. Sci Rep Niigata Univ, Ser E 11:35-42
Kawaguchi R, Burgess JG, Sakaguchi T, Takeyama H, Thornhill RH, Matsunaga T (1995) Phylogenetic analysis of a novel sulfate-reducing magnetic bacterium, RS-1, demonstrates its membership of the δ-Proteobacteria. FEMS Microbiol Lett 126:277-282
Kirchhausen T (2000) Three ways to make a vesicle. Nature Rev Mol Cell Biol 1:187-198
Kirschvink JL (2000) A grand unified theory or biomineralization. *In*: Biomineralization: From Biology to Biotechnology and Medical Application. Bäuerlein E (ed) Wiley-VCH, Weinheim, Germany, p 139-149
Kirschvink JL, Kobayashi-Kirschvink A, Woodford BJ (1992) Magnetite biomineralization in the human brain. Proc Natl Acad Sci USA 89:7683-7687
Kobayashi A, Kirschvink JL (1995) Magnetoreception and electromagnetic field effects: sensory perception of the geomagnetic field in animals and humans. Adv Chem Ser 250:367-394
Lowenstam HA (1981) Minerals formed by organisms. Science 211:1126-1131
Maher BA (1990) Inorganic formation of ultrafine-grained magnetite. *In*: Iron Biominerals. Frankel RB, Blakemore RP (eds) Plenum Press, New York, p 179-192
Maher BA, Taylor RM (1988) Formation of ultra-fine grained magnetite in soils. Nature 336:368-370
Mandernack KW, Bazylinski DA, Shanks WC, Bullen TD (1999) Oxygen and iron isotope studies of magnetite produced by magnetotactic bacteria. Science 285:1892-1896
Mann S (1986) On the nature of boundary-organized biomineralization. J Inorg Chem 28:363-371
Mann S, Frankel RB (1989) Magnetite biomineralization in unicellular organisms. *In*: Biomineralization: Chemical and Biochemical Perspectives. Mann S, Webb J, Williams RJP (eds) VCH Publishers, New York, p 389-426
Mann S, Frankel RB, Blakemore RP (1984a) Structure, morphology and crystal growth of bacterial magnetite. Nature 310:405-407
Mann S, Moench TT, Williams RJP (1984b) A high resolution electron microscopic investigation of bacterial magnetite. Implications for crystal growth. Proc R Soc London B 221:385-393
Mann S, Sparks NHC, Blakemore RP (1987a) Ultrastructure and characterization of anisotropic inclusions in magnetotactic bacteria. Proc R Soc London B 231:469-476
Mann S, Sparks NHC, Blakemore RP (1987b) Structure, morphology and crystal growth of anisotropic magnetite crystals in magnetotactic bacteria. Proc R Soc London B 231:477-487
Mann S, Sparks NHC, Frankel RB, Bazylinski DA, Jannasch HW (1990) Biomineralization of ferrimagnetic greigite (Fe_3S_4) and iron pyrite (FeS_2) in a magnetotactic bacterium. Nature 343:258-260
Mann S, Sparks NH, Walker MM, Kirschvink JL (1988) Ultrastructure, morphology and organization of biogenic magnetite from sockeye salmon, *Oncorhynchus nerka*: implications for magnetoreception. J Exp Biol 40:35-49
Manson MD (1992) Bacterial chemotaxis and motility. Adv Microbiol Physiol 33:277-346
Matitashvili EA, Matojan DA, Gendler TS, Kurzchalia TV, Adamia RS (1992) Magnetotactic bacteria from freshwater lakes in Georgia. J Basic Microbiol 32:185-192
Matsunaga T, Nakamura C, Burgess JG, Sode K (1992) Gene transfer in magnetic bacteria: transposon mutagenesis and cloning of genomic DNA fragments required for magnetite synthesis. J Bacteriol 174:2748-2753
Matsunaga T, Sakaguchi T, Tadokoro F (1991) Magnetite formation by a magnetic bacterium capable of growing aerobically. Appl Microbiol Biotechnol 35:651-655
Matsunaga T, Tsujimura N (1993) Respiratory inhibitors of a magnetic bacterium *Magnetospirillum* sp. AMB-1 capable of growing aerobically. Appl Microbiol Biotechnol 39:368-371
Matsunaga T, Tsujimura N, Okamura Y, Takeyama H (2000) Cloning and characterization of a gene, *mpsA*, encoding a protein associated with intracellular magnetic particles from *Magnetospirillum* sp. strain AMB-1. Biochem Biophys Res Commun 268:932-937
McKay DS, Gibson Jr. EK, Thomas-Keprta KL, Vali H, Romanek CS, Clemett SJ, Chillier XDF, Maechling CR, Zare RN (1996) Search for past life on Mars: possible relic biogenic activity in Martian meteorite ALH84001. Science 273:924-930
Meldrum FC, Heywood BR, Mann S, Frankel RB, Bazylinski DA (1993a) Electron microscopy study of magnetosomes in a cultures coccoid magnetotactic bacterium. Proc R Soc London B 251:231-236
Meldrum FC, Heywood BR, Mann S, Frankel RB, Bazylinski DA (1993b) Electron microscopy study of magnetosomes in two cultured vibrioid magnetotactic bacteria. Proc R Soc London B 251:237-242
Moench TT (1988) *Biliphococcus magnetotacticus* gen. nov. sp. nov., a motile, magnetic coccus. Antonie van Leeuwenhoek 54:483-496

Moench TT, Konetzka WA (1978) A novel method for the isolation and study of a magnetic bacterium. Arch Microbiol 119:203-212

Moskowitz BM (1995) Biomineralization of magnetic minerals. Rev Geophys Supp 33:123-128

Moskowitz BM, Frankel RB, Bazylinski DA (1993) Rock magnetic criteria for the detection of biogenic magnetite. Earth Planet Sci Lett 120:283-300

Moskowitz BM, Frankel RB, Bazylinski DA, Jannasch HW, Lovley DR (1989) A comparison of magnetite particles produced anaerobically by magnetotactic and dissimilatory iron-reducing bacteria. Geophys Res Lett 16:665-668

Nakamura C, Burgess JG, Sode K, Matsunaga T (1995a) An iron-regulated gene, *magA*, encoding an iron transport protein of *Magnetospirillum* AMB-1. J Biol Chem 270:28392-28396

Nakamura C, Kikuchi T, Burgess JG, Matsunaga T (1995b) Iron-regulated expression and membrane localization of the MagA protein in *Magnetospirillum* sp. strain AMB-1. J Biochem 118:23-27

Nakamura C, Sakaguchi T, Kudo S, Burgess JG, Sode K, Matsunaga T (1993) Characterization of iron uptake in the magnetic bacterium *Aquaspirillum* sp. AMB-1. Appl Biochem Biotechnol 39/40:169-176

Neilands JB (1984) A brief history of iron metabolism. Biol Metals 4:1-6

Noguchi Y, Fujiwara T, Yoshimatsu K, Fukumori Y (1999) Iron reductase for magnetite synthesis in the magnetotactic bacterium *Magnetospirillum magnetotacticum*. J Bacteriol 181:2142-2147

Nyquist LE, Bogard DD, Shih C-Y, Greshake A, Stoffler D, Eugster O (2001) Ages and geologic histories of Martian meteorites. Space Sci Rev 96:105-164

O'Brien W, Paoletti LC, Blakemore, RP (1987) Spectral analysis of cytochromes in *Aquaspirillum magnetotacticum*. Curr Microbiol 15:121-127

Okamura Y, Takeyama H, Matsunaga T (2001) A magnetosome specific GTPase from the magnetic bacterium *Magnetospirillum magneticum* AMB-1. J Biol Chem 276:48183-48188

Okamura Y, Takeyama H, Sekine T, Sakaguchi T, Wahyudi AT, Sato R, Kamiya S, Matsunaga T (2003) Design and application of a new cryptic-plasmid-based shuttle vector for *Magnetospirillum magneticum*. Appl Environ Microbiol 69:4274-4277

Okuda Y, Denda K, Fukumori Y (1996) Cloning and sequencing of a gene encoding a new member of the tetratricopeptide protein family from magnetosomes of *Magnetospirillum magnetotacticum*. Gene 171:99-102

Okuda Y, Fukumori Y (2001) Expression and characterization of a magnetosome-associated protein, TPR-containing MAM22, in *Escherichia coli*. FEBS Lett 491:169-173

Palache C, Berman H, Frondel C (1944) Dana's System of Mineralogy. Wiley, New York

Paoletti LC, Blakemore RP (1986) Hydroxamate production by *Aquaspirillum magnetotacticum*. J Bacteriol 167:73-76

Paoletti LC, Blakemore RP (1988) Iron reduction by *Aquaspirillum magnetotacticum*. Curr Microbiol 17:339-342

Peck JA, King JW (1996) Magnetofossils in the sediments of Lake Baikal, Siberia. Earth Planet Sci Lett 140:159-172

Penninga I, deWaard H, Moskowitz BM, Bazylinski DA, Frankel RB (1995) Remanence curves for individual magnetotactic bacteria using a pulsed magnetic field. J Magn Magn Mater 149:279-286

Petersen N, von Dobeneck T, Vali H (1986) Fossil bacterial magnetite in deep-sea sediments from the South Atlantic Ocean. Nature 320:611-615

Pósfai M, Buseck PR, Bazylinski DA, Frankel RB (1998a) Reaction sequence of iron sulfide minerals in bacteria and their use as biomarkers. Science 280:880-883

Pósfai M, Buseck PR, Bazylinski DA, Frankel RB (1998b) Iron sulfides from magnetotactic bacteria: structure, compositions, and phase transitions. Am Mineral 83:1469-1481

Pósfai M, Cziner K, Marton E, Marton P, Frankel RB, Buseck PR, Bazylinski DA (2001) Crystal size distributions and possible biogenic origin of Fe sulfides. Eur J Mineral 13:691-703

Proksch RB, Moskowitz BM, Dahlberg ED, Schaeffer T, Bazylinski DA, Frankel RB (1995) Magnetic force microscopy of the submicron magnetic assembly in a magnetotactic bacterium. Appl Phys Lett 66:2582-2584

Rogers FG, Blakemore RP, Blakemore NA, Frankel RB, Bazylinski DA, Maratea D, Rogers C (1990) Intercellular structure in a many-celled magnetotactic procaryote. Arch Microbiol 154:18-22

Sakaguchi T, Burgess JG, Matsunaga T (1993a) Magnetite formation by a sulphate-reducing bacterium. Nature 365:47-49

Sakaguchi H, Hagiwara H, Fukumori Y, Tamaura Y, Funaki M, Hirose S (1993b) Oxygen concentration-dependent induction of a 140-kDa protein in magnetic bacterium *Magnetospirillum magnetotacticum* MS-1. FEMS Microbiol Lett 107:169-174

Schleifer K-H, Schüler D, Spring S, Weizenegger M, Amann R, Ludwig W, Kohler M (1991) The genus *Magnetospirillum* gen. nov., description of *Magnetospirillum gryphiswaldense* sp. nov. and transfer of *Aquaspirillum magnetotacticum* to *Magnetospirillum magnetotacticum* comb. nov. Syst Appl Microbiol 14:379-385

Schopf JW (1993) Microfossils of the Early Archean Apex chert: new evidence of the antiquity of life. Science 260:640-646

Schopf JW, Kudryavtsev AB, Agresti DG, Wdowlak TJ, Czaja AD (2002) Laser-Raman imagery of Earth's earliest fossils. Nature 416:73-76

Schopf JW, Packer BM (1987) Early Archean (3.3-billion to 3.5-billion-year-old) microfossils from Warrawoona Group, Australia. Science 237:70-73

Schüler D, Bäuerlein E (1996) Iron-limited growth and kinetics of iron uptake in *Magnetospirillum gryphiswaldense*. Arch Microbiol 166:301-307

Schüler D, Bäuerlein E (1998) Dynamics of iron uptake and Fe_3O_4 mineralization during aerobic and microaerobic growth of *Magnetospirillum gryphiswaldense*. J Bacteriol 180:159-162

Schüler D, Spring S, Bazylinski DA (1999) Improved technique for the isolation of magnetotactic spirilla from a freshwater sediment and their phylogenetic characterization. Syst. Appl. Microbiol. 22:466-471

Schultheiss D, Schüler D (2003) Development of a genetic system for *Magnetospirillum magnetotacticum*. Arch Microbiol 179:89-94

Segall, JE., Block SM, Berg HC (1986) Temporal comparisons in bacterial chemotaxis. Proc Natl Acad Sci USA 83:8987-8991

Short KA, Blakemore RP (1986) Iron respiration-driven proton translocation in aerobic bacteria. J Bacteriol 167:729-731

Short KA, Blakemore RP (1989) Periplasmic superoxide dismutases in *Aquaspirillum magnetotacticum*. Arch Microbiol 152:342-346

Silverman M, Simon M (1974) Flagellar rotation and the mechanism of bacterial motility. Nature 249:73-74

Sparks NHC, Mann S, Bazylinski DA, Lovley DR, Jannasch HW, Frankel RB (1990) Structure and morphology of magnetite anaerobically-produced by a marine magnetotactic bacterium and a dissimilatory iron-reducing bacterium. Earth Planet Sci Lett 98:14-22

Spormann AM, Wolfe RS (1984) Chemotactic, magnetotactic and tactile behavior in a magnetic spirillum. FEMS Lett 22:171-177

Spring S, Amann R, Ludwig W, Schleifer K-H, Petersen N (1992) Phylogenetic diversity and identification of non-culturable magnetotactic bacteria. Syst Appl Microbiol 15:116-122

Spring S, Amann R, Ludwig W, Schleifer K-H, Schüler D, Poralla K, Petersen, N (1994) Phylogenetic analysis of uncultured magnetotactic bacteria from the alpha-subclass of Proteobacteria. Syst Appl Microbiol 17:501-508

Spring S, Amann R, Ludwig W, Schleifer K-H, van Gemerden H, Petersen N (1993) Dominating role of an unusual magnetotactic bacterium in the microaerobic zone of a freshwater sediment. Appl Environ Microbiol 59:2397-2403

Spring S, Bazylinski DA (2000) Magnetotactic bacteria. *In*: The Prokaryotes. Dworkin M et al. (eds) Springer-Verlag, Inc., New York, at http://www.springer-ny.com

Spring S, Lins U, Amann R, Schleifer K-H, Ferreira LCS, Esquivel DMS, Farina M (1998) Phylogenetic affiliation and ultrastructure of uncultured magnetic bacteria with unusually large magnetosomes. Arch Microbiol 169:136-147

Spring S, Schleifer K-H (1995) Diversity of magnetotactic bacteria. Syst Appl Microbiol 18:147-153

Stolz JF, Chang S-BR, Kirschvink JL (1986) Magnetotactic bacteria and single domain magnetite in hemipelagic sediments. Nature 321:849-850

Stolz JF, Lovley DR, Haggerty SE (1990) Biogenic magnetite and the magnetization of sediments. J Geophys Res 95:4355-4361

Suzuki H, Tanaka T, Sasaki T, Nakamura N, Matsunaga T, Mashiko S (1998) High resolution magnetic force microscope images of a magnetic particle chain extracted from magnetic bacteria AMB-1. Jpn J Appl Phys 37:L1343-L1345

Tamegai H, Fukumori Y (1994) Purification, and some molecular and enzymatic features of a novel *ccb*-type cytochrome *c* oxidase from a microaerobic denitrifier, *Magnetospirillum magnetotacticum*. FEBS Lett 347:22-26

Tamegai H, Yamanaka T, Fukumori Y (1993) Purification and properties of a "cytochrome a_1"-like hemoprotein from a magnetotactic bacterium, *Aquaspirillum magnetotacticum*. Biochim Biophys Acta 1158:237-243

Taoka A, Yoshimatsu K, Kanemori M, Fukumori Y (2003) Nitrate reductase from the magnetotactic bacterium *Magnetospirillum magnetotacticum* MS-1: purification and sequence analysis. Can J Microbiol 49:197-206
Taylor BL (1983) How do bacteria find the optimal concentration of oxygen? Trends Bioch Sci 8:438-441
Thomas-Keprta KL, Bazylinski DA, Kirschvink JL, Clemett SJ, McKay DS, Wentworth SJ, Vali H, Gibson Jr. EK, Romanek CS (2000) Elongated prismatic magnetite (Fe_3O_4) crystals in ALH84001 carbonate globules: potential Martian magnetofossils. Geochim Cosmochim Acta 64:4049-4081
Thomas-Keprta KL, Clemett SJ, Bazylinski DA, Kirschvink JL, McKay DS, Wentworth SJ, Vali H, Gibson Jr. EK, McKay MF, Romanek CS (2001) Truncated hexa-octahedral magnetite crystals in ALH84001: presumptive biosignatures. Proc Natl Acad Sci USA 98:2164-2169
Thomas-Keprta KL, Clemett SJ, Bazylinski DA, Kirschvink JL, McKay DS, Wentworth SJ, Vali H, Gibson Jr.EK, McKay MF, Romane CS (2002) Magnetofossils from ancient Mars: a robust biosignature in the Martian meteorite ALH84001. Appl Environ Microbiol 68:3663-3672
Thornhill RH, Burgess JG, Sakaguchi T, Matsunaga T (1994) A morphological classification of bacteria containing bullet-shaped magnetic particles. FEMS Microbiol Lett 115:169-176
Torres de Araujo FF, Pires MA, Frankel RB, Bicudo CEM (1986) Magnetite and magnetotaxis in algae. Biophys J 50:385-378
Towe KM, Moench TT (1981) Electron-optical characterization of bacterial magnetite. Earth Planet Sci Lett 52:213-220
Van Ho A, Ward DM, Kaplan J (2002) Transition metal transport in yeast. Annu Rev Microbiol 56:237-61
Wahyudi AT, Takeyama H, Matsunaga T (2001) Isolation of *Magnetospirillum magneticum* AMB-1 mutants defective in bacterial magnetite particle synthesis by transposon mutagenesis. Appl Biochem Biotechnol 91-93:147-154
Wahyudi AT, Takeyama H, Okamura Y, Fukuda Y, Matsunaga T (2003) Characterization of aldehyde ferredoxin oxidoreductase gene defective mutant in *Magnetospirillum magneticum* AMB-1. Biochem Biophys Res Commun 303:223-229
Wakeham SG, Howes BL, Dacey JWH (1984) Dimethyl sulphide in a stratified coastal salt pond. Nature 310:770-772
Wakeham SG, Howes BL, Dacey JWH, Schwarzenbach RP, Zeyer J (1987) Biogeochemistry of dimethylsulfide in a seasonally stratified coastal salt pond. Geochim Cosmochim Acta 51:1675-1684
Waleh NS (1988) Functional expression of *Aquaspirillum magnetotacticum* genes in *Escherichia coli* K12. Mol Gen Genet 214:592-594
Walker MM, Diebel CE, Haugh CV, Pankhurst PM, Montgomery JC (1997) Structure and function of the vertebrate magnetic sense. Nature 390:371-376
Weiner S, Dove PM (2003) An overview of biomineralization processes and the problem of the vital effect. Rev Mineral Geochem 54:1-29
Wiltschko R, Wiltschko W (1995) Magnetic Orientation in Animals. Springer-Verlag, New York
Woese CR (1987) Bacterial evolution. Microbiol Rev 51:221-271
Yamazaki T, Oyanagi H, Fujiwara T, Fukumori Y (1995) Nitrite reductase from the magnetotactic bacterium *Magnetospirillum magnetotacticum*; a novel cytochrome cd_1 with Fe(II):nitrite oxidoreductase activity. Eur J Biochem 233:665-671
Yoshimatsu K, Fujiwara T, Fukumori Y (1995) Purification, primary structure, and evolution of cytochrome *c*-550 from the magnetic bacterium, *Magnetospirillum magnetotacticum*. Arch Microbiol 163:400-406
Zavarzin GA, Stackebrandt E, Murray RGE (1991) A correlation of phylogenetic diversity in the Proteobacteria with the influences of ecological forces. Can J Microbiol 37:1-6
Zhang C, Liu S, Phelps TJ, Cole DR, Horita J, Fortier SM (1997) Physiochemical, mineralogical, and isotopic characterization of magnetite-rich iron oxides formed by thermophilic iron-reducing bacteria. Geochim Cosmochim Acta 61:4621-4632
Zhang C, Vali H, Romanek CS, Phelps TJ, Lu SV (1998) Formation of single domain magnetite by a thermophilic bacterium. Am Mineral 83:1409-1418
Zhulin IB, Bespelov VA, Johnson MS, Taylor BL (1996) Oxygen taxis and proton motive force in *Azospirillum brasiliense*. J Bacteriol 178:5199-5204

9 Mineralization in Organic Matrix Frameworks

Arthur Veis

The Feinberg School of Medicine
Department of Cell and Molecular Biology
Northwestern University
Chicago, Illinois 60611 U.S.A.

INTRODUCTION

The primordial earth surface exposed minerals comprised mainly of carbonates, silicates and smaller amounts of phosphates. Weathering eventually led to the dissolution of the surface rock and the leaching of their components into the rivers, lakes and oceans. There, complex chemistry depending upon the temperature, pH, pressure, and atmospheric carbon dioxide content led to the reprecipitation of the dissolved minerals into new forms as part of the sedimentary rock. The minerals themselves could be transformed by passive diagenesis to further structures. When primitive organisms emerged, they added a very significant component to the processing of the dissolved mineral constituents in the marine environment, where both prokaryotic and eukaryotic organisms have the ability to produce mineralized skeletal elements and mineralized fecal pellets that also accumulate in the marine sediments. The increasing diversity of plant and animal life has continually accelerated the dynamic relationship between the composition of the earths crust and the living world. The resulting minerals of biogenic origin comprise a surprisingly large portion of the earths crust, and represent a huge reservoir of sequestered carbonate, silicate and phosphate ions. The majority of the carbonates appear as calcite or aragonite, produced principally by plankton, invertebrates and so on, while the phosphates, as carbonated hydroxyapatite, are the products of vertebrates as well as the weathering of igneous rock.

This chapter is focused on the mechanisms of formation of the vertebrate mineralized structures; bone, tooth enamel, tooth dentin and otoliths. Although other Chapters deal explicitly with the invertebrate and bacterial systems, we need to consider if there are any general considerations that apply to all biogenic mineralization systems. We shall begin from that perspective and inquire as to the way by which living organisms can organize their mineral phases into such complex and specific structures.

BIOMINERALIZATION: GENERAL ASPECTS

The term "Biomineralization" implies that a mineral phase that is deposited requires or is occasioned by the intervention of a living organism. This can happen in two basic ways, either the mineral phase develops from the ambient environment as it would from a saturated solution of the requisite ions, but requires the living system to nucleate and localize mineral deposition, or the mineral phase is developed under the direct regulatory control of the organism, so that the mineral deposits are not only localized, but may be directed to form unique crystal habits not normally developed by a saturated solution of the requisite ions. Moreover, the shape, size and orientation of the crystals may be controlled by the cells involved. However, it must be emphasized that all of the biomineral deposited may not be crystalline. In a very famous paper (Lowenstam 1981) and two extended elaborations (Lowenstam and Weiner 1982, 1989) the first type of mineralization was called "biologically induced" mineralization and the second "(organic) matrix-mediated" mineralization. Single-celled organisms and protoctists such as algae may deposit biologically-induced mineral either intra- or extra (inter)-cellularly. The majority of eukaryote matrix-mediated mineralization is extracellular.

An important generalization that can be made with respect to matrix-mediated mineralization is that the mineralization takes place within defined and restricted compartments or spaces. In essence, this compartmentalization of mineral deposition is crucial to the entire process, as it demands that the construction of the compartment or space take place first, and that mechanisms exist for regulation of the flux of organic components and the mineral ions into the localized space. This is very different from the absorption of mineral ions from solution onto a preformed surface of organic polymers with subsequent mineral crystal growth limited only by the further uptake of ions from solution. In matrix-mediated mineralization the physical size and shape of the space may be crystal volume and shape limiting. A variety of means are used to construct compartment boundaries. The compartments do not have to be completely sealed. The compartment walls may contain pores, and channels with limited access may serve as well as sites for mineral deposition.

One of the most compelling examples of a compartment structure is illustrated in Figure 1, a compilation of data obtained by several investigators on the structure of the shells of the bivalves, *Pinctada radiata, Monodonta labio, Callistoma unicum.* Sectioned perpendicular to the shell surface, it can be seen in Figure 1A that the most heavily mineralized portion of the shell in *Pinctada* is like a brick wall, with layers of mineral plates of aragonite stacked on top of each other, but clearly separated by an intervening material of non-mineral origin. Looking at a plane parallel to the shell surface, Figure 1B, in a mature shell growth zone layer, scanning electron micrographs clearly show the plates to be hexagonal and separated by wall material. After demineralization with EDTA in such a way that the organic matter was not disrupted, Figure 1C, the compartment walls were clearly visualized. Figure 1D shifts the view to the earliest growth layer, adjacent to the shell forming cells. It is clear that the compartment walls have been constructed, and that the mineral begins to form at a single central point near the base of the compartment. The highly hydrated matrix filling the compartment contains a fine network of matrix macromolecules, revealed by transmission electron micrographs (not shown here). The aragonite plates grow in size until they fill each compartment. The stack of plates may contain an organic core at the center of each plate, Figure 1E. Thus, there may be a channel for the transport of the required mineral ions into each compartment until the tablets of mineral fill the entire compartment. The fine filamentous matrix within each compartment is occluded in the mineral phase as the compartment fills, as shown by electron micrographs of partially demineralized shell matrix, Figure 1F. The most important point established in these views of the growing region of the shell is that the structural matrix is indeed formed first and has mechanical integrity; the second point is that the matrix within the compartment contains other macromolecules that may be involved with crystal nucleation. The structural macromolecular framework may not have nucleation capability. In the mollusc shell, Weiner and Traub (1984) described the shell organic matrix itself as a complex of several layers. The polysaccharide β-chitin is the main structural component. Initially, it was reported that the next layer was silk fibroin, upon which acidic macromolecules were deposited. These were proposed to be the molecules directing the apposition of the aragonite crystals. More recently, cryo-TEM studies of *Atrina serrata* shell indicated that the silk fibroin was not present as a defined layer. Rather, prior to mineralization, the silk was in a hydrated gel state, and as shown in Figure 2, it is likely that aragonite crystal growth takes place within the hydrated layer, while the Asp-rich acidic protein is at the chitin–gel interface (Levi-Kalisman et al. 2001).

An example of mineralization in channel spaces is the mineralization within sea urchin teeth. Figure 3 shows a microCT scan of a *Lytechinus variegatus* tooth cross section (Fig. 3B) in an area where the mineral is being deposited in the structures known as the primary plates and the corresponding section stained directly for the organic matrix (Fig. 3A). The

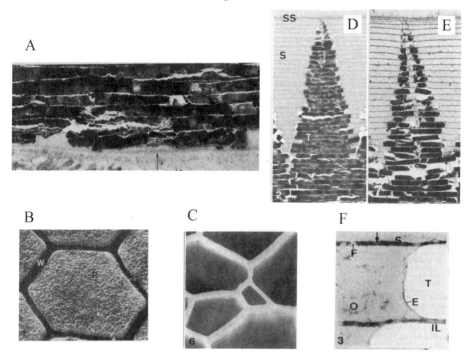

Figure 1. Mineral placement in various invertebrate shells. (A) *Pinctada radiata*, scanning electron micrograph, perpendicular to shell surface showing brick tablet structure. (from Bevelander and Nakahara 1969; 1980 with permission of Springer-Verlag). (B) Parallel to surface, showing hexagonal shape of prism tablets (P), and space for matrix walls (W) between prisms (from Nakahara et al. 1980 with permission of the Malacological Society of Japan). (C) The demineralized shell, with matrix retained, same view as in B. The shape of the insoluble matrix defines the shape of the prism compartment (from Nakahara et al. 1980 with permission of the Malacological Society of Japan). (D) A view of the growth region as the mineral begins to fill in the preformed compartments of *Monodonta labia*. Growth initiates at the surface sheet, SS. The more mature surface sheets, S, are evident. The stack of prisms, T, grow in area but not width as they mature. (E) A partially demineralized view of the growing stacks in *Callistoma unicum*, showing that they may have a central channel for the transfer of ions and organic components into each compartment. Mineral crystallization begins at this point within the compartment. (F) Electron micrograph of the contents of the matrix of *Monodonta labia* adjacent to an empty space which had been occupied by an aragonite tablet, T, in a growth compartment. Note the structural organic matrix of the wall, S, the apparent fenestrations in the wall, F, and organic deposits, O, in the matrix fluid. The envelope, E, covering the crystal prism probably contains proteins occluded in the crystal, as shown by further transmission electron micrographs. (Figures D–F from Nakahara 1982 and used with permission of D. Rideal Publishing Co).

cellular network forms a set of syncitia in which multinucleated layers of cellular material are bounded by cell membrane. The mineral crystals of the primary plates, calcite in this case, grow in the channels defined directly by the cell membranes. The intimate layering of cell contents and calcite crystals has made it very difficult to determine the nature of the macromolecules that may be associated specifically with the inter-membranous mineral phase, as compared to the contents of the cells themselves. Nevertheless, it has been possible to determine that there are several likely mineral-related protein components (Veis et al. 1986; Katijima et al. 1996; Veis et al. 2002). From our own studies, not yet complete (Barss et al., unpublished results, 2003), it appears that at least one of the important

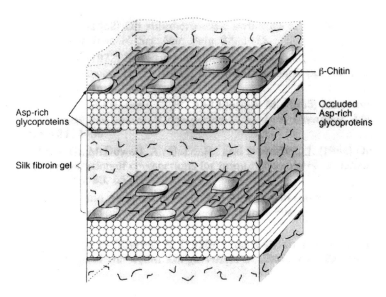

Figure 2. Schematic of the structure of the demineralized nacreous layer of *Atrina serrata*, (From Levi-Kalisman et al. 2001 with permission of Elsevier) showing the relationship between the structural β chitin and the silk fibroin gel. The Asp-rich proteins, thought to be part of the mineral nucleating system are at the chitin, silk fibroin interface. There are also Asp-rich proteins occluded in the gel, and presumably, the mineral phase.

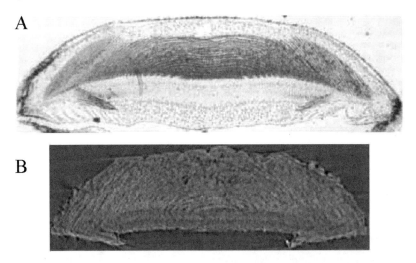

Figure 3. Mineralization in the sea urchin tooth takes place in inter-membrane spaces. A cross section view of a developing, partially mineralized zone in a *Lytechinus variegatus* tooth. (A) A histological section of an equivalent region stained with toluidine blue to show the cellular contents. The cell nuclei are seen as intensely stained spots. In these sections the mineral appears to have a granular surface and is clearly in the spaces between adjacent cell membranes. At this light microscopy level, the fluid-filled inter-membrane space appears empty. (B) The distribution of mineral in the plane of the section as revealed by microCT scanning. Lighter areas represent the high density mineral. (Work in progress, S. Stock and A. Veis). The keel of the tooth has not yet formed, but the mineral is forming within the spaces between adjacent cell surfaces.

proteins is intercalated into the cell membrane, and has a role comparable to that of the Asp-rich protein of the mollusk shell at the interface between the silk fibroin gel and the crystal surface. The sea urchin tooth is a continuously growing structure, with its size and mineral content increasing from the aboral plumula to the adoral incisal edge. All stages of growth and mineralization are present in a single tooth at all times, making this a very interesting system to study. The three dimensional reconstruction of the mineral distribution made possible by microCT scanning (Stock et al. 2002) indicates that the crystals in the plane of the inter-membrane space are quite large, while their thickness remains small. The components of the inter-cell membrane compartments must enter via transport from the cells through the cell membrane. Labeling with tetracycline shows that calcite deposition takes place in all parts of the teeth, even in the oldest adoral zone. Although it is far from proven at this point, it is reasonable to hypothesize that the presumed membrane proteins provide the interactive sites for mineral crystal initiation within the inter-membrane plate channel space in this matrix-mediated mineralization system. In effect, the syncitium of cells provides the structural matrix. There are no defined polymeric organic matrix structural compartment walls, as evident in the bivalve shells.

A quite different system and organization is seen in the structure of the avian eggshell, one of the two systems in which the mineral phase in a vertebrate is calcium carbonate derived rather than a calcium phosphate. The mineral phase deposition takes place within a very well defined zone during the passage of the egg through the oviduct. In addition to secreting albumen, the epithelial cells of the oviduct also secrete a variety of proteins and proteoglycans that organize into distinct layers and tightly regulate the shell structure. As shown schematically in Figure 4 (Gautron et al. 2001), the inner shell membrane is not mineralized, but consists of type I and X collagens, several glycoproteins, and

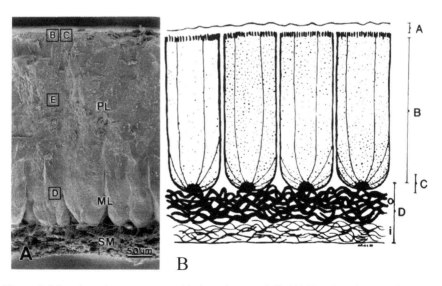

Figure 4. Mineral-matrix arrangement with the avian egg shell. (A) Scanning electron micrograph across a fractured avian shell (from Gautron et al. 2001 with permission of The American Society for Biochemistry & Molecular Biology). Internal shell membrane, SM; Mammillary layer , ML; Pallisade layer, PL; Surface cuticle, [B][C]. (B) Schematic of the shell cross-section. The mammillary layer, C, rests on a collagenous fiber net, D, the mammillary knobs are evident. These are the sites of mineral nucleation to form the palisade layer, B, which is capped by a cuticle membrane, A (from Wu et al. 1992a with permission of Elsevier).

proteoglycans. Dense aggregates called mammillary knobs form on the shell membrane surface. These are mainly mammillan, a keratan sulfate proteoglycan, and they act as nidi for crystal growth. The calcite crystals that make up the bulk of the shell grow in a complex matrix of glycoproteins and proteoglycans (Soledad Fernandez et al. 2001) which guides their orientation into parallel arrays to form what is known as the palisade layer. Crystal elongation is stopped by the formation of a final cuticle comprised of still other proteins, localized in the fluid of the oviduct. This is a dynamic process and regulation is kinetic in the sense that during the passage of the egg through the oviduct one sees both topographic localization of the components of the oviduct fluid, and the temporal secretion of the shell-forming macromolecules. It is not important to consider the details of these processes here, the main point is that within this open system where the egg traverses the restricted space of the tubular oviduct in about 22 hours, the epithelial cells regulate the composition of the secreted proteins and proteoglycans available at every point, and thus closely control the shell mineralization. It is noteworthy that the type I collagen within the inner shell membrane, the principal membrane constituent, does not mineralize. The calcium carbonate crystals are initiated with respect to the sulfated mammillan.

The three examples given above do not represent all possibilities for matrix-mediated biomineralization, but were chosen to emphasize the wide range of situations that nature has used. The mollusk builds its shell layer by layer, creating fixed compartments that can be filled by injections of the proper components in the correct ratios. Once the spaces are filled, the mineralized matrix remains through the life of the animal. A finer control is achieved by the echinoderm. The teeth are used to scrape algae and other organisms off of hard rock surfaces; hence their incisal edge is worn away. The teeth, as in rodent incisors, respond with continual growth. Each tooth remains "alive" with viable cells throughout from the initial cell divisions in the plumula to their disappearance as they are removed as the incisal edge is worn off. The matrix is never entirely mineralized, and the cellular compartments may always be capable of metabolic response and some repair. The mineral phase is in direct apposition to the cell membrane. The avian egg shell is closer to the mollusk situation although the mineralized compartment is prepared in assembly line fashion along the oviduct with varying compartment walls (e.g., the inner shell matrix is distinct in composition from the shell cuticle). In each case, however, the scenario is the same: a structural interface is created in a delimited space; a second set of macromolecules interacts with the structural interface and creates sites for the specific deposition of the mineral phase; and, finally, macromolecules either bind to the mineral at certain crystal faces to inhibit growth or the compartment walls themselves impede further crystal growth. The selective pressures determined by the ultimate use of the mineralized tissue must have determined the strategy selected during evolution. As we shall now explore, the vertebrate mineralized tissues were designed to fit a variety of particular requirements, thus all bone in a given animal, for example, is not the same in architecture, in composition or in metabolic activity. Nevertheless, all bone uses a similar repertoire of macromolecular components in meeting its functional requirements. The size, shape, orientation and composition of the mineral crystals in different vertebrate mineralized tissues are obviously crucial to the particular function of a given tissue, but our emphasis will be on the organic matrix components that control those properties.

THE COLLAGEN MATRIX

The structural matrices of vertebrate bone, dentin and cementum are comprised essentially of a network of type I collagen fibrils, built in distinct hierarchical arrays from type I collagen molecules. The ways in which the collagen fibrillar arrays are constructed have a direct relationship to the arrangement of the mineral phase, a carbonated apatite.

The designation "type I" collagen indicates that several different collagens exist, and, in fact, "collagen" is now used to denote members of a large family of proteins that share certain common features. More than 40 vertebrate genes have been identified as members of the collagen family (Pace et al. 2003), and there are related genes in invertebrates as distant as sponges, sea urchins, and the byssus threads of mussels (Lucas et al. 2002; Boot-Handford and Tuckwell 2003). The key feature defining a collagen is the presence of continuous or interrupted sequences in the individual polypeptide chains of repeated $(GXY)_n$ amino acid sequence domains. (In the single letter abbreviations for amino acids, G is Glycine while X and Y may be any amino acid.) These domains on one polypeptide chain interact with similar domains on two additional chains to form compound triple-helical units with characteristic folding, the "collagen-fold." In the collagen fold domains, the X and Y chain positions can accommodate any amino acids except cysteine and tryptophan. In the vertebrate collagens, X is frequently P (proline) and Y is P or O (hydroxyproline). The 40 different chains can assemble into homotrimeric or heterotrimeric triple-helical structures, with the requirement that the lengths of the $(GXY)_n$ domains on each chain must match the lengths on the partner chains so that stable triple-helical segments can be formed. As of this time, 27 different combinations of chains have been recognized as assembled into distinct molecules.

The molecular forms have been grouped into several broad classes: fibrillar collagens, types I, II, III, V, XI, XXIV and XXVII, which are all characterized by a long uninterrupted triple helix containing 300 or more GXY repeats; network forming collagen, type IV, with shorter interrupted triple-helix segments joined by more flexible, hinged chain regions (Blumberg et al. 1987, 1988) so that three-dimensional nets can be formed; and the remaining FACIT or fibril-associated collagens with interrupted triple helix (Olsen et al. 1989) that may be involved in linking other extracellular matrix components to the fibrillar matrix or in regulating fibril size or other properties. In all cases the triple-helical domains are bounded by "noncollagenous" (NC) sequence domains that do not fold into triple-helical conformations, although in the three-chain assembly interactions of both covalent and non-covalent nature may form between NC domains. The collagens of primary importance to this discussion of the vertebrate mineralized tissues are the fibrillar collagens. At the protein level, collagens I, II, and III are the major fibrillar collagens, while collagen types V, XI, XXIV and XXVII are present in minor amounts. All of the fibrillar collagens are related in terms of their gene structure as well as amino acid structure. The exons of the fibrillar collagen genes within the triple helical regions all appear to have been derived by gene duplication from a primordial 54 base pair (18 amino acid, $[GXY]_6$ coding) gene. Type I, V and XI collagens form heterotrimers, with types I and XI having two chains, designated COL1A1 and COL1A2, and COL11A1 and COL11A2, respectively. Type V forms heterotrimers built from COL5A1, COL5A2, COL5A3 chains. Type II molecules are homotrimers comprised of COL2A1 chains, and similarly type III is comprised of three COL3A1 chains. Types XXIV and XXVII are probably homotrimers constructed from COL24A1 and COL27A1 chains, respectively.

Phylogenetic analyses show the major and minor fibrillar collagens separate into three clades, Figure 5. The organization of the exons of different sizes within the triple helical regions is different in detail in each case but leads to the same type of major triple helix. However, a major difference can be seen in Figure 6 (Välkkilä et al. 2001; Pace et al. 2003) relating to the NC domains flanking the major helix. Clearly, the composite N-terminal NC domains of the three lineages are quite different; the minor fibrillar collagens all have very large and complex N-terminal NC domains as compared to the much smaller N-terminal NC domains of the major fibrillar collagens. On the other hand, although they are different and have been used to distinguish the phylogenetic relationships of the fibrillar collagens, the C-terminal NC domains are similar in

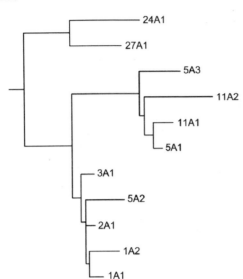

Figure 5. Unrooted phylogenetic tree for the fibrillar collagen genes showing three clades. (from Pace et al. 2003 with permission of Elsevier).

		N-Propeptide			C-Propeptide	
	Exons	Nucleotides	AA	Exons	Nucleotides	AA
Clade 1						
COL1A1	6	534	178	5	801	267
COL1A2	6	501	167	5	831	277
COL2A1	8	600	200	5	864	288
COL3A1	6	528	176	5	873	291
Clade 2						
COL5A1	14	1707	554	5	807	269
COL11A1	15	1770	590	5	901	267
Clade 3						
COL27A1	3	1745	583	5	726	242
COL24A1	3	1530	510	5	705	235

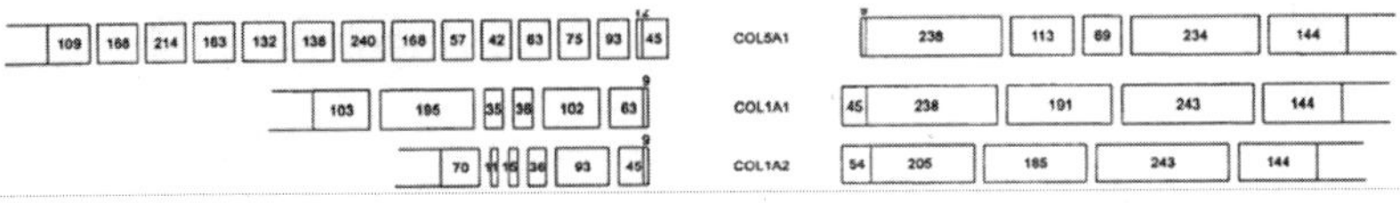

Figure 6. Composite of gene structures of fibrillar collagen N-propeptide and C-propeptide domains, derived from the data of Välkkilä et al. (2001) and Pace et al. (2003) with permission of Elsevier. The very major differences in the N-propeptide domains and N-telopeptide domains, linked to extracellular organization, in contrast to the similar structures of the C-propeptide domains, related to intracellular molecular assembly, are evident. The N-propeptide differences are the basis for the division into distinct clades.

structure. This must relate to the function of the C-propeptide (NC1) domains in the assembly of the trimeric molecules, to be discussed below.

Molecular assembly of the Type I collagen molecule

The COL1A1 and COL1A2 terminology relates to the polypeptide chain that is the product of the translation of the entire final processed messenger RNA. Since each chain is itself a multi-domain structure that is post-translationally processed into the final functional units during the initial three chain association into molecules, the subsequent export into the extracellular space, and the assembly into tissue specific fibrillar arrays, a different nomenclature was developed. The individual biosynthetic products, with all domains intact, are designated as pro-alpha chains, pro-$\alpha1$(I) and pro-$\alpha2$(I), respectively. The normal type I procollagen heterotrimer (with all domains intact) has the formula ([pro $\alpha1$(I)]$_2$ pro $\alpha2$(I)). The homotrimer [pro $\alpha1$(I)]$_3$ also is formed, but only a small fraction of the collagen assembles in this fashion. The potential homotrimer [pro $\alpha2$(I)]$_3$ does not appear to exist naturally.

Figure 7 provides a schematic overview of the pro $\alpha1$(I) and pro-$\alpha2$(I) chain domain arrangements and domain sizes. The central domain of 338 GXY amino acid repeats that fold into the uninterrupted triple helix structure is identical in length in each chain, but the sequences (G is always in the proper position, but X and Y differ between chains) are not. Regions or subdomains with special reactivity or properties exist even within the uninterrupted triple helix and these areas are also denoted specifically. Lysine residues (K) that ultimately participate in cross-linking reactions are depicted by the heavy arrows at residues K87 in both chains and at residue K930 in the $\alpha1$ chain. K933 in the $\alpha2$ chain can also participate in cross-linking. The remaining domains differ in sequence and size. Assembly of the ribosomes and translation of the mRNA begins in the cytosol. Following elaboration of the signal peptide sequences translation is paused until the signal peptides find their receptors on the endoplasmic reticulum (ER). After the signal peptides insert the nascent chains into the ER cisternal space, chain synthesis and elongation resume within the restricted ER space and environment, Figure 8. The growing chains remain in the unfolded state until elongation is essentially completed. As they grow the separate chains are subject to a number of interactions and post-translational modifications that modulate the chain properties and prepare them for their ultimate functionality. One important interaction is with an ER-resident chaperone, heat-shock protein Hsp 47, which reacts specifically with the nascent $\alpha1$ N-propeptide, and perhaps other parts of the chain, to prevent premature interchain interaction and random collagen-fold formation (Hu et al. 1995). Koide et al (2002) have shown that, within the ER, HSP 47 binds to the newly formed triple helix, stabilizing the structure until helix formation is complete and is ready for export to the Golgi. Chaperone proteins have the function of preventing misfolding of nascent protein chains. HSP 47 is uniquely a chaperone for collagen. Although it has been known for a long time that full length collagen chains properly in register can form triple-helix in either direction, i.e., N $\Rightarrow$ C or N $\Leftarrow$ C (Veis and Cohen 1960; Drake and Veis 1964; Frank et al. 2003), *in vivo* the triple helical folding is in the C $\Rightarrow$ N direction (Bruckner et al. 1981, Bachinger et al. 1980), following completion of elongation and registration of the C-propeptides. Recent studies show that this registration is a very complex process.

The principal modification during chain elongation is that of the hydroxylation of selected proline residues in the Y positions of the repetitive GXY sequences. In this potential helical domain, the sequence GPP occurs frequently. The second (Y-position) P is subject to the action of the enzyme prolyl hydroxylase that places a hydroxyl group on the 4-position of the P pyrrolidine ring. This crucial reaction is both tissue and species specific and regulates the stability of the final triple helix. An unhydroxylated chain will not join into the three chain triple helical structure of collagen at body temperature. That

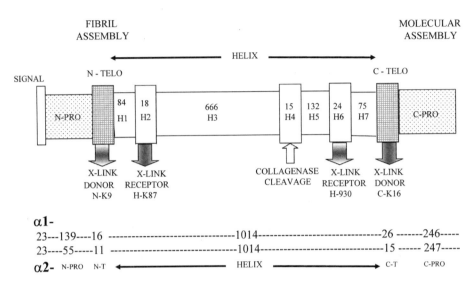

Figure 7. Schematic representation of the domains of the Type I collagen α1 and α2 chains. The block arrows represent the positions of the lysine (K) resides involved in cross-linking reactions. The shading within the arrows is to emphasize that the N-telopeptide at K^{9N} reacts with the similarly shaded K930 in H-6 on an adjacent, properly staggered molecule, and the K^{16C} of the C-telopeptide reacts with K87 in H-2.

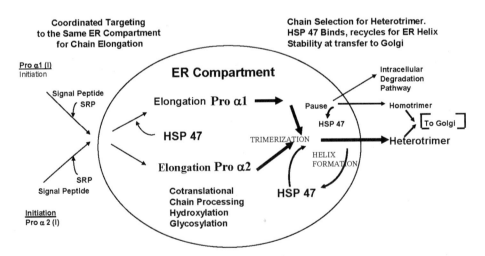

Figure 8. Schematic representation of the synthesis of type I collagen, with emphasis on the processing within the ER compartment. Key points are the brining of the different mRNAs into the same ER compartment, the central but multiple roles of the chaperone protein HSP 47 in stabilizing the chains as they fold into triple helix and in preventing premature interactions until triple helix can begin. At the point of export of completed molecules note that in addition to the major pathway of heterotrimer formation movement to the Golgi, a lesser amount of α1(I) folded homotrimer also moves on for secretion whereas misfolded, or unfolded chains are moved into the lysosomal pathway for degradation.

is, the unhydroxylated collagen molecule is intrinsically unstable at body temperature, and hydroxylation to a certain level is required to permit formation of the three-chain structure. The collagens of cold-water fish, for example, are lower than that of land animals living at higher ambient temperatures. The mechanism of the stabilization is an area of current debate (Burjanadze 1979; Burjanadze and Bezhitadze 1992; Burjanadze and Veis 1997; Leikina et al. 2002) but the hydroxylation reaction is well understood.

The extended 15 triplet GPP rich domain within the N-propeptide of the nascent elongating α-chain probably begins to be hydroxylated immediately; therefore premature helix formation must be inhibited. This is probably accomplished through interactions of the nascent α1(I) N-propeptide chains with Hsp47 (Sauk et al. 1994; Hu et al. 1995). Molecular assembly may take place while the chains are still anchored to the endoplasmic reticulum membrane (Beck et al. 1996; Gura et al. 1996). An important fact is that hydroxylation of all potential substrate sequences in a single chain generally does not occur (Bornstein 1967). The rate of chain elongation is faster and of higher fidelity than the post-translational enzymatic reactions, introducing hydroxylation microheterogeniety into any collagen tissue. The critical stability of the triple-helix is reached before all of the post-translational hydroxylation reactions are completed. Chains in the triple helical conformation are no longer substrates for prolyl hydroxylase so that hydroxylation ceases. This important physiological mechanism assures that the final triple-helical collagen molecule is held to a marginally stable state at the host body temperature. As triple helix forms Hsp47 binds and stabilizes the structure while it is still being processed in the ER compartment. The Hsp47 is released at the time that the completed molecule is passed on to the Golgi compartment for secretion. The marginal stability of the collagen at body temperature enables the turnover and remodeling of collagen, and may be especially important in bone, since bone collagen is remodeled at a much faster rate than soft tissue collagens. The cut ends resulting from a proteolytic cleavage within the helix destabilize it, and render it more susceptible to further proteolytic degradation.

Selected Lysine (K) residues in the Y-position on nascent collagen chains are also subjected to hydroxylation (Kivirikko et al.1973; Brownell and Veis 1975), followed by a second post-translational modification, variable glycosylation to either galactosyl-hydroxylysine or glucosyl-galactosyl-hydroxylsine (Butler and Cunningham 1966). Like the P hydroxylation, the extent of hydroxylation of the K's is variable, and the subsequent extent and nature of the glycosylation of the K is variable. Moreover, the extent of K hydroxylation and glycosylation is different in hard and soft tissues in a given animal. This has a very important effect on the ultimate tissue properties (Morgan et al. 1970) as explained in detail below. Like prolyl hydroxylase, the enzyme carrying out this reaction, lysyl hydroxylase, acts only on non-helical collagen chains, moreover, there are three lysyl hydroxylases, PLOD1(procollagen-lysine, 2-oxoglutarate, 5-dioxygenase 1), PLOD2 and PLOD3 (Sipilä et al. 2000). Of these PLOD 1 and PLOD 3 hydroxylate the lysines in the helix region, and possibly regulate their subsequent glycosylation. PLOD 2 and perhaps a fourth PLOD enzyme isoform are involved in the hydroxylation of the telopeptide Lys (Bank et al. 1999; Van Der Slot et al. 2003). The hydroxylation of the telopeptide lysines is not found in soft tissues, it is a bone and dentin specific process. Interestingly, PLOD 1 has the ability to hydroxylate K87 and K930 as well as other Lys residues in the helix in the type I collagen helix region, whereas PLOD 3 does not hydroxylate K87 and K930. These two helix region Lys residues are involved in the cross-linking reactions.

Secretion

Following synthesis, post-translational modification, chain selection and triple-helical folding, the completed molecules are passed from the ER compartment into the Golgi apparatus and packaged into secretory vesicles. In the vesicles, which finally bud off from the Golgi, the individual molecules appear to be aligned in loose bundles with the same axial orientation. The procollagen molecules within the secretory vesicles retain the bulky propeptide extensions at both ends and these inhibit fibrillar packing within the vesicle. The parallel, aligned collagen aggregates are similar to, but less densely packed than the segment-long-spacing (SLS) form (Bruns et al. 1979; Hulmes et al. 1983). The secretory process has been described at the electron microscopic level by several beautiful studies (Weinstock and Leblond 1974; Birk and Linsenmeyer 1994; Prockop and Hulmes 1994).

The mineralized tissues appear to handle the exocytosis of the procollagen from the odontoblasts and osteoblasts quite differently from the soft tissue fibroblasts. According to Birk and Linsenmeyer (1994), in tendon the contents of the vesicles are exocytosed into confined spaces in extracytoplasmic channels where they fuse or merge into filaments within the restricted channel space. Osteoblasts and odontoblasts do not show the kind of channels demonstrated in tendon, rather the secretory vesicles appear to release their contents directly into the open pericellular space (Weinstock and Leblond 1974; Rabie and Veis 1995). There, the procollagen molecules are trimmed and assemble into fibrils. It is not clear whether the aligned packets of parallel procollagen molecules aggregate to form fibrils directly or disaggregate and then reassemble into fibrils.

Fibrillogenesis

The procollagen molecules, which can be readily isolated from fibroblast tissue culture supernatant, are very soluble so long as they retain both N- and C-propeptide domains. *In vivo*, different procollagen peptidases specific for the N- and C- propeptides are present and cleave the propeptides from the main triple helical section. The order of cleavage is important, it is not possible to form filaments with collagens containing the C-propeptide (Kadler et al. 1987, 1990), and thus, the C-propeptide must be excised first. Molecules with the N-propeptide attached still can form axially extended structures (Lenaers et al. 1971; Pierard et al. 1987), as seen in Dermatosparaxis (Levene 1966; Lenaers et al. 1971), but the aggregation is generally restricted to thin sheets in staggered arrays, with the bulky N-propeptides on either side of the collagen sheets (Hulmes et al. 1989).

When both propeptides have been cleaved one is left with rod-like collagen I monomers with an uninterrupted triple helical segment that consists of 1014 residues, 338 GXY triplets in each polypeptide chain, trimmed by the short telopeptide regions at each end (see Fig. 7). The molecules are about 296 nm in length and 1.5 nm in diameter. The type I monomers are semi-flexible as clearly demonstrated by electron micrographs of rotary shadowed collagen monomers (Fig. 9) (Veis 1982; Hofmann et al. 1984), by viscoelasticity measurements (Nestler et al.1983; Amis et al.1985), and by electric birefringence and dynamic light scattering solution measurements (Bernengo et al. 1983). The electron microscopy and viscoelasticity studies, in particular, demonstrated that there were several points of higher flexibility. The major triple helix stability arises from GPP and GPO rich regions (Kuznetsova and Leikin 1999), while sequence regions with a lower content of P and O are more flexible (Malone and Veis 2003). These regions of lesser stability are of crucial importance for the collagen structure. They represent the sites of helix region-telopeptide interactions that are most important for establishing the axial repeat order in the fibril and the regions where cross-linkages can form. When both propeptides are properly excised, the collagen forms fibrillar structures in which all of the molecules are essentially axially oriented. The *in vitro* assembly of collagen monomers is a path dependent process. While the detailed mechanism of assembly is debatable (Veis and George 1994; Kuznetsova and Leikin 1999) the result is the formation of fibrils indistinguishable in packing arrangement from those formed *in vivo*. *In vitro* the kinetics of assembly and fidelity of the packing order are crucially dependent upon the intactness of the telopeptide regions.

Mineralization in Organic Matrix Frameworks 261

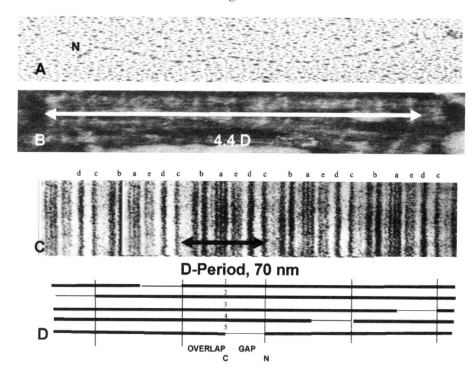

Figure 9. Electron micrographs of type I collagen in various states. (A) Rotary shadowed monomer. Note that the individual chain is not entirely a rigid rod, but has some kinks and flexibility. In this EM the N-terminus is on the left. The sharp kink is typical of the N-terminal region. In solution a collagen monomer always has a shorter end-to-end length than its contour length. The difference in length between free molecule and extended length in either native fibril or SLS aggregate shows the stabilizing role of the intermolecular interactions in stiffening the triple helix. (B) Collagen molecules aggregated in side-by-side, axially aligned segment-long-spacing form, SLS. The elongated molecules have a length 4.4 times the D-periodicity noted in native fibrils. (C) A section of the band pattern of a native collagen fibril, doubly positively stained with phosphotungstate and uranyl acetate. The length of the repeat D-period, ~70 nm, is denoted by the double arrow. The major stained bands are denoted by electron microscopists as the a-, b-, c-, d-, and e-bands, although the bands may be complexes of bands as in the a-band. (D) The molecular packing in a native fibril in the Petruska-Hodge model, arranged to correspond to the band pattern of the native fibril. Each molecule has the length shown in (A) or (B), 4.4D. The numbers, 1,2,3,4,5, are the relative axial positions of the domains within the D-period.

Fibril structure

Figure 9 shows electron micrographs of the collagen monomer (Fig. 9A), the parallel-packed SLS form (Fig. 9B) and the structure of a typical fibril (Fig. 9C). The SLS aggregate (Fig. 9B), has an asymmetric banding pattern, in a combination positive/negative staining procedure, showing 52 or 53 clearly demarcated bands. This asymmetry of the band pattern confirms the identical orientation of the molecules within the SLS aggregate, with all of the molecules in the same N to C orientation. The stain pattern is directly related to the sequence of charged amino acid side chains. The length of the SLS corresponds to the contour length of the monomers, 290 nm, as demonstrated in Fig. 9A. On the other hand, the fibril (Fig. 9C) has a periodicity of 67 nm. The problem of packing rigid rods ~ 300 nm long to achieve a ~67 nm periodic structure (D-periodic) was solved (in two dimensions) by Petruska and Hodge (1964; Hodge 1989) as the "quarter stagger-overlap" packing arrangement. Each molecule is about 4.4 times the D-period length. Each molecule can be considered as divided into 5 consecutive segments, numbered 1, 2, 3, 4, 5 in the N to C direction. Segments 1, 2, 3 and 4 are 1-D in length. Segment 5, containing the C-terminal telopeptide as depicted, is 0.4-D in length (Fig. 9D). The minimum possible cross section of a D-periodic fibril would contain 5 molecules in the overlap zone and 4 in the gap region. Thus, each D-period has a minimum of 5 molecular sections from 5 neighboring D-staggered molecules. In this two-dimensional representation, the 5 sections in the D-period correspond to the full contents of a complete molecule. However, the "gap" or "hole" section of ~0.6 D has 4/5 the density of the "overlap" and the contents of molecular sections 1, 2, 3 and 4 only. The Petruska-Hodge model ruled out the possibility of linear end-to-end aggregation between molecules and emphasized the overlap zone side-by-side molecular interactions.

The telopeptides, at the ends of each molecule, are at the boundaries of the gap regions. The consequences of this structure are enormous. As soon as one moves from the 5 molecule limiting microfibril to larger diameter assemblies and packs the molecules or limiting fibrils in register to duplicate the band pattern in a normal collagen fibril (from 80 to 1000 Å in diameter) one realizes that the structure is filled with in-register defects which can form channels throughout a fibril. These empty channels are, in fact, where the mineral crystals in bone, dentin and cementum form. However, in fully mineralized mature bone or dentin all the mineral does not fit in the gap channels, there is always some mineral deposited between fibrils and between large fibers. That is, the mineral can have three placements: in the periodic gap spaces and related channels (intrafibrillar), in the pores between microfibrils (intrafibrillar), and in the spaces between packed fibrils, a hierarchical arrangement. The chemistry of the gap zone is most interesting. The telopeptides bounding the gap are chemically reactive and participate in the cross-linking reactions. The gap zone itself may be a site for the binding of noncollagenous proteins and other molecules, themselves with potential biologic activity.

A single two-dimensional layer of molecules in D-periodic array within a fibril is shown in Figure 10. Each "point" along each chain represents the correct amino acid in the sequence of each of the three chains. The letters in larger font are the positions of the ionic groups, Arg, Lys, His, Asp and Glu. Wherever the clusters or patches of ionizable groups appear it is not possible to have the GPP or GPO triplets, and these sequence regions have a lower intrinsic triple helical stability than those rich in P and O. It is possible that salt bridges between adjacent positively and negatively charged residues can increase local stability (Venugopal et al. 1994; Long et al. 1995) in compensation for the lack of P and O. It is evident in Figure 10 that the charged groups are clustered, not distributed randomly along the fibril length, and that the charged clusters correspond very well with the positive stain band pattern. In Figure 10 the basic residues in the e-band region of the gap zone are emphasized in a larger, bold font and the acidic residues are in larger font only. The adjacent + − pairs are obvious, but all of the groups are not paired, as is the case throughout each molecule, giving rise to a distinctive net charge pattern. These sequences with excess charge may be regions of heightened molecular flexibility.

When the molecules aggregate side by side in a 0 D-stagger (SLS form) or in D-staggered fibrils, intermolecular interactions potentiate the helix stability and stiffness so that the molecules within the fibrils or SLS exhibit nearly their full extension in the axial direction. The melting temperature of non-crosslinked collagen fibrils is on the order of 20°C higher than the monomer in solution, indicating that the intermolecular packing energy contributes to the helix stability and molecular stiffness. On the other hand, the denaturation temperature of an unhydroxylated collagen monomer is about 15°C lower

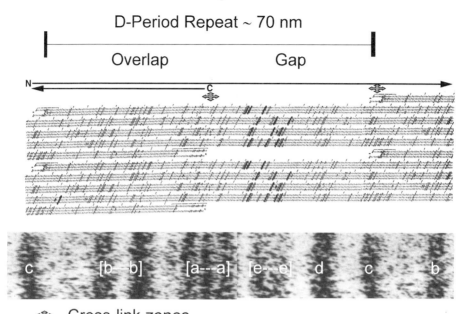

Figure 10. The detailed structure of the hole and overlap regions based on the Petruska-Hodge model. The three chains of each molecule are shown, the oblique lines evident are the locations of acidic and basic amino acid residues, showing the non-uniform disposition of the charged side chains. The lower chain density within the hole region is evident in this two dimensional representation. Note that the composition of five chains in a D-period in the structure includes the sequences of a single complete molecule. The crosses are where cross-linking domains are present, at the boundaries of the gap region. A section of an electron micrograph of a native fibril D-period is shown at the bottom, with the electron microscopic designations of the a, b, c, d, and e bands shown. According to Traub et al. (1992b) the e-band is the locus of the first mineral crystals in turkey tendon mineralization.

than that of a fully hydroxylated monomer, indicating that the hydroxylated GPO triplets also add to the overall molecular stability in an aqueous environment.

All type I collagen fibrils exhibit a distinctive X-ray diffraction pattern indicative of the long-range triple helical organization. However, the coherence length of the crystalline regions of tendon fibers, sampled in the meridional (axial) direction, is only ~ 120 Å, significantly shorter than the molecular length. Thus, there must be more disordered regions interspersed between the most highly crystalline zones. Similarly, the diffraction pattern along the equator shows a diffuse equatorial scattering indicating only a liquid-like short-range order in the direction perpendicular to the fibril axis (Miller 1976; Woodhead-Galloway et al. 1976). NMR data on fibrillar collagen showed that individual chains retained azimuthal mobility even within the fibrils, indicating that segments of the molecules were free to rotate about their long axis (Jelinski et al. 1980; Torchia 1982), probably correlating with the less ordered segments between the most highly crystalline regions. The problem of the three dimensional packing of fibrils so as to permit both crystalline segments, short range disorder and chain rotational freedom (Fraser et al. 1983; Wess et al. 1995) was studied by Hulmes et al. (1995) who proposed that the D-period could be considered as containing four structurally significant segments, the N-terminal junction at the gap-overlap interface, the remaining overlap (five molecular segments), the C-terminal gap-overlap interface, and the remaining gap, with 4 molecular segments. The "internal" gap and overlap regions were proposed to have 80 % and 60 % liquid-like radial disorder, respectively, while both gap-overlap interface regions containing the N- and C-telopeptides had the highest degree of crystallinity.

The basic units of assembly that exhibit all of the features of the quarter-stagger array are of two types, a microfibril structure in which the molecules assemble in a symmetric (Veis et al. 1967; Smith 1968) or asymmetric (Veis and Yuan 1975; Trus and Piez 1980) compressed microfibril array. In the microfibril models, the thin or "limiting" microfibrils coalesce laterally to form larger fibrils and fibers, with lateral registration of gap and overlap regions. An alternate proposition was direct assembly into a quasi-hexagonal pattern (Hulmes and Miller 1979) without the involvement of a microfibril intermediate. Current studies favor the quasi-hexagonal model for the final packed structure although dissolution of collagen fibrils results in the appearance of thinner and thinner fibrils apparently unwinding from the larger fibril. That is, it appears as if a fibril is composed of smaller fibril assemblies. Thus, in spite of many years of study, agreement has not been reached on an assembly mechanism.

Cross-linking stabilization

In spite of the beautiful structural arrangement described above, fibrils based on non-covalent interactions alone are quite weak, and the fibrils would have a low tensile strength. Thus, a system of covalent cross-linkages has been developed to provide the appropriate functional strength. The most crystalline regions, as noted above, are in the overlap zones at either edge of the "hole" zone. It is at that point the covalent cross-linkages between the telopeptides and their helix receptors bind the molecules within a fibril together to form a stable polymeric network. The cross-linkages are of several varieties, but essentially involve reactions between the K residues in the telopeptides and specific helix domain K residues, matched in the 4-D staggered pairing of telopeptide and helix domains. However, at the N-telopeptide (segment 1) – helix K930 (segment 5) cross-link sites, and at the C-telopeptide (Segment 5) – helix K87 (segment 1) sites the geometry is different because of the differences in structures and sizes of the two telopeptides (Malone et al. 2003; Malone and Veis 2003).

Figure 11 (top) schematically depicts a cross-section slice through a quasi-hexagonally packed fibril at the 1-5 overlap level, the N-telopeptide-helix K930 cross-link site. At this level, in three dimensions, the model has a mix of two intermolecular axial staggers, 1-D and 2-D. The 1-D stagger planes of near neighbor molecules are in layers, but because of the triclinic unit cells and the quasi-hexagonal packing each layer is offset from its adjacent layer and this creates a second set of planes in which each molecule is staggered from its nearest neighbors in that plane by 2-D. Each unit cell (outlined in Fig. 11) is equivalent to a whole molecule in composition. If the unit cell was considered to form a compressed microfibril, the order within a microfibril would not be the symmetric 1-2-3-4-5 wrapping of the Smith (1968) or Trus and Piez (1980) model, but would have a staggered wrapping preserving the 1 and 2-D steps. The specific azimuthal orientation of near neighbor microfibrils at a particular axial overlap level must be the same. However, the order may persist only over small sets of microfibrils (Hulmes et al. 1981), and over relatively short distances along the fibril axis. If the cross-section packing had been drawn through the fibril at the level of the C-telopeptide and associated cross-linkages the larger C-telopeptide (Orgel et al. 2000) and its cross-linkage network arrangement requires a different packing arrangement, depicted in Figure 11 (bottom).

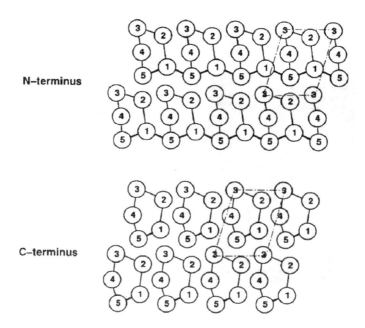

Figure 11. Packing arrangements within the cross-linking zones of collagen fibrils, according to Orgel et al. (2001). The X-ray data on which these drawings were made emphasize the different packing structure in the fibrils at the N- and C-telopeptide cross-linking domains within a fibril. Reproduced with permission of Elsevier.

The X-ray data indicate that the molecular axes are tilted by ~1.3° with respect to the fibril axis, and this corresponds to one molecular diameter per 234 residue D- period to maintain molecular continuity. Thus, the surface of the "hole" or "channel" changes in a uniform way in the C-telopeptide to N-telopeptide channel, (N→C fibril) direction. The low density holes or channels in the gap zones can be thought of as discrete spaces, but the high liquid-like disorder in the gap zone can also be thought of as permitting distortions allowing the vacant molecular spaces to become filled by solvent, other non-collagenous matrix molecules or crystals larger than a single molecular space. This is an important consideration, since small plate-like crystals of carbonated hydroxyapatite are thought to occupy these intrafibrillar spaces in bone and dentin. The enhanced density of packing at the N-telopeptide and C-telopeptide levels is clear, but the packing and azimuthal orientations of the chains in these more dense and crystalline regions need not be the same. Moreover, upon being filled by mineral or protein ligands, the relative chain orientations may change. Thus, the collagen fibril structure, even after initial assembly, is dynamic and can be modulated by mineralization, cross-linking, interaction with other extracellular matrix (ECM) molecules, and by mechanical forces. Intrafibrillar mineralization is accompanied by expulsion of the water from the fibrils (Lees 2003).

Cross-linkage chemistry

As shown by studies of lathyrism (Levene 1966), a disorder in which cross-linkages are prevented from forming, an assembled collagen fibril stabilized only by the packing interactions with neighboring molecules has little tensile strength. Stress in the axial direction simply causes the molecules to slide along each other and slip apart. Thus, the cross-linkages are required for the development of tensile strength. Once again the differing physiological requirements for the collagen fibril systems are met by a variety of tissue specific cross-linkage densities and chemistries. This discussion will be confined to type I collagen.

Intermolecular cross-linking takes place specifically between the K^{16C} of the $\alpha 1(I)$ C-telopeptides of one molecule with a K87 ($\alpha 1$ or $\alpha 2$) in the helical region opposed to it in the N-terminal sequence of an adjacent 4-D staggered molecule. Similarly, cross-linkages may form between the K^{9N} of the $\alpha 1(I)$ N-telopeptide of one molecule and a 4-D staggered K930 of the $\alpha 1(I)$ C-terminal helical sequence in an adjacent molecule. The cross-link chemistry, unlike that of most proteins, is based on the formation of a Schiff base between the ε-amino group of a lysine in the helical sequence and an aldehyde created from the oxidative deamination of a lysine in the telopeptide. The cross-linking takes place only after the fibrils are established in the extracellular matrix, and requires post-secretory chemical modification. The first step in this process is the enzymatic oxidative deamination of the telopeptide ε-NH_2 group and conversion of the ε-CH_2 to an aldehyde, –CHO. The derivatized lysine is called "allysine." The aldehyde then reacts with the appropriate helix K to form a Schiff's base cross-linkage, telo-CH=NC-helix. Schiff's bases are perfectly good covalent bonds creating the initial cross-linked fibril network, but they are quite reactive. The chemistry of the cross-linking is seriously affected by the presence or absence of additional pre-secretory lysine hydroxylation, and possibly glycosylation, in either or both the partners in the cross-linking reaction. Four hydroxylation conditions are possible at each cross-link site:

- telo K- helix K [$\Delta^{6,7}$dehydrolysinonorleucine (ΔLN)]
- telo-HOK- helix K [$\Delta^{6,7}$dehydro-5-hydroxy-lysinonorleucine (ΔLHN)]
- telo K- helix HOK [$\Delta^{6,7}$dehydrohydroxylysinonorleucine (ΔHLN)]
- telo-HOK- helix HOK [$\Delta^{6,7}$dehydro-5-hydroxy-hydroxylysinonorleucine(ΔHLHN)]

The hydroxylated Schiff's bases are unstable and rearrange to more stable form:

- ΔLHN $\Rightarrow$ lysino-5-ketonorleucine, -C-C-CO-C-N-C-C-C- (LKN)
- ΔHLHN $\Rightarrow$ hydroxylysino-5-ketonorleucine, -C-C-CO-C-N-C-C(OH)-C- (HLKN)

Within the close packed cross linking zones, in their very specific intermolecular and intrafibrillar arrays adjacent LKN groups can condense and form a complex pyridinium compound, called lysylpyridinoline (LP), joining 3 polypeptide chains on at least 2 molecules. LKN adjacent to a HLKN can form hydroxylysylpyridinoline (HLP). These are very stable compounds, resistant to disruption by acid or base. However, they are sensitive to ultraviolet radiation; consequently they cannot be found at significant levels in skin. They are present in bone matrix collagens (Robins and Duncan 1987) and cartilage collagen (Farquharson et al. 1996).

Another tri-functional cross-link is the pyrrole formed from a telopeptide allysine aldehyde and a bifunctional ketonorleucine (HKLN) (Kuypers et al. 1992). The major locus of that bond is in the N-telopeptide region, with the majority as $\alpha 2(I)$OHK933 × $\alpha 1(I)K(OHK)^{9N}$ × $\alpha 2(I)K(OHK)^{5N}$ and a smaller amount as $\alpha 1(I)$OHK930 × $\alpha 1(I)K(OHK)^{9N}$ × $\alpha 2(I)K(OHK)^{5N}$. A C-telopeptide derived pyrrolic cross-link, $\alpha 1(I)$OHK87 × $\alpha 1(I)K(OHK)^{16C}$ × $\alpha 1(I)K(OHK)^{16C}$ has also been found. It was estimated that 85% of the pyrrolic cross-linkages in bone collagen are involved with the N-telopeptide (Brady and Robins 2001). The placement of the cross-linkages involved in bone is not entirely clear because the methodologies of demineralization, hydrolysis and analysis used by different investigators all affect the stability of these complex bonds in different ways. Thus, while Hanson and Eyre (1996) found a predominant participation of $\alpha 1(I)K(OHK)^{9N} \times \alpha 2(I)K(OHK)^{5N}$ in the pyrrole cross-links, Brady and Robins (2001) reported that $\alpha 1(I)K(OHK)^{9N} \times \alpha 1(I)K(OHK)^{9N}$ was the predominant form.

In contrast to the pyrroles, the pyridinoline based cross-links are mainly derived from the C-telopeptide regions (Yamauchi et al. 1981, 1988, 1989,1992; Katz et al. 1989, 1992; Otsubo et al. 1992). Virtually all of the available K and OHK in the C-telopeptides were found to be quantitatively converted to the aldehyde form, and these aldehydes stoichiometrically cross-linked to K87 on both $\alpha 1$ and $\alpha 2$ chains of adjacent molecules. Moreover, the telopeptide K in bone was completely hydroxylated, present as OHK. However, the ratio of cross-linked $\alpha 1$-chains to $\alpha 2$-chains was 3 to 1 rather than 2 to 1, suggesting that there was some specificity of structure with regard to the azimuthal orientation in the packing of the bone collagen fibrils. It is important to recall here that the C-telopeptide of $\alpha 2(I)$ is devoid of K and hence cannot participate in cross-linking. However, the $\alpha 2(I)$ K87 can be present as OHK87 and participate in a cross linkage.

Otsubo et al. (1992) and Yamauchi et al. (1989, 1992) showed that the cross-linking in dentin was related to the extent of mineralization. Collagen is secreted from the odontoblasts into an adjacent nonmineralized layer called predentin. Predentin subsequently mineralizes to become the dentin. No collagen is secreted directly into the dentin. The predentin cross-linking was all in the form of DHLN, and at the level of ~ 2 moles of cross-link per mole of collagen. All the cross-linkages were between C-telo-K^{16C} and helix-K^{87}, so that molecules must be linked to both nearest neighbors as allowed by the quasi-hexagonal packing model formation of an extended system of the appropriate 1-5-1-5- etc. cross-links throughout a fibril. The surprising feature of these data was that in the mineralized dentin, the DHLN cross link content was less than half that of the predentin, 0.86 cross-links per mole of collagen. Moreover, 0.08 moles/mole collagen were HLN and 0.10 mole/mole collagen were in the form of pyridinoline (LP+HLP, PYR)). Further, the predentin DLHN cross-links do not mature to PYR, and it is evident that about half of the bonds are disrupted during mineralization. Perhaps the precursor form of the cross-linkages, ΔHLHN, may exist in the predentin and their maturation is overtaken by the mineralization (15-20 h) before the Amadori reaction has had time to occur, so that *in situ* the bonds are still labile and reversible. In support of this idea, the free aldehyde content is greater in dentin than in predentin. The mineralization of the dentin matrix may reorganize the fibril network and provide enough energy to disrupt the C-telo to helix bonding network. Any collagen differences between dentin and predentin must be the result of post-secretory processing, tissue maturation and mineralization. The insertion of the mineral phase can change both the fibril organization and chemistry.

More recently, in addition to the pyridinoline tri-functional cross-linking related to the C-telopeptide to helix bonds described above, a set of N-telopeptide to helix pyridinolines were found in bone. The main N-telo related pyridinoline involved $\alpha 1(I)K^{9N} - \alpha 2(I)K^{5N} - \alpha 1(I)OHK(K)930$ (Hanson and Eyre 1996). The chemical properties of this cross-linkage are such that the N-telo pyridinoline fluorescence, the main procedure used to detect its presence, was quenched on chromatography, leading to the underestimation of its presence. Pyridinolines linking two $\alpha 1(N)$ telopeptides to helix were a minor component. The cross-link ratio of HLP to LP differed between N-telopeptide and C-telopeptide sites, and between the individual interchain combinations. Cross-linked N-telopeptides accounted for two-thirds of the total lysylpyridinoline in bone.

It seems paradoxical that the domains with the highest chain packing densities and crystallinity (Hulmes et al. 1995) within bone and dentin collagen fibrils are so readily modified by deposition of the mineral phase. As pointed out earlier, the gap or hole zone spaces within which the mineral crystals may nucleate and grow are not uniform in properties, and, in fact, are the regions of maximum chain disorder. The higher collagen matrix crystallinity at the boundaries of the gap zones does not prohibit the dynamic changes in cross-linking and chemistry that take place during introduction of mineral crystals within the collagen framework.

COLLAGEN MATRIX MINERALIZATION

The SIBLING protein family

Many studies have shown that a collagen fibril matrix by itself does not have the capacity to induce mineralization from a solution of calcium and phosphate ions at the appropriate pH, degree of saturation, and temperature. The matrix-regulated models of mineral induction place the focus of mineral nucleation squarely on the organic matrix, and, in particular, on the presence of proteins or polysaccharides with the ability to interact and localize on the structural matrix, and, when bound, induce the nucleation of the mineral deposition (Lowenstam 1981; Veis and Sabsay 1982). Since these proteins may have more than one function in the mineralization process, e.g., crystal nucleation, crystal growth regulation, it is difficult to give them a meaningful generic name. For convenience here, we can think of them as "accessory" proteins. In general, the bone, dentin and cementum proteins are acidic in nature, rich in glutamic acid, aspartic acid, and, in dentin, exceptionally rich in serine, frequently as phosphoserine. Taking advantage of many recent advances in the cloning, sequencing and chromosomal localization of these proteins, Fisher et al. (2001) considered the set of acidic mineralized tissue matrix proteins as belonging to a group with similar biochemical properties and related genetic features and called them the SIBLING (Small Integrin-Binding-Ligand-N-linked Glycoprotein) family. The gene relationships are not clear, although most, in the human, are closely linked in the same chromosomal region, 4q21 and have a similar exon-intron organization. They all contain one or more consensus sequences for phosphorylation by casein kinase II, suggesting that their phosphorylation is important to their function. Fisher has made the cautionary point that, *in vivo*, members of the SIBLING family may have regulatory functions different from direct involvement in mineralization. Indeed, all of these molecules have several distinct sequence domains and potentially may be multifunctional molecules within the extracellular matrix. The story may not end there. There are other genes that encode secretory Ca-binding phosphoproteins including the enamel matrix proteins (EMPs), milk caseins and salivary proteins. These genes are clustered on human chromosome 4q13 and appear to be a related gene family. It has been argued that the SIBLING protein genes, clustered on human chromosome 4q21, only 15 Mb distal to the EMP cluster, are related to it and arose from the same common ancestor by gene duplication (Kawasaki and Weiss 2003).

The first acidic phosphoproteins of a mineralized matrix to be identified were in dentin (Veis and Schlueter 1963,1964; Schlueter and Veis 1964) and described in detail by Veis and Perry (1967). The major phosphoprotein was subsequently named "phosphophoryn" or phosphate carrier protein (Dimuzio and Veis 1978a,b), because it was more than 50 residue % serine of which 85-90% was phosphorylated. The phosphophoryn (PP) could have just as easily been named for its very remarkably high content of aspartic acid, which in bovine dentin accounted for 40-45 residue % (Stetler-Stevenson and Veis 1983). It is perhaps unfortunate that PP was the first of the SIBLINGS isolated and described, as it is the most unique member of the family. The other members generally have less serine (and consequently lesser phosphorylation), less aspartic acid and more glutamic acid. However, PP is present in the teeth of all vertebrate species examined (Rahima and Veis 1988), and may be very primitive in origin. Odontodes and teeth were the first vertebrate mineralized tissues (Donoghue and Sansom

2002). Table I lists the SIBLINGS, their origin and relationships. The majority of these proteins are in bone. *In vitro* studies of the mineral nucleating ability of SIBLINGS osteocalcin, osteopontin, bone sialoprotein (BSP) and PP. Hunter et al. (1996) suggested that BSP and PP could be nucleating agents, whereas osteopontin might function as an inhibitor of hydroxyapatite formation.

PP is initially synthesized as a larger precursor molecule, dentin sialophosphoprotein (DSPP). The DSPP molecule has never been isolated intact, because it is processed into three portions. The N-terminal portion corresponds to the molecule known as DSP, dentin sialoprotein. The C-terminal portion is PP. A central connecting portion is cleaved out by proteases (not yet defined) and the central peptide has not been isolated from tissues. The C-terminal PP domain is highly conserved among species and has unusual sequences of $(DSS)_n$ repeats, with n as large as 13 (George et al. 1996: Veis et al. 1998). These sequences can be phosphorylated by the casein kinases and the phosphate and carboxylate groups can create an extended template-like structure with the potential ability to bind Ca^{2+} ions in an organized array. Ser residues located in acidic sequence environments are generally good substrates for phosphorylation by tissue casein kinases (Wu et al. 1992; George et al. 1993; Sfeir and Veis 1996). PP contains many Ser residues at consensus sites which could be phosphorylated by either casein kinase II or casein kinase I and are consequently highly phosphorylated and markedly enhance the number of anionic, negatively charged residues, creating patches which can bind multivalent cations such as Ca^{2+}. In solution, the PP and similar proteins have a high capacity for cation binding, hence they form complexes which remain soluble and by reducing the net ion activity coefficient, inhibit the precipitation of Ca ions. On the other hand, Crenshaw and colleagues (Lussi et al. 1988, Linde et al. 1989; Saito et al. 1997, 1998, 2000) showed that when immobilized to a collagen or other polymer surface, bound PP will induce crystal formation. DMP1, another dentin acidic matrix protein with a lesser content of both phosphoserine and aspartic acid, but richer in glutamic acid, can initiate hydroxyapatite crystal nucleation when deposited on a glass surface (He et al. 2003).

In addition to its potent Ca ion binding capacity, the PP strongly binds to fibrillar collagen (Stetler-Stevenson and Veis 1986, 1987). When bound, the PP enhanced the ability of the fibril network to bind and retain calcium ions. However, the binding of Ca ion was complex in that two binding affinity constants were found. The binding of PP to the collagen was also complex, indicating regions of differing PP binding capacity. Although immuno-electron microscopy demonstrated that PP was bound predominantly at the e-band of turkey tendon collagen fibrils, it was evident that several binding domains of lower affinity were present (Traub et al. 1992a). As shown in Figure 10, the e-band is near the middle of the gap region. Direct visualization of the binding of PP to monomeric collagen (Dahl et al. 1998; Veis et al. 2000) by rotary shadowing electron microscopy showed that the highest affinity site for PP interaction was between helix residues 680-720 along a monomer. However, this required very high collagen to PP ratios. As the relative content of PP was increased additional binding sites were revealed. This was true as well for the collagen assembled into fibrils. An analysis of the charge distribution along the fibril surface suggested that there would be charged group constellations favoring binding at the gap zone boundaries as well as in the e-band site (Veis et al. 2000) (see Fig. 10). Saito et al. (2000) not only showed that the binding of PP to collagen was crucial for nucleation of apatite deposition, but also that the binding site involved the collagen telopeptide domains which would be at the gap region edges. More work is necessary to clarify the nature, location and number of matrix protein binding sites on a collagen fibril, and which telopeptides might be involved. No comparable studies have been made for the other SIBLING proteins, but they may also bind to collagen fibrils in specific fashion. The enamel matrix proteins are discussed below.

Mineral placement

The geologically produced igneous apatites are large and well-crystallized. The unit cell, ideally $Ca_{10}(PO4)_6(OH)_2$ in composition, is a right rhombic prism. In the basal plane, the a and b dimensions are equivalent at 9.432Å. The c-axis spacing is 6.881Å. Crystal growth is preferentially in the c-axis direction. Hydroxyapatite crystals grown from supersaturated solutions of calcium phosphate or via biological processes rarely have such a composition or size. More generally the crystals are of submicroscropic size, and are non-stoichiometric. They may be calcium-, hydroxyl- and phosphate deficient, and most frequently have substitutions of divalent and trivalent cations for Ca^+, F^- for OH^- and CO_3^{2-} for phosphate. The general formula for the non-stoichiometric hydroxyapatite is $Ca_{10-x}H_x(PO_4,CO_3)(OH)_{2-x}$. From the very earliest electron microscopic and x-ray diffraction studies it was demonstrated that the crystals of apatite within a bone or dentin matrix are small and plate or needle-like, oriented so that the c-axes of the apatite crystals are near parallel with the collagen fibril axis. Many studies show bone apatite crystals to range between 250-350 Å in their largest dimension with a thickness of 25-50 Å. Further, the small, discrete crystals are placed so as to yield the same ~67 nm periodicity as the collagen fibrils (Glimcher et al. 1957; Glimcher 1976). These, and similar, observations led to the general conclusion that the crystals were mainly limited to deposition within the gap regions of the collagen fibrils. One of the most convincing studies in this respect was that of White et al. (1977) who showed by low angle X-ray scattering and neutron diffraction in mineralizing turkey tendon that the first deposits of mineral indeed were likely to fill the gap region of the collagen fibrils. The turkey tendon system was especially favorable for that study, as the tendon collagen fibrils were highly oriented and thus gave excellent diffraction patterns. Traub et al. (1992b) used transmission electron microscopy to examine the first traces of mineral density at the mineralization front in turkey tendon and determined that the initial mineral density was located near the center of the gap zone, at the e-band position. As discussed below, it is less likely that this is the case in the mineralization of the collagen of bone or dentin.

Whatever the order and placement of the crystals, the primary mechanical properties of the bone matrix-mineral composite are determined by the collagenous matrix. The continuity of the bone and its shape and tensile strength are defined by the collagen matrix. Demineralization of a long bone, removing the mineral with acid or a Ca ion chelating agent such as EDTA (ethylene diamine tetraacetic acid) under conditions where degradation of the crosslinked collagen is inhibited, results in a structure that retains the shape and external volume of the original bone. It also retains high tensile strength, but becomes flexible and is easily bent. On the other hand, following deproteinization under conditions that do not support changes in the mineral crystal size the mineral phase looks to have the shape and volume of the original bone, but is friable so that under minor stress it disintegrates into the individual small crystallites. This demonstrates the mineral particles are not fused, but are independent of each other and have no macroscopic continuity or compressive strength. Different bones have different macroscopic physical properties because of the way in which the collagen fibril network is organized in three dimensions. At the meso scale, within a single collagen fibril bundle, the relationships of both intra- and inter-fibrillar crystals to an individual collagen fibril are probably similar in nature.

The three-dimensional network of fibrils in a particular tissue is assembled into a defined array by the cells (osteoblasts, odontoblasts, cementoblasts, fibroblasts) that secrete the molecules of bone, dentin, cementum and tendon, respectively. Initially the network is in non-mineralized form as osteoid or predentin. As discussed earlier there are three types of space within these arrays, two of which are intrafibrillar, within a single fibril, while the third is between the fibrils. The gap region space, may have the gaps aligned to form channels through the fibril over significant distances (Katz and Li 1973a, b), but there are also pores between molecules within the fibrils, even in the overlap zones. Because of the higher chain packing density in the cross-link regions and the balance of the overlap, the pores themselves are not uniform in terms of their length in the axial direction. The liquid-like disorder in the radial direction within the overlap zone, but excluding the higher crystallinity zones containing the cross-linkages, compounds the problem further. The literature is replete with discussions about the space available for insertion of the non-collagenous proteins and micro anions and cations necessary to initiate crystallization, into the matrix, using a variety of physical properties; sonic velocity, tissue anisotropy, density, dielectric constant and composition among others. Adult wet bovine cortical bone was estimated to be 25% by volume water, 32% organic matrix (mostly collagen) and 43% mineral (Lees 1981). Of the mineral 28% by volume was in the gap zone, 58% was radially intrafibrillar (in the pores), and 14% was radially extrafibrillar. If the mineral is nucleated in the gap zones it is clear from these mineral volume estimates that, once initiated, crystallization moves along the fibril surface outside of the holes or channels. The situation can be drastically different depending on the type of bone. The design of a particular bone or tooth is directly related to its function. Consider the differences between a rat incisor and a rat molar tooth. The rat incisor tooth is a continuously growing structure designed for puncturing and tearing. It must be rigid and resistant to bending, yet not brittle or easily cracked. The incisor wears at the incisal end by abrasion; it is self-sharpening as the incisal edge is continuously renewed. This is achieved by having an outer layer of hard, brittle enamel lying on the less mineralized and more compliant dentin. The dentin at the incisal edge is more easily removed by abrasion leaving the enamel sharpened at the incisal edge. The enamel is abraded at a slightly lower rate than the dentin. The permanent, non-renewing rat molar tooth, on the other hand, is used for mastication. It has a high compressive strength and a hard, non-deformable, enamel surface structure, with good wear resistance provided by the underlying dentin which also contributes to the compressive strength. Similarly, compact, cancellous lamellar and membrane bone have different physical properties within the same animal to match their differing requirements for toughness, compressive and tensile strength. Their physical parameters such as the Young's modulus and tensile strength may be very different. For example, deer antler bone has a higher toughness than compact bone, but a lower stiffness. Different mineralized structures also have varying mineral content, and the extent of mineralization in a particular mineralized tissue is an age dependent property (see Currey 2001).

Figure 12 (Lees and Prostak 1988) shows the organic matrix/mineral weight ratios in a variety of animals. The rostral bones of the toothed whale, *Mesoplodon densirostris*, are hypermineralized, with a mineral content of 87% by weight. The collagen is largely comprised of thin fibrils in tubular networks, aligned in longitudinal register. The mineral rods are enclosed within the tubular network. In this case, it is not clear that the collagen has any intrafibrillar gap or pore mineral (Zylberberg et al. 1998). The toughest and most compliant bone is in the deer antler, with an organic matrix/mineral weight ratio of 0.6 g organic matrix/g mineral. In this case the collagen matrix represents > 53% of the bone volume. The highest compressive strength but stiffest bone is in the rostrum and other hyperdense bones. Although difficult to quantify, there is a clear relationship between toughness and organic matrix content (Burr 2002).

Compact lamellar bone is quite complex in structure. The lamellae are clearly visible in SEM micrographs of fracture surfaces of bone. (Fig. 13A). The lamellae have an organized, layered substructure in which the collagen is organized into five subarrays of parallel fibrils. The fibrils within adjacent sublamellae are rotated by approximately 30° to form what has been called a rotated plywood structure (Weiner et al. 1991, 1997).

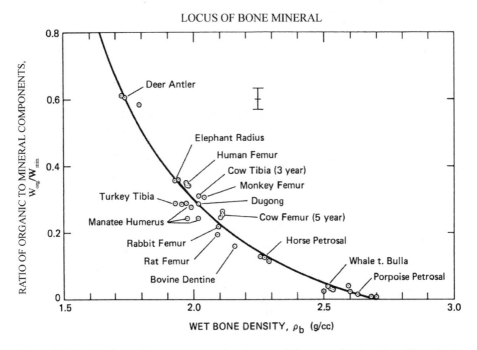

Figure 12. The ratio of organic components to mineral content in bone as a function of total bone density (from Lees and Prostak 1988 with permission of Taylor and Francis). Tissue toughness decreases as the bone density increases; hardness is inversely related to bone density.

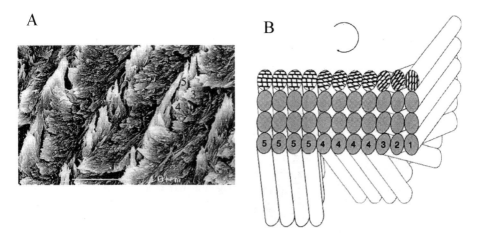

Figure 13. The "rotated plywood" structure of lamellar bone. (A) Scanning electron micrograph of a fracture surface, oblique to the lamellar boundary plane in human lamellar bone. Five non-uniform sublayers are numbered. (B) A schematic of the five sublayer model for bone, showing the relative thickness and orientation of the fibrils within each layer. The cross-hatched upper layer of fibrils depicts the orientation of platy mineral crystals within each fibril (from Weiner et al. 1999 with permission of Elsevier).

Weiner et al. (1999) noted that in each subarray the individual fibrils were either thick or thin. Layers of the thick fibrils were rotated at about 30°, but the arrays of the narrower fibrils were at about 70° to an adjacent array of the thicker fibrils. The narrow fibril arrays were essentially separators at boundaries between the thick fibril arrays; thin fibril arrays were not adjacent to each other (Fig. 13B). In addition to the changing orientation of the collagen fibrils the crystal c-axes are not directly aligned with the underlying fibril axes. Electron microscopic, small angle scattering (SAXS) and acoustic microscopy data showed that c-axes of the crystals are oriented along the external long axis of the bone whereas the collagen fibrils were aligned at 30° to that direction (Fratzl et al. 1992, 1996; Turner et al. 1995).

As pointed out above, all of the mineral in a mature bone cannot fit into the hole zones of the collagen fibrils. Even in the supposedly simple model system of the turkey tendon mineralization, where nucleation was demonstrated to be related to the gap zone (Traub et al. 1992b), the platy crystals that developed *in vivo* were larger than the gap zone space (Traub et al. 1989), thus, if they nucleated within the gap zone they would have had to grow out of the gap, into the intrafibrillar pore space, or out of the fibril to interfibrillar locations. Landis (1995a,b) took advantage of high voltage electron microscopy and tomographic imaging techniques to reconstruct the placement of mineral within a mineralizing zone of turkey tendon, and in the bone of a normal mouse and a mouse strain (*oim/oim*) that exhibits the properties of osteogenesis imperfecta, a genetic disorder leading to bone fragility. In these and subsequent studies (Landis and Hodgens 1996; Siperko and Landis 2001) it was evident that in the avian tendon and normal bone the platy crystals did initiate in the collagen gap zones and penetrate in register across the gap channels as proposed by Traub et al. (1989), but they also grew out of the gaps in the axial direction and extended in the pore spaces in the overlap zones, separating the collagen layers. Similar crystal initiation and growth were seen in *oim/oim* mouse bone, but the collagen fibrils were more disordered, and the initial crystallites were smaller and did not retain their registration with the fibrils as they grew. The current scheme is illustrated in Figure 14 and described in the accompanying figure legend. The mineral phase develops in all three locations as predicted much earlier by Katz and Li (1973a,b).

The effect of mineralization on collagen cross-linking, and the compaction of the collagen fibril structure (Lees 2003) indicate that there are changes in matrix architecture during mineralization. To follow this in detail Beniash et al. (2000) selected rat incisor dentin as the model system. Cryosectioning (−100 to −115°C) was used to retain the *in situ* structure, followed by vitrification of water in the sections by plunging them into liquid ethane at liquid nitrogen temperatures. This process prevented dimensional changes due to ice crystal growth. Sections were obtained at a slightly oblique angle so that the entire range of collagen structures could be seen from the first secreted collagen fibrils near the odontoblast surface through the proximal, central and distal portions of the predentin and into the dentin. The cryosections were stained with uranyl acetate, or anti-PP antibodies conjugated to 1.4 nm gold particles. Mineralization began in the distal portion of the predentin at the mineralization front. PP deposition, as detected by the gold-conjugated anti-PP antibody, began at the mineralization front and continued into the mineralized dentin, as shown earlier (Rabie and Veis 1995). The fibril diameter and packing density increased in traversing from proximal to distal portions of the predentin. In the distal zone entering into the dentin the collagen fibrils became essentially contiguous and filled all the space available. Thus, much of the non-collagenous components of the initially deposited predentin matrix and their associated water of hydration were removed as the collagen network became more densely packed. The highly hydrated proteoglycans of the predentin appear to be excluded from the dentin; they are not present in mineralized dentin. The collagen uranyl acetate banding pattern

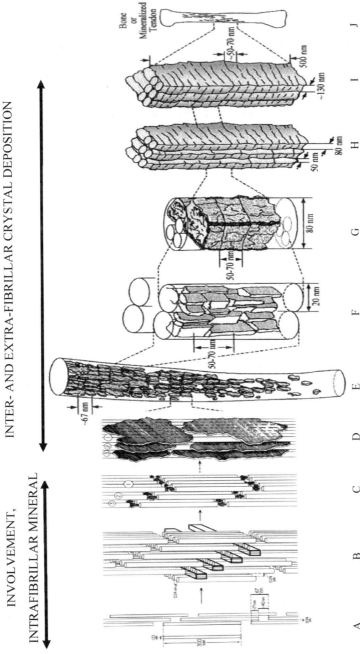

Figure 14. A combined schematic of the hierarchal development of mineralized bone as proposed by Siperko and Landis (2001) and Landis (1996) (rearranged and combined with the permission of the author and Elsevier). At the left, the model depicts the assembly of molecules into the native staggered array, and the initial placement of the mineral within the hole spaces and fibrils. Beginning with D the mineral overgrows the hole and intrafibrillar deposits and begins to fuse to larger plates. The plates retain their axial periodic features all the way though stage I. Finally, there is surface mineralization not directly related to the underlying collagen hole-overlap structure.

changed significantly from proximal to distal predentin, from a D-periodicity of 67 to 71 nm. In contrast to the localization of the PP at the e-band in calcifying turkey tendon (Traub et al. 1992a), immunogold labeling showed the PP to bind at the gap-overlap boundaries rather than at the e-band, as predicted by Veis et al. (2000) and Saito et al. (2000). These data all lead to the conclusion that both the collagen fibril packing and detailed molecular relationships such as cross-linking within the individual fibrils change in the vicinity of the mineralization front as the PP-binds and the collagen begins to mineralize. PP is absent from the bone matrix and it is not yet clear which of the SIBLING proteins may carry out that function. The most probable candidate appears to be bone sialoprotein, BSP, which clearly enhances the binding of mineral to collagen matrices, via a sequence of eight contiguous glutamic acid (E) residues (Goldberg et al. 1996; Tye et al. 2003). The regions on the collagen to which the BSP may bind have not been determined, nor has the conformation of the acidic domains within the BSP been examined. However, BSP binding to collagen may be mediated by decorin binding to both collagen and BSP (Hunter et al. 2001). Both of the two poly-E sequences in BSP are required for hydroxyapatite nucleation, suggesting that there may be a conformation that pairs these domains when Ca ions are bound. A similar situation is evident in DMP 1 (He et al. 2003) in which two acidic, E-rich domains fold together to form the calcium binding, crystal nucleating domain. No such domains are present in PP so that a different mechanism of nucleation may be involved.

There are a number of paradoxes in considering the properties of bone and dentin. On the one hand, each is essentially a two component system, an organic matrix of mainly type I collagen fibrils packed at the molecular level into very similar molecular arrays, and a crystal phase of calcium hydroxyapatite. On the other hand, the mesoscale organization of the collagen fibril matrix varies from on species to another, from one tissue to another in the same animal, and even from one portion of a bone or tooth to another. Likewise, the mineral phase is disposed differently in different bones, the crystal size varies and the mineral composition is not fixed. The calcium, hydroxyl and phosphate groups can all be replaced with other ions to some extent. Further, in the same bone, the crystal-collagen relationships may vary with age of the host animal. The relevant cells obviously guide and control all aspects of the tissue organization and composition. Bone and cementum differ from dentin in this respect. Bone is in a constant state of high metabolic activity and is being remodeled throughout life; the adult male human skeleton is entirely replaced about every seven years. Dentin, once put into place is not remodeled and generally, in the absence of pathology is not subject to controlled removal. However, mature dentin is not an inert structural entity. The odontoblasts remain alive throughout life and continue synthesis of dentin matrix at a low level. The primary dentin is complete at the time in development of the closure of the tooth apex (final production of the root). Following that, production of new dentin (secondary dentin) continues at a low rate, and with a somewhat different structure. Although the outer appearance of a tooth may be identical between a 20 year old and a 70 year old person, the secondary dentin production reduces the inner space available for the dental pulp in the older individual. The blood, and hence nutrient, supply to the pulp fibroblasts and dentin-lining odontoblasts diminishes.

The discussion of bone mineral placement has focused on compact bone, because that has been the easiest to study. However, the cancellous bone, in the end-regions of a bone adjacent to the cartilage of the joints, is not compact, but is an open network of mineralized collagen fibers with only 7-15% bone and the balance marrow. The mineralized collagen fibrils of the network, the trabeculae, may have very different microarchitectures within a bone. In general, however, the trabeculae are aligned in the direction of principle stress within the bone. Small angle X-ray scattering has shown that the collagen fibers and mineral crystals within the trabeculae are predominantly aligned parallel to the trabeculae. Banse et al. (2002) have examined the structure of bone trabeculae in relation to the resultant mechanical properties of the bone. They determined the amount of mineralized bone, osteoid, and collagen, and the nature of the collagen cross-links in relation to the bone volume fraction, trabecular thickness, number, spacing and length. The major correlations found were between the bone volume fraction and the ratio of immature (e.g., hydroxylysinonorleucine (HLN)) cross-links to mature cross-links (hydroxylysylpyridinoline (HP) and pyrrole (PY)), and in the ratio of mature cross-links to each other, PY/HP. The higher the PY/HP the thicker were the trabeculae. The important message of these results is to emphasize the close relationship between the orderliness of the collagen fibril network in the mineralizing matrix and the placement of the mineral within the matrix. Considering that the cross-linking chemistry is dynamic and changes with the cross-link maturation and tissue mineralization state, it will be extremely important to correlate these very diverse properties and reactions in terms of the functional mechanical properties of various types of bone. Ultrastructural studies of bone in the future must be correlated with the collagen cross-linkage chemistry and cross-link distribution. Another example can be seen in dentin. The dentinal tubules, cell processes that run through the dentin from the predentin to the dentino-enamel junction, are surrounded in many species, particularly prominently in man, with a highly mineralized collar of peritubular dentin. The organic components of the peritubular dentin are different from that of the bulk intertubular dentin, and there is relatively little collagen (Weiner et al. 1999). There are clearly positional differences in properties, such as hardness and elastic properties related to the disposition of the tubules and extent of the peritubular dentin (Kinney et al. 2002). Thus, bone and dentin are complex composites and their physical and chemical properties have to be considered at various levels: nanoscale, microscale and mesoscale organizations.

TOOTH ENAMEL

Cellular compartmentation

Vertebrate teeth are covered with a dense mineralized layer of a collagen-free carbonated apatite, the enamel. In humans mature enamel is about 95% carbonated apatite, about 2% organic matrix and 2% water. Immature enamel, however, is highly proteinaceous and the mineral phase increases and hardens slowly during tooth development, a process that begins early in fetal life. After the rostral-caudal orientation of the embryo is set, the cells forming the neural crest migrate in the head direction, as they advance and expand cellular outgrowths bud off laterally forming the branchial arches. The first branchial arch divides to form the maxilla and mandible and define the oral cavity. The teeth form along the developing mandible and maxilla in specific locations specified by a number of homeobox genes. The first sign of a tooth is a localized thickening of the oral epithelium that begins to invaginate into the underlying oral mesenchyme. In response to the epithelial thickening mesenchymal cells begin to condense and present a surface around which epithelial cells fold. The patterns of epithelial folding and mesenchymal cell condensation determine the ultimate shape of the tooth. The layer of epithelial cells folded in contact with the mesenchyme becomes the inner enamel epithelium (IEE), the cells which will ultimately differentiate to become the ameloblasts and construct the enamel. The condensed mesenchymal cells, the dental papilla, in contact which the IEE will differentiate to ultimately become the odontoblasts and the remainder of the papilla will become the dental pulp cells. It is beyond the scope of this article to consider the many signaling events based on the epithelial-mesenchymal interactions during the differentiation and cell maturation processes that control the rates of differentiation. It will suffice to say here that the cells of the IEE form a continuous layer or sheet in which the preameloblasts have a cuboidal appearance. As they differentiate to their mature form the preameloblasts elongate and their nuclei recede proximally away from the dentino-enamel junction (DEJ) producing highly polarized secretory cells. At the DEJ, the ameloblasts form a secretory structure similar to dentinal tubules in function but not shape. These asymmetric, tapered cell extensions, called the Tomes' processes, define spaces into which a number of specialized proteins and the mineral ions are secreted. The enamel mineral is nucleated and the crystals grown within in these limited chambers defined by the surfaces of adjacent Tomes' processes at the apical ends of the ameloblasts. Mineralization is initiated at the DEJ, and the mineral forms as thin ribbon-like hydroxyapatite crystals in small bundles called prisms. As the crystals grow in length and the enamel layer thickens the ameloblasts recede from the DEJ, however, each cell retains the closed Tome's process space and the same bundle of apatite crystals, the same prisms, continue to elongate in the direction of the outer enamel surface. Each enamel prism, made up of thin ribbon-like crystals is thus the product of a single secretory cell. The initial ribbons grow in length and may be continuous from the DEJ to the tooth outer surface. When the enamel has reached its full thickness, the ameloblasts cease their secretory production of enamel and enter a post-secretory, resorptive phase, cleaning up in a sense, the enamel surface before dying and sloughing off the tooth surface. The ameloblasts, and hence the prisms, follow a sinuous path to the outer tooth surface and the prisms form interlocking structures that contribute to the strength of the enamel, but this structure is dependent on the location within the tooth, the tooth function, and the species of animal producing the enamel.

Enamel matrix proteins

The mature ameloblasts secrete a complex mixture of proteins, glycoproteins and proteolytic enzymes into the extracellular Tomes' process compartments, and these proteins control the growth and form of the mineral phase. At the early stage of apatite development the majority of the proteins are amelogenins, essentially hydrophobic proteins that aggregate and fill the inter-Tomes' process spaces. There are, however, other proteins and proteases present. The enamel matrix proteins, as noted earlier, fit into a phylogenetic clade rising by gene duplication from the same primordial ancestor gene as the SIBLING proteins. Ameloblastin and enamelin are closely grouped on human chromosome 4q13 while amelogenin, more distant in structure and function, resides on the X and Y chromosomes (Kawasaki and Weiss 2003). The genes for amelogenin (AMEL), ameloblastin (AMBN) and enamelin (ENAM) have all been sequenced in a variety of species and found to be different, but highly conserved across species and the amino acid sequences have been determined from protein data as well. It is important to note that all of these proteins are essentially removed during the formation of the enamel, so that the mature enamel is only 1-2% protein. Thus, the mineralizing compartments are undoubtedly changing in content during the entire process of mineral development. This is clearly reflected in the amelogenin content. When isolated from teeth undergoing mineralization, the amelogenin fraction is very heterogeneous, so that many amelogenin-related peptides can be seen on SDS-gel electrophoresis. Most of these are degradation products of the full-length protein. However, the amelogenin gene is subject to alternative splicing. In the mouse up to nine peptides corresponding to specific gene splice products have been identified (Hu et al. 1997). AMBN and ENAM proteins have also been found to be degraded during their existence in the mineralizing enamel matrix.

The amelogenin-related peptides account for more than 90% of the initial enamel matrix. Crystal growth begins in this amelogenin-rich milieu. Full-length amelogenins have between 173 and 210 amino acids, depending on the species, with a molecular weight approximately 27 kDa. The molecule has essentially three domains. The first 50 amino-acid terminal sequence of the secreted protein is highly conserved in most species and contains an internal YxxxxYxxxxxxWYxxxxxxxYxxYxY motif, where W is tryptophane and Y is tyrosine as in the human sequence:

MPLPPHPGHPGYINFS*YEVLTPLKWYQSMIRPPYSSYGYEMPG**GW**LHHQ-

The underlined GW is a conserved proteolytic cleavage site between G45 and W46, The S* represents a single phosphorylation site. The cleaved N-terminal 45-mer peptide is designated as the TRAP peptide. Another highly conserved sequence is at the C-terminus. The final 29 residues, corresponding to exon 6d and 7,

PLPPMLPDLTEAWPSTDKTKREEVD

is polar and acidic. This acidic tail is a very important feature of the intact amelogenin molecule. The splice product peptide comprised from the amino acids encoded by exons 2, 3, 5, 6d,7, 59 amino acids in length, is known as the "leucine rich amelogenin peptide" LRAP. The hydrophilic 14 residue peptide encoded by exon 4 is usually not expressed, the long peptide sequence encoded by exon 6a,b,c in the large molecular weight amelogenin is quite hydrophobic.

Ameloblastin (also called amelin or sheathlin) is the second most abundant protein in the secretory enamel matrix. The nascent AMBN has a $M_r \approx 68$ kDa but this is rarely found in the enamel matrix since, like the amelogenin, it is proteolytically processed rapidly to a series of peptides with apparent molecular weights of 52, 40, 37, 19, 17, 16, 15, 14 and 13 kDa, all of which have been detected on gels (Brookes et al. 2001). In the rat incisor, the 68, 52, 40, 37 and 13 kDa forms are readily soluble and can be extracted in simulated enamel fluid. The 19, 17 and 16 kDa proteins were only partly soluble in the enamel fluid and required phosphate buffers to wash them from the mineral surface. The 15 and 14 kDa peptides and the final part of the 17 and 16 kDa peptides were obtained only after dissociative extraction in SDS-containing buffer, apparently because they formed an insoluble aggregate *in vivo* and were bound to the mineral phase. *In vivo* the AMBN proteins were completely absent from mature enamel. The function of the AMBN is not known, but the aggregated peptides may be accumulated at the growing prism surfaces and prism boundaries, hence the name sheathlin.

Enamelin is present in still smaller amount in the secretory enamel, but it is strongly bound and represents the major protein of the mature enamel, bound to the enamel crystals. It cannot be dissolved without dissolution of the mineral phase (Termine et al. 1980). Mouse enamelin has been cloned (Hu et al. 1998) and is postulated to have a sequence of 1236 amino acids, a pI of 9.4 and a calculated molecular mass of 137 kDa without consideration of any post-translational modifications. It is immediately processed upon secretion into the Tomes' process compartment to lower mass fragments. It is processed stepwise from the carboxyl terminus to products of 155, 142, and 89 k apparent molecular mass, each of which retains the amino-terminus (Fukae et al. 1996). The 89 kDa form accumulates but is further processed to 32 and 25 kDa products. The 32 kDa form is glycosylated and phosphorylated and has an isoelectric point of pH 6.4, and accounts for about 1% of the enamel matrix protein (Tanabe et al. 1990). The enamelin degradation products are present in the secretory enamel, but disappear as the enamel matures. The enamelin is not present in the prism sheaths but is restricted to the rod and interrod crystallites. The functions of the enamelin cleavage products are not known (Uchida et al 1998).

The final enamel matrix protein of note is tuftelin, a protein localized to the DEJ, potentially nucleating crystallization of the enamel rods. There have been several different descriptions of the tuftelin proteins, first described by Deutsch et al. (1989) and there was significant disagreement in the data (Deutsch et al. 1998; MacDougall et al. 1998). Tuftelin mRNA has been detected in some non-dental tissues (MacDougall et al. 1998). Human tuftelin cDNA has been sequenced and shown to have an open reading frame of 1170 bp, encoding a 390 amino acid protein with a calculated M_r = 44.3 kDa and isoelectric point of pH 5.7. The tuftelin gene contains 13 exons and alternatively spliced mRNAs have been detected (Mao et al. 2001). An intact protein corresponding to the cDNAs cloned as tuftelin, has not yet been isolated from the enamel matrix.

ENAMEL MINERALIZATION

The spaces defined by the Tomes' processes are initially filled with the ill-defined mixture of matrix proteins and degradation products described above, but are richest in the amelogenins (> 90%). Early electron microscopic studies of the nature of the enamel matrix suggested that the matrix was organized into periodic substructures of globular particles (Ronnhölm 1962; Travis and Glimcher 1964). That idea was challenged in many studies over the years, but was finally substantiated when the amelogenin was sequenced, cloned and the recombinant protein was prepared in sufficient quantity for study. In a brilliant series of studies led by Alan Fincham (Fincham et al. 1994, 1995, 1998; Moradian-Oldak et al. 1994, 1995, 1998a,b, 2000) it was shown that the amelogenins showed a temperature-dependent, pH dependent self-association to form globular units called "nanospheres" about 20 nm in diameter. The nanospheres aggregated to form an extensive gel-like network, and it was proposed that, in the tooth the enamel apatite crystals grow within this network to ultimately fill the entire space. The assembly of the nanospheres, and their coalescence into larger fused aggregates depends on the integrity of the amelogenin molecules. The most amino-terminal domain, the "A-domain," appears to be necessary for the assembly into uniform size nanospheres, while the "B-domain" at the carboxyl-terminal region (lacking residues 157-173 in mouse recombinant amelogenin rM179) blocked their fusion to larger, more irregular aggregates (Moradian-Oldak et al. 2000). Deletion of the A-domain in transgenic mice delayed the initial formation of aprismatic enamel in the molars and caused severe structural abnormalities in the incisor enamel. B-domain deletion yielded a less drastic phenotype (Dunglas et al 2002). The amelogenins and most other proteins are degraded as mineralization proceeds to make space for the hydroxyapatite. In the end, the proteins that do remain, principally the enamelins, are accumulated at the surfaces of the apatite crystals. It is not clear as to the function of the residual protein in the mechanical properties of the enamel, but it has been suggested that they do moderate crack propagation and fracture in the otherwise brittle, highly crystalline apatite.

Since it is removed, the function of the amelogenin rich matrix is not related directly to the strength or toughness of the final product, in distinct contrast to the collagen of bone. However, there must be a very specific role for the amelogenins in providing the environment in which the crystals grow. In mouse molar teeth enamel mineral organization is not uniform throughout the tooth but is distinctly less organized at the region of the DEJ (Diekwisch et al. 1995). The enamel crystals begin as very thin ribbons with elongated hexagonal cross sections representing the a,b planes. The crystals grow in their c-axis direction. Figure 15 shows the earliest crystals forming in porcine enamel. The organization of the ameloblast layer essentially determines the packing of the enamel rods, probably via asymmetric secretion of matrix proteins through the adjacent Tomes' processes. In different teeth, and in teeth from different species, the sizes and organization of the enamel rods differ greatly. The materials properties of the different enamels are determined by the layering and intertwining of the apatite rods. This process

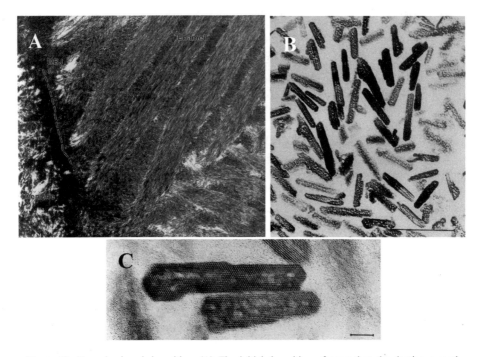

Figure 15. Enamel mineral deposition. (A) The initial deposition of enamel at the dentino-enamel interface (DEJ) in a mouse mandibular molar (Diekwisch et al. 1995). There is a very striking difference between the small, collagen-associated crystal organization in the dentin, and the giant, elongated apatite crystals within the enamel rods. Each rod or prism, contains many aligned apatite ribbons, and each prism represents the mineral deposited in a single Tomes' process compartment by a single ameloblast. In the lower right and upper left one can see how the prisms change orientation and begin to interweave. (B) Low-resolution electron microscopic cross-section of crystals found in early secretory porcine enamel. The thin, hexagonal shape, and the matrix filled space are evident. (C) A high-resolution electron micrograph of a single growing porcine enamel crystal cross-section. The stepwise growth along the elongated edges of the crystal emphasize the much greater rate of growth at crystal ends. The bar = 10 nm. (Figures B and C from Miake et al. 1993). All images reproduced with permission of Springer-Verlag.

is under direct cellular control and relates to the organization of the Tomes' process chambers. Thus it is difficult to consider the enamel as much of a protein composite material in comparison with the matrix-dominated underlying dentin. In spite of the paucity of organic matrix in the final enamel, Fincham et al. (1999) proposed that enamel formation was a typical matrix-regulated process in which 1) the Tomes' process and dentin at the DEJ delineate the initial space; 2) secreted proteins assemble in that space and form a supramolecular structural framework; 3) mineral ions are conveyed via the ameloblasts into the delimited space at a concentration sufficient for precipitation; 4) apatite crystals are nucleated by the macromolecules near the DEJ and grow into elongated prisms, with their subsequent growth, morphology and orientation all controlled by the matrix organization; and 5) the matrix is degraded and crystal growth ceases, the degradation products are removed as the crystals mature and harden. The removal of the matrix is unique to the enamel with respect to the other matrix regulated mineralization systems.

OTOLITHS

The inner ear in humans is comprised from a bony labyrinth which forms a complex cavity. This cavity is filled with a membranous labyrinth, a fluid filled assembly of interconnected sacs and channels. The fluid filling the membranous labyrinth is called the endolymph. The central part of the labyrinth is the vestibule which has two membranous chambers, the utricle and saccule. Each of these membranes contains patches of epithelial hair cells projecting into the gelatinous endolymph. These hairs support masses of aragonitic calcium carbonate, the otoliths, which function in both hearing and gravitational orientation. The patches of epithelial hair cells are associated with the vestibular nerves in a complex called the macula. The otoliths of mouse and teleost fish are similar and have been studied in biochemical detail. The principal structural glycoprotein of the matrix in vestibular endolymph is otoconin-90 (Oc90). The Oc90 is a potentially highly glycosylated protein of 453 amino acid residues, with two domains of strong homology to secretory phospholipase A_2. The aragonitic otoconia form on the Oc90-rich matrix gel. However, the Oc90 mRNA is not expressed in the sensory macular epithelia related to the otoconia, but is produced elsewhere in the non-sensory epithelia (Wang et al. 1998). Borelli et al. (2001) compared the proteins of the otolith with the endolymph protein. Although the total organic content of the mineral was about 0.2% of the total weight, the mineral related proteins were 23% collagen, 29% proteoglycan and 48% other proteins. The endolymph contained all of the protein components of the mineralized matrix, plus others, in larger amounts. Thus, the proteins accumulated in the mineral were selectively retained. The distribution of proteins in the gel-like endolymph was also non-uniform in terms of total concentrations, but not in relative content. Alcian blue staining, which reveals anionic components, suggests that the proteins and proteoglycans do form an anionic environment from which the calcium carbonate does crystallize. Glycogen may also be present (Pisam et al. 2002). Much more work remains to define the mineral deposition processes to form the otoliths. The saccular arrangement of the compartments in which the otoliths form, with the endolymph constituents delivered by epithelial cells not directly associated with the macula is reminiscent of the process by which the avian eggshell develops. Perhaps it would be of interest to compare the components of eggshell and otolith systems to determine any similarities that might relate to the formation of carbonate rather than phosphate containing mineral phases.

A mechanistic model

Each of the systems discussed here have very specific attributes and control points for the assembly of its specific biomineralized tissue, yet they also have many points in common. In a sense one may consider that all use a similar overall strategy but each case demands the use of different local tactics to achieve structures with different properties. Figure 16 summarizes the overall scheme we propose for the mineralization of bone and dentin. However, in different mineralizing systems in the vertebrate and invertebrate worlds, the variety of tactics which lead to the enormous diversity of biominerals in different species, and within the same animal, can be thought of as merely the result of substituting different components, at different concentrations, along each of the arrows. Thus chitin in the mollusc shell could substitute for the collagen in bone. The whole train of events is orchestrated by the cells that produce or transport the components that comprise the mineralized tissue. As complex as Figure 16 may appear, there are a host of additional factors that have been ignored. At any given point in the lifetime of an animal the cells are forced to respond to numerous biochemical inputs including nutrients, growth factors, and hormones, and to mechanical stresses. Bone, dentin and cementum are living tissues, always responding to these factors and always changing in subtle ways to meet their functional responsibilities. Only enamel mineral is fixed in amount and

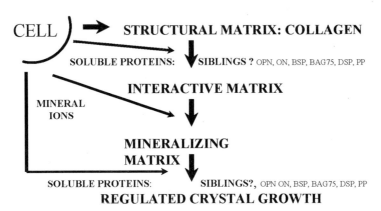

Figure 16. A generalized scheme for the construction of all matrix-mediated mineralized tissues, with emphasis on the bone and dentin systems. The three key steps, for which all of the players are not known, are the deposition of the structural matrix, the addition of structure-interactive-mineral interactive molecules onto the structural matrix and consequent nucleation of crystal formation, and finally, the regulation of crystal growth rate and crystal size by the further intervention of acidic macromolecules which may or may not be the same as the structure interactive molecules. It is postulated that similar schemes, depending on the nature of the structural materials, are used in all matrix-regulated mineralization systems.

structure after synthesis, once the ameloblasts have produced the enamel and have been removed by apoptosis. In the teleosts, the otoliths continue to grow by daily increments, so that in a given species the diurnal increments can be used to analyze age, growth rates and environmental effects.

The bone, dentin and cementum apatites have a similar structural collagen matrix, and similar accessory proteins that together limit the crystals to small size and platy shape and prevents the crystals from coalescing to larger sizes. They are thus organic polymer matrices with their physical properties determined by the arrangement of the polymer matrix and modified by the placement, arrangement and amount of the reinforcing matrix. The same main components can thus produce a varied spectrum of final mineralized tissues, and, as living tissues they are under the surveillance of functioning cells with the capability of tissue repair. The tooth enamel is distinctly different; it is essentially an impure crystalline array with the protein components present as minor constituents in the mature, non-living structure. The crystals or crystal prisms are macroscopic in size, at least in length, and cannot be repaired except by external supply of the requisite ions in the oral cavity and strict adherence to physical chemically determined solubility product equilibria as modified by the environmental pH. Nevertheless, the variations in structures of enamels from different species and in different teeth are under cellular control during the process of tooth formation and are essentially determined by the structures of the Tomes' process spaces and the genetically specified pathway of ameloblast migration during amelogenesis. The mechanical properties of the enamel are determined by the organization and interweaving of the elongated enamel prisms as they follow the sinuous path of the ameloblasts from the dentino-enamel junction to the outer tooth surface.

ACKNOWLEDGMENT

I am pleased to acknowledge that all of the work from my laboratory cited in this chapter has been supported by two grants from the National Institutes of Health: Grant AR13921 from the National Institute of Arthritis and Musculoskeletal and Skin Diseases; and Grant DE01374 from the National Institute of Dental and Craniofacial Research.

REFERENCES

Amis EJ, Carriere CJ, Ferry JD, Veis A (1985) Effect of pH on collagen flexibility determined from dilute solution viscosity measurements. Int J Biol Macromol 7:130-134

Bachinger HP, Bruckner P, Timpl R, Prockop DJ, Engel J (1980) Folding mechanism of the triple helix in type-III collagen and type-III pN-collagen. Role of disulfide bridges and peptide bond isomerization. Euro J Biochem 106:619-632

Bank RA, Robins SP, Wijmenga C, Breslau-Siderius LJ, Bardoel AF, Van der Sluijs HA, Pruijs HE, TeKoppele JM (1999) Defective collagen crosslinking in bone, but not in ligament or cartilage, in Bruck syndrome: indications for a bone-specific telopeptide lysylhydroxylase on chromosome 17. Proc Natl Acad Sci USA 96:1054-1058

Banse X, Devogelaer JP, LaFosse A, Sims TJ, Grynpas M, Bailey AJ (2002) Cross-link profile of bone collagen correlates with structural organization of trabeculae. Bone 31:70-76

Beck K, Boswell BA, Ridgway CC, Bachinger HP (1996) Triple helix formation of procollagen type I can occur at the rough endoplasmic reticulum membrane. J Biol Chem 271:21566-21573

Beniash E, Traub W, Veis A, Weiner S (2000) A transmission electron microscope study using vitrified ice sections of predentin: Structural changes in the dentin collagenous matrix prior to mineralization. J Struct Biol 132:212-225

Bernengo JC, Ronziere MC, Bezot P, Bezot C, Herbage D, Veis A (1983) A hydrodynamic study of collagen fibrillogenesis by electric birefringence and quasielastic light scattering. J Biol Chem 258:1001-1006

Bevelander G, Nakahara H (1969) An electron microscope study of the formation of the nacreous layer in the shell of certain bivalve mollusks. Calcif Tiss Res 3:84-92

Bevelander G, Nakahara H (1980) Compartment and envelope formation in the process of biological mainerlization. In: The Mechanisms of Biomineralization in Animals and Plants. Omori M, Watabe N (eds) Tokai University Press, Tokyo, p 19-27

Birk DE, Linsenmeyer TF (1994) Collagen fibril assembly, deposition, and organization into tissue specific matricies. In: Extracellular Matrix Assembly and Structure. Yurchenko PD, Birk DE, Mecham RP (eds) Academic Press, New York, p 91-128

Blumberg B, MacKrell AJ, Fessler JH (1988) Drosophila basement membrane procollagen alpha 1(IV). II. Complete cDNA sequence, genomic structure, and general implications for supramolecular assemblies. J Biol Chem 263:18328-18337

Blumberg B, MacKrell AJ, Olson PF, Kurkinen M, Monson JM, Natzle JE, Fessler JH (1987) Basement membrane procollagen IV and its specialized carboxyl domain are conserved in Drosophila, mouse, and human. J Biol Chem 262:5947 - 5950.

Boot-Handford RP, Tuckwell DS (2003) Fibrillar collagen: the key to vertebrate evolution? A tale of molecular incest. Bioessays 25:142-151

Borelli G, Mayer-Gostan N, De Pontual H, Boeuf G, Payan P (2001) Biochemical relationships between endolymph and otolith matrix in the trout (*Oncorhynchus mykiss*) and turbot (*Psetta maxima*). Calcif Tissue Int 69:356-364

Bornstein P (1967) The incomplete hydroxylation of individual prolyl residues in collagen. J Biol Chem 242: 2572-2574

Brady JD, Robins SP (2001) Structural characterization of pyrrolic cross-links in collagen using a biotinylated Ehrlich's reagent. J Biol Chem 276:18812-18818

Brookes SJ, Kirkham J, Shore RC, Woor SR, Slaby I, Robinson C (2001) Amelin extraction, processing and aggregation during rat incisor amelogenesis. Arch Oral Biol 46:201-208

Brownell AG, Veis A (1975) Intracellular location of the glycosylation of hydrosylysine of collagen. Biochem Biophys Res Comm 63:371-377

Bruckner P, Eikenberry EF, Prockop DJ (1981) Formation of the triple helix of type I procollagen in cellulo. A kinetic model based on cis-transisomerization of peptide bonds. Euro J Biochem 118:607-613

Bruns RR, Hulmes DJ, Therrien SF, Gross J (1979) Procollagen segment-long-spacing crystallites: their role in collagen fibrillogenesis. Proc Natl Acad Sci USA 76:313-317

Burjanadze TV (1979) Hydroxyproline content and location in relation to collagen thermal stability. Biopolymers. 18:931-938

Burjanadze TV, Bezhitadze MO (1992) Presence of a thermostable domain in the helical part of the type I collagen molecule and its role in the mechanism of triple helix folding. Biopolymers. 32:951-956

Burjanadze TV, Veis A (1997) A thermodynamic analysis of the contribution of hydroxyproline to the structural stability of the collagen triple helix. Connect Tiss Res 36:347-365

Burr DB (2002) The contribution of the organic matrix to bones material properties. Bone 31:8-11

Butler WT, Cunningham LW (1966) Evidence for the linkage of a disaccharide to hydroxylysine in tropocollagen. J Biol Chem 241:3882-3888

Currey JD (2001) Bone strength: what are we trying to measure? Calcif Tissue Int 68:205-210

Dahl T, Sabsay B, Veis A (1998) Type I collagen-phosphophoryn interactions: Specificity of the monomer-monomer binding. J Struct Biol 123:162-168

Deutsch D. (1989) Structure and function of enamel gene products. Anat Rec 224:189-210

Deutsch D, Palmon A, Dafni L, Mao Z, Leytin V, Young M, Fisher LW (1998) Tuftelin-aspects of protein and gene structure. Eur J Oral Sci 106:315-323

Deutsch D, Palmon A, Fisher LW, Kolodny N, Termine JD, and Young MF (1991) Sequencing of bovine enamelin (tuftelin), a novel acidic enamel protein. J Biol Chem 266:16021-16028

Diekwisch TGH, Berman BJ, Gentner S, Slavkin HC (1995) Initial enamel crystals are not spatially associated with mineralized dentine. Cell Tissue Res 279:149-167

Dimuzio MT, Veis A (1978b) Phosphophoryns—major non-collagenous proteins of the rat incisor dentin. Calcif Tiss Res 25:169-178.

Dimuzio MT, Veis A (1978a) The biosynthesis of phosphophoryns and dentin collagen in the continuously erupting rat incisor. J Biol Chem 253:6845-6852

Donoghue P C J, Sansom IJ (2002) Origin and early evolution of vertebrate skeletonization. Microsc Res Tech 59:352-372

Drake MP, Veis A (1964) Interchain interactions in collagen-fold formation. I. The kinetics of renaturation of γ-gelatin. Biochemistry 3:135-145

Dunglas C, Septier D, Paine ML, Zhu DH, Snead ML, Goldberg M (2002) Ultrastructure of forming enamel in a mouse bearing a transgene that disrupts the amelogenin self-assembly domains. Calcif Tissue Int 71:155-166

Farquharson C, Duncan A, Seawright E, Whitehead CC, Robins SP (1996) Distribution and quantification of pyridinium cross-links of collagen within the different maturational zones of the chick growth plate. Biochim Biophys Acta 1290:250-256

Fincham AG, Leung W, Tan J, Moradian-Oldak J (1998) Does amelogenin nanosphere assembly proceed through intermediary-sized structures? Connect Tissue Res 38:237-240

Fincham AG, Moradian-Oldak J, Diekwisch TGH, Lyaruu DM, Wright JT, Bringas PJr, Slavkin HC (1995) Evidence for amelogenin nanospheres as functional components of secretory-stage enamel. J Struct Biol 115:50-59

Fincham AG, Moradian-Oldak J, Simmer JP, Sarte PE, Lau EC, Diekwisch T, Slavkin HC (1994) Self-assembly of a recombinant amelogenin protein generates supramolecular structures. J Struct Biol 112:103-109

Fincham AG, Moradian-Oldak J, Simmer JP (1999) The structural biology of the developing dental enamel matrix. J Struct Biol 126:270-299

Fisher LW, Torchia DA, Fohr B, Young MF, Fedarko NS (2001) Flexible structures of SIBLING proteins, bone sialoprotein, and osteopontin. Biochem Biophys Res Commun 280:460-465

Frank S, Boudko S, Mizuno K, Schulthess T, Engel J, Bachinger HP (2003) Collagen triple helix formation can be nucleated at either end. J Biol Chem 278:7747-7750

Fraser RDB, MacRae TP, Miller A, Suzuki E (1983) Molecular conformation and packing in collagen fibrils. J Mol Biol 167:495-521

Fratzl P, Groschner M, Vogl G, Plenk H Jr., Eschberger J, Fratzl-Zelman N, Koller K, Klaushofer K (1992) Mineral crystals in calcified tissues: a comparative study by SAXS. J Bone Min Res 7:329-334

Fratzl P, Paris O, Klaushofer K, Landis WJ (1996) Bone mineralization in an osteogenesis imperfecta mouse model studied by small-angle X-ray scattering. J Clin Invest 97:396-402

Fukae M, Tanabe T, Murakami C, Dohi N, Uchida T, Shimizu M (1996) Primary structure of porcine 89 kDa enamelin. Adv Dent Res 10:111-118

Gautron J, Hincke MT, Mann K, Panheleux M, Bain M, McKee MD, Solomon SE, Nys Y (2001) Ovocalyxin-32, a novel chicken eggshell matrix protein, isolation, amino acid sequencing, cloning, and immunocytochemical localization. J Biol Chem 276:39243-39252.

George A, Sabsay B, Simonian PAL, Veis A (1993) Characterization of a novel dentin matrix acidic protein. Implications for biomineralization. J Biol Chem 268:12624-12630

George A, Bannon L, Sabsay B, Dillon JW, Malone J, Veis A, Jenkins NA, Gilbert DJ, Copeland NG (1996) The carboxyl terminal domain of phosphophoryn contains unique extended triplet amino acid repeat sequences forming ordered carboxyl-phosphate interaction edges which may be essential in the biomineralization process. J Biol Chem 271:32869-32873

Glimcher MJ (1976) Composition, structure and organization of bone and other mineralized tissues and the mechanism of calcification, *In:* Handbook of Physiology-Endocrinology VII, American Physiological Society, Williams and Wilkins Co., Baltimore, MD. p 25-116

Glimcher MJ, Hodge AJ, Schmitt FO (1957) Macromolecular aggregation states in relation to mineralization: the collagen hydroxyapatite system as studied *in vitro*. Proc Natl Acad Sci USA 43:860-967

Goldberg HA, Warner KJ, Stillman MJ, Hunter GK (1996) Determination of the hydroxyapatite-nucleating region of bone sialoprotein. Connect Tissue Res 35:385-392

Gura T, Hu G, Veis A (1996) Posttranscriptional aspects of the biosynthesis of type I collagen pro-alpha chains. The effects of posttranslational modifications on synthesis pauses during the elongation of the pro α1(I) chain. J Cellular Biochem 61:194-215

Hanson DA, Eyre DR (1996) Molecular site specificity of pyridinoline and pyrrole cross-links in type I collagen of human bone. J Biol Chem 271:26508-26516

He G, Dahl T, Veis A, George A (2003) Nucleation of apatite crystals *in vitro* by self-assembled dentin matrix protein 1. Nat Mater 2:552-558

Hincke MT, Gautron J, Tsang CP, McKee MD, Nys Y (1999) Molecular cloning and ultrastructural localization of the core protein of an eggshell matrix proteoglycan, ovocleidin-116. J Biol Chem 274:32915-32923

Hodge AJ (1989) Molecular models illustrating the possible distributions of "holes" in simple systematically staggered arrays of type I collagen molecules in native-type fibrils. Connect Tiss Res 21:137-147

Hofmann H, Voss T, Kuhn K, Engel J (1984) Localization of flexible sites in thread-like molecules from electron micrographs. Comparison of interstitial, basement membrane and intima collagens. J Mol Biol 172:325-343

Hu C-C, Ryu OH, Quian Q, Zhang CH, Simmer JP (1997) Cloning, characterization and heterologous expression of murine exon 4-containing amelogenin mRNAs. J Dent Res 76:641-647

Hu C-C, Simmer JP, Bartlett JD, Quian Q, Zhang C, Ryu OH, Xue J, Fukae Uchida T, MacDougall M (1998) Murine enamelin: cDNA and derived protein sequences. Connect Tissue Res 39:47-62

Hu G, Gura R, Sabsay B, Sauk J, Dixit SN, Veis A (1995) Endoplasmic Reticulum Protein Hsp47 Binds Specifically to the N-Terminal globular Domain of the Amino-Propeptide of the Procollagen I α1(I)-Chain. J. Cellular Biochem 59:350-367

Hulmes DJ, Bruns RR, Gross J (1983) On the state of aggregation of newly secreted procollagen. Proc Natl Acad Sci USA 80:388-392

Hulmes DJ, Jesior JC, Miller A Berthet-Colominas C, Wolff C (1981) Electron microscopy shows periodic structure in collagen fibril cross sections. Proc Natl Acad Sci USA 78:3567-3571

Hulmes DJ, Kadler KE, Mould AP, Hojima Y, Holmes DF, Cummings C, Chapman JA, Prockop DJ (1989) Pleomorphism in type I collagen fibrils produced by persistence of the procollagen N-propeptide. J Mol Biol 210:337-345

Hulmes DJS, Miller A (1979) Quasi-hexagonal packing in collagen fibrils. Nature 282:878-880

Hulmes DJS, Wess TJ, Prockop DJ, Fratzl P (1995) Radial packing, order, and disorder in collagen fibrils. Biophys J 68:1661-1670

Hunter GK, Hauschka PV, Poole AR, Rosenberg LC, Goldberg HA (1996) Nucleation and inhibition of hydroxyapatite formation by mineralized tissue proteins. Biochem J 317:59-64.

Hunter GK, Poitras MS, Underhill TM, Grynpas MD, Goldberg HA (2001) Induction of collagen mineralization by a bone sialoprotein-decorin chimeric protein. J Biomed Mater Res 55:496-502

Jelinski LW, Sullivan CE, Torchia DA (1980) ^{2}H NMR study of molecular motion in collagen fibrils. Nature 282:878-880

Kadler KE, Hojima Y, Prockop DJ (1990) Collagen fibrils *in vitro* grow from pointed tips in the C- to N-terminal direction. Biochem J 268:339-343

Kadler KE, Hojima Y, Prockop DJ (1987) Assembly of collagen fibrils de novo by cleavage of the type I pC-collagen with procollagen C-proteinase. Assay of critical concentration demonstrates that collagen self-assembly is a classical example of an entropy-driven process. J Biol Chem 262:15696-15701

Katz EP, Billings E, Yamauchi M, David C (1992) Computer simulations of bone collagen cross-linking patterns. *In:* Chemistry and Biology of Mineralized Tissues. Slavkin H, Price P. (eds) Elsevier Science Publishers B.V., New York, p 61-67

Katz EP, Li ST (1973a) The intermolecular space of reconstituted collagen fibrils. J Mol Biol 73:351-369

Katz EP, Li ST (1973b) Structure and function of bone collagen fibrils. J Mol Biol 80:1-15

Katz EP, Wachtel E, Yamauchi M, Mechanic GL (1989) The structure of mineralized collagen fibrils. Connect Tissue Res 21:149-158

Kawasaki K, Weiss KM (2003) Mineralized tissue and vertebrate evolution: The secretory calcium-binding phosphoprotein gene cluster. Proc Natl Acad Sci (US) 100:4060-4065

Kinney JH, Oliveira J, Haupt DL, Marshall GW, Marshall SJ (2002) The spatial arrangement of tubules in human dentin. J Materials Sci: Materials in Med 12:743-751

Kitajima T, Tomita M, Killian CE, Akasaka K, Wilt FH (1996) Expression of spicule matrix protein gene SM30 in embryonic and adult mineralized tissues of sea urchin *Hemicentrotus pulcherrimus*. Development Growth & Differentiation 38:687-695

Kivirikko KI, Ryhanen L, Anttinen H, Bornstein P, Prockop DJ (1973) Further hydroxylation of lysyl residues in collagen by protocollagen lysyl hydroxylase *in vitro*. Biochemistry 12:4966-4971

Koide T, Takahara Y, Asada S, Nagata K (2002) Xaa-Arg-Gly triplets in the collagen triple helix are dominant binding sites for the molecular chaperone HSP47. J Biol Chem 277:6178-6182

Kuypers R, Tyler M, Kurth LB, Jenkins ID, Horgan DJ (1992) Identification of the loci of the collagen-associated Ehrlich chromogen in type I collagen confirms its role as a trivalent cross-link. Biochemical J 283:129-136

Kuznetsova N, Leikin S (1999) Does the triple helical domain of type I collagen encode molecular recognition and fiber assembly while telopeptides serve as catalytic domains? Effect of proteolytic cleavage on fibrillogenesis and on collagen-collagen interaction in fibers. J Biol Chem 274:36083-36088

Landis WJ, Hodgens KJ (1996) Mineralization of collagen may occur on fibril surfaces: Evidence from conventional and high-voltage electron microscopy and three-dimensional imaging. J Struct Biol 117:24-35

Landis WJ, Song MJ, Leith A, McEwen L, McEwen B (1993) Mineral and organic matrix interaction in normally calcifying tendon visualized in three dimensions by high voltage electron microscopic tomography and graphic image reconstruction. J Struct Biol 110:39-54

Landis WJ, Song MJ (1991) Initial mineral deposition in calcifying tendon characterized by high voltage electron microscopy and three-dimensional graphic imaging. J Struct Biol 107:116-127

Landis WJ (1996) Mineral characterization in calcifying tissues: Atomic, molecular, and macromolecular perspectives. Connect Tissue Res 34:239-246

Landis WJ (1995a) Tomographic imaging of collagen-mineral interaction: Implications for osteogenesis imperfecta. Connect Tissue Res 31:287-290

Landis WJ (1995b)The strength of a calcified tissue depends in part on the molecular structure and organization of its constituent mineral crystals in their organic matrix. Bone 16:533-544

Lees S, Prostak K (1988) The locus of mineral crystallites in bone. Connect Tissue Res 18:41-54

Lees S (1981) A mixed packing model for bone collagen. Calcif Tiss Int 33:591-602

Lees S (2003) Mineralization of Type I collagen. Biophys J 85:204-207

Leikina E, Mertts MV, Kuznetsova N, Leikin S (2002) Type I collagen is thermally unstable at body temperature. Proc. Natl. Acad. Sci. USA 99:1314-1318

Lenaers A, Ansay M, Nusgens BV, Lapiere CM (1971) Collagen made of extended α-chains, procollagen, in genetically-defective dermatosparaxic calves. Eur J Biochem 23:533-543

Levene CI (1966) Plant toxins and human disease. Collagen and lathyrism. Proc Roy Soc Med 59:757-758

Levi-Kalisman Y. Falini G. Addadi L. Weiner S (2001) Structure of the nacreous organic matrix of a bivalve mollusk shell examined in the hydrated state using cryo-TEM. J Struct Biol 135:8-17

Linde A, Lussi A, Crenshaw MA (1989) Mineral induction by immobilized polyanionic proteins. Calcif Tiss Int 44:286-295

Long GC, Thomas M, Brodsky B (1995) Atypical G-X-Y sequences surround the interruptions in the repeating tripeptide pattern of basement membrane collagen. Biopolymers 35:621-628

Lowenstam HA (1981) Minerals formed by organisms. Science 211:1126-1131

Lowenstam HA, Weiner S (1983) Mineralization by organisms and the evolution of biomineralization. In: Biomineralization and biological metal accumulation. Westbroek P, deJong EW (eds) D. Reidel Pub Co, Dordrecht, Holland, p 191-203

Lowenstam HA, Weiner S (1989) On Biomineralization, Oxford University Press, New York

Lucas JM, Vaccaro E, Waite JH (2002) A molecular, morphometric and mechanical comparison of the structural elements of the byssus from Mytilus edulis and Mytilus galloprovincialis. J Exptl Biol 205:1807-1817

Lussi A, Crenshaw MA, Linde A (1988) Induction and inhibition of hydroxyapatite formation by rat dentine phosphoprotein *in vitro*. Arch Oral Biol 33:685-691

MacDougall M, Simmons D, Dodds A, Knight C, Luan X, Zeichner-David M, Zhyang C, Ryu OH, Quian Q, Simmer JP, Hu C-C (1998) Cloning, characterization and tissue expression pattern of mouse tuftelin cDNA. J Dent Res 77:1970-1978

Mao Z, Shay B, Hekmati M, Fermon E, Taylor A, Dafni L, Heikinheimo K, Lustmann J, Fisher LW, Young M, Deutsch D (2001) The human tuftelin gene: cloning and characterization. Gene 279:181-196

Miake Y, Shimoda S, Fukae M, Aoba T (1993) Epitaxial overgrowth of apatite crystals on the thin ribbon precursor at early stages of porcine enamel mineralization. Calcif Tissue Res 53:249-256

Miller A (1976) Molecular packing in collagen. In: Biochemistry of Collagen. Ramachandran GN, Reddi AH (eds) Plenum Pub Corp, New York, p 85-136.

Moradian-Oldak J, Lau EC, Diekwisch T, Slavkin HC, Fincham AG (1995) A review of the aggregation properties of a recombinant amelogenin. Connect Tissue Res 35:125-130

Moradian-Oldak J, Leung W, Fincham AG (1998b) Temperature and pH dependent supramolecular assembly of amelogenin molecules: a dynamic light-scattering analysis. J Struct Biol 122:320-327

Moradian-Oldak J, Paine ML, Lei YP, Fincham AG, Snead ML (2000) Self-assembly properties of recombinant engineered amelogenin proteins analyzed by dynamic light scattering and atomic force microscopy. J Struct Biol 131:27-37

Moradian-Oldak J, Simmer PJ, Lau EC, Sarte PE, Slavkin HC, Fincham AG (1994) Detection of monodisperse aggregates of a recombinant amelogenin by dynamic light scattering. Biopolymers 34:1339-1347

Moradian-Oldak J, Tan J, Fincham AG (1998a) Interaction of amelogenin with hydroxyapatite crystals: An adherence effect through amelogenin self-association. Biopolymers 46:225-238

Morgan PH, Jacobs HG, Segrest JP, Cunningham LW (1970) A comparative study of glycopeptides derived from selected vertebrate collagens. A possible role of the carbohydrate in fibril formation. J Biol Chem 245:5042-5048

Nakahara H (1982) Calcification of gastropod nacre. In: Biomineralization and biological metal accumulation. Westbroek P, deJong EW (eds) D. Reidel Pub Co, Dordrecht, Holland, p 225-230

Nakahara H, Kaeki M, Bevelander G (1980) Fine structure and amino acid composition of the organic envelope in the prismatic layer of some bivalve shells. Venus Jap J Malac 39:167-177

Nestler H, Hvidt S, Ferry J, Veis A (1983) Flexibility of collagen determined from dilute solution viscoelastic measurements. Biopolymers 22:1747-1758

Olsen BR, Gerecke D, Gordon M, Green G, Kimura T, Konomi H, Muragaki Y, Ninomiya Y, Nishimura I, Sugrue S (1989) A new dimension in the extracellular matrix. In: Collagen, Vol. 4. Olsen BR, Nimni ME (eds) CRC Press, Boca Raton, Fl. p 1-19

Orgel JP, Miller A, Irving TC, Fischetti RF, Hammersley AP, Wess TJ (2001) The *in situ* supermolecular structure of type I collagen. Structure 9:1061-1069

Orgel JP, Wess TJ, Miller A (2000) The *in situ* conformation and axial location of the intermolecular cross-linked non-helical telopeptides of type I collagen. Structure 8:137-142

Otsubo K, Katz EP, Mechanic GL, Yamauchi M (1992) Cross-linking connectivity in bone collagen fibrils: the COOH-terminal locus of free aldehyde. Biochemistry 31:396-402

Pace JM, Corrado M, Missero C, Byers PH (2003) Identification, characterization and expression of a new fibrillar collagen gene, COL27A1. Matrix Biol 22:3-14

Petruska JA, Hodge AJ (1964) A subunit model for the tropocollagen macromolecule. Proc Natl Acad Sci USA 51:871-876

Pierard GE, Le T, Hermanns JF, Nusgens BV, Lapiere CM (1987) Morphometric study of cauliflower collagen fibrils in dermatosparaxis of the calves. Coll Relat Res 6:481-492

Pisam M, Jammet C, Laurent D (2002) First steps of otolith formation in zebrafish: role of glycogen? Cell Tissue Res 310:163-168

Prockop DJ, Hulmes DJS (1994) Assembly of collagen fibrils de novo from soluble precursors: Polymerization and copolymerization of procollagen, pN-collagen and mutated collagens. In: Extracellular Matrix Assembly and Structure, Yurchenko PD, Birk DE, Mecham RP (eds). Academic Press, New York, p 47-90

Rabie AM, Veis A (1995) An immunocytochemical study of the routes of secretion of collagen and phosphophoryn from odontoblasts. Connect Tissue Res 31:197-209

Rahima M, Veis A (1988) Two classes of dentin phosphophoryn from a wide range of species contain immunologically cross-reactive epitope regions. Calcif Tiss Int 42:104-112

Robins SP, Duncan R (1987) Pyridinium crosslinks of bone collagen and their location in peptides isolated from rat femur. Biochim Biophys Acta 914:233-239

Ronnhölm E (1962) The amelogenesis of human teeth as revealed by electron microscopy. II. J Ultrastruct Res 6:249-303

Saito T, Arsenault AL, Yamauchi M, Kuboki Y, Crenshaw MA (1997) Mineral induction by immobilized phosphoproteins. Bone 21:305-311

Saito T, Yamauchi M, Abiko Y, Matsuda K, Crenshaw MA (2000) *In vitro* apatite induction by phosphophoryn immobilized on modified collagen fibrils. J Bone Min Res 15:1615-1619

Saito T, Yamauchi M, Crenshaw MA (1988) Apatite induction by insoluble dentin collagen. J Bone Min Res 13:265-270

Sauk JJ, Smith T, Norris K, Ferreira L (1994) Hsp47 and the translation-translocation machinery cooperate in the production of alpha 1(I) chains of type I procollagen. J Biol Chem 269:3941-3946

Schlueter RJ, Veis A (1964) The macromolecular organization of dentine matrix collagen. II. Periodate degradation and carbohydrate cross-linking. Biochemistry 3:1657-1665

Sfeir C, Veis A (1996) The membrane associated kinases which phosphorylate bone and dentin matrix phosphoproteins are isoforms of cytosolic CKII. Connect Tissue Res 35:215-222

Siperko LM, Landis WJ (2001) Aspects of mineral structure in normally calcifying avian tendon. J Struct Biol 135:313-320

Sipilä L, Szatanik M, Vainionpaa H, Ruotsalainen H, Myllyla R, Guenet JL (2000) The genes encoding mouse lysyl hydroxylase isoforms map to chromosomes 4,5, and 9. Mammalian Genome 11:1132-1134

Smith JW (1968) Molecular pattern in native collagen. Nature 219:157-158

Soledad Fernandez M, Moya A, Lopez L, Arias JL (2001) Secretion pattern, ultrastructural localization and function of extracellular matrix molecules involved in eggshell formation. Matrix Biol 19:793-803

Stetler-Stevenson WG, Veis A (1983) Bovine dentin phosphophoryn: composition and molecular weight. Biochemistry 22:4326-4335

Stetler-Stevenson WG, Veis, A (1986) Type I collagen shows a specific binding affinity for bovine dentin phosphophoryn. Calcif Tissue Int 38, 135-141

Stetler-Stevenson, W.G. and Veis, A. (1987) Bovine dentin phosphophoryn: calcium ion binding properties of a high molecular weight preparation. Calcif Tissue Int 40, 97-102

Stock SR, Barss J, Dahl T, Veis A, Almer JD (2002) X-ray absorption microtomography (microCT) and small beam diffraction mapping of sea urchin teeth. J Struct Biol 139:1-12

Tanabe T, Aoba T, Moreno EC, Fukae M, Shimazu M (1990) Properties of phosphorylated 32 kd nonamelogenin proteins isolated from porcine secretory enamel. Calcif Tissue Int 46:205-215

Termine JD, Belcourt AB, Christner PJ, Conn KM, Nylen MU (1980) Properties of dissociatively extracted fetal tooth matrix proteins. Principal molecular species in developing bovine enamel. J Biol Chem 20:9760-9768

Torchia DA (1982) Solid-state NMR studies of molecular motion in collagen fibrils. Methods Enzymol 82:174-186

Traub W, Arad T, Weiner S (1989) Three-dimensional ordered distribution of crystals in turkey tendon collagen fibers. Proc Nat Acad Sci USA 86:9822-9826

Traub W, Arad T, Weiner S (1992b) Growth of mineral crystals in turkey tendon collagen fibers. Connect Tissue Res 28:99-111

Traub W, Jodaikin A, Arad T, Veis A, Sabsay B (1992a) Dentin phosphophoryn binding to collagen fibrils. Matrix 12:197-201

Travis DF, Glimcher MJ (1964) The structure and organization of, and the relationships between the organic matrix and the inorganic crystals of embryonic bovine enamel. J Cell Biol 23:477-497

Trus BL, Piez KA (1980) Compressed microfibril models of the native collagen fibril. Nature 286:300-301

Turner CH, Chandran A, Pidaparti RMV (1995) The anisotropy of osteonal bone and its ultrastructural implications. Bone 17:85-89

Tye CE, Rattray KR, Warner KJ, Gordon JA, Sodek J, Hunter GK, Goldberg HA (2003) Delineation of the hydroxyapatite-nucleating domains of bone sialoprotein. J Biol Chem 278:7949-7955

Uchida T, Murakami C, Wakida K, Dohi N, Iwai Y, Simmer JP, Fukae M, Satoda T, Takahashi O (1998) Sheath proteins: Synthesis, secretion, degradation and fate in forming enamel. Eur J Oral Sci 106:308-314

Välkkilä M, Melkoniemi M, Kvist L, Kuivaniemi, Tromp G, Ala-Kokko L (2001) Genomic organization of the human COL3A1 and COL5A2 genes: COL5A2 has evolved differently than the other minor fibrillar collagen genes. Matrix Biol 20:357-366

Van Der Slot AJ, Zuurmond AM, Bardoel AF, Wijmenga C, PruijsHE, Sillence DO, Brinkmann J, Abraham DJ, Black CM, Verzijl N, DeGroot J, Hanemaaijer R, TeKoppele JM, Huizinga TW, Bank RA (2003) Identification of PLOD2 as telopeptide lysyl hydroxylase, an important enzyme in fibrosis. J Biol Chem. (in press)

Veis A, Barss J, Dahl T, Rahima M, Stock S (2002) Mineral related proteins of the sea urchin teeth. *Lytechinus variegatus*. Microsc Res Technique 59:342-351

Veis A (1982) Collagen fibrillogenesis. Connect Tissue Res. 1:11-24

Veis A, Cohen J (1960) Reversible transformation of gelatin to collagen structure. Nature 186:720-721

Veis A, Perry A (1967) The phosphoprotein of the dentin matrix. Biochemistry 6:2409-2416

Veis A, Sabsay B (1982) Bone and tooth formation. Insights into mineralization strategies. *In:* Biomineralization and Biological Metal Accumulation. Westbroek P, deJong EW (eds) D. Reidel Pub. Co., Dordrecht, Holland, p 273-284

Veis A, Schlueter RJ (1964) The macromolecular organization of dentine matrix collagen. I. characterization of dentine collagen. Biochemistry 3:1650-1656

Veis A, Dahl T, Sabsay B (2000) The specificity of phosphophoryn-collagen I interactions. *In:* Proc. 6th Int. Conf. Chemistry & Biology of Mineralized Tissues, Vittel, France. Goldberg M, Robinson C, Boskey, A (eds) Orthopedic Research Soc., Rosemont, IL. p 169-173

Veis A, George A. (1994) Fundamentals of interstitial collagen self-assembly. *In:* Extracellular Matrix Assembly and Structure. Yurchenko PD, Birk DE, Mecham RP (eds) Academic Press, New York, p 15-45

Veis A, Schlueter R (1963) Presence of phosphate-mediated cross-linkages in hard tissue collagens. Nature 197:1204

Veis A, Wei K, Sfeir C, George A, Malone J (1998) The properties of the $(DSS)_n$ triplet repeat domain of rat dentin phosphophoryn. Eur J Oral Sci 106 (Suppl 1):234-238

Veis A, Yuan L (1975) Structure of the collagen micro-fibril. A four-strand overlap model. Biopolymers 14:895-900

Veis DJ, Albinger TM, Clohisy J, Rahima M, Sabsay B, Veis A (1986) Matrix proteins of the teeth of the sea urchin *Lytechinus variegatus*. J Exptl Zool 240:35-46

Venugopal MG, Ramshaw JAM, Braswell E, Zuh D, Brodsky B (1994) Electrostatic interactions in collagen-like triple-helical peptides. Biochemistry 33:7948-7956

Wang Y, Kowalski PE, Thalmann I, ornitz DM, Mager DL, Thalmann R (1998) Otoconin-90, the mammalian otoconial matrix protein, contains two domains of homology to secretory phospholipase A_2. Proc Natl Acad Sci USA 95:1534-15350

Weiner S, Traub W (1992) Bone structure: from angstroms to microns. FASEB J 6:879-885

Weiner S, Traub W (1984) Macromolecules in mollusk shells and their functions in biomineralization. Phil Trans R Soc London Ser B 304:412-438

Weiner S, Veis A, Beniash E, Arad T, Dillon JW, Sabsay B, Siddiqui F (1999) Peritubular dentin formation: crystal organization and the macromolecular constituents in human teeth. J Struct Biol 126:27-41

Weiner S, Arad T, Sabanay I, Traub W (1997) Rotated plywood structure of primary lamellar bone in the rat: orientations of the collagen fibril arrays. Bone 20:509-514

Weiner S, Arad T, Traub W (1991) Crystal organization in rat bone lamellae. FEBS Let. 285:49-54

Weiner S, Traub W, Wagner HD (1999) Lamellar bone: structure-function relations. J Struct Biol 126:241-255

Weinstock M, Leblond CP (1974) Synthesis, migration, and release of precursor collagen by odontoblasts as visualized by radioautography after (^{3}H) proline administration. J Cell Biol 60:92-127

Wess TJ, Hammersley A, Wess L, Miller A (1995) Type I collagen packing, conformation of the triclinic unit cell. J Mol Biol 248:487-493

White SW, Hulmes DJ, Miller A, Timmins PA (1977) Collagen-mineral axial relationship in calcified turkey leg tendon by X-ray and neutron diffraction. Nature 266:421-425

Woodhead-Galloway J, Machin PA (1976) Modern theories of liquids and the diffuse equatorial x-ray scattering from collagen. Acta Crystallogr 32:368-372

Wu T-M, Fink DJ, Arias JL, Rodriguez JP, Heuer AH, Caplan AI (1992) The molecular control of avian eggshell mineralization. *In:* Chemistry and Biology of Mineralized Tissues. Slavkin H, Price P (eds) Elsevier Science Pub, NY, p 133-141

Wu CB, Pelech SL, Veis A (1992) The *in vitro* phosphorylation of the native rat incisor dentin phosphophoryns. J Biol Chem 267:16588-16594

Yamauchi M, Banes AJ, Kuboki Y, Mechanic GL (1981) A comparative study of the distribution of the stable crosslink, pyridinoline, in bone collagens from normal, osteoblastoma, and vitamin D-deficient chicks. Biochem Biophys Res Commun 102:59-65

Yamauchi M, Chandler GS, Katz EP (1992) Collagen cross-linking and mineralization. *In:* Chemistry and Biology of Mineralized Tissues. Slavkin H, Price P (eds) Elsevier Science Publishers B.V., New York p 39-46

Yamauchi M, Katz EP, Otsubo K, Teraoka K, Mechanic GL (1989) Cross-linking and stereospecific structure of collagen in mineralized and nonmineralized skeletal tissues. Connect Tissue Res 21:159-67

Yamauchi M, Young DR, Chandler GS, Mechanic GL (1988) Cross-linking and new bone collagen synthesis in immobilized and recovering primate osteoporosis. Bone 9:415-418

Zylberberg L, Traub W, de Buffrenil V, Allizard F, Arad T, Weiner S (1998) Rostrum of a toothed whale: ultrastructural study of a very dense bone. Bone 23:241-247

10 Silicification: The Processes by Which Organisms Capture and Mineralize Silica

Carole C. Perry

Division of Chemistry, Interdisciplinary Biomedical Research Centre
The Nottingham Trent University
Clifton Lane, Nottingham, NG11 8NS, United Kingdom

INTRODUCTION

Silicification is widespread in the biological world and occurs in bacteria, single-celled protists, plants, invertebrates and vertebrates. Minerals formed in the biological environment often show unusual physical properties (e.g., strength, degree of hydration) and often have structures that exhibit order on many length scales. The minerals are formed from an environment that is undersaturated with respect to silicon and under conditions of around neutral pH and low temperature ca. 4–40°C. Formation of the mineral may occur intra- or extra-cellularly and specific biochemical locations for mineral deposition that include lipids, proteins and carbohydrates are known. The significance of the cellular machinery cannot be over emphasized and it is with advances in experimental techniques (cell biology and materials characterization) and advances in understanding (including the ability to design laboratory experiments to mimic the biological environment) that much progress has been made in the field in recent years. In most cases the formation of the mineral phase is linked to cellular processes and if we understand this process the knowledge so gained could be used to good effect in designing new materials for biomedical, optical and other applications. The study of living organisms could result in wealth generation/creation. It should be noted that although significant advances have been made in the last ten years, new questions have arisen and there are many areas requiring exploration.

This contribution will place emphasis on the systems for which most is known, namely sponges and diatoms, however, it should be borne in mind that many other organisms from single-celled species such as choanoflagellates (Mann et al. 1982) and radiolaria through to higher plants and molluscs such as the limpet (Mann et al. 1986) make use of silica and form species-specific structures that show structural organization on several length scales. Examples of some of the silicified structures observed for organisms that will be little discussed in this chapter are given in Figure 1. Table 1 provides definitions for the general terminology used in this contribution. At the end of the chapter, the bibliography lists some excellent reviews which can be used to provide more information on the topics discussed as well as providing an excellent source of further literature.

STRUCTURAL CHEMISTRY OF SILICA

Silica and silicates are extensively used in industry and medicine. The materials find use in paints, foods, medicines, adhesives, detergents, chromatography materials, catalysts and photonic materials (reviewed by Iler 1979). Silica may be produced at high temperature, via aqueous processing or by largely non-aqueous routes such as the low-temperature sol-gel process (reviewed in Brinker and Scherrer 1990; Hench and West 1990). Whatever the eventual use of the silica, it is its structure that determines its

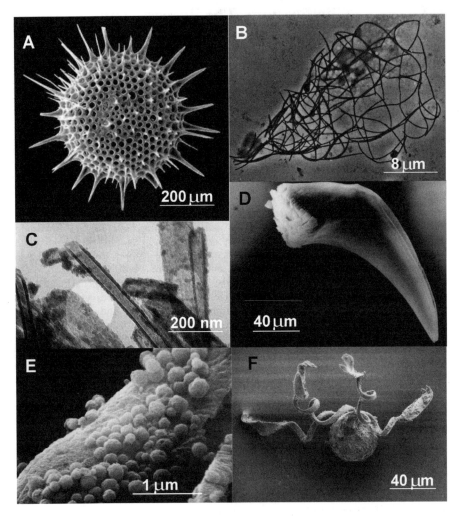

Figure 1. Electron microscope images of biosilicas from (A) Radiolarian, (B) choanoflagellate (courtesy of Professor Steve Mann, Bristol University), (C) silica tubes that surround goethite (iron containing) crystals in limpet teeth, (D) limpet tooth with iron containing mineral removed, (E) silica coated particles on the surface of the spore elators of *Equisetum arvense*, (F) spore and elators from *Equisetum arvense*.

properties. By structure we mean order and organization on length scales from angstroms to the size of the final object, morphology, surface area, porosity and surface functionality. The essential building block is the SiO_4 tetrahedron although other structural units such as the SiO_6 octahedron are also used. These units can be put together in a wide range of patterns to yield both porous and non-porous crystalline materials including silica based zeolite materials. As well as crystalline silica and silicates there exist an extremely diverse range of amorphous materials such as disordered precipitates, gels, glasses and shaped objects (spheres, screws, hollow tubes, etc; Stöber et al. 1968; Yang et al.1997; Miyaki et al. 1999) produced using a wide range of synthesis conditions.

Silicification: Organism Mineralization of Silica

Table 1. General terminology used in this chapter

Si: the chemical symbol for the element and the generic term used when the nature of the specific silicon compound is not known.

Si(OH)$_4$: silicic acid, or more correctly orthosilicic acid, the fundamental building block used in the formation of silicas.

SiO$_2$·nH$_2$O or SiO$_{2-x}$(OH)$_{2x}$·2H$_2$O: amorphous, hydrated, polymerized material produced from orthosilicic acid.

Oligomerization: the formation of dimers and small oligomers from orthosilicic acid by removal of water. e.g., 2Si(OH)$_4$ → (HO)$_3$Si–O–Si(OH)$_3$ + H$_2$O

Polymerization: the mutual condensation of silicic acid to give molecularly coherent units of increasing size.

Organosilicon compound: must contain silicon covalently bonded to carbon within a distinct chemical species

Silanol: hydroxyl group bonded to silicon atom

Silicate: a chemically specific ion having negative charge (e.g., SiO$_3^{2-}$), term also used to describe salts (e.g., sodium silicate Na$_2$SiO$_3$) or the mineral beryl (Be$_3$Al$_2$Si$_6$O$_{18}$) that contains the Si$_6$O$_{18}^{12-}$ ion.

Opal: the term used to describe the gem-stone and often used to describe the type of amorphous silica produced by biological organisms. The two are similar in structure at the molecular level (disordered or amorphous) but at higher levels of structural organization are distinct from one another. The mineral opal contains amorphous silica particles of regular size packed in a well ordered array which may extend over many mm or cm in length. This is not the case for silica produced in the biological environment as will be discussed below. By the terminology of Jones and Segnit (1971), this is opal-A.

The amorphous materials, in contrast to their crystalline analogs exhibit no long-range order and are built up from SiO$_4$-tetrahedra with variable Si–O–Si bond angles and Si–O bond distances. These materials are covalent inorganic polymers of general formula [SiO$_{n/2}$(OH)$_{4-n}$]$_m$ (Mann et al. 1983). This formula where n = 0 to 4 and m is a large number, indicates the variation in the number of silanol functional groups within the condensed structure, Figure 2. There can also be a similar variation in the extent of hydration. The formation of "regular" amorphous silica structures such as Stöber spheres requires great care during synthesis (Stöber et al. 1968), Figure 3a and some of the more novel amorphous silica structures (length scale for organization not defined) are produced under extreme conditions of acidity (silica screws; Yang et al. 1997), Figure 3b. The effect of organic templates (including ammonium ions and substituted ammonium ions) is well documented for the synthesis of zeolites (Lewis et al. 1996).

SILICA CHEMISTRY IN AQUEOUS AND NON-AQUEOUS ENVIRONMENTS

The simplest soluble form of silica is the monomer orthosilicic acid, which is silicon tetrahedrally coordinated to four hydroxyl groups and has the formula Si(OH)$_4$ (N.B. this species has never been isolated). It is only weakly acidic with a pK$_a$ of 9.8 (Iler 1979) and is found almost universally at concentrations of a few ppm. In water, at 25°C, it is stable for long periods of time at levels below ca. 100 ppm. However, once the concentration exceeds the solubility of the amorphous solid phase, at 100–200 ppm, it undergoes autopolycondensation reactions. This process has been studied and reported by many authors (reviewed in Iler 1979; Tarutani 1989; Perry and Keeling-Tucker 2000), and can be divided into three distinct stages:

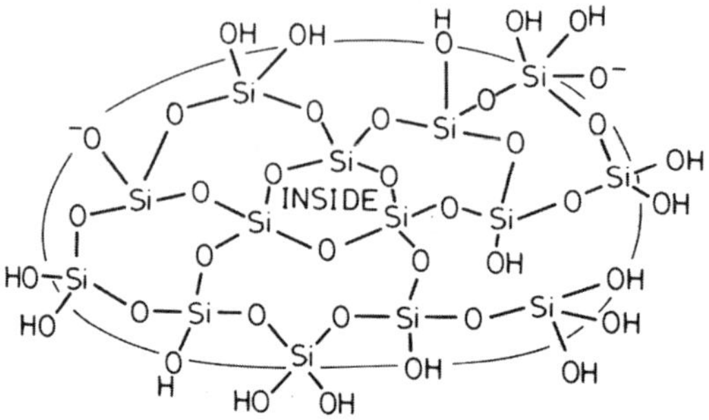

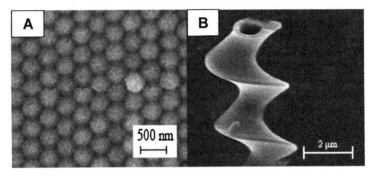

Figure 2. Schematic to show the range of functionalities possible for fundamental silica particles.

Figure 3. Scanning electron microscope images of (A) Stöber spheres, (B) silica screw (courtesy of Professor Geoff Ozin).

1. Polymerization of monomers to form stable nuclei of a critical size.
2. Growth of the nuclei to form spherical particles.
3. Aggregation of particles to form branched chains or structural motifs.

The term polymerization must be considered in its broadest sense, meaning the "...mutual condensation of silicic acid to give molecularly coherent units of increasing size..." (Iler 1979), as the product does not retain the general formula of the precursors. An alternative term is oligomerization, which is particularly appropriate for the early stages of the reaction where discrete, identifiable units are formed.

There are various mechanisms proposed for the condensation reaction. The simplest involves the condensation of two molecules of silicic acid with the release of water, Figure 4. This process results in no net change in pH for the reaction system.

$$Si(OH)_4 + Si(OH)_4 \leftrightarrow (HO)_3Si-O-Si(OH)_3 + H_2O$$

An alternative mechanism invokes a bimolecular collision between an ionized and an un-ionized silicic acid molecule (e.g., see Harrison and Loton 1995; Perry and Keeling-Tucker 2000)

$$Si(OH)_4 + Si(OH)_3O^- \leftrightarrow (HO)_3Si-O-Si(OH)_3 + OH^-$$

This process, although favored in terms of the nucleophilic attack of a negatively charged oxygen atom on the electropositive silicon center generates hydroxyl ions and

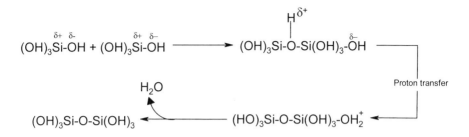

Figure 4. Mechanism of the condensation of two silicic acid molecules under conditions where neither contain Si–O⁻ groups.

would, therefore, increase the pH of the reaction system in the absence of a buffer of some sort. At ca. neutral pH where silica is formed in biological organisms (or down to ca. pH 5 as proposed for silicification in diatoms; Vrieling et al. 1999) the proportion of ionized silicic acid molecules is very small (ca. 0.002% at pH 5 and ca. 0.18% at pH 7).

As the particles get progressively bigger, pK_a values for the removal of an hydrogen atom from a silanol group decrease such that at ca. pH 7, all particles with nanometer dimensions will carry a negative charge (even at ca. pH 5, ~2% of all silanol groups will be ionized). Thus charge will be present on the fundamental particles used to build up the silica structures observed in diatoms, sponges, etc. which will require some means of neutralizing the negative charge on the surface of each particle in order for a flocculated and/or aggregated material to form.

As a rule, silicic acid molecules condense in such a way as to produce a sufficient number of siloxane (Si–O–Si) bonds with oligomers cyclizing during the early stages of the process to give rings containing predominantly 3–6 silicon atoms linked by siloxane bonds, for examples see Figure 5. Almost all silicas except those formed under conditions of extreme acidity contain high proportions of cyclic species.

Once the cyclic species dominate, monomers and dimers and other small oligomers react preferentially with these (Tarutani 1989), as the oligomers have a higher density of ionized silanol groups. A process known as Ostwald ripening also occurs where smaller, more soluble particles, dissolve and release silicic acid that re-deposits onto the larger particles. At ca. neutral pH as the particles bear a negative charge on the surface they grow as isolated sol particles, Figure 6, until the levels of soluble silica reach the solubility of amorphous silica. If salts are present or under acidic conditions (e.g., within the diatom silica deposition vesicle), however, the surface negative charge that causes repulsion between individual particles is reduced or eliminated and the particles aggregate to form dense three-dimensional networks, Figure 6.

There are many studies of the kinetics of silica formation and the orders of reaction reported range from first to fifth (see references in Iler 1979; Perry and Keeling-Tucker 2000). Despite the differences in reported orders of reaction for silica deposition, it is universally acknowledged that orthosilicic acid concentration, temperature, pH and the presence of additional chemical species all affect the rate of oligomerization. The first two factors always cause a concomitant increase in the rate. The effect of pH is that, in general, silica prepared under acidic conditions is built up from extended networks and silica formed under basic conditions is built up from highly branched networks with pH

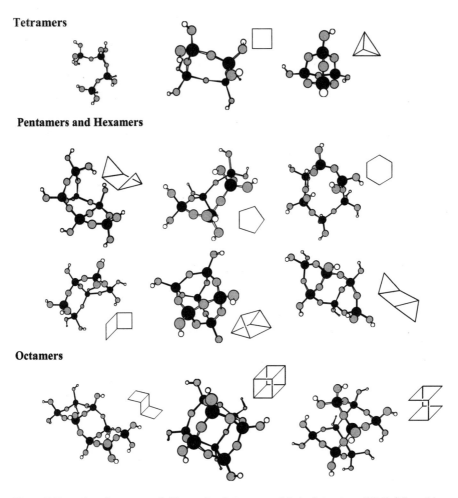

Figure 5. Examples of structures of silicates showing energy minimized structures (MM2 followed by application of the AM1 program) and idealized shape. The energy minimization was performed with all silanols protonated although the extent of deprotonation will vary according to the pH of the system. Included are five- and six-membered rings that are thought to exist in the approximately neutral solutions relevant to biological silica precipitation.

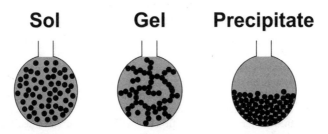

Figure 6. Schematic of a sol, gel and precipitate containing the same number of particles.

affecting the relative rates of condensation, and particle growth and aggregation, Figure 6. The presence of cations can modify the rates of the various condensation reactions (Iler 1979; Harrison and Loton 1995), the smaller the cation, the larger the hydration sphere for a unicharged metal ion, the faster the reaction (i.e., condensation is faster in the presence of Li^+ than Na^+ than K^+). The effect of cations derived from ammonium ions is less clear but these ions are able to reduce the solubility of any siliceous material formed in their presence. Cations present in the reaction system also affect speciation. One major difference is in the effect of "traditional" metal ions, such as sodium, (found in the biological environment) as opposed to alkylammonium ions, which are often used in chemical synthesis and are also found in the biosilica extracts from diatoms, see later in this chapter. In the presence of ions such as $[NR_4^+]$ (R = methyl, ethyl, propyl etc.) cages made of double cycles of 3–5 SiO_4 tetrahedra are stabilized and can be observed in solution (Engelhardt et al. 1982; Knight et al. 1986) with $[N(CH_3)_4]^+$ particularly favoring the formation of the cubic octamer $[Si_8O_{20}]^{8-}$. In the presence of sodium ions, the transverse bonds are readily broken to yield simple $[Si_4O_{12}]^{4-}$ cycles (Jolivet 2000), Figure 5.

As will become obvious later, some researchers in this field utilize non-ionic silicic acid precursors in their *in vitro* studies of silicification and a brief introduction and indication of the differences in the chemistry involved are given here (Brinker and Scherrer 1990; Hench and West 1990).

Silicon alkoxides contain a central silicon atom attached via ether linkages to organic moieties including methyl, ethyl, phenyl and other functional groups. Alkyl and other organic groups may also be directly bonded to the central silicon atom. The alkoxy groups are generally hydrolyzed when the compounds are placed in an aqueous environment, particularly under conditions of acid or base catalysis, but organic groups directly bonded to the silicon atom are not hydrolyzed. The molecular reactions that occur are given below:

$$\equiv Si-OR + H_2O \rightarrow \equiv Si-OH + ROH \qquad \text{Hydrolysis}$$

If tetramethoxysilane, a typical precursor, is considered then hydrolysis leads to the following species: $Si(OCH_3)_3OH$, $Si(OCH_3)_2(OH)_2$, $Si(OCH_3)(OH)_3$, $Si(OH)_4$. The reaction conditions will determine the relative proportions of these species but they may all be involved in condensation reactions, the simplest form of which is described below

$$\equiv Si-OH + \equiv Si-OR \rightarrow \equiv Si-O-Si\equiv + ROH \qquad \text{Condensation}$$

$$\equiv Si-OH + \equiv Si-OH \rightarrow \equiv Si-O-Si\equiv + H_2O$$

A very wide range of species can result owing to the different reactivities of hydrolyzed and partially hydrolyzed species. The presence of acid and base modifies the rates and relative rates of the hydrolysis and condensation reactions and affects the silica structures that develop. Factors known to affect hydrolysis and condensation include; (1) the chemical identity of the precursor, catalyst and solvent, (2) the relative amount of water present in the system, (3) the temperature, (4) the pH, (5) stirring, and (6) the absolute amounts of any of the components in the precursor mixtures and the order in which they are added one to the other. Gel structures (acid hydrolysis leads to largely linear structures) and particulate materials (base catalysis leads to highly branched structures) can be formed, Figure 6. As an example of the effect of precursor identity on form, reaction of tetramethoxysilane leads to a structures built up from mainly three-membered ring structures and reaction of tetraethoxysilane leads to structures built up mainly from four-membered rings, see Figure 5. The different building blocks used to make the silica-based materials can affect particle sizes, surface chemistry and porosity, all features that must be controlled in the generation of genetically invariant biologically produced materials.

The processes of oligomerization and aggregation in the formation of silicas can be modified by the presence of additivies (Iler 1979). Additives are able to affect all aspects of silica deposition from the kinetics through to the macroscopic properties including surface area, pore structure, particle size and aggregation. This makes the use of additives, from simple inorganic or organic molecules to complex organic polymers, a very powerful tool for the controlled production of silica. It should be noted that surfactants and small molecules such as ammonium-based cations are used in the production of mesoporous materials (e.g., Kresge et al. 1992) and silica-based zeolites (e.g., Lewis et al. 1996), respectively. Some of the most proficient systems at using this type of regulation are biological organisms. Their ability to tailor-make inorganic materials for a specific use is unsurpassed and has become the focus of much research effort in the search for ways to produce materials with modified properties for use by industry and academe alike.

THE STRUCTURAL CHEMISTRY OF BIOSILICAS

Biological organisms are able to tailor-make biominerals that exhibit structural and physical properties that are distinct from materials produced in the laboratory. Biosilicas can exhibit controlled particle size distributions, hydration and aggregation characteristics. Control of these structural features tends to be attributed to the interaction of the mineral as it forms with an organic matrix, but the fact that the biomineralization environment is likely to contain small organic and/or inorganic chemical species cannot be ignored.

Biological organisms form much more silica than man (Gigatons per annum as opposed to Megatons produced by man). As stated in the introduction, the silica is formed from an aqueous environment vastly under-saturated with respect to silicic acid at ca. atmospheric pressure (unless produced deep in the oceans) and at temperatures ranging from ca. 4°C to 40°C. Both gel-like and particulate siliceous materials are formed in confined spaces where membranes (lipid based) or extracellular matrices (carbohydrate or protein based) govern the location of precipitation. All biosilicates are thought to form by synchronous cooperative assembly (organic and inorganic phases form simultaneously). The biosilicate growth space contains biopolymers which may be fixed on surfaces or solubilized, metal ions, anions and other small organic molecules. The organic phase, whether intracellular or extracellular, may play important roles in (1) the formation of primary particles by altering supersaturation requirements and promotion of oligomer formation (catalysis), (2) the formation of networks and morphological features, and (3) in the stabilization and protection of the mineral structure from the environment. It is likely that the precise pathway followed during mineralization and the relative importance of the different stages will vary for each silicifying organism (Perry 1989). Biosilicas are not stoichiometric minerals and their density, hardness, solubility, viscosity and composition may vary considerably. The question of structure templating in these materials is not known but is highly likely.

Silicified organisms are very beautiful and have complex structural features, for examples see the many figures distributed throughout this chapter and the text edited by Simpson and Volcani (1981). The rationale behind this is largely unknown. The material used to construct the silicified skeletons is amorphous to X-rays and electrons at the angstrom level. No evidence has yet been found for short range order extending above 1 nm (ca. three Si–O bond lengths) although some early studies did suggest that very small crystallites of cristobalite might be present within biosilicas (Frondel 1962). We believe these findings were probably due to sample contamination. The fundamental unit that is used to build up silicified structures is a polymeric, probably particle-like object, as depicted in Figure 2 and it is difficult to obtain detailed structural information on the nature of these particles and how they are arranged within the silicified structures. Detailed structural studies of siliceous biominerals are, however, an extremely important starting point in studies of biosilicification. The structure of a particular mineral provides, in its make-up, information that may point to the path/mechanism by which it was formed. If we could correctly "read" the information that structures provide we may proceed to a much deeper understanding of the biosilicification process for the full range of organisms.

A wide range of techniques is now available to help us understand the structure of these complex amorphous materials. Examples of these techniques and the information they provide are described below.

Techniques for the study of biosilica structure

Small-angle and wide-angle X-ray diffraction techniques that can detect order at the nm level (i.e., to look for small crystalline regions within a structure) have been applied to both diatoms (Vrieling et al. 2000) and the primitive plant *Equisetum arvense* (Holzuter et al. 2003) and have found no evidence for crystallinity although small differences in instrument response to the technique have been observed by the authors for analysis along the length and across the fibers of the siliceous sponge from the Hexactinellida family, N.B. no crystallinity was detected. Other biosilicified materials should also be examined by this technique. Recent studies of model silica-protein composites and diatoms by ultrasmall angles (USAXS) have shown that structures on different length scales (including pore structures) could be identified by this technique (Vrieling et al. 2002).

Electron microscopy has been extensively used to look at the sizes of fundamental particles and their interconnection one to another. Figure 7 provides structural information on particle size and orientation for small regions of silica that were removed from plant hairs where a range of particle sizes and arrangements were observed. The technique can be used to follow structural changes with time and gel-like and particulate structures can be observed within the same organism at different development times,

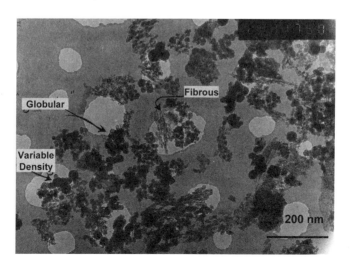

Figure 7. Transmission electron microscope image of silica from hairs on the lemma of *Phalaris canariensis* L. Examples of fibrillar, globular and other structural arrangements all obtained from a single cell hair sample.

Figure 8 (Perry and Fraser 1991) More information on the changes in structure with time will be found in the section on diatom silica structures later in the chapter. The technique can also give information on the degree of hydration by comparison of sample stability under the electron beam. Biosilicas, in comparison to silicas produced in the laboratory are much more stable under the electron beam. An extension of these studies would be to use CRYO transmission electron microscopy to look at the fundamental particles present in silicifying organisms. The technique has the advantage that contrast is enhanced and small chains, and rings of particles that are ca. 2 nm in diameter can be visualized. An example is shown in Figure 9 that was obtained from a model experiment to follow silica formation (Harrison and Loton 1995).

Electron microscopy has also widely been used to "look at" microscopic structures, see the many figures distributed throughout the chapter. The technique, in combination with energy dispersive X-ray analysis can be used to locate the mineral within the organism and to look for any co-localized species including metal ions. Maps of element localization can be produced and Figure 10 shows an example of maps produced for nettle hairs, a silicifying organism that also mineralizes calcium). The combination of

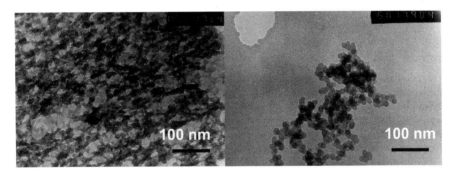

Figure 8. Transmission electron microscope images of gel-like and globular silica from the primitive plant *Equisetum arvense* at different stages of development.

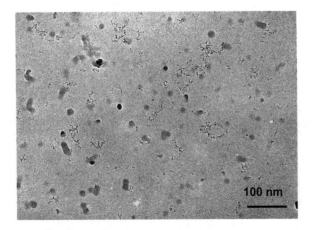

Figure 9. Cryo transmission electron microscope image of small (~2 nm diameter) silica particles arranged in chains and rings.

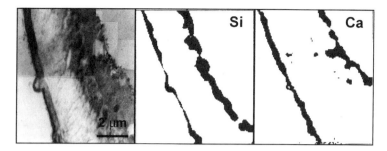

Figure 10. Transmission electron microscope image of a section through a silicified hair from the common nettle *Urtica dioica* together with silicon and calcium elemental maps obtained using energy dispersive X-ray analysis.

electron microscopy and EDXa can also be used to investigate levels of hydration for individual structural types within a mixed sample by use of chemicals containing heavy metals that react with silanol groups or water (Perry et al. 1990). Further applications of elemental analysis involve the use of the proton microprobe where very high energy focused proton beams are used to provide elemental compositions down to the part per million level at a spatial resolution of ca. one micron. The technique has been used to follow the silicification of plant hairs in relation to changes in cell ion composition (e.g., Perry et al. 1984), spore germination in the primitive plant *Equisetum arvense* where silicon is concentrated in the germinating tip of the germination tube and in diatoms (Watt et al. 1991). Other observational techniques that have been used to good effect include the atomic force microscope and Figure 11 shows a comparison of a scanning electron microscope image and an AFM image on the same diatom sample. The AFM has a range of viewing modes and Figure 12 shows two different images, height and deflection images obtained from the same area of sample. This method has been able to show that the frustules of different diatom species are built up from aggregates of particles with very narrow size distribution (Crawford et al. 2001).

Surface area and porosity measurements can be used to differentiate between silica structures that are built up from distinguishable particles (e.g., some plant silicas with

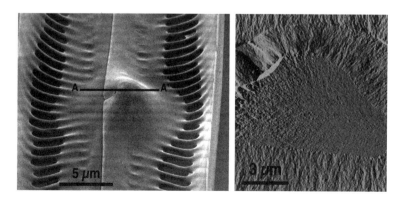

Figure 11. A comparison of a scanning electron microscope image and an atomic force microscope image for the diatom *Pinnularia viridis* valve in cross-section. Published with permission of the Journal of Phycology from Crawford et al. 2001.

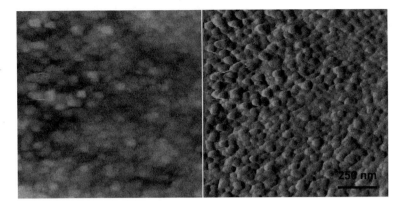

Figure 12. A comparison of a height image (left hand side) and a deflection image (right hand side) for a valve of *Pinnularia viridis* in cross-section. Published with permission of the Journal of Phycology from Crawford et al. 2001.

surface areas of several hundred meters squared per gram when the organic matrix has been removed) and those where particles are poorly distinguished (e.g., some sponges where surface areas are virtually zero). Measures of surface area can give information on pore structures that are open to the adsorbate, often nitrogen, as well as providing an indication of voids that could be occupied by the organic matrix within the structure (Perry 1989).

Infrared spectroscopy can be used to provide information on the internal structure of the siliceous phase. Functional groups such as Si–OH and Si–O–Si exhibit characteristic vibrational frequencies, the exact positions of which in the spectrum give information on the strength of bonding within the structure and the strength of interaction with other external species including the organic matrix. Information on extent of hydroxylation and the extent of hydrogen bonded interactions can be obtained. Moreover the relationship between the intensity of antisymmetric stretching modes and symmetric stretching modes seems to be characteristic of particular silica types (Kamatani 1971; Perry 1989; Schmidt et al. 2001). More recent studies using this technique have predicted that as many as one in four oxygens at the apices of the individual silica tetrahedra may be in the form of an O–H group, indicating that biogenic silica is not as highly condensed as perhaps thought (Frohlich 1989).

Solid state ^{29}Si NMR spectroscopy which looks at the local environment around the silicon nucleus can be used to find out local structural information about silicon atoms in the mineralized structures. A range of single pulse and multiple pulse methods can provide information on speciation and on the extent of interaction between the inorganic phase and any organic material present. These structural studies identify biological silicas as being disordered materials containing only tetrahedral structural units with differences in residual hydroxyl functionality. Information on the range of bond angles surrounding, for example a silicon atom attached to four oxygen atoms can be obtained and this information used to give an indication of the range of cyclic and other structures that have been used in the generation of the structure (Perry 1989; Perry et al. 2003; Schmidt et al. 2001). Using this method it has been possible to determine that plant silicas are most like silica gels with average Si–O–Si bond angles of less than 145° and they exhibit extensive hydration. Silica from diatoms and sponges have average Si–O–Si bond angles of ca. 150° and are much less heavily hydrated. The use of chemical reagents that can selectively react with silanol groups in conjunction with NMR spectroscopy is able to provide evidence on accessibility of such functional groups to the external environment. For example, reaction of silanol groups in a plant silica sample with trimethylchlorosilane (Mann et al. 1983) provided evidence that many of the silanol groups in the intact plant hairs were probably protected against dissolution by reaction with the plant cell wall biopolymers as they were unavailable to react with the reagent. Use of NMR experiments designed to look at the relaxation behavior of selected species can provide circumstantial evidence for the extent of interaction between the siliceous phase and any organic phase (Perry 1989). Of the siliceous materials investigated to date (sponge, diatom, plant and limpet teeth), the strongest interactions between inorganic and organic phases is proposed to exist for diatoms.

Solution ^{29}Si NMR spectroscopy has, until now, found little use in the study of biosilicification owing to poor sensitivity and low natural abundance of the nucleus. Recent experiments using enriched silicon sources have started to probe the internal reaction environment of a silicifying diatom *Navicula pelliculosa* (Kinrade et al. 2002). Figure 13. shows the presence of silicic acid detected within the diatom. During some of

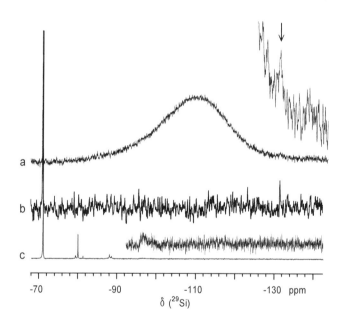

Figure 13. Reprinted with permission of the Royal Society of Chemistry from Kinrade et al. 2002. (a) Silicon-29 ^{1}H-decoupled NMR spectrum (149 MHz) of a synchronized culture of *Navicula pelliculosa* following 6 h feeding on ^{29}Si-enriched (75 atom%) silicon at 298 K. The spectrum was acquired over 9.3 h at 278.2 K using π/2 (16 μs) pulses with a cycling period of 11 s. The dominant feature centered at −110 ppm arises from Si-containing components in the NMR probe. The sample itself yielded a sharp orthosilicic acid peak at −71.0 ppm and a weak signal at −131.5 ppm with a half-height peak width of 25 Hz. (See vertical expansion). (b) The last of four successive 3.3 h silicon-29 ^{1}H-decoupled NMR spectra (99.4 MHz) acquired of another synchronized *Navicula pelliculosa* culture, following 6 h exposure to ^{29}Si (99.8 atom%) at 298 K in a medium also enriched (99.6 atom%) in ^{15}N. It was acquired at 273.5 K using π/2 (14 μs) pulses with a cycling period of 12 s. The strong orthosilicate peak at −71.0 ppm appeared in every spectrum. However, the peak at −131.5 ppm occurred only in the fourth spectrum and was considerably narrower than in Figure 13a, the half-height line-width being only 6 Hz. The signal accounted for ca. 5-10% of the detected Si. (c) Silicon-29 ^{1}H-decoupled NMR spectrum (99.4 MHz) at 273.5 K of the nutrient medium in Figure 13b immediately following the addition of 16 ppm ^{29}Si. No peak is apparent in the vicinity of −131.5 ppm. (See vertical expansion.)

the spectral accumulations an additional signal for silicon in five or six fold coordination was observed. Further experimentation using both solution and solid state NMR spectroscopy of silicon alone and in conjunction with other NMR active nuclei (^{1}H, ^{13}C, $^{14/15}$N) may prove beneficial in our search for the mechanisms by which silicas with controlled structure, form and chemistry are produced in the biological environment.

All of the above-mentioned techniques have many applications in the study of biomineral structures and provide us with clues as to the mechanisms by which such well controlled amorphous gel and particulate structures are formed. We must remember that the chemistry involved in the formation of such well-ordered amorphous materials is regulated by the organisms concerned. We will now consider the two classes of mineralizing organisms that live in water (freshwater and marine) and for which there is a significant amount of knowledge concerning both the mineral phase and the process of mineral formation.

SILICA FORMATION IN SPONGES

Introduction to sponges including structural chemistry

Sponges (the Porifera) are multi-cellular (differentiated) sedentary filter-feeder organisms that occur in both fresh water and marine environments. In the marine environment they can be found in both shallow waters and deep sea environments up to 300 meters where the environment for precipitation will/could be very different as concentrations of dissolved species, temperature, pressure and light levels differ from those found in shallow waters. They possess a well constructed and complex network of water connecting channels and choanocyte chambers, which are lined with the flagellated choanocyte cells (Brusca and Brusca 1990). Sponges can process their own volume of water in ~5 seconds (reviewed in Vogel 1977) thus supplying their cells with oxygen and food and allowing for the elimination of toxic gases. The network of water containing channels will be the route by which silicon, probably as orthosilicic acid will enter the organism.

There are three classes of sponges, two of which produce silicified spicules as their body support (the Hexactinellida and the Demospongiae) and one of which (the Calcarea) that produces calcium carbonate spicules. Spicules are secreted in specialized cells known as sclerocytes. Silicified spicules in the Demospongiae are deposited within a membrane bounded compartment around an axial organic filament but no organic axial filament is found in spicules from Calcarea. The locus of formation is different for these two classes of sponges with the siliceous spicules being deposited intracellularly and the calcium carbonate spicules being produced extracellularly around a number of sclerocyte cells (reviewed in Simpson 1984). The concentration of calcium in seawater is ca. 10 mM and the concentration of silicon is of the order of 2.5% of this value so it is perhaps surprising that silicon is used in the production of a mineral framework! One reason may be the high concentration of polyphosphate found in/with silicified sponges (probably produced by symbiotic bacteria that are not found in calcareous sponges) that is absent in calcareous sponges and would be counterproductive to the production of a carbonate containing calcium mineral (Müller et al. 2003). It is worth noting that work of Müller and others on model "primmorph" systems where silicon provided in the form of sodium hexafluorosilicate at a concentration of 60 micromolar promoted spicule formation but silicon provided as tetraethoxysilane at the same concentration did not promote spicule formation. However, no detailed information on how the silicon containing solutions were prepared has been provided in any of the publications describing this effect (Krasko et al. 2000).

Species may contain many different spicule types, often separated into megascleres and microscleres with the former appearing to have a more important structural role. An example of a spicule is shown in Figure 14. The megascleres, or spicules that have a defined structural role, can vary hugely in size with the smallest only being a few microns in width and length and the largest being ca. 8 mm in diameter and 3 meters in length (Levi et al. 1989). The microsceleres are usually significantly smaller than the megasceleres and are often more ornate. The shapes, sizes and numbers of spicule types secreted by a species is of paramount importance in sponge taxonomy as they are assumed to not vary within a species. However, a note of caution should be added here as studies of spicule formation under silicon-poor and silicon-rich conditions have found that "the concentration of silicic acid in seawater modulates the phenotypic expression of the various spicule types genetically available in a sponge species" (Maldonado et al. 1999). A decline in Mesozoic reef-building sponges was explained by silicon limitation.

In the Demospongiae the spicules are covered with "spongin" a collagenous protein material that also serves to cement the structure together, more about this later. In the Hexactinellida, for which much less is known, the organic content of the living sponge is significantly lower than for the other group of siliceous sponges and the network of spicules is connected one to another via additional silica deposits. No studies of the silcifying process itself have been conducted on this system. More is known for the Demospongiae and in particular for the sponges, *Tethya aurantia*, *Suberites domuncula* and *Geodia cydonium*.

Silicateins

Investigations into the nature of an intra-silica matrix have been performed on several sponges For example, the sponge *Tethya aurantia* possesses silica spicules several millimeters in length that constitute 75% of the dry weight of the organism (Shimizu et al. 1998). Demineralization of the spicules with aqueous buffered HF yields a central protein filament which can be further dissociated into three similar sub-units, referred to as silicatein (for silica protein) alpha, beta and gamma. The amino acid composition of the purified extracts contains 20 mol% acidic residues, 16–20 mol% serine/threonine and ~15 mol% glycine. The more abundant alpha subunit of the sponge filament proteins is 218 amino acids in length and has been identified as a novel member

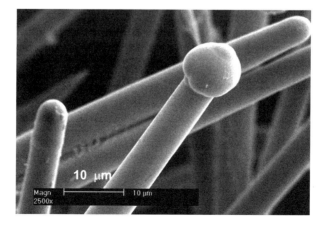

Figure 14. Spicules from *Suberites domuncula* with kind permission from Professor WEG Müller.

of the cathepsin L subfamily of papain-like cysteine proteases (Shimizu et al. 1998). If the amino acid sequence of human cathepsin L and silicatein alpha are compared then identical amino acids are found at 50% of the positions and amino acids with genetic or functional similarities are found at 75% of the positions. Similar proteins isolated from the sponge *Suberites domuncula* (Krasko et al. 2000) have 70% identical and 79% similar (identical plus physico-chemically related amino acids) with the *T. aurantia* silicatein polypeptide. In contrast to a cathepsin isolated from the sponge *G. cydonium* (Krasko et al. 1997) the silicatein sequences have two potential transmembrane helices whose role is as yet not understood. The silicatein proteins have distinct hydrophobic domains on their molecular surfaces and the macroscopic silicatein filaments are suggested to form as a result of hydrophobic interactions between subunits. Hydrophobic domains are not found on the surface of human cathepsin L, that is known to function as a monomer. An additional feature of the silicateins extracted from these sponges is a distinctive clustering of hydroxy amino acids (serine) (SSRCSSSS or Ser–Ser–Arg–Cys–Ser–Ser–Ser–Ser or OH–OH–N^+–SH–OH–OH–OH–OH), which strengthens the idea that these proteins are molecular templates for biosilicification. The proteases catalyze hydrolysis reactions (the reverse of condensation), and it is proposed that the silicateins catalyze condensation reactions.

Activity of silicateins. As silicateins are highly homologous to members of a family of hydrolytic enzymes, they have been tested for catalytic activity during silicification. The active site for cathepsin L has an asparagine (amide side chain), an histidine (basic side chain) and a cysteine group (sulphydryl side chain). Two of the three amino acids at the active site were exactly conserved in silicateins from the different sponge systems with the third, cysteine, being replaced by serine (hydroxyl side chain). In vitro studies using silicon alkoxides as the silicon source found that silicateins (as individual units or present in filaments as extracted from the sponge spicules) could catalyze the hydrolysis and subsequent condensation reactions to form silica at neutral pH (efficient reaction of the alkoxides normally requires acid or base catalysis) and ambient temperature (Cha et al. 1999). It should be noted that estimates of catalytic activity were made based on the amount of silica that could be collected by centrifugation after a certain time period which could relate to extent of aggregation as much as to extent of condensation as the collection by centrifugation really only identifies silicas moieties that are "beyond a certain size." The filaments were found to exhibit both catalytic activity and a "template-like activity" in which the small nanometer sized silica particles were found to "pattern" the underlying organic proteinaceous filament. The catalytic behavior was dependent on the three dimensional conformation of the protein (subunits and/or filaments) as thermal denaturation prior to the silicification assay removed all catalytic activity.

Genetic engineering techniques have been used to produce the silicatein alpha subunit from a recombinant DNA template cloned in bacteria (Cha et al. 1999; Zhou et al. 1999; Krasko et al. 2000). Structural variants of the silicatein protein produced using site directed mutagenesis (Zhou et al. 1999) have shown that specific serine and histidine side chains are required at the suspected active site for optimal activity in the catalysis and formation of silica. The proposed mechanistic action involves the active site hydroxyl and imidazole nitrogen acting in concert. It is suggested that the formation of a hydrogen bond between these two groups will enhance the nucleophilicity of the hydroxyl group thus facilitating its attack on the silicon of an alkoxide to form a transitory protein-silicon covalent intermediate, see Figure 15. Hydrolysis of this intermediate generates a reactive silanol species that is used in subsequent condensation reactions while regenerating the hydrogen bonded pair at the protein's active site. This mechanism although aesthetically pleasing is unlikely as alkoxycompounds are not found in the natural environment.

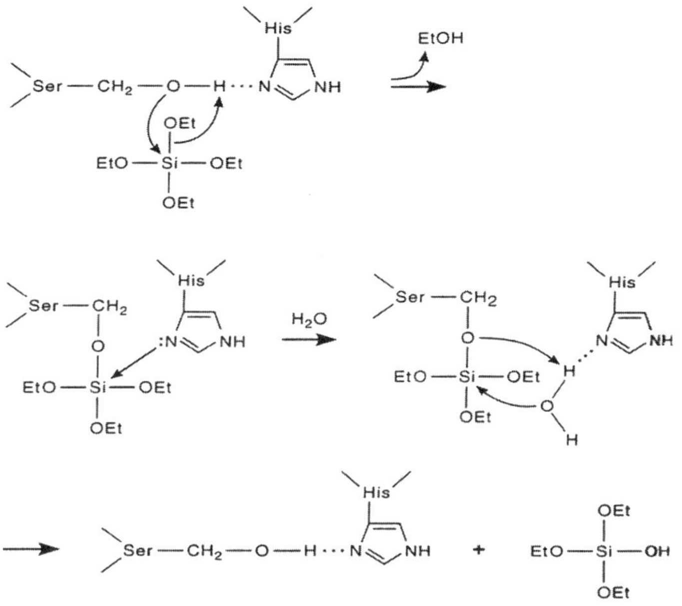

Figure 15. Mechanism of action of silicatein showing the key role of the amino acids present at the active site that are required for activity of the enzyme. Published with permission from Zhou et al. 1999 and Wiley-VCH.

Collagen, spiculogenesis and the effect of silicon concentration

Collagen is also required for the formation of the functional skeleton in sponges, particularly the Demospongiae where the spicules are glued together with a collagenous "cement" made of microfibrils (Garrone 1969). The Hexactinellida are different in that their spicules are largely held together with further deposits of silica. Collagens are extracellular matrix proteins that have a triple helical structure that readily form fibrils. Collagens typically contain a (Gly–X–Y)$_n$ (where X and Y are other amino acids) repeating sequence. Examples in the sponge world are n = 24 for *Suberites domuncula* (Schroder et al. 2000) and *E. Muelleri* with n = 79 (Exposito and Garrone 1990).

The formation of siliceous spicules requires the synthesis of both silica and collagen and has been found to only take place if silicon and iron (in available form) are present in the surrounding medium (Müller et al. 2003). The optimum concentration of silicon is between 5 and 100 micromolar, although this is species dependent (Jewell 1935) and detailed studies by Maldonado et al (1999) have shown that not only is silicon essential for spicule formation but that it also modifies the rich diversity of spicule types produced by a particular organism. As an example, Figure 16 shows examples for *Crambe crambe* (Schmidt) grown under different silicon concentrations. In explanation, for most natural populations of C Crambe only one spicule type is observed, namely small needles called styles (Maldonado et al. 1999), however, specimens in some localities also produce larger styles, C-shaped spicules (called isochelae) and centrally-branched aster-like desmas. An explanation for this variability has been obtained by laboratory studies using different concentrations of silicic acid in the growth medium (natural concentrations, ~0.03–4.5 micromolar), and elevated concentrations (~30 and 100 micromolar). The sponges reared in the control or "natural silicon concentration" cultures produced simple styles only but

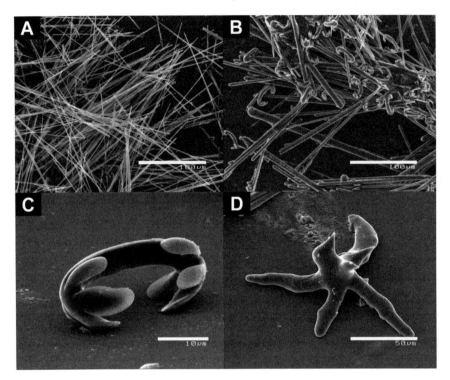

Figure 16. Examples of spicules formed under different concentrations of silicic acid. (A) Control (3 micromolar), only one needle-like spicule type called a style. (B) (30 micromolar) Three spicule types: small styles, Large styles and small C-shaped spicules called isochelae. (C) Larger version of an isochelae showing teeth. (D) Aster-like desma with a developed centrum from samples grown in 100 micromolar silicic acid. Scale bars for (A) and (B) are 100 microns, for (C) 10 microns, and for (D) 50 microns. Modified from Maldonado et al. 1999 and published with permission of Nature Publishing Group.

sponges grown in higher concentrations of silicic acid produced an abundance of at least three different spicule types although some of the spicules appeared deformed for the sponges grown at the highest concentration of silicic acid, perhaps suggesting problems with silicon transport under these conditions. The authors postulated that the changing composition of sea water, in respect of silicon concentration, caused, in part by the rise of the importance of diatoms in geological time around the Cretaceous-Tertiary boundary (Lowenstam and Weiner 1989) may have been a factor in the loss of many varieties of siliceous sponges from shallow waters where silicon concentrations were dramatically reduced. It is also of note that the sponge *Suberites domuncula* grows best at a silicon concentration of 60 micromolar (Simpson et al. 1985). Silicon concentration is also found to affect expression of a gene encoding collagen (the structural protein) and a morphogen, myotrophin (Krasko et al. 2000). Figure 17 shows the relative expression of both collagen and silicatein in the model primmorph system.

The formation of the sponge skeleton is mediated by an Fe^{2+} dependent enzyme (Holvoet and Van de Vyver 1985) and an iron chelator, 2,2′-bipyridine, when added to the sponge medium can inhibit skeletogenesis (as above). Model experiments on *Suberites domuncula* have shown that iron concentrations of 10 micromolar and above stimulate the formation of spicules and stimulate DNA and protein, including ferritin,

Transcript levels for silicatein   Transcript levels for collagen-1

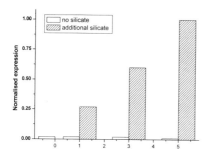
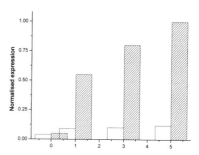

Figure 17. Levels of transcription in the model primmorph system for silicatein and collagen after 1, 3 and 5 days growth in the presence and absence of additional silicate at a concentration of 60 micromoles per liter. The relative degree of expression is correlated to that seen for maximal expression (after 5 days in the presence of silicate) (Krasko et al. 2000).

synthesis (Krasko et al. to be published). Silicatein which is required for the spicule axial filament and an integrin/scavenger receptor thought to be one of/the docking molecule for the collagen matrix into which the spicules are embedded were not upregulated in the presence of the iron supplement (Müller et al. 2003) suggesting that the role for iron is in proliferation and morphogenesis of cells rather than directing the synthesis of silicatein.

Model studies: primmorphs: studies of the effect of silicon and germanium on spiculogenesis

A model cell system has been developed to study spiculogenesis and other cellular processes in sponges with application to higher animals (Müller et al. 1999). The model primmorphs are obtained after the following procedure. Sponge tissue samples are transferred into calcium and magnesium free artificial seawater containing EDTA where they dissociate into single cells. These are transferred into natural seawater supplemented with antibiotics and 0.1% (v/v) of Marine broth 2216 (Difco). The dissociated cells retain their proliferation potency. Starting from single cells, primmorphs of at least 1 mm in diameter, with an average of 2 to 5 mm are formed after two days that have an external layer of epithelial cells and different cells in the interior of the primmorph. For a given growth period, primmorphs grow larger in silicon supplemented medium and in the presence of iron(III) (Müller et al. 2003). This system has been used to show that silicon from silicate at a concentration of sixty micromoles per liter upregulates the expression of collagen and silicatein and the primmorphs are able to form spicules of similar dimensions to those observed in the natural sponge (Krasko et al. 2000). This system has also been used to show that germanium at sixty micromoles per liter has no effect on silicatein expression and does not interfere with silicate-induced silicatein expression. Thus, where germanium is found to inhibit spicule formation (Simpson et al. 1979) and modify spicule type (Simpson et al. 1985) its effect must be felt at the level of the enzyme mediated silica deposition stage.

Use of this system in other areas of this field of research may enable a fuller picture of many of the processes involved in silicification to be obtained and a deeper understanding of, e.g., transport, concentration of the element and the maintenance of "soluble" pools, mechanisms of polymerization, particle-particle interactions, the role of membranes and biopolymers etc. in this "animal" system to be understood. Figure 18 contains a postulated mechanism for the formation of a functional skeleton in siliceous sponges including the stage that is most likely to be affected by germanium (Krasko et al. 2000)

Diatoms: introduction and structural information

Diatoms are single celled algae of the class Bacillariophyceae and there are more than 100,000 species known today. They are found in both sea water and fresh water and are a dominant component of the phytoplankton. They did not exist before the Permian-Triassic extinction of ~250 Ma ago and are implicated in the change of pattern of location of many sponge types as their blooms drastically reduced the silicon levels of the oceans where sponges had previously dominated. They have been studied since the advent of microscopy in the early 18th century and the studies performed on them have largely been dictated by the techniques available. As time has gone on the process of formation has been explored progressively by more advanced microscopic techniques (e.g., electron microscopy, atomic force microscopy) and by chemical, biochemical and cell biology techniques. The latter has been as important in the study of these organisms as in the study of silicified sponges discussed above.

The diatom cell wall, or frustule has two halves with the upper half called the epitheca and the lower half the hypotheca. The upper and lower surfaces of the frustule are known as valves and the structures extending on the sides and overlapping the two halves of the cell are the girdle bands (Fig. 19). The valves may be perforated with numbers of pores, known as areolae and both the pore system and the number of girdle bands is species-specific. Diatoms also vary in the extent of their silicification. There are two general classes of diatoms, the centrics and the pennates, thus defined according to their general symmetry (the centrics are radially symmetric, the pennates have an axis of symmetry) with the pennates commonly exhibiting a central slit known as the raphe fissure through which

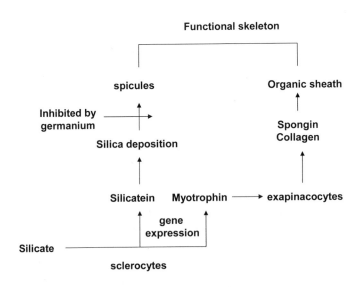

Figure 18. A possible mechanism for the formation of a functional skeleton in a siliceous sponge such as *Suberites domuncula*. Silicate is assumed to cause the expression of silicatein and myotrophin in the sclerocytes, the cells involved in silicification. N.B. Silicified spicules and collagen are cemented together to form the functioning skeleton (Krasko et al. 2000).

Silicification: Organism Mineralization of Silica 311

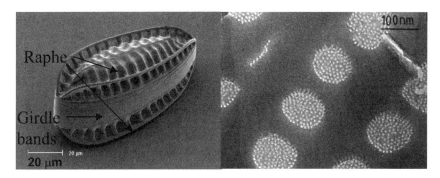

Figure 19. Field emission scanning electron microscope image of the diatom *Surirella sp.* to show the two valves, the raphes and girdle bands. Transmission electron microscope image of the diatom *Navicula pelliculosa* showing examples of the areolae. With the kind permission of Professor Rick Wetherbee.

mucilage is extruded to assist in movement, Figure 19. The silica from which these structures are built up is protected from the external environment by an organic matrix. Although the silica is defined as being "amorphous" at the angstrom level, microfibrillar and hexagonal columnar arrangements have been observed during the development of certain diatoms (Li and Volcani 1985). At maturity certain structures are built up from ~5 nm particles (Perry 1989) and small aggregates around 40 nm (Crawford et al. 2001), Figure 20. Structural studies that have used selective dissolution with sodium hydroxide at high pH have shown that the silica used to build up the diatom frustules varies within the frustule and also between species. The silica in diatoms has been proposed to exhibit proton buffering activity in the oceans, thus enabling the efficient conversion of inorganic bicarbonate to carbon dioxide (Milligan and Morel 2002).

The diatom reproductive cycle, Figure 21 results in the formation of two daughter cells, each retaining half of the original frustule with a new valve being formed within the original cell structure prior to separation of the daughter cells and completion of the girdle bands. At this point it is important to note that the first stage in the process of

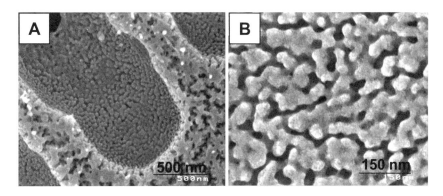

Figure 20. Scanning electron microscope image of ca. 40 nm particles visible within the diatom *Pinnularia sp.* cell wall after treatment with mild alkali. With the kind permission of Professor Rick Wetherbee.

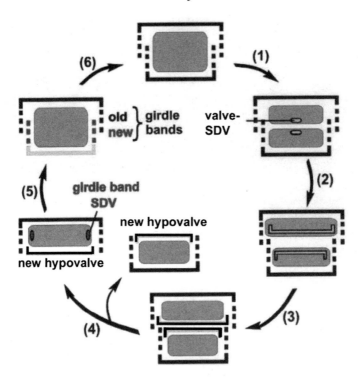

Figure 21. The diatom reproductive cycle. Modified from Kröger and Wetherbee (2000) and used with permission of Urban & Fischer Verlag. Diatom cells are represented schematically in cross section. The frustule is composed of two halves, the epitheca and the hypotheca. Each theca consists of a valve together with several girdle bands. The extent of silicification and the number of girdle bands are species-specific. Stage (1) Division of the cytoplasm and the formation of a silica deposition vesicle within each daughter protoplast. (2) Growth of the silica deposition vesicles and concomitant formation of the new valves. (3) Exocytosis of the newly formed valve. (4) Separation of the daughter cells and the formation of silica deposition vesicles for the girdle bands. (5) Consecutive formation and secretion of the girdle bands. (6) DNA replication. Note, in the early stages of development the silicalemma is represented as thick black lines but at maturity it is no longer visible nor separable from the silica skeleton.

reproduction involves the cytoplasm of the enlarged original or mother cell dividing to provide two localities for the formation of one of each of the required epitheca or hypotheca. Each of the separated cytoplasmic regions "knows" which half of the frustule it is supposed to make! The new valves are formed within silica deposition vesicles of unknown origin that are bounded by a membrane called the silicalemma, the exact nature of which is unknown. It is pertinent to note that any entry of silicic acid and other biochemical components to the developing valve has to take place through the main body of the cytoplasm and for the silicic acid, through the existing valves as the two developing valve structures of the daughter cells are opposed to one another within the mother cell.

The first stage in silicification: transport of the raw ingredients into the cell!

Demonstrations that diatoms could take up silicon from the environment were presented in the mid 50's (Lewin 1955, 1957). In the late 90's evidence that silicic acid is the form in which silicon is generally sequestered were published (Del Amo and Brzezinski 1999) although it was noted that specific diatoms could also take up the ionized form as well. In marine diatoms, uptake is linked to the transport of sodium ions in a 1Si:1Na ratio (Bhattacharyya and Volcani 1980). Silicic acid uptake is specifically induced just prior to cell wall silicification (Sullivan 1977). More recent studies have defined and sequenced a family of five silicic acid transporters (SIT 1–5) (Hildebrand et al. 1997, 1998) from one specific diatom, *Cylindotheca fusiformis*. These transporters have highly conserved hydrophobic regions, a signature amino acid sequence for sodium symporters and a long hydrophilic carboxy-terminus (Hildebrand et al. 1997; Hildebrand 2000). The carboxy-terminal segment in all five SITs had a very high probability to form a coiled-coil structure, suggesting that this portion of the protein interacts with other proteins, It is proposed that this end of the molecule may be involved with control of activity and localization (Hildebrand et al. 1998). The most significant amount of variation between the five SIT proteins was in the C-terminus region suggesting that the different SIT proteins might be found in distinct locations and perform different roles during the formation of the silicified cell wall. Exact information on specific locations for the individual SITs is at present unavailable so this hypothesis cannot be validated.

In most diatom species silicic acid transport and cell wall silica deposition are temporally coupled (Chisholm et al. 1978) and the model that satisfies the experimental data is that of "internally controlled uptake" that requires a feedback mechanism operating through saturation or near saturation of intracellular pools (Conway and Harrison 1977). The rate of silicic acid uptake is largely independent of external silicic acid concentration, but for exceptions see below, and depends on the rate of utilization or cell wall deposition. More recently it has been proposed that regulation of uptake may not be determined by the absolute levels of silicic acid in the intracellular pools but by the relative ratio of silicic acid to silicic acid binding components (Hildebrand 2000). According to this model, silicic acid for the cell wall is drawn from internal pools as needed with flux from the pools connecting the silicification process with uptake. In this model, uptake does not drive silicification. Although the components of this model are consistent with one another, there is likely to be some effect of external silicon concentration on uptake and transport as the overall extent of silicification for a diatom species can be affected by external silicic levels (Martin-Jezequel et al. 2000) in much the same way as has been found for sponges.

Internal silicon pools

Intracellular pools of soluble silicon were identified in the mid 1960's by Werner (Werner 1966). There have been difficulties in measuring the soluble silicon pool due to problems in the estimation of cell water volume and hence water content and thus widely differing measures of cell silicon have been reported. Reliable estimates from Hildebrand (Hildebrand 2000) where the assumption has been made that the soluble silicon, measured as orthosilicic acid is evenly dispersed throughout the cell, gives a range of 19–340 mM for a range of diatoms, Figure 22. For an individual diatom these values can also change in a regular manner during the course of cell wall synthesis (Hildebrand and Wetherbee 2003). The solubility limit for orthosilicic acid is of the order of 2 mM and may drop to ca. 0.5 mM (Harrison and Loton 1995) in the presence of specific positively charged ions including ammonia/amine counterions thus the concentrations measured are all in vast excess of this and questions arise as to the location of the silicic acid and to how such high levels can be maintained?

Experiments performed on *Nitzschia alba* and *Cylindrotheca fusiformis* (Mehard et al. 1974) showed that silicic acid was able to move around the cell freely as experiments with radiolabelled silicic acid co-purified with a range of cell organelles and subcellular fractions. Three possibilities for the maintenance of high solution concentrations are, silicic acid is stabilized as the monomer or as small oligomers by complexation (the use

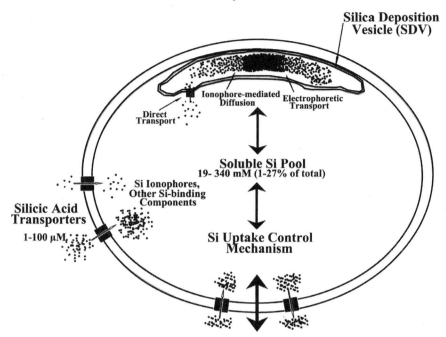

Figure 22. Model of silicic acid uptake, its intracellular transport and its deposition within the silica deposition vesicle. With the kind permission of Professor Mark Hildebrand.

of ionophores or other complexing agents including catechol (1,2-dihydroxybenzene) and its derivatives), the packaging of the silicic acid within small membrane bounded vesicles, or that the internal chemistry of the diatom has a composition as yet unknown that is able to "cope with" such high concentrations of silicic acid! The experimental methods that have been used to obtain estimates of internal silicic acid concentrations have utilized dilution (helps to depolymerize colloidal and oligomeric species) and prolonged heating at the boiling point of water (again, aiding depolymerization), thus when the molybdate method for orthosilicic acid detection is used, perhaps the concentrations of orthosilicic acid appear higher than they really are in the living organism? A different approach to the measurement of cellular silicon concentrations has been to use isotopically labelled silicon 29 and use ^{29}Si NMR spectroscopy to identify soluble silicon species inside the diatom (Kinrade et al. 2002), Figure 13. Some published data suggest that hypervalent silicon may be present within silicifying cells of *Navicula pelliculosa*. Complexation to give hypervalent silicon containing species is a possible chemical strategy by which silicic acid may be retained in solution until required for cell wall synthesis. The importance of "silicon transport vesicles" (Schmid and Schulz 1979) in the preservation of soluble silicon pools is somewhat doubtful as microscopic evidence suggests that there are an insufficient number to provide the silicic acid required for cell wall synthesis (Li and Volcani 1985) and there would still be a requirement for maintenance of readily soluble silicon, perhaps by complexation of the silicic acid within the vesicles.

The silica deposition environment

For diatoms, the silicon deposition environment is within the silica deposition vesicle whose structure and composition are unknown largely due to problems with its extraction from the siliceous valve (hypotheca or epitheca) formed within the shaped vesicle. However, some information on the chemical environment present in the SDV is known as a result of the use of a fluorescent, cationic, lipophilic dye rhodamine 123 that can accumulate within the SDV that indicates precipitation under acidic conditions around pH 5 (Vrieling et al. 1999). The molecules, other than silica within the SDV are unknown.

Armed with the knowledge that the formation of species-specific biominerals other than silica requires the services of biopolymers including proteins and carbohydrates a search has been carried out to find similar regulatory molecules for silicifying organisms including diatoms. For the production of silicified structures, the "control" molecules need to regulate nucleation (location and time of deposition in relation to the organism/cell cycle), growth and cessation of growth. Early observations came from the study of silicifying diatoms where a marked increase in protein concentration in the cell wall was found during silica deposition with carbohydrate incorporation only occurring after silicification (Coombs and Volcani 1968). Nakajima and Volcani also identified the presence of three unusual amino acids, namely 3,4-dihydroxyproline, ε-N-trimethyl-δ-hydroxylysine and its phosphorylated derivative from the cell walls of different diatoms including *Cylindrotheca fusiformis* (Nakajima and Volcani 1969, 1970). They were not able to proceed further with their investigations as they could not isolate the proteins containing these species. Other experiments designed to chemically inhibit protein synthesis during valve development produced morphological abnormalities, suggesting the requirement for *de novo* protein synthesis in biosilicification (Blank and Sullivan 1983). Although the SDV cannot be isolated, researchers have found ways to explore the identity of biopolymers associated with the silica phase. Protein containing extracts have been isolated from various freshwater and marine diatoms and their amino acid composition determined (Hecky et al. 1973; Swift and Wheeler 1992). Treatment of "cleaned" diatoms with buffered HF to dissolve the silica phase released material enriched in serine/threonine (ca. 25 mol%), glycine (ca. 25 mol%) and acidic residues (ca. 20 mol%) (Swift and Wheeler 1992). (It is perhaps notable that none of these researchers identified the unusual amino acids found by Nakajima and Volcani in their extracts). The model proposed by Hecky et al. (1973) envisaged that the serine/threonine-enriched protein would form the inner surface of the silicalemma and, as such, would present a layer of hydroxyls, facing into the SDV, onto which orthosilicic acid molecules could condense. This initial layer of geometrically constrained orthosilicic acid molecules would promote condensation with other orthosilicic acid molecules. Carbohydrates were also found in extracts from the diatoms and this was variously assigned a range of roles including action as a physical buffer between the organism and the aquatic environment, providing resistance to chemical and bacterial degradation and functioning as an ion exchange medium (Hecky et al. 1973).

More recently extremely sophisticated structural studies have been performed on a range of isolates from the diatom cell wall of *Cylindrotheca fusiformis* and three classes of proteins involved with the diatom cell wall identified. Of these groups of molecules, the frustulins with molecular weights ~75–200 kDa are mainly involved in forming a protective coat around the diatom and will not be discussed further (Kröger et al. 1994; Kröger and Sumper 2000). The second family of proteins are known as HEPs (hydrogen fluoride extractable proteins) or pleuralins (Kröger et al. 1997). The apparent molecular weights of these molecules are 150–200 kDa and they are highly acidic molecules with a structure comprising a signal peptide, a proline-rich (> 65%) region, a C-terminal domain and repeats of proline (> 22%)-serine (> 11%)-cysteine (> 11%) aspartate (> 9%) rich domains. The molecules have been sequenced and antibodies raised against the PSCD-domains so that experiments to investigate the cellular localization of these molecules could be performed. Immunoelectronmicrosopy studies showed that for the specific pleuralin, HEP200 investigated, it is localized only where the epitheca overlaps the hypotheca. It is important to note that the molecule could not be localized during the developmental stages leading to silicification of the girdle bands suggesting that the role of this family of molecules is not with silicification but that it may be involved in establishing a reversible connection between the epitheca and hypotheca, Figure 23. During cell reproduction, this connection must be broken for the new hypotheca and epitheca to form but when these new valves have been formed the connection must again be closed so that the mechanical stability of the daughter cell can be maintained.

The third group of proteins, the low molecular weight (ca. 4–17 kDa (although these molecular weights will have been revised recently when a milder extraction agent was used (Kröger et al. 2002) silaffins (1A, 1B and 2) can also be released from the diatom cell wall using hydrogen fluoride and the sil–1–gene has been sequenced (Kröger et al. 1999, 2001). Structural information obtained for the sil–1 encoded polypeptide displays a modular primary structure that comprises an N-terminus signal peptide, an acidic region having high negative charge and a series of repeat units with a high content of post-translationally modified lysine and serine. Although the sil–1 polypeptide contains several different functional domains, only the repeat units containing high levels of post-translationally modified lysine and serine are found in the polypeptides isolated from diatoms, Figure 24. The structure of these low molecular weight species varies according to the extraction method used (Kröger et al. 2002). The use of a milder extractant, ammonium fluoride, previously used by others (Swift and Wheeler 1992; Harrison 1996) has yielded the most comprehensive information on this family of molecules (Kröger et al. 2002) with both lysine and serine showing post-translational modifications. A range of detailed structural methods including high resolution NMR (proton and multinuclear) and mass spectrometric techniques were required to provide unequivocal evidence for the proposed structures (Kröger et al. 1999, 2001). The chemical modifications of lysine involve the ε-amino group and give rise to ε-N-dimethyllysine or δ-hydroxy-ε-N,N,N-trimethyllysine residues or a modification that has a methyl group on the ε-amino group together with an oligomeric structure consisting of N-methylpropylamine units where between 5 and 10 methylpropylamine groups are attached to the lysine residue (Kröger et al. 1999, 2001). The ε-N-trimethyl-δ-hydroxylysine residue or its phosphorylated derivative as identified by

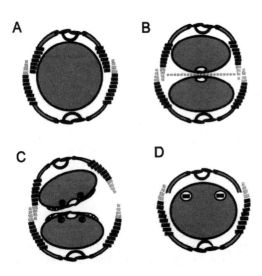

Figure 23. A model scheme showing the overlap of girdle bands and their separation during cell wall development. Published with permission of Urban & Fischer Verlag from Kröger and Wetherbee 2000. (A) intact diatom, (B) cleavage of daughter cells, (C) separation of daughter cells, and (D) beginning of interphase. Pleuralins are involved in the opening and closing of the two theca during reproduction and the development of daughter cells.

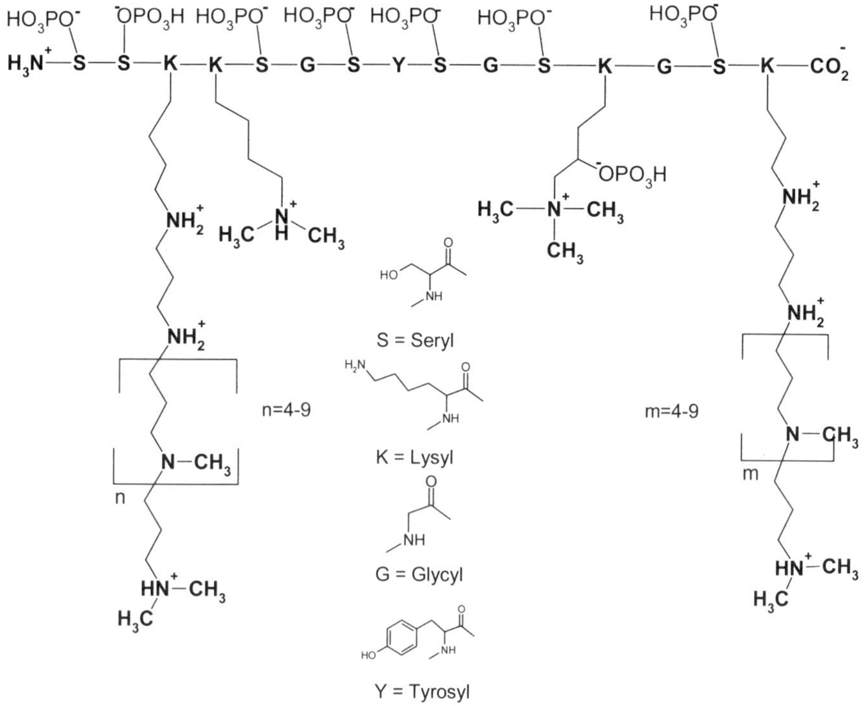

Figure 24. A schematic structure of Nat-Sil-1A, after Kröger et al. 2002.

Nakajima and Volcani have not been identified in the highly detailed study performed by Kröger, Sumper and coworkers. The other form of poststranslational modification involves the serine residues and these, together with the hydroxyl group on the modified lysine residues are thought to be phosphorylated (Kröger et al. 2002). The polypeptide has a high positive charge arising from the number of amino groups but it also has a high negative charge (each phosphorylated amino acid will contribute one unit of negative charge) such that the polypeptide will have little overall net charge.

When similar extraction procedures were carried out on a series of other diatoms from a broad range of species (Kröger et al. 2000) a further class of molecules were identified whose molecular weights were less than 3.5 kDa and with the exception of *Cylindrotheca fusiformis*, appear to be either one of or the main silaffin type component of all the other diatoms investigated. Again, electrospray mass spectrometry helped in the identification of the compounds which were found to be linear polyamines composed of N-methylaminopropyl or aminopropyl units. An example structure is shown in Figure 25. The largest number of polyamines was isolated from *Nitzschia angularis* with molecular weights between 0.6–1.25 kDa (Kröger et al. 2000).

What are the roles for the various extracted biopolymers (sponges and diatoms)?

The formation of silica based materials with specific shape and form requires manipulation and control of the polymerization process of an amorphous covalently bonded polymeric material. When the mineral structure of biosilicas is examined by microscopy (electron and atomic force microscopy) all structures seem to be built up from

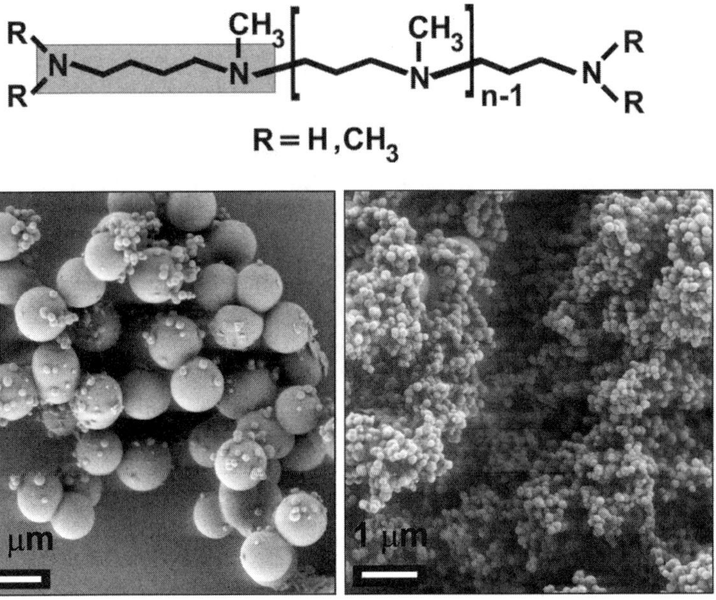

Figure 25. A generalized structure of the polyamines and scanning electron microscope images of silica structures formed in the presence of low molecular weight polyamines (left hand side) and higher molecular weight polyamines (right hand side). From Kröger et al. 2000 and used with permission of the PNAS.

particles that have defined shape and size (e.g., Perry 1989; Perry and Fraser 1991; Wetherbee et al. 2000) and in some instances, degree of surface hydration (Perry et al. 1990). Fundamental particles are of the order of 3–5 nm in diameter and for diatoms, spherical objects of the order of ca. 40 nm (exact size ranges are species-specific) have been detected during the formation of the diatom cell wall by scanning electron microscopy and can still be identified in mature walls using atomic force microscopy. Figure 11 shows examples of silica particles visible by these two different methods. For sponges, precipitation occurs around a proteinaceous filament within a membrane bounded compartment and for diatoms, precipitation again occurs in a membrane bounded compartment in the presence of polypeptide/polyamine/proteins. Investigations of silicification in higher plants have shown that here too, silica deposits are closely associated with proteinaceous material (Harrison 1996; Perry and Keeling-Tucker 1998, 2000, 2003).

In some organisms, the precise location of isolated proteinaceous material has been identified (see section on pleuralins) and a functional role for these molecules proposed that is not directly involved with silica deposition. For all other biosilica proteinaceous materials, the precise role that the isolates play in silicification is not known *in vivo*. For this reason, *in vitro* studies of silicification have been performed using the biosilica extracts in an attempt to understand the role of the isolated biomolecules.

In the model studies of silicification silicon sources have included alkoxysilanes (Zhou et al. 1999; Cha et al. 1999; Perry and Keeling-Tucker 2000), functionalized alkoxysilanes (Cha et al. 1999), metastable solutions of silicic acid (e.g., Kröger et al. 1999), solutions of sodium silicate (Swift and Wheeler 1992) and solutions of a catecholato complex of silicon (Perry and Keeling-Tucker 1998, 2003) Experimental approaches used to monitor the effect of biosilica extracts on silicification include kinetic studies of the early stages of oligomerization, measurement of the mass of centrifugable material obtained after a certain reaction time, the ability of the biopolymer extracts to bind silica and structural studies of rhe silica produced. Very early studies utilizing isolates from diatoms (Swift and Wheeler 1992) did not find any effect on the rate of polymerization of supersaturated solutions of orthosilicic acid at neutral pH contrary to the findings of more recent experiments.

Studies on the effect of silicatein α extracted from sponge spicules on polysiloxane synthesis at neutral pH have shown a ca. 10-fold increase in the amount of harvestable silica compared to experiments run in the presence of the denatured silicatein or proteins such as bovine serum albumin and trypsin (Cha et al. 1999). These studies did not report on whether the protein was occluded within the silica. A catalytic role for the silicatein in the polycondensation of the siloxanes was proposed in a manner analogous to the action of serine- and cysteine-based proteases. The suggested mechanism was that of a general acid/base catalyst with a specific requirement for both serine and histidine amino acids at the active site (Zhou et al. 1999). A scaffolding or structure-directing activity was also reported, as the polymerized silica was able to form a layer following the contours of the underlying protein filament (as extracted from the sponge), behavior that was not observed for the reaction performed in the presence of cellulose or silk fibroin fibers (Cha et al. 1999).

Experiments performed using low molecular weight silaffins (Kröger et al. 1999) in the presence of phosphate buffers have shown that these molecules accelerate the silica condensation process and aid in flocculation, the biopolymer extract being occluded in the silica formed. Silaffins were able to cause a metastable solution of silicic acid to spontaneously aggregate, with small ~50 nm diameter spherical particles being produced in the presence of mixtures of silaffins (1A, 1B and 2) with particles ~700 nm in diameter being produced in the presence of a solution containing a purified silaffin 1A. The silaffins, in the presence of a phosphate buffer, were found to be effective at promoting silicification at sub-neutral pH values down to pH 4. N.B. a pH value of around 5 has been measured for the lumen of the silica deposition vesicle (Vrieling et al. 1999). The provision of highly positively charged side chains has been thought to be very important in enabling the silaffins to promote condensation and aggregation under mildly acidic conditions. Most recently, phosphorylated silaffins, isolated when milder extraction methods were used to remove the siliceous phase have been examined for their silica forming ability (Kröger et al. 2002). NatSil–1A is able to precipitate silica from a monosilicic acid solution in the absence of additional phosphate (change of buffer systems) and the activity of silaffin–1A can be totally restored if an excess of phosphate is added to the system. As before, 400–700 nm diameter silica spheres were formed in a short reaction time. The authors conclude that the numerous phosphate groups in natSil–1–A serve as an intrinsic source of anions required for silica formation by diatoms. Further, they provided evidence that the zwitterionic structure of natSil–1–A (polyamine moieties and phosphate groups) form ionic strength dependent aggregates via electrostatic interactions. They hypothesize that silaffin self-assembly provides a template for silicic acid polycondensation and have shown through a detailed microscopic study the formation of silica spheres from an initial "plastic" silica–natSil–1–A phase. They believe that this "plasticity" may be important in the molding of diatom silica for all structural components other than those that require patterning on the nanoscale to produce nanoscale pores, for example. Other *in vitro* experiments have been conducted separately on the polyamines extracted from *Nitzschia angularis* (the species examined to date for which the largest number of polyamines have been found). These have also been investigated for their ability to promote silica formation (Kröger et al. 2000). A range of spherical silica particles were formed, the size and distribution of sizes depending on the pH of the reaction mixture and the molecular weight of the polyamine fraction used. For experiments conducted at a fixed pH (pH 5) smaller sized particles were found when lower molecular weight polyamines (600–700 Da) were used. If pH is increased from ca. 5 to 8 then particle sizes fall from ca. 700 nm to 50 nm, all other experimental conditions remaining constant, Figure 25. It will be interesting to see what effect the combination of natSil–1–A and polyamines will have on silica formation, on its rate of precipitation and the form of the silica deposited.

How to go from molecular level chemical control to macroscopic control?

In all biomineralized organisms, chemical control at the molecular control is manifest in structural regulation at the visible level. It is still a mystery how three-dimensional structures are produced from information contained in chemical form (Pickett-Heaps et al. 1990). In all the studies described above, no mention has been made of the processes by which specific features in the diatom (or for that matter the sponge) are formed. An excellent review on this subject, amongst others by Hildebrand and Wetherbee (2003) is a good starting point for this subject. The entire cell components seem to be involved in formation of the silicified frustule with its many innate structural features including raphe, areolae, girdle bands etc. The process of silicification,, in conjunction with all the other stages of cell division are tightly coordinated. In order to study the relationships between cell organelles and silica deposition electron microscopy has proved invaluable as it is a technique that is able to visualize all components within the cell, albeit only in a static mode and in two dimensions. Microtubules, microfilaments and mitochondria have all been shown to occupy specific locations prior to and during silicification (Hildebrand and Wetherbee 2003 and references therein). The pattern of deposition in valves is generally from the outside in (Cox 1999) but there are examples for both centric and pennate diatoms where the deposition pattern is the opposite way round (Cox 1999; Schmid and Volcani 1983; Crawford and Schmid 1986). Pore patterns are also under cellular control and mathematical models based on the maintenance of optimal packing have been used to explain observed pore patterns (Longuet-Higgins 2001). Observation of diatom structures in detail suggests that a substantial amount of molding is involved in their formation. Evidence for this comes from the study of frustules where some surfaces appear smooth and some rough. The smooth surfaces are proposed to arise from silica being shaped directly by the silicalemma, or the outer membrane structure of the silica deposition vesicle. The shape of the silica deposition vesicle is not only governed by the chemical composition and structure of those molecules proposed to be involved in silica deposition. Microtubules, derived from actin and microfilaments are also involved (Chiappino and Volcani 1977; van der Meene and Pickett-Heaps 2002). Their role is in the determination of size and in manipulation of the shape of the silica deposition vesicle, however, much remains to be discovered concerning the relationship between the "chemistry" involved in silica precipitation and its subsequent manipulation within the silica deposition vesicle to produce objects that have both form and function. A model has been proposed by Sumper (2002) to explain the nanopatterning of silica in diatoms via phase separation. The polyamines isolated from a range of diatoms (Kröger et al. 2000) are methylated and exhibit amphiphilic properties. The model postulates the existence of repeated phase separation processes within the silica deposition vesicle which produce emulsions of microdroplets and, subsequently, nanodroplets or micelles consisting of a polyamine-containing organic phase. It is proposed that the contact sites between polyamine droplets and the aqueous phase that contains the hydrophilic silicic acid promotes silicic acid polymerization, the resulting silica occupying the spaces between the droplets. Modification of the size and packing of the droplets will then lead to different patterns of silica being observed.

Progress towards understanding biosilicification at the molecular level in other systems

Other silicifying systems are also being studied to increase our understanding of how silica that is produced in specific shape and form is produced. Silica is found in many higher plants and can occur in cell walls, cell lumens, intracellular spaces, roots, leaves and other spaces. The point of entry of the silicic acid (or other precursor) is through the roots and for rice, specifically through the lateral roots and not through the root hairs (Ma et al. 2001). N.B. transport processes in plants will of necessity be more complex than in diatoms as additional modes of transport are needed to enable long-distance transport across specialized tissues and compartments from the root to the site of deposition (Richmond and Sussman 2003). Siliceous remains or phytoliths are important objects in species identification and in the study of the use of ancient sites by man. Progress in the understanding of silicification is not so far advanced as for sponges and diatoms but studies in the early 90's identified that in common with other mineralizing organisms, plant biosilicas contained (within the mineral) biomolecules that were proposed to be involved in silica formation (Harrison 1996). Studies have been made on the primitive group, the Equisetaceae and for single cell plant hairs from the outside seed coating of *Phalaris canariensis*, the plant grown to obtain grain for canaries (e.g., Harrison 1996) ! Both of these plant systems show evidence of regulated silica production in the formation of different structural motifs in different regions of their silicified structures (Perry and Fraser 1991; Perry et al. 1984).

Extraction of biopolymers intimately associated with the silica has been performed utilizing a range of concentrated acids and mild heating in sequence to remove cell wall materials not intimately associated with the siliceous mineral phase prior to dissolution of the silica with aqueous, buffered HF (Harrison 1996). Specific glycosylation patterns were not lost on dissolution of the mineral phase using this procedure and xylose and glucose were the monosaccharides found in highest abundance. Three biopolymer extracts were obtained with distinctive chemical composition. The most readily released is enriched in serine/threonine (25 mol%), glycine (20 mol%) and acidic residues (25 mol%). The second extract has lower levels of glycine and hydroxyl-containing residues, with the glycine being largely replaced by proline and contains a ca. two-fold increase in the levels of lysine. A third, insoluble material is also obtained which contains even higher levels of lysine (up to 26 mol%), high levels of proline and high levels of aliphatic amino acids. We have proposed that this latter material may be involved in the regulation of nucleation. The other two groups of materials that contain higher levels of hydroxylated amino acids (such as serine) are proposed to be involved in the regulation of particle growth. These extracts have also been used in model studies where the emphasis was on quantitation of the effect of the extracts on the rates of the early stages of silicic acid condensation as well as on the structure of the silicas that are produced (Perry and Keeling-Tucker 1998, 2000, 2003). An acceleration of the initial stages of the condensation process was observed and particle growth within the early stages of experiments was larger than for control systems that might be expected if the initial condensation rates had increased. At later stages of reaction, the biopolymer phase was found to limit particle growth and a narrower distribution of particles sizes was observed. Other unusual structures were observed and for the first time, crystalline silica has been formed from an aqueous solution at ca. neutral pH and at room temperature. The lath-like crystalline structures observed suggested a "soft" or perhaps layered material but the diffraction information (electron diffraction) suggested that the material formed was quartz. When the organic material embedded/ incorporated in the silica produced in the model reaction was released and analyzed its composition suggested a β-sheet type conformation. It is possible that the biopolymers provide a nucleating structure from which the silica phase continues to develop although it must be noted that crystalline silica is not observed in the many silicified organisms examined to date and must be prevented by the organism as a crystalline material is inherently more difficult to mold than an amorphous covalent polymer.

The completion of genome mapping for silicon accumulating plant species including rice (Goff et al. 2002; Yu et al. 2002) will make it possible to identify which genes are silicon dependent and perhaps also to identify any genes involved with the mineralization process. It will also be possible to look for synergy or not with other cellular processes including lignin synthesis.

Unanswered questions, a selection

In diatoms, the silica deposition vesicle is the location where silica is produced and molded to generate species-specific structures. It provides a specialized microenvironment that requires transport of silicic acid (most likely) and all required organic moieties across the silicalemma in order to generate the silicified structures that are then molded by the "expandable reaction vessel" in conjunction with other cytoskeletal structures such as microtubules and microfilaments. Energy is required for the process and mitochondrial centers are located close to points of silicification. The composition of the membrane is not known, nor is the mechanism by which it is able to expand and maintain its integrity during frustule formation. When the daughter cells separate with their new valve components (N.B. the system knows to make the correct "half" of the frustule!) it is thought that the silicalemma becomes all or part of an organic casing surrounding the silicified components of the mature wall (Pickett-Heaps et al. 1990). Other components that may be present in the outer coating include sulphated polysaccharides and protein or proteoglycans (McConville et al. 1999) and mucilage (Hoagland et al. 1993). A range of techniques including electron microscopy, atomic force microscopy and biochemical separation procedures have been used to try to elucidate the structure and chemical composition of the silicalemma but with little success and much work is required in this area. The availability of genomic sequences for diatoms will be invaluable in this study as specific genes can be "switched off" and the consequences of this investigated so that structural information specific to the silicalemma, its composition, its three dimensional structure and function can be elucidated. N.B. the first steps in gene manipulation have been achieved (Zaslavskaia et al. 2001) but much remains to be done. Gene mutation techniques and techniques that enable the "addition" of specific genes will both be needed in order to unequivocally establish the functions encoded by specific genes.

Similarly in sponges, the availability of genomic sequence data will enable the identification of genes specifically involved in all stages of the biosilicification process and by mutation and removal/insertion of specific genes and a measure of the organisms response to external stimuli (e.g., silicon in some form). Eventually these experiments should be possible for plants although it is likely that more immediate success in the field will be achieved by the route, extract protein, find function with model *in vitro* studies, clone gene etc.

Although there is now more information on the likely chemical environment in which silica is formed in nature, the relationship between biochemical structure and their precise function *in vivo* are unknown. We still do not know if the identified molecules are the only ones involved in the formation of nanoscale silica structures. We also do not know how such silica particles might be manipulated to produce structures on higher length scales. Probes to investigate reactions *in situ* are required (for example see Shimizu et al. 2001) as are approaches that will look at the whole cell system and find the relationships between chemistry at the molecular level and cytoskeletal behavior. Patterning of silica is still an area that is little understood.

Where to from here?

Silica formation in the natural environment requires the uptake, transport, condensation, growth, precipitation and molding of species involving silicic acid. Significant progress has been made in respect of transport and in respect of the role(s) played by biomacromolecules in the controlled formation of silica structures. However, much is not understood and there are many unresolved controversies concerning uptake, transport and the specific role of silica associated biomolecules such that there is plenty of scope for scientists wishing to enter the field and to make a contribution to the understanding of the processes required to make and mold an amorphous mineral that is able to withstand pressure (sponges are found at depths of ~300 m), dissolution (for all organisms found in water) and predation (water- and land-based species) by a range of animals. The information obtained in this search will, however, assist us in our understanding of the essentiality of silicon to life processes and in the generation of new materials with specific form and function for industrial application in the twenty-first century. A new approach to this area has been the use of a combinatorial phage peptide display library to select for peptides that bind silica (Naik et al. 2002). The information so gained from these studies may help us in our search for biomolecules necessary for the production of organized silicified structures in a wide range of biological organisms.

ACKNOWLEDGMENTS

The author would like to thank ESRC, AFRC, BBSRC, Ineos Chemicals, The EU and AFOSR for their interest and funding of research programs in biological and biomimetic silica chemistry. Members of my research group (David Belton, Neil Shirtcliffe, and Jian Xu) are thanked for help with preparation of figures.

BIBLIOGRAPHY

Bauerlein E (2003) Biomineralization of unicellular organisms: an unusual membrane biochemistry for the production of inorganic nano- and microstructures. Angewandte Chemie 42(6):614–641 (a review of biomineralization—iron oxide, calcium carbonate and silica—pertaining to single cell organisms)

Hildebrand M, Wetherbee R (2003) Components and control of silicification in diatoms. *In:* Silicon Biomineralization: Biology-Biochemistry-Molecular Biology-Biotechnology, Progress in Molecular and Subcellular Biology. Vol 28. Müller WEG (ed) Springer, New York, p 11–57 (a review on the "state of the art" in all aspects of diatom biomineralization research)

Jones JB, Segnit ER (1971) Nature of opal. I. Nomenclature and constituent phases. J Geol Soc Aust 18:57–68

Kröger N, Sumper M (2000) The biochemistry of silica formation. *In:* Biomineralization from biology to biotechnology and medical application. Bauerlein E (ed) Wiley-VCH, Weinheim, Germany, p 151–170 (a review on the role of the organic matrix in diatom silicification

Perry CC (2001) Silica in Encyclopedia of Life Sciences http:/www.els.net London, Nature Publishing Group, London (précis on silica in biological organisms)

Perry CC, Keeling-Tucker T (2000) Biosilicification: the role of the organic matrix in structure control. J Biol Inorg Chem 5:537–550 (a review on the specifics of silica chemistry, the organic matrix in sponges, diatoms and plants, *in vitro* reactions and molecular modeling)

Simpson TL, Volcani BE (eds) (1981) Silicon and Siliceous Structures in Biological systems, Springer-Verlag, New York (a book containing many wonderful images of silicified organisms) specific chapters of relevance are chapter 2 on silicon in the cellular metabolism of diatoms by Sullivan and Volcani, Chapter 6 on the siliceous components of the diatom cell wall and their morphological variation by Crawford, Chapter 7 on cell wall formation in diatoms, morphogenesis and biochemistry, Chapter 16 on the form and distribution of silica in sponges by Hartmann and Chapter 17 on ultrastructure and deposition of silica in sponges by Garrone, Simpson and Pottu-Boumendil. The book, although dated contains much relevant information to the study of silicification in diatoms and sponges as well as providing information on silicification in other organisms.

REFERENCES

Bhattacharyya P, Volcani BE (1980) Sodium-dependent silicate transport in the apochlorotic marine diatom *Nitzschia alba*. Proc Natl Acad Sci USA 77:6386–6390

Blank GS, Sullivan CW (1983) Diatom mineralization of silicic acid. VI. The effects of microtubule inhibitors on silicic acid metabolism in *Navicula saprophila*. J Phycol 19:39-44

Brinker CJ, Scherrer GW (1990) Sol-Gel Science—The Physics and Chemistry of Sol-Gel Processing, Academic Press, London

Brusca RC, Brusca GJ (1990) Invertebrates. Sinauer Associates Publishers, Sunderland, UK

Cha JN, Shimizu K, Zhou Y, Christiansen SC, Chmelka BF, Stucky GD, Morse DE (1999) Silicatein filaments and subunits from a marine sponge direct the polymerization of silica and silicones in vitro. Proc Natl Acad Sci USA 96:361–365

Chiappino ML, Volcani BE (1977) Studies on the biochemistry and fine structure of silica shell formation in diatoms. VIII. Sequential cell wall development in the pennate *Navicula pelliculosa*. Protoplasma 93:205–221

Chisholm SW, Azam F, Eppley RW (1978) Silicic acid incorporation in marine diatoms on light:dark cycles: use as an assay for phased cell division Limnol Oceanogr 23:518–529

Conway HL, Harrison PJ (1977) Marine diatoms grown in chemostats under silicate or ammonium limitation. IV. Transient response of *Chaeteroceros debilis*, *Skeletonema costatum* and *Thalossiosira gravida* to a single addition of the limiting nutrient. Mar Biol 43:33–43

Coombs J, Volcani BE (1968) Studies on the biochemistry and fine structure of silica-shell formation in diatoms. Chemical changes in the wall of *Navicula pelliculosa* during its formation. Planta 82:280–292

Cox EJ (1999) Variation in patterns of valve morphogenesis between representatives of six biraphid diatom genera. J Phycol 35:1297–1312

Crawford RM, Schmid AM (1986) Ultrastructure of silica deposition in diatoms. In: Biomineralization in lower plants and animals. Vol 30. Leadbetter BS, Riding R (eds) System Soc. Oxford Univ Press, Oxford, p 291-314

Crawford SA, Higgins MJ, Mulvaney P, Wetherbee R (2001) Nanostructure of the diatom frustule as revealed by atomic force and scanning electron microscopy. J Phycol 37:543–554

Del Amo Y, Brzezinski MA (1999) The chemical form of dissolved Si taken up by marine diatoms. J Phycol 35:1162–1170

Engelhardt G, Hoebbel D, Tarmale M, Samoson A, Lippmaa E (1982) Si-29 NMR investigations of the anion structure of crystalline tetramethylammonium-aluminosilicate and aluminosilicate solutions. Z Anorg Allg Chem 484:22–32

Exposito JY, Garrone R (1990) Characterisation of a fibrillar collagen gene in sponges reveals the early evolutionary appearance of two collagen families. Proc Natl Acad Sci USA 87:6669–6673

Frondel C (1962) The System of Mineralogy of DANA, 7th Edition. Vol 3. Wiley, New York

Garrone R (1969) Collagene, spongine et squelette mineral chez l'eponge *Haliclona rosea*. J Microscop 8:581–598

Goff SA, Ricke D, Lan TH, Presting G, Wang R, Dunn M, Glazebrook J, Sessions A, Oeller P, Varma A et al. (2002) A draft sequence of the rice genome (*Oryza sativa L. ssp. Japonica*) Science 296:92–100

Harrison CC (1996) Evidence for intramineral macromolecules containing protein from plant silicas. J Phytochem 41:37–42

Harrison CC, Loton N (1995) Novel routes to designer silicas-studies of the decomposition of $(M^+)_2[Si(C_6H_4O_2)_3] \cdot xH_2O$, the importance of M^+ identity on the kinetics of oligomerization and the structural characteristics of the silicas produced. J Chem Soc Faraday Trans 91:4287–4297

Hecky RE, Mopper K, Kilham P, Degens ET'(1973) The amino acid and sugar composition of diatom cell walls. Marine Biology 19:323–331

Hench LL, West JK (1990) The sol-gel process. Chem Rev 90:33–72

Hildebrand M (2000) Silicic acid transport and its control during cell wall silicification in diatoms. In: Biomineralization: from biology to biotechnology and medical application. Baeuerlein E (ed) Wiley-VCH, Weinheim p 171-188

Hildebrand M, Volcani BE, Gassmann W, Schroeder JI (1997) A gene family of silicon transporters. Nature 385:688–689

Hildebrand, M, Dahlin K, Volcani BE (1998) Characterization of a silicon transporter gene family in *Cylindrotheca fusiformis*: sequences, expression analysis, and identification of homologs in other diatoms. Molec General Genetics 260:480–486

Hildebrand M, Wetherbee R (2003) Components and control of silicification in diatoms. In: Silicon Biomineralization: Biology-Biochemistry-Molecular Biology-Biotechnology. Progress in Molecular and Subcellular Biology Vol. 28. Müller WEG (eds) Springer, New York, p 11–57

Hoagland KD, Rosowski JR, Gretz MR, Roemer SC (1993) Diatom extracellular polymeric substances: function, fine structure, chemistry and physiology. J Phycol 29:537–566

Holvoet S, Van de Vyver G (1985) Effects of 2,2'-bipyridine on skeletogenesis of *Ephydatia fluviatilis*. In: New perspectives in sponge biology. Rutzler K (ed) Smithsonian Institution Press, Washington DC p 206–210

Holzuter G, Narayama K, Gerber T (2003) Structure of silica in *Equisetum arvense*. Anal Bioanal Chem 10.1007/S00216-003-1905-2

Iler RK (1979) The Chemistry of Silica. Plenum Press, New York

Jewell ME (1935) An ecological study of the fresh water sponge of northern Wisconsin. Ecol Monogr 5:461–504

Jolivet JP (2000) Metal Oxide Chemistry and Synthesis from Solution to Solid State. John Wiley, Chichester

Jones JB, Segnit ER (1971) Nature of opal. I. Nomenclature and constituent phases. Jour Geol Soc Aust 18:57–68

Kamatani A (1971) Physical and chemical characteristics of biogenous silica. Mar Biol 8:89–95

Kinrade SD, Gillson A-ME, Knight CTG (2002) Silicon-29 NMR evidence of a transient hexavalent silicon complex in the diatom *Navicula pelliculosa*. J Chem Soc Dalton 3:307–309

Knight CTG, Kirkpatrick RJ, Oldfield E (1986) The unexpectedly slow approach to thermodynamic equilibrium of the silicate anions present in aqueous tetramethylammonium silicate solution. J Chem Soc Chem Commun 11:66–67

Krasko A, Gamulin V, Seack J, Steffen R, Schröder HC, Müller WEG (1997) Cathepsin, a major protease of the marine sponge *Geodia cydonium*: purification of the enzyme and molecular cloning of cDNA. Molec Marine Biol and Biotechnol 6:296–307

Krasko A, Batel R, Schröder HC, Müller WEG (2000) Expression of silicatein and collagen genes in the marine sponge *Suberites domuncula* is controlled by silicate and myotrophin. Europ J Biochem 267:4878–4887

Kresge CT, Leonowicz ME, Roth WJ, Vartuli JC, Beck JS (1992) Ordered mesoporous molecular sieves synthesized by a liquid-crystal template mechanism. Nature 359:710–712

Kröger N, Bergsdorf C, Sumper M (1994) A new calcium binding glycoprotein family constitutes a major diatom cell wall component. The EMBO Journal 13:4676–4683

Kröger N, Deutzmann R, Bergsdorf C, Sumper M (2000) Species-specific polyamines from diatoms control silica morphology. Proc Natl Acad Sci USA 97:14133–14138

Kröger N, Deutzmann R, Sumper M (1999) Polycationic peptides from diatom biosilica that direct silica nanosphere formation. Science 286:1129–1132

Kröger N, Deutzmann R, Sumper M (2001) Silica-precipitating peptides from diatoms, the chemical structure of silaffin-1a from *Cylindrotheca* fusiformis. J Biol Chem 276:26066-26070

Kröger N, Lehmann G, Rachel R, Sumper M (1997) Characterization of a 200-kDa diatom protein that is specifically associated with a silica-based substructure of the cell wall. Eur J Biochem 250:99–105

Kröger N, Lorenz S, Brunner E, Sumper M (2002) Self-assembly of highly phosphorylated silaffins and their function in biosilica morphogenesis. Science 298:584–586

Kröger N, Sumper M (2000) The biochemistry of silica formation. *In:* Biomineralization from biology to biotechnology and medical application. Bauerlein E (ed) Wiley-VCH, Weinheim, Germany, p 151–170

Kröger N, Wetherbee R (2000) Pleuralins are involved in theca differentiation in the diatom *Cylindrotheca fusiformis*. Protist 151:263-273

Lewin JC (1955) Silicon metabolism in diatoms III. Respiration and silicon uptake in *Navicula pelliculosa*. J Gen Physiol 39:1–10

Lewin JC (1957) Silicon metabolism in diatoms IV. Growth and frustule formation in *Navicula pelliculosa*. Can J Microbiol 3:427–433

Lewis DW, Wilcock DJ, Catlow CRA, Thomas JM, Hutchings GJ (1996) De novo design of structure-directing agents for the synthesis of microporous solids. Nature 382:604–606

Levi C, Barton JL, Guillemet C, le Bras E, Lehuede P (1989) A remarkably strong natural glassy rod—the anchoring spicule of the monraphis sponge. J Mat Sci Lett 8:337–229

Li CW, Volcani BE (1985) Studies on the biochemistry and fine structure of silica shell formation in diatoms VIII, morphogenesis of the cell wall in a centric diatom *Ditylum brightwelli*. Protoplasma 124:10–29

Longuet-Higgins MS (2001) Geometrical constraints on the development of a diatom. J Theoret Biol 210:101–105

Lowenstam HA, Weiner S (1989) On Biomineralization. Oxford University Press, New York

Ma JF, Goto S, Tamai K, Ichii M (2001) Role of root hairs and lateral roots in silicon uptake by rice. Plant Physiol 127:1773–1780

Maldonado M, Carmona MC, Uriz MJ, Cruzado A (1999) Decline in Mesozoic reef-building sponges explained by silicon limitation. Nature 401:785–787

Mann S, Williams RJP (1982) High resolution electron microscopy studies of the silica lorica in the choanoflagellate *Stephanoeca diplocostata* Ellis. Proc Roy Soc Lond B216:137–146

Mann S, Perry CC, Williams RJP, Fyfe CA, Gobbi GC, Kennedy GJ (1983) The characterisation of the nature of silica in biological systems. J Chem Soc Chem Commun 168–170

Mann S, Perry CC, Webb J, Luke B, Williams RJP (1986) Structure, morphology, composition and organization of biogenic minerals in limpet teeth. Proc Roy Soc Lond B227:179–190

Martin-Jezequel V, Hildebrand M, Brzezinski MA (2000) Silicon metabolism in diatoms: implications for growth. J Phycol 36:821–840

McConville MJ, Wetherbee R, Bacic A (1999) Subcellular localization and composition of the wall and secreted extracellular sulphated polysaccharides/proteoglycans of the diatom *Stauroneis amphioxys* Gregory. Protoplasma 206:188–200

Mehard CW, Sullivan CW, Azam F, Volcani BE (1974) Role of silicon in diatom metabolism IV. Subcellular localization of silicon and germanium in *Nitzschia alba* and *Cylindrotheca fusiformis*. J Physiol 30:265–272

Milligan AJ, Morel FMM (2002) A proton buffering role for silica in diatoms. Science 297:1848–1850

Miyaki F, Davis SA, Charmant JPH, Mann S (1999) Organic-crystal templating of hollow silica fibers. Chem Material 11:3021–3024

Müller WEG, Wiens M, Batel, Steffen R, Borojevic R, Custodio MR (1999) Establishment of a primary culture from a sponge: Primmorphs from *Suberites domuncula*. Marine Ecol Progr Ser 178:205–219

Müller WEG, Kraski A, Le Pennec G, Steffen R, Wiens M, Shokry MAA, Müller IM, Schroder HC (2003) Molecular mechanism of spicule formation in the demosponge *Suberites domuncula*: silicatein-collagen-myotrophin. *In*: Prog Molec Subcell Biol. Vol 33. Müller WEG (ed) Springer, New York, p 195–221

Naik RR, Brott LL, Clarson SJ, Stone MO (2002) Silica-precipitating peptides isolated from a combinatorial phage display peptide library. J Nanosci Nanotech 2:95–100

Nakajima T, Volcani BE (1969) 3,4-dihydroxyproline: a new amino acid in diatom cell walls. Science 164:1400–1401

Nakajima T, Volcani BE (1970) Eta-N-trimethyl-L-gamma hydroxylysine phosphate and its nonphosphorylated compound in diatom cell wall.s Biochem Biophys Res Commun 39:28–33

Mizutani T, Nagase H, Fujiwara N, Ogoshi H (1998) Silicic acid polymerization catalyzed by amines and polyamines. Bull Chem Soc Jpn 71:2017–2022

Naik RR, Brott LL, Clarson SJ, Stone MO (2002) Silica-precipitating peptides isolated from a combinatorial phage display peptide library. J Nanosci Nanotech 2:95–100

Perry CC (1989) Chemical studies of biogenic silica. *In:* Biomineralization, Chemical and Biological Perspectives. Mann S, Webb J, Williams RJP (eds) VCH, Weinheim, p 233–256

Perry CC, Mann S, Williams RJP (1984) Structural and analytical studies of silicified macrohairs from the lemma of the grass *Phalaris canariensis* L. Proc Roy Soc Lond B222:427–438

Perry CC, Moss EJ, Williams RJP (1990) A staining agent for biological silica. Proc Roy Soc Lond B241:47–50

Perry CC, Fraser MA (1991) Silica deposition and ultrastructure in the cell wall of *Equisetum arvense*: the importance of cell wall structures and flow control in biosilicification. Phil Trans R Soc Lond B 334:149–157

Perry CC, Keeling-Tucker T (1998) Crystalline silica prepared at room temperature from aqueous solution in the presence of intrasilica bioextracts. J Chem Soc Chem Commun 2587–2588

Perry CC, Keeling-Tucker T (2000) Biosilicification: the role of the organic matrix in structure control. J Biol Inorg Chem 5:537–550

Perry CC, Keeling-Tucker T (2003) Model studies of colloidal silica precipitation using biosilica extracts from *Equisetum telmateia*. Colloid Polym Sci 281:652-664

Perry CC, Belton D, Shafran KL (2003) Studies of biosilicas: structural aspects, chemical principles, model studies and the future. *In:* Silicon Biomineralization: Biology-Biochemistry-Molecular Biology-Biotechnology. Progress in Molecular and Subcellular Biology vol 28. Müller WEG (ed) Springer, New York, p 269–299

Pickett-Heaps J, Schmid A-MM, Edgar LA (1990) The cell biology of diatom valve formation. *In:* Progress in pycological research, Vol 7. Round FE, Chapman DJ (eds) Biopress, Bristol, p 1-168

Richmond KE, Sussman M (2003) Got silicon? The non-essential beneficial plant nutrient. Curr Opin Plant Biol 6:268–272

Schmid A-MM, Schulz D (1979) Wall morphogenesis in diatoms: deposition of silica by cytoplasmic vesicles. Protoplasma 100:267–288

Schmid A-MM, Volcani BE (1983) Wall morphogenesis in *Coscinodiscus wailesii*. I Valve morphology and development of its architecture. J Phycol 19:387–402

Schmidt M, Botz, R, Rickert D, Bohrmann G, Hall SR, Mann S (2001) Oxygen isotopes of marine diatoms and relations to opal-A maturation. Geochim Cosmochim Acta 65:201–211

Schroder HC, Krasko A, Batel R, Skorokhod A, Pahler S, Kruse M, Müller IM, Müller WEG (2000) Stimulation of protein (collagen) synthesis in sponge cells by a cardiac myotrophin-related molecule from Suberites domuncula. FASEB J 14:2022–2031

Shimizu K, Cha J, Stucky GD, Morse DE (1998) Silicatein α: Cathepsin L-like protein in sponge biosilica. Proc Natl Acad Sci USA 95:6234–6238

Shimizu K, del Amo Y, Brzezinski MA, Stucky GD, Morse DE (2001) A novel fluorescent silica tracer for biological silicification studies. Chem Biol 8:1051–1060

Simpson TL (1984) The cell biology of sponges. Springer, Berlin Heidelberg New York

Simpson TL, Gil M, Connes R, Diaz JP, Paris J (1985) Effects of germanium (Ge) on the silica spicules of the marine sponge *Suberites domuncula*: transformation of the spicule type J Morphol 183:117–128

Simpson TL, Refelo LM, Kaby M (1979) Effects of germanium on the morphology of silica deposition in a freshwater sponge. J Morphol 159:343–354

Simpson TL, Volcani BE (eds) (1981) Silicon and Siliceous Structures in Biological Systems. Springer, New York

Stöber W, Fink A, Bohn, (1968) Controlled growth of monodisperse silica spheres in the micron size range. Colloid Interface Sci 26:62–69

Sullivan CW (1977) Diatom mineralization of silicic acid (II) Regulation of silicic acid transport rates during the cell cycle of *Navicula pelliculosa*. J Phycol 13:86–91

Sumper M (2002) A phase separation model for the nanopatterning of diatom biosilica. Science 295:2430–2433

Swift DM, Wheeler AP (1992) Evidence of an organic matrix from diatom biosilica. J Phycol 28:202–209

Tarutani T (1989) Polymerization of silicic acid: a review. Anal Sci 5:245–252

Van der Meene AML, Pickett-Heaps JD (2002) Valve morphogenesis in the centric diatom *Proboscia alalta* Sundstrom J Phycol 38:351–363

Vogel S (1977) Current-induced flow through living sponges in nature. Proc Natl Acad Sci USA 74:2069-2071

Vrieling EG, Gieskes WWC, Beelen TPM (1999) Silica deposition in diatoms: control by the pH inside the silica deposition vesicle. J Phycol 35:548–559

Vrieling EG, Beelen TPM, van Santen RA, Gieskes WWC (2000) Nano-scale uniformity of pore architecture in diatomaceous silica: a combined small and wide angle X-ray scattering study. J Phycol 36:146–159

Vrieling EG, Beelen TPM, van Santen RA, Gieskes WWC (2002) Mesophases of (bio)polymer-silica particles inspire a model for silica biomineralization in diatoms. Angew Chem Int Ed 41:1543–1546

Watt F, Grime GW, Brook AJ, Gadd GM, Perry CC, Pearce RB, Turnau K, Watkinson SC (1991) Nuclear microscopy of biological specimens. Nucl Instrum Meth Phys Res B54:123–143

Werner D (1966) Silicic acid in the metabolism of *Cyclotella cryptica*. Arch Mikrobiol 55:278–308

Wetherbee R, Crawford S, Mulvaney P (2000) The nanostructure and development of diatom biosilica. *In:* Biomineralization; from Biological to Biotechnology and Medical Applications. Baeuerlein E (ed), Wiley VCH, Weinheim, Germany, p 189–206

Yang H, Coombs N, Ozin GA (1997) Morphogenesis of shape and surface patterns in mesoporous silica. Nature 386:692–695

Yu J, Hu S, Wang J, Wong GK, Li S, Liu B, Deng Y, Dai L, Ahou Y, Zhang X et al. (2002) A draft sequence of the rice genome (*Oryza sativa L. ssp. indica*). Science 296:79–92

Zaslavskaia LA, Lippmeier JC, Kroth PG, Grossman AR, Apt KE (2001) Transformation of the diatom *Phaeodactylum tricornutum* with a variety of selectable markers and reporter genes. J Phycol 36:379–386

Zhou Y, Shimizu K, Cha JN, Stucky GD, Morse DE (1999) Efficient catalysis of polysiloxane synthesis by Silicatein α requires specific hydroxy and imidazole functionalities. Angew Chem Int Ed 38:780–782

11 Biomineralization and Evolutionary History

Andrew H. Knoll

Department of Organismic and Evolutionary Biology
Harvard University
Cambridge, Massachusetts, 02138 U.S.A.

INTRODUCTION

The Dutch ethologist Niko Tinbergen famously distinguished between proximal and ultimate explanations in biology. Proximally, biologists seek a mechanistic understanding of how organisms function; most of this volume addresses the molecular and physiological bases of biomineralization. But while much of biology might be viewed as a particularly interesting form of chemistry, it is more than that. Biology is chemistry with a history, requiring that proximal explanations be grounded in ultimate, or evolutionary, understanding. The physiological pathways by which organisms precipitate skeletal minerals and the forms and functions of the skeletons they fashion have been shaped by natural selection through geologic time, and all have constrained continuing evolution in skeleton-forming clades. In this chapter, I outline some major patterns of skeletal evolution inferred from phylogeny and fossils (Figure 1), highlighting ways that our improving mechanistic knowledge of biomineralization can help us to understand this evolutionary record (see Leadbetter and Riding 1986; Lowenstam and Weiner 1989; Carter 1990; and Simkiss and Wilbur 1989 for earlier reviews).

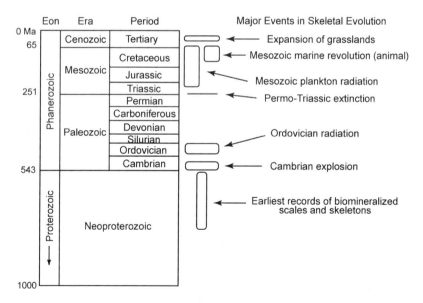

Figure 1. A geologic time scale for the past 1000 million years, showing the principal time divisions used in Earth science and the timing of major evolutionary events discussed in this chapter. Earlier intervals of time—the Mesoproterozoic (1600–1000 million years ago) and Paleoproterozoic (2500–1600 million years ago) eras of the Proterozoic Eon and the Archean Eon (> 2500 million years ago)—are not shown. Time scale after Remane (2000).

My discussion proceeds from two simple observations:
1. Skeletal biomineralization requires energy and so imposes a metabolic cost on skeleton-forming organisms.
2. Mineralized skeletons have evolved in many clades of protists, plants and animals.

If both statements are true, then clearly for many but not all eukaryotic organisms the biological benefits of biomineralization must outweigh its costs. But cost-benefit ratio is not static. It will change through time as a function of environmental circumstance. Thus, if mineralized skeletons confer protection against predators, the energy expended on biomineralization should vary through time as a function of predation pressure. And insofar as the metabolic cost of skeleton formation reflects the physical chemistry of an organism's surroundings, cost-benefit can also change as a function of Earth's environmental history.

THE PHYLOGENETIC DISTRIBUTION OF MINERALIZED SKELETONS

Figure 2 shows the phylogenetic relationships among eukaryotic organisms. The tree is incomplete and parts of it remain contentious. Nonetheless, it provides an essential roadmap for studies of skeletal evolution. The three principal classes of skeletal biominerals—calcium carbonates, calcium phosphates, and silica—all occur widely but discontinuously on the tree. This suggests either that gene cassettes governing biomineralization spread from group to group via horizontal transfer or that skeletal biomineralization evolved independently in different groups using, at least in part, component biochemical pathways that evolved early within the Eucarya.

Carbonate skeletons

Five of the eight major clades in Figure 2 contain species with mineralized skeletons, and all five include organisms that precipitate calcite or aragonite. Accepting this phylogenetic distribution, we can ask how many times carbonate skeletons originated within the eukaryotic domain. To arrive at an answer, we need to draw conclusions about skeletal homology, and we should also differentiate between skeletal structures and the underlying biochemical processes that govern their formation. Skeletons in two taxa can be accepted as independently derived if they share no common ancestor that was itself skeletonized. Thus, foraminiferans and echinoderms unambiguously record two separate origins of $CaCO_3$ skeletons—carbonate biomineralization evolved within the forams well after their differentiation from other cercozoans and, so, long after the opisthokont and cercozoan clades diverged. In contrast, the question of common ancestry for echinoderm and ascidian skeletons is less clear cut. In principle, the two clades could have shared a calcified deuterostome ancestor, although skeletal anatomies in the two groups—spicules in ascidians and the remarkable stereom of echinoderms—differ sharply. A conservative estimate is that carbonate skeletons evolved twenty-eight times within the Eucarya. This includes at least twenty origins among metazoans (as many as eight within the Cnidaria alone; e.g., Romano and Cairns 2000), four within the Plantae (two each in green and red algae), and one each within the cercozoans, haptophytes, heterokonts, and alveolates (the calcified cysts of some dinoflagellates). Given the diversity of problematic skeletons in Lower Cambrian rocks (e.g., Bengtson and Conway Morris 1992) and the likelihood that differing patterns of skeletal precipitation and discontinuous stratigraphic distributions reflect multiple origins of carbonate biomineralization in coralline demosponges (Reitner et al. 1997), the actual number could be significantly higher.

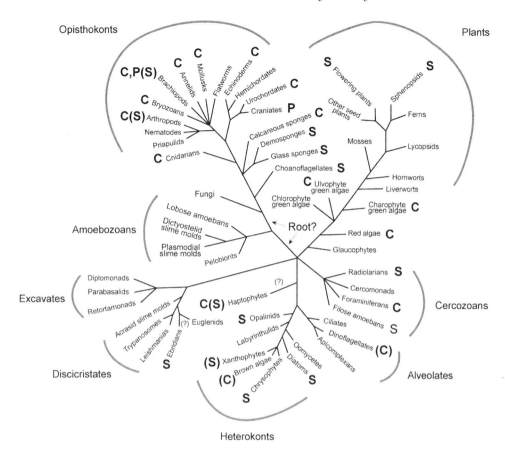

Figure 2. Molecular phylogeny of eukaryotic organisms, showing the phylogenetic distribution of mineralized skeletons. C = calcium carbonate minerals; P = calcium phosphate minerals; S = opaline silica. Letters in parentheses indicate minor occurrences. Phylogeny based principally on Mishler et al. (1994) for the Plantae; Giribet (2001) for opisthokonts; and Baldauf (2003) for the Eucarya as a whole. Data on skeletons principally from Lowenstam and Weiner (1989).

Phylogeny thus points toward repeated innovations in the evolution of carbonate skeletons, but this leaves open the question of homology in underlying molecular process. As pointed out by Westbroek and Marin (1998), skeleton formation requires more than the ability to precipitate minerals; precipitation must be carried out in a controlled fashion in specific biological environments. Skeletal biomineralization requires directed transport of calcium and carbonate, molecular templates to guide mineral nucleation and growth, and inhibitors that can effectively stop crystal growth. All cells share the ability to bind Ca^{2+} ions and regulate calcium concentrations (e.g., MacLennan et al. 1997; Sanders et al. 1999), and both photoautotrophic and heterotrophic eukaryotes regulate their internal inorganic carbon chemistry using carbonic anhydrase and other enzymes (e.g., Aizawa and Miyachi 1986). Thus, the biochemical supply of ions required for calcite or aragonite precipitation appears to be an ancient feature of eukaryotes. Organisms in which carbonate formation has been well studied also synthesize acidic proteins and glycoproteins that provide templates for mineralization (Weiner and Addadi 1997), guiding what in many cases appears to be essentially diagenetic crystal growth from amorphous precursors (Gotliv et al. 2003). Thus, many animals and protists with no skeleton-forming common ancestors nonetheless share the underlying capacity to synthesize and localize carbonate-inducing biomolecules, although homologies among these molecules remain uncertain.

Biochemical similarities extend, as well, to the molecules that inhibit $CaCO_3$ mineralization. Marin et al. (1996, 2000) proposed that organisms sculpt mineralized skeletons via "anti-calcifying" macromolecules that locally inhibit crystal growth. Because spontaneous calcification of cell and tissue surface may have been a problem in highly oversaturated Proterozoic oceans (Grotzinger 1989; Knoll et al. 1993), Marin et al. (1996) reasoned that molecular inhibitors evolved early as anti-calcification defenses and were later recruited for the physiological control of skeleton growth. [Even today, molecular inhibitors are necessary to retard spontaneous calcification of internal tissues in vertebrates; Luo et al. (1997).] Immunological comparison of anti-calcifying molecules in mollusks and cnidarians supports the hypothesis that this biochemistry already existed in the last common ancestor of cnidarians and bilaterian animals, if not before that.

Thus, in the view of Westbroek and Marin (1998), the many origins of calcareous skeletons reflect multiple independent cooptations of molecular and physiological processes that are widely shared among eukaryotic organisms. Skeletons that are not homologous as *structures* share underlying physiological pathways that are. This powerful idea has to be correct at some level. The details—how much biochemical innovation and diversity underlies carbonate skeletonization across clades—await an expanded program of comparative biological research.

Silica skeletons

Skeletons of opaline silica also enjoy a wide distribution among eukaryotes, occurring in five or (depending on the still debated phylogenetic relationships of ebridians) six of the eight great eukaryotic clades. Biologically, however, SiO_2 skeletons differ in distribution from those made of $CaCO_3$. Silica biomineralization is limited to intracellular precipitation. Thus, microscopic scales and skeletons are abundant in cercozoans (radiolarians and their relatives, as well as the siliceous scales of euglyphid amoebans), ebridians, and heterokonts (reaching their apogee in the exquisite frustules of diatoms). In animals, however, siliceous skeletons are limited to sponge spicules and minor occurrences such as the opalized mandibular blades of boreal copepods (Sullivan et al. 1975) and the micron-scale silica tablets formed intracellularly in the epidermis of some brachiopod larvae (Williams et al. 2001). Within Plantae, the carbonate biomineralization of marine and freshwater algae is replaced by silica phytolith mineralization in the epidermis of some vascular plants, especially grasses, sedges, and the sphenopsid genus *Equisetum* (Rapp and Mulholland 1992). Accepting that choanoflagellates and sponges could share a common ancestor that precipitated silica, we can estimate that silica skeletons evolved at least eight to ten times in eukaryotic organisms. Much has been learned in recent years about silica precipitation by diatoms (e.g., Kröger and Sumper 1998; Kröger et al. 2000, 2002; Zurzolo and Bowler 2001). Frustule formation requires active silica transport to local sites of precipitation (silica deposition vesicles), where opal deposition is mediated by enzymes. The vesicular spaces for mineral formation appear to be modifications of the Golgi-vesicle apparatus shared almost universally by eukaryotes. On the other hand, little is known about intracellular silica transport. Various molecules have been shown to promote the polymerization of silicic acid molecules—polyamines and silaffins in diatoms (Kröger et al. 2000), silicatein and collagen in sponges (Krasko et al. 2000). To date, however, we do not know enough about the comparative biology of these molecular processes to know the extent to which organisms that produce siliceous skeletons mirror carbonate precipitators in the recruitment of common biochemistries.

Phosphate skeletons

Calcium phosphate mineralization is widespread in the form of amorphous granules used in phosphate storage, gravity perception and detoxification (Gibbs and Bryan 1984; Lowenstam and Weiner 1989). In contrast, phosphatic skeletons are limited to animals and within this kingdom occur prominently only in lingulate brachiopods, vertebrates, and several extinct clades (see below). The absence of wider usage may reflect the metabolic extravagance of deploying nutrient phosphate for skeletons, especially in protists and plants. Phosphatic skeletons may best justify their cost when ontogeny requires constant remodeling, as it does in vertebrates.

Although phosphatic skeletons differ in evolutionary trajectory from those made of calcium carbonate, the two likely share much of the biochemical machinery underlying mineralization. As noted above, calcium transport is ubiquitous in eukaryotes, and so is the basic molecular biology of phosphate control. As evidenced by the capacity of nacre to induce bone formation by human osteoblasts (Lopez et al. 1992), $CaCO_3$ and phosphate precipitating animals also share at least some of the signaling molecules that direct mineralization. Given that vertebrates and mollusks characteristically precipitate different minerals, key features of their skeletal biochemistry must differ—but what those features are and how many of them separate bone from clam shells remains to be seen.

THE GEOLOGIC RECORD OF SKELETONS

Biomineralization in Precambrian oceans

The antiquity of eukaryotic clades. In Figure 3, the eukaryotic tree is trimmed with dates that mark (1) the earliest geological records of constituent taxa and (2) the earliest records of mineralized skeletons within clades. The oldest evidence for eukaryotic biology comes from C_{28}–C_{30} steranes preserved in 2700 million year old shales from northwestern Australia (Brocks et al. 1999). The chemical structures of inferred precursor molecules are inconsistent with biosynthesis by any of the few bacterial groups known to produce sterols. Nonetheless, nothing else is known about the biology of source organisms, which could have been stem group eukaryotes only distantly related to extant clades. Protistan body fossils are rare and poorly characterized in rocks older than ca. 1800–1600 million years, when both problematic macrofossils and unambiguously eukaryotic microfossils enter the record. The macrofossils include centimeter-scale helices, tubular structures, and articulated filaments up to 7 or 8 cm long (Walter et al. 1990; Grey and Williams 1990; Rasmussen et al. 2002), all plausibly eukaryotic but not easily mapped onto the phylogeny in Figure 3. Organically preserved microfossils include forms up to several hundred microns in diameter whose walls display closely packed polygonal plates, branching processes, spines and other ornamentation, and ultrastructures (imaged by TEM) found today only in eukaryotes (Javaux et al. 2001, 2003). Such fossils are protistan, but they also remain problematic at any finer taxonomic scale.

Large populations of filamentous fossils, preserved in ca. 1200 million year old cherts from northern Canada, display patterns of cell differentiation and division that relate them to extant bangiophyte red algae (Butterfield 2000). Florideophyte reds began to diversify no later than 600 million years ago (Xiao et al. in press), but calcified red algae enter the record only during the Ordovician Period (Riding et al. 1998). Similarly, green algae occur in 700–800 million year old shales from Spitsbergen and are possibly recorded by phycoma-like microfossils in rocks older than 1000 million years

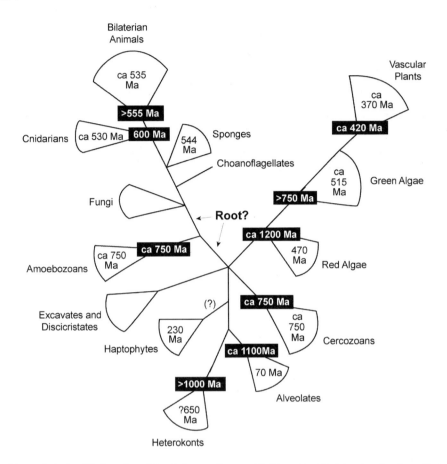

Figure 3. A simplified version of the phylogeny illustrated in Figure 2, with dates showing the earliest geological records of major clades (black boxes) and the earliest known records of biomineralized scales and skeletons within these clades (dates within wedges). First appearances of clades are shown only if they constrain the timing of subsequent divergences on a given branch. Thus, the oldest record of fungi is not shown because its Cambrian age postdates the earliest appearance of cnidarians at 600 million years—fungal divergence from other opisthokonts must have predated the first appearance of recognizable cnidarians.

(Butterfield et al. 1994), yet calcareous dasyclad and caulerpalean greens occur only in Cambro-Ordovician and younger carbonates (Wray 1977). *Vaucheria*-like heterokonts occur in Siberian shales cut by 1000 million year old intrusions (German 1990) and in 700–800 million year old shales from Spitsbergen (Butterfield 2002)—the latter are particularly compelling, as they include most phases of the vaucherian life cycle. In contrast, siliceous scales possibly formed by heterokonts appear only in ca. 610–650 million year old rocks and well skeletonized chrysophytes and diatoms not until the Mesozoic Era (see below).

Rounding out the picture of eukaryotic clade divergence, cercozoan fossils occur in ca. 750 million year old rocks (Porter et al. 2003), well before radiolarians and forams rose to biogeochemical importance. Lobose amoebans, an early branch of the amoebozoan clade, occur in the same rocks (Porter et al. 2003). Biomarkers ascribed to alveolates have been found in shales as old as 1100 million years (Summons and Walter 1990), but animals are unknown before 600 million years ago (Xiao and Knoll 2000). Save for the late appearance of animals, this paleontological picture of Proterozoic clade divergence is consistent with recent molecular clock estimates (e.g., Wang et al. 1999; Yoon et al. 2002).

The overall pattern, then, appears to be one of early eukaryotic clade differentiation, with later—and in some cases much later—evolution of biomineralized skeletons within clades. Preservational biases have undoubtedly influenced this pattern; skeletal fossils can be recognized in the rock record only when both mechanical and solubility thresholds for preservation and morphological thresholds for identification have been crossed. Nonetheless, the inferred evolutionary pattern must be at least broadly correct, as detailed in the following sections.

The earliest records of mineralized skeletons. Today, calcium carbonate and silica leave the oceans largely as skeletons. This could not have been the case early in Earth history; nonetheless, as expected when chemical weathering continuously introduces calcium and silica into the oceans, limestone, dolomite, and chert are common in Proterozoic sedimentary successions. Proterozoic and Phanerozoic successions differ not so much in the abundance of carbonates and silica as in the facies distributions of preserved deposits. Prior to the radiation of skeleton-forming eukaryotes, $CaCO_3$ and SiO_2 both accumulated preferentially along the margins of the oceans—in tidal flats and coastal lagoons, where evaporation drove precipitation (Maliva et al. 1989; Knoll and Swett 1990). Bacteria undoubtedly facilitated this deposition—the microscopic textures preserved in silicified stromatolites include both encrusted filaments and radially oriented crystal fans nucleated within surface sediments (e.g., Knoll and Semikhatov 1998; Bartley et al. 2000)—but the overall distribution of Proterozoic carbonates and silica reflects primary physical controls on precipitation. In this regard, it is worth noting that surface textures of Proterozoic sandstone beds record the widespread distribution of microbial mats in environments where no stromatolites accreted (e.g., Hagadorn and Bottjer 1997; Noffke et al. 2002); stromatolites formed where carbonate was deposited, not the reverse.

At present, the oldest inferences of protistan biomineralization come from vase-shaped tests preserved in 742±6 million year old rocks of the Chuar Group, Grand Canyon, Arizona (Porter and Knoll 2000; Porter et al. 2003). These fossils are veneered by pyrite (or iron oxides after pyrite) with regularly arranged ovoid holes, interpreted as insertion sites for mineralized scales in originally proteinaceous tests (Figure 4B). The morphology, inferred organic construction, and scale distribution of these fossils closely (and uniquely) resemble the tests of euglyphid filose amoebans (Porter et al. 2003). Further evidence of silica biomineralization is provided by a remarkable assemblage of small (10–30 µm) scales preserved in early diagenetic chert nodules from ca. 630–650 million year old rocks of the Tindir Group, northwestern Canada (Figure 4A; Allison and Hilgert 1986). These variously ornamented microfossils resemble the siliceous scales of chrysophytes, albeit at a larger size.

Horodyski and Mankiewicz (1990) described a possible instance of early carbonate skeletonization in silicified dolomites from the Pahrump Group, California, that are at least broadly correlative with Chuar rocks in the Grand Canyon. The fossils consist of sheet-like cellular structures originally encrusted by finely crystalline calcium carbonate and now preserved texturally in early diagenetic chert nodules. In some specimens, cell walls are preserved as clear chert, sharply differentiated from the carbonate-textured silica within and without. Horodyski and Mankiewicz (1990) proposed that the Pahrump fossils record thalloid algae with carbonate impregnated cell walls. More likely, however, the fossils

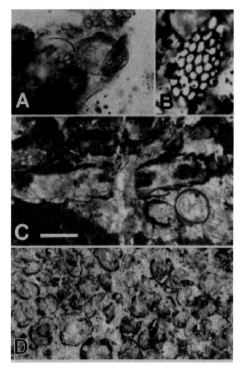

Figure 4. Evidence for skeletal biomineralization in Neoproterozoic rocks: (A) chrysophyte-like scales from ca. 650–630 million year old cherts of the Tindir Group, northwestern Canada; (B) *Melicerion*, test of a scale-forming filose amoeban from rocks just below a 742±6 million year ash bed in the Chuar Group, Grand Canyon, Arizona; (C) *Cloudina*, thin section photograph showing skeletal structure, from 548–543 million year old reefs of the Nama Group, Namibia; and (D) *Namacalathus*, also from the Nama Group, Namibia—note the goblet shaped individual in the upper right part of the image. (See text for references.) Scale bar (found in C) = 35 μm for (A), = 40 μm for (B), = 600 μm for (C), and = 1.5 cm for (D).

record unmineralized thalli whose relatively decay resistant walls were cast by encrusting carbonates soon after burial. Cyanobacterial sheaths preserved in similar fashion occur widely if sporadically in Neoproterozoic rocks around the world (e.g., Knoll and Semikhatov 1998). Ca. 600 million year old rocks of the Doushantuo Formation, China, contain extraordinary fossils of multicellular red and green algae preserved in anatomical detail, and these include florideophyte reds interpreted as stem group members of the (now) skeleton-forming Corallinales. Despite their superb preservation, however, these fossils show no evidence for the precipitation of $CaCO_3$ skeletons (Xiao et al. in press).

Early animals are best known from so-called Ediacaran fossils, problematic remains found globally as casts and molds in latest Proterozoic (575–543 million year old) storm deposits and turbidites (Narbonne 1998). Ediacaran remains and associated trace fossils contain little evidence for mineralized skeletons, prompting the widespread belief that pre-Cambrian animals were exclusively soft-bodied. Coeval carbonates, however, tell a different story. Microbial reefs in Namibia and elsewhere contain abundant and moderately diverse assemblages of calcified metazoans. *Cloudina* is a broadly cylindrical fossil built of funnel-like, apically flaring tubes set one within the next—ontogenetically, but not necessarily phylogenetically, reminiscent of pogonophoran worm tubes (Figure 4C; Grant 1990). The systematic relationships of these fossils are poorly known, but occasional evidence for budding supports a broadly cnidarian interpretation. Two distinct size classes, 0.5–2 and up to 7 mm in cross-sectional diameter, have been distinguished as different species (Germs 1972). Preserved walls are thin (a few tens of microns), but they commonly retain their rounded cross-sections in sediments, displaying only minimal evidence of compaction. Shell hash sometimes shows evidence of brittle fracture, although folded walls also occur (Conway Morris et al. 1990). Collectively, these observations suggest that *Cloudina* tubes consisted principally of organic materials, with only weak biomineralization. Nonetheless, because *Cloudina* fossils differ in diagenetic behavior from encompassing (originally) aragonitic cements, Grant (1990) proposed they precipitated skeletons of Mg calcite—not the aragonite favored by the physical chemistry of surrounding seawater.

Namacalathus constitutes a second calcified animal that was common in terminal Proterozoic reefs (Figure 4D; Grotzinger et al. 2000). Goblet-shaped fossils up to 2 cm across again suggest a broadly cnidarian biology, although *Namacalathus* continues the theme of problematic relationships established by Ediacaran remains and *Cloudina*. As in *Cloudina*, walls were thin and somewhat flexible, suggesting light and perhaps discontinuous calcification, as by spicules or granules. Many specimens are preserved by void-filling cement, but locally unique overgrowth of outer surfaces by euhedral calcite crystals may suggest original calcitic biomineralization.

Thin calcified plates also occur in terminal Proterozoic platform carbonates of Namibia. A general resemblance to Squamariacean red algae and the possible preservation of conceptacles led Grant et al. (1991) to interpret these fossils as rhodophytes, but like much else in terminal Proterozoic paleontology, this interpretation remains uncertain. More promising, Wood et al. (2002) recently described well skeletonized modular fossils that grew in fissures within Namibian reefs. Christened *Namapoikia*, these remains resemble chaetitid sponges or simple colonial cnidarians; petrology suggests originally aragonitic walls. Finally, latest Proterozoic strata from Siberia and Mongolia contain simple calcareous tubes called anabaritids and cambrotubulids, as well as clusters of siliceous spicules precipitated by early sponges (Brasier et al. 1997).

The diversity of skeletal organisms in terminal Proterozoic successions remains to be documented fully, but it will never be high. On the other hand, the *total* diversity of latest Proterozoic animals was also low, and so the *proportional* representation of skeleton-formers seems not too different from that of the Cambrian Burgess Shale or, for that matter, modern oceans. By the end of the Proterozoic Eon, animals had evolved the physiological pathways required to make skeletons of calcite, aragonite, and silica. Nascent predation may have favored the evolution of mineralized skeletons—Bengtson and Yue (1992) have reported possible predator borings in well-preserved *Cloudina* specimens from China. In light of this, it is surprising that the most heavily mineralized skeletons thus far known from terminal Proterozoic rocks are the modular fossils found in cryptic environments (fissures) within Namibian reef systems.

It is possible to elaborate on individual claims for Proterozoic biomineralization because there are so few of them. Other than the microscopic remains of siliceous scales, there is little evidence to suggest that skeletons played important functional roles before the latest Proterozoic Eon and none to suggest that they loomed large in Proterozoic carbonate and silica cycles. Skeletal biomineralization was part of the initial diversification of animals, but skeleton-forming metazoans began their rise to biological and biogeochemical prominence only during the next, explosive phase of diversity increase.

Biomineralization and the Cambrian explosion

The great watershed in skeletal evolution was the Cambrian Period, when mineralized skeletons appeared for the first time in many groups (see Cloud 1968; Lowenstam and Weiner 1989 for comprehensive reviews of early thinking on this subject). Various hypotheses propose that the Cambrian "explosion" was precipitated by environmental changes favoring skeletal biomineralization. There is, however, no geological evidence to support this conjecture, and, in any case, it isn't clear what physical event would facilitate the broadly simultaneous evolution of carbonate, phosphatic, siliceous, and agglutinated skeletons in marine animals and protists. It has also been proposed that Cambrian rocks document an explosion of fossils, made apparent by robust skeletons, rather than a true diversification of animals in the oceans. Again, the growing record of latest Proterozoic fossils and trace fossils discourages this idea. More likely, as Bengtson (1994) has argued, the Cambrian diversification of mineralized skeletons was part and parcel of a broader Cambrian radiation of animal diversity, with increases in predation pressure favoring the evolution of protective armor.

Exceptionally well preserved Cambrian fossils such as those of the Burgess (Briggs et al. 1994) and Chengjiang (Chen and Zhou 1997) biotas show that, then as now, most animals did not precipitate mineralized skeletons; only about 15% of Burgess species formed skeletons that would be preserved in conventional fossil assemblages (Bengtson and Conway Morris 1992). Nonetheless, many clades did evolve mineralized skeletons (Figure 5), and these gave rise to the predominant fossil record of the past 500 million years. Skeletal diversity increased tremendously in the sponges, even though this clade must have differentiated from other animals before Ediacaran fossils first entered the record. The Cambrian diversity of siliceous spicules matches or exceeds anything known from subsequent periods (Bengtson et al. 1994; Dong and Knoll 1996), and the carbonate-precipitating archaeocyathids produced the first massively mineralized skeletons known from the geological record. Indeed, by virtue of their skeletal evolution, archaeocyathids qualify as the first metazoan reef formers, although many Cambrian build-ups remained predominantly microbial (Riding and Zhuravlev 1998; Rowland and Shapiro 2002).

Anabaritids, small calcitic tubes with three-fold symmetry found in basal Cambrian rocks, may have been cnidarian, but for the most part, Cambrian rocks contain only

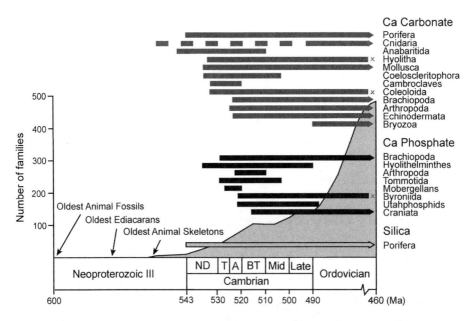

Figure 5. Summary of early skeleton evolution in metazoans, showing the diversity, skeletal mineralogy, and stratigraphic distribution of Cambro-Ordovician animals. Skeleton data modified from Bengtson and Conway Morris (1992); family diversity from Sepkoski (1982).

scattered records of cnidarian (or presumptive cnidarian) biomineralization (Debrenne et al. 1990). Thus, in contrast to their importance in subsequent periods, cnidarian skeletons are not a major feature of the Cambrian fossil record. Bilaterian skeletons are also scarce in lowermost Cambrian deposits, but over the first 25 million years of the period, they increased markedly in abundance and diversity. Earliest Cambrian bilaterians are dominated by extinct groups whose relationships to younger animals remain uncertain. By the end of the Early Cambrian, however, faunas included recognizable stem group members of many extant phyla and classes (Figure 5; Budd and Jensen 2000). Among protists, radiolarians first appear in Cambrian rocks (Dong et al. 1997; Won and Below 1999; Won and Iams 2002), as do agglutinated, but not calcified foraminiferans (McIlroy et al. 2001). Skeletons assigned to the dasyclad green algae have been reported from Cambrian carbonates in Siberia, but, by and large, algal skeletons are not a conspicuous feature of Cambrian rocks (Johnson 1966).

Functionally, Early Cambrian skeletons included spicules, sclerites, tubular and conical exoskeletons, bivalved shells, internal skeletons, and mineral-impregnated cuticle that molted as organisms grew—of the 178 distinct skeletal architectures recognized in marine animals by Thomas and Reif (1993), 89 are known to have appeared within the first 20 million years of the period, and by the Middle Cambrian fully 146 (80%) were present (Thomas et al. 2000). Craniate chordates were part of this mix (Shu et al. 1999), and of these the conodonts, at least, formed dermal bone in the form of tooth-like structures containing dentine (Donoghue et al. 2000). Other major innovations in the vertebrate skeleton postdated the Cambrian.

Many Early Cambrian skeletons are phosphatic, and in the past, this gave rise to speculation that phosphatic mineralization was favored by the chemistry of Early Cambrian oceans. Detailed investigations of preserved shell microstructure, however, show that the phosphate in most Early Cambrian fossils originated by the early diagenetic replacement of originally calcitic, aragonitic, or organic skeletons (e.g., Runnegar 1985; Kouchinsky 2000). Of course, some animals precipitated phosphatic skeletons, and this list includes a number of extinct clades in addition to lingulate brachiopods and craniates. Despite this, the proportional representation of phosphatic skeletons in Early Cambrian ecosystems was not very different from that of younger oceans (Bengtson and Runnegar 1992).

In summary:

(1) The evolution of mineralized skeletons in Cambrian organisms was not instantaneous. Only a few taxa occur in lowermost Cambrian rocks; skeletal abundance and diversity increased dramatically over the succeeding 25 million years.

(2) The body plans characteristic of most extant bilaterian phyla took shape over this same interval, although most Early and Middle Cambrian animals were stem and not crown group members of persistent phyla and classes. Skeletons evolved as principal components of some body plans, constraining the subsequent course of evolution in these clades. That so many different groups evolved mineralized skeletons during this time lends support to the hypothesis that animals deployed "off-the-shelf" biochemistry in skeletal evolution.

(3) All major skeletal biominerals were exploited during this radiation, as were most known skeletal architectures. The diversity of minerals employed in early skeletal evolution discourages the idea that changing ocean chemistry drove Cambrian evolution. Likewise, the fact that protists (including skeletonized protists) diversified along side animals, limits explanations that rely exclusively on developmental genetics (Knoll 1994). The diverse skeletons of Cambrian organisms share only one principal feature in common—they would have protected their owners against the bilaterian animal predators

that took shape during the Cambrian explosion. Thus, ecology must loom large in hypotheses to explain Cambrian biomineralization (Bengtson 1994). This does not preclude a role for physical change in early animal evolution, but does limit its specific application to the evolution of skeletons and skeletal chemistry. In general, Cambrian diversification likely reflects the interplay between genetic possibility and environmental opportunity, including new ecological opportunities introduced by both late Proterozoic increase in atmospheric oxygen tensions and strong carbon cycle perturbation at the Proterozoic-Cambrian boundary (Knoll 2003; Amthor et al. 2003).

An aside on later Cambrian oceans. Interestingly, mineralized skeletons are more apparent in Lower Cambrian sedimentary rocks than they are in many Middle and, especially, Upper Cambrian successions. Much Early Cambrian diversity was lost during major extinctions near the end of the age (Zhuravlev and Wood 1996). The massively calcified archaeocyathids dwindled to a handful of species, and trilobites underwent extensive turnover. Many of the small shelly taxa known from Lower Cambrian rocks disappeared as well, although the role of changing taphonomy in producing this pattern remains poorly known; Porter (2003) has shown that the diversity history of small shelly fossils closely matches the stratigraphic distribution of the phosphatized limestones in which they occur most abundantly. Phosphatized limestones diminished in abundance through later Cambrian time, taking with them the preservational window through which small shelly fossils are viewed.

Moderately mineralized trilobite carapaces and lingulate brachiopods are common in Upper Cambrian deposits, and viewed under the microscope, carbonate sands commonly contain abundant echinoderm debris. But diverse mollusks, calciate brachiopods, and well-preserved echinoderms—all known from older rocks and staples of the younger Paleozoic record—are not common. In contrast, stromatolites and thrombolites are widespread (Kennard and James 1986; Rowland and Shapiro 2002). Trace fossils and exceptional assemblages like the Burgess Shale tell us that diverse animals lived in later Cambrian oceans, and that quite a few of them were large—but large animals did not precipitate robust skeletons.

Rowland and Shapiro (2002) asked why metazoan-built reefs disappeared at the end of the Early Cambrian, a question that can be generalized to include all carbonate precipitating eukaryotes. In answer, they hypothesized that high temperatures and high P_{CO2} associated with a later Cambrian supergreenhouse inhibited carbonate biomineralization, especially by organisms with only limited capacity to buffer the fluids from which they precipitate skeletons. In terms of the framework presented here, this means that the environmentally dictated cost of precipitating carbonate skeletons is thought to have been high for all organisms and prohibitively high for groups such as calcareous sponges, cnidarians, and algae.

Rowland and Shapiro's (2002) innovative idea deserves careful testing by both biologists and Earth scientists. Geologists need to sharpen tools for paleoenvironmental reconstruction in the deep past, and they need to develop better data on the paleolatitudinal distribution of early skeleton-forming organisms. Berner and Kothavala's (2001) model of atmospheric carbon dioxide levels through Phanerozoic time is consistent with the Rowland/Shapiro scenario, as is the widespread distribution of carbonaceous shales in later Cambrian successions. Nonetheless, empirical constraints on Middle and Late Cambrian environmental conditions remain limited. Biologists need to provide better experimental data on the responses of skeletal physiology to elevated temperature and CO_2, measured across a phylogenetically wide range of organisms. Experiments prompted by current threats of global warming point the way (e.g., Langdon et al. 2000; Zondervan et al. 2001; see below).

The Ordovician radiation of heavily calcified skeletons

Although Cambrian evolution produced diverse body plans, species diversity within major clades remained low as the Ordovician Period began. But all this was about to change. During renewed Ordovician diversification, the global diversity of families and genera preserved as fossils increased three- to four-fold (Figure 5; Sepkoski 1982; Miller 1997), crown group members of major invertebrate phyla and classes came to dominate marine ecosystems, and the marine carbonate and silica cycles both came under substantial biomineralogical control.

Animal taxa that radiated during the Ordovician Period include sclerosponges and other new experiments in demosponge calcification; both tabulate and rugose corals; cephalopods, bivalves and gastropods; bryozoans; calciate brachiopods; crinoids and other echinoderms; and fish with dermal armor—establishing the fauna that would dominate oceans for the rest of the Paleozoic Era (Sepkoski and Sheehan 1986). Metazoan reefs returned to the oceans, skeletonized red and green algae became widespread, and radiolarians expanded dramatically, firmly establishing biological control over the marine silica cycle (Maliva et al. 1989; Racki and Cordey 2000). Moreover, both maximum size (e.g., Runnegar 1987) and skeletal mass increased in many groups.

Again, it is helpful to frame these observations in terms of evolutionary costs and benefits. The physiological cost of precipitating such massive skeletons must have been high, although, if, as noted above, Late Cambrian environmental conditions made the costs of carbonate skeletons prohibitive, then relaxation of those conditions might have contributed to Ordovician radiation. In any event, the phylogenetically broad evolution of robust skeletons as part of an overall increase in biological diversity once again suggests that an upward ratcheting of predation pressure contributed to the observed evolutionary pattern; important new predators included nautiloid cephalopods and starfish. (Increase in skeleton mass is a biomechanical requirement of increased body size but cannot explain why groups like corals and bryozoans evolved robust carbonate skeletons *de novo*.) Miller (1997) pointed out that Ordovician diversification took place in the context of increasing tectonic activity and by implication, therefore, increased nutrient flux to the oceans. Increasing nutrient status would certainly help to explain increased body size and predation in Ordovician oceans (see Vermeij 1995 and Bambach 1999 for arguments why this should be so).

Permo-Triassic extinction and its aftermath

The end of an era. For most of the Permian Period, marine biology looked much as it had in for the preceding two hundred million years. Large, well-skeletonized sponges, rugose and tabulate corals, calciate brachiopods, mollusks, and stenolaemate bryozoans characterized benthic communities, while radiolarians contributed a skeletal component to the plankton. Calcified algae played important roles in carbonate build-ups on continental shelves and platforms (Samankassou 1998; Wahlman 2002), and—in contrast to most earlier Paleozoic reefs—so did non-skeletal cementstones, precipitated at least in part under microbial influence in reef cavities and on the seafloor (Grotzinger and Knoll 1995; Weidlich 2002). Evidently, Permian seawater was highly oversaturated with respect to calcium carbonate minerals, a circumstance related in part to low sea levels and, consequently, limited shelf area for skeleton-forming benthos (Grotzinger and Knoll 1995). Despite the broad Paleozoic continuity of the marine biota, several important innovations in skeletal evolution did occur after the Ordovician radiation. By the Devonian Period, foraminiferans had evolved the capacity to precipitate calcitic tests, allowing benthic forams, especially the large, symbiont-bearing fusulinids, to emerge as major carbonate producers in Late Paleozoic seaways (Culver 1993). And vertebrates evolved phosphatized post-cranial skeletons, providing, among other things, biomechanical support for the colonization of land (Clack and Farlow 2002).

It was not to last. Mass extinction at the end of the Permian Period devastated Paleozoic ecosystems. More than half of all invertebrate families, and perhaps ninety percent or more of existing species, perished in the oceans (Raup 1979; Erwin et al. 2002), and both tetrapods and vascular plants suffered substantial losses on land (Maxwell 1992; Looy et al. 2001). Recent geochronologic data suggest a rapid pulse of extinction (Jin et al. 2000), precipitated by some discrete event or cascade of events (Bowring et al. 1998). Debate continues, however, about both the physical trigger(s) for the extinction and the physiological mechanisms that induced mortality on such a vast scale.

The reason for considering end-Permian mass extinction in a discussion of biomineralization is that extinction and survival at the Permo-Triassic boundary show a pattern of strong selectivity with respect to skeletal physiology. Knoll et al. (1996) asked whether catastrophically high carbon dioxide levels could have furnished a kill mechanism for Permo-Triassic mass extinction. The standard geological approach to such a problem is to search in boundary rocks for geochemical evidence of elevated CO_2, but Knoll and colleagues looked, instead, for physiologically informative pattern in the fossil records of animals that did or did not survive the extinction. Consulting the large experimental literature on hypercapnia in both tetrapods and marine invertebrates, they articulated paleontologically recoverable traits of physiology, anatomy, functional biology, and ecology that correlate with vulnerability to or tolerance for elevated CO_2 levels. From these data, they developed a set of predictions about extinction and survival at the Permo-Triassic boundary.

Within the marine realm, animals with active circulation and elaborated gills (or lungs) for gas exchange compensate for elevated P_{CO2} better than organisms that lack these features. Also, animals that normally experience high internal P_{CO2}, including infauna and metazoans capable of exercise metabolism at high rates, tolerate experimental increases in carbon dioxide differentially well. And among animals that form skeletons of calcium carbonate, species that can exert physiological control over the pH balance of the fluids from which minerals are precipitated fare better than those with limited buffering capacity. Based on these features, Knoll et al. (1996) divided the Late Permian marine fauna into two groups, with high and low expected survival, respectively. As predicted, 81% of genera in the "vulnerable" group disappeared at the boundary, while only 38% of nominated "tolerant" genera became extinct (Figure 6A). The selectivity is even more pronounced when skeletal physiology is considered alone (Figure 6B; Bambach and Knoll, in preparation). Fully 88% of genera disappeared in groups that formed massive carbonate skeletons but had limited capacity to buffer fluids (sponges, cnidarians, bryozoans, calciate brachiopods). In contrast, groups that built skeletons of materials other than calcium carbonate lost only about 10% of their genera. Intermediate taxa—groups like mollusks that have a substantial commitment to carbonate skeletons but some physiological capacity to regulate the chemistry of internal fluids—lost 53% of their genera. Of these, those genera considered vulnerable on the basis of the other criteria noted above went extinct at twice the rate of those otherwise considered tolerant.

To a first approximation, then, Late Permian extinction in the oceans is about biomineralization. Specifically, it appears that the metabolic cost—and for some organisms, even the metabolic feasibility—of building $CaCO_3$ skeletons changed catastrophically at the Permo-Triassic boundary. Most proposed explanations for end-Permian mass extinction do not predict this. Indeed, while oxygen starvation and nutrient collapse, two commonly favored kill mechanisms, can explain the prevalence of small

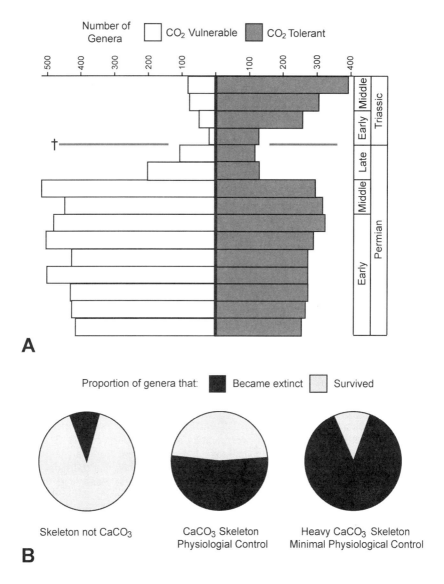

Figure 6. Summary of extinction and survival associated with Permo-Triassic mass extinction, as seen from the perspective of anatomy and physiology. (A) Permian and Early to Middle Triassic diversity of genera within groups predicted to be vulnerable to physiological disruption by rapid increase in carbon dioxide levels (white bars—includes calcifying sponges, corals, brachiopods, bryozoans, echinoderms, and calcareous foraminiferans) and groups predicted to be relatively tolerant to hypercapnic stress (shaded bars—includes mollusks, arthropods, chordates, and protists without $CaCO_3$ skeletons). Note that both absolute genus richness and proportional diversity change across the P-Tr boundary (marked by the cross). Pre-boundary Late Permian diversity decline reflects both true extinction at the end of the Middle Permian and the Signor Lipps effect, a backward smearing of last appearances caused by record loss. Redrawn from Knoll et al. (1996). (B) The relationship between skeletal physiology and Permo-Triassic extinction (Bambach and Knoll, in prep).

sizes in Early Triassic animals, they make biological predictions about extinction selectivity that run counter to the pattern actually observed (Bambach and Knoll in preparation). Moreover, catastrophic carbon dioxide increase provides two distinct kill mechanisms—direct physiological inhibition of metabolism and climate change associated with greenhouse enhancement—so CO_2 increase at the boundary is consistent with both terrestrial and marine patterns of extinction. Other, as yet unarticulated, environmental perturbations may account for observed patterns of extinction equally well, but for now, carbon dioxide deserves a prominent place on the short-list of plausible kill mechanisms. Accepting this, we can ask what sort of trigger could introduce lethal amounts of CO_2 into the atmosphere and surface oceans. Knoll et al. (1996) favored catastrophic overturn of anoxic deep waters, but this idea has not generated widespread enthusiasm among oceanographic modelers. Oxidation of methane released catastrophically from hydrates on the continental shelf and slope might also be suggested, but isotopic constraints on the volume of methane that could have been released suggest that by itself this mechanism cannot provide the necessary trigger (Berner 2002). The same problem applies to carbon dioxide release associated with extensive continental basalt eruption at the Permo-Triassic boundary. The wild card that remains is bolide impact, an attractive mechanism for which empirical evidence remains controversial (Becker et al. 2001; Kaiho et al. 2001; Verma et al. 2002).

Delayed rediversification of the marine fauna. The Early Triassic aftermath of end-Permian mass extinction poses its own set of questions, some of which potentially provide new perspective on the extinction event itself. Given the magnitude of biological disturbance, it is not surprising that basal Triassic rocks are paleontologically depauperate. What is surprising is that the exceptional features of the earliest Triassic biosphere persisted for four to six million years; true ecosystem recovery really didn't begin until the Middle Triassic (Erwin 2001).

Early Triassic marine diversity was lower than at any previous time since the Cambrian Period. Characteristically, Early Triassic faunas consist of small, lightly skeletonized survivors, many with apparently cosmopolitan distributions (Schubert and Bottjer 1995; Rodland and Bottjer 2001). The gastropods, bivalves, and ostracodes found in local abundance in Lower Triassic carbonates are commonly only 1–2 mm long (Lehrmann et al. 2003); ammonoid cephalopods and calciate brachiopods were similarly small and thin walled. Echinoderm debris formed bioclastic sands, but individual crinoid columnals and skeletal plates are tiny. Calcified red and green algae were important carbonate producers in both Late Permian and later Triassic seaways, but with one high paleolatitude exception (Wignall et al. 1998), neither is known from Lower Triassic rocks (Flügel 2002). Similarly, all Paleozoic corals disappeared by the end of the Permian, and the scleractinian corals that would rise to ecological prominence in Mesozoic and Cenozoic ecosystems appear only in Middle Triassic rocks (Stanley 2003), apparently evolving carbonate skeletons *de novo* in at least two distinct actinarian stocks (Romano and Cairns 2000). Thus, no metazoan or algal reefs are known from Lower Triassic rocks. In contrast, microbialites and precipitated carbonates less obviously formed under microbial influence are relatively widely distributed (Schubert and Bottjer 1992; Baud et al. 1997; Kershaw et al. 2002; Lehrmann et al. 1993).

The paucity of skeletal remains doesn't mean that carbonates are rare in Lower Triassic successions; in fact Lower Triassic limestones and dolomites are widespread and regionally thick. Facies development, however, resembles that of Neoproterozoic basins, as well as Upper Cambrian successions that have similarly limited contributions from skeletal algae and animals—carbonate mudstones, generated at least in part by whitings, and oolites occur along with the microbialites noted in the previous paragraph.

CO_2-induced decrease in the saturation state of sea water must, therefore, have been transient, pointing to more directly physiological limitations on calcification.

During the Anisian Age of the Middle Triassic, large, robustly skeletonized floras and faunas reappeared globally. Ecological networks were reestablished in the oceans, but in novel ways which insured that subsequent evolution would travel new paths (Bambach et al. 2002). New reefs were, at first, dominated by *Tubiphytes*, an unusual fossil formed by (?)bacterially induced carbonate encrustation of algae and perhaps other non-skeletal organisms (Senowbari-Daryan and Flügel 1993). Soon, however, scleractinian corals emerged along with both coralline and siliceous sponges as major framework builders (Flügel 2002; Stanley 2003). Biological reorganization was seeded in the plankton as well as the benthos, paving the way for one last momentous change in skeletal evolution.

Two Mesozoic revolutions

A revolution of marine animals. In 1977, Geerat Vermeij outlined what he called the "Mesozoic marine revolution," a change of guard in skeletal faunas, driven by step-wise increase in the abundance and mechanical strength of shell crushing predators. Vermeij's ecological research had convinced him that in marine gastropods, both shell strength and probability of shell damage varied biogeographically with the crushing strength of claws sported by predatory crabs. His insight was to observe that gastropod shell strength also varied in time: beginning in the Cretaceous Period and accelerating toward the present, gastropod taxa with stronger, more highly ornamented, and more readily remodeled shells replaced earlier evolved forms. Other marine groups from coralline red algae to echinoids also changed in ways that reduced their vulnerability to shell-crushing predators. Among bivalves, infaunal groups capable of avoiding predators by burrowing diversified markedly during the Cretaceous and Cenozoic, while epifaunal taxa withdrew to protected environments or disappeared entirely (Stanley 1975).

What predators radiated during this interval? The list includes crabs, lobsters, stomatopods, shell crushing sharks and rays, teleost fish, and shore birds, not to mention naticid and other gastropods that drill through shells to reach their prey (Vermeij 1977). Bambach (1993) has argued that Mesozoic and Cenozoic marine animals were larger and more nutritious than their Paleozoic forebears—Devonian New Englanders would doubtfully have prepared trilobite boils and brachiopod bakes—and that this evolutionary change in the nutritional quality of prey might have facilitated predator evolution. The problem, also noted by Bambach, is one of timing. The shift toward larger and more energetic animals was set in place by selective extinction at the Permo-Triassic boundary and subsequent recovery, but Vermeij's marine revolution took hold only in mid-Cretaceous times, some 150 million years later. Thus, while overall changes in the prey biota may have been necessary for Mesozoic marine revolution, they were probably not sufficient. That so many distinct classes of predator should have diversified at the same time suggests, once again, that some enabling environmental change began 100 million years ago. Bambach (1993, 1999) proposed that the radiation of flowering plants brought more nutrients to continental shelves and platforms. The nutritional quality of angiosperms (N:C) is, on average, higher than that of ferns and other seed plants (Midgley et al. 2002). Moreover, the specialized chemical defenses that flowering plants produce tend to be synthesized in lower concentrations than the more general feeding deterrents characteristic of gymnosperms and ferns (Rhoades 1979), further enhancing the food quality of angiosperm debris. At the same time, Cretaceous and Cenozoic orogenesis increased the mean elevation of continents, increasing weathering rates (Huh and Edmond 1999) and so fueling increased productivity in coastal oceans. Furthermore, both flowering plants (especially sea grasses) and kelps radiated in late Cretaceous and

Cenozoic oceans, providing new in situ resources for animals. In consequence, marine communities may have come to support a new tier of top predators, including shell crushing animals that drove adaptive change in the skeletal biology of prey organisms.

A revolution of marine phytoplankton. A second Mesozoic marine revolution took place in the planktonic realm. For more than three billion years, primary production in the oceans had been dominated by organisms bearing chlorophyll a (cyanobacteria) or chlorophyll a+b (green algae). Beginning at the time of Middle Triassic recovery from end-Permian events, however, phytoplankton containing chlorophyll a+c began to spread, and by the end of the era, three groups—the dinoflagellates, diatoms, and coccolithophorids—would come to dominate the oceans (Falkowski et al. in press). The reasons for this photosynthetic change of guard remain a subject for debate, although the distinct trace metal requirements of Chl a+c algae relative to greens and cyanobacteria may have favored evolutionary replacement in increasingly well oxidized Mesozoic oceans (Quigg et al. in press). Whatever its causes, the phytoplankton revolution stands as a major episode in the history of biomineralization because two groups, the coccolithophorids and the diatoms, rose to biogeochemical dominance in the marine carbonate and silica cycles, respectively.

Dinoflagellates were the first Chl a+c group to diversify. Preserved biomarker molecules suggest that the dinoflagellates existed in Neoproterozoic and Paleozoic oceans (Summons and Walter 1990; Moldowan and Talyzina 1998), although morphologically diagnostic microfossils have not been identified to date. Unambiguous dinoflagellate fossils enter the record in modest diversity during the Middle Triassic (Fensome et al. 1996), in concert with a major increase in dinoflagellate biomarkers in sedimentary organic matter (Moldowan et al. 1996). Species diversity increased by an order of magnitude during the second half of the Jurassic Period, mostly due to the radiation of a single family, the Gonyaulacaceae (Fensome et al. 1996). Cretaceous radiation of a second family, the Peridiniaceae, brought the group to its late Cretaceous diversity acme. Biomineralization within the group, however, never expanded beyond calcareous cysts in a few lineages.

Coccolithophorids similarly show modest Triassic origins and later Mesozoic expansion to a late Cretaceous zenith (Siesser 1993). Unlike the dinoflagellates, however, coccolithophorids evolved calcitic skeletons made of minute scales called coccoliths (Young et al. 1999). The functional biology of coccoliths remains uncertain, although the apparent light dependence of calcification offers some support to hypotheses that tie calcification to CO_2 supply within cells (e.g., Anning et al. 1996). [One can generalize to state that our understanding of skeletal function, and, therefore, of the evolutionary benefits associated with skeletal biomineralization, varies inversely with size.] The ecological expansion of coccolithophorids, joined in the Middle Jurassic by planktonic foraminiferans with calcitic tests, changed the nature of carbonate deposition in the oceans. For the first time, a pelagic carbonate factory delivered carbonate sediments directly to the deep sea floor. Seafloor cementstones, common in later Triassic carbonate platforms, exit the record as coccolithophorids expand, suggesting that coccolithophorid evolution led to a global decrease in the saturation state of seawater with respect to carbonate minerals.

The third Chl a+c group, the diatoms, radiated later than the others, but expanded to unparalleled ecological success. Uncertainty plagues Rothpletz' (1896) report of diatoms in Middle Jurassic rocks, but unambiguous frustules occur in Lower Cretaceous marine rocks from many localities (Harwood and Nikolaev 1995). By late in the Cretaceous Period the group had both radiated in the oceans (Harwood and Nikolaev 1995) and colonized non-marine environments (Chacón-Baca et al. 2002). The fossil record suggests that diatom diversity has continued to expand up to the present (Barron 1993).

Nasselarian radiolaria also diversified in the Cretaceous and Tertiary, as did two additional silica precipitating groups. Silicoflagellates are chrysophyte algae with an internal skeleton of siliceous tubes. Ebridians also form internal skeletons of silica rods; they have sometimes been grouped with dinoflagellates, although recent ultrastructural data suggest euglenid affinities (Hargraves 2002). Regardless of the expanded diversity of silica-precipitating protists, however, diatoms account quantitatively for silica removal from modern seawater, limiting silica concentrations in surface waters to only a few parts per million (Wollast and Mackenzie 1983). The "diatomization" of the marine silica cycle left its mark on evolutionary pattern in other silica users. Cenozoic radiolarians, for example, show a pattern of decreasing use of silica in tests (Harper and Knoll 1974). Siliceous sponges, which had been major reef builders in Jurassic seas, largely vacated shelf environments (Maliva et al. 1989), and the morphological diversity of siliceous spicules decreased (Maldonado et al. 1999). Interestingly, when grown experimentally in waters with presumed pre-Cretaceous concentrations of dissolved silica, modern Mediterranean sponges precipitate spicules with morphologies not seen as fossils since the Mesozoic (Maldonado et al. 1999).

What, if any, links connect the two Mesozoic biological revolutions? Perhaps changing nutrient status and patterns of oceanographic circulation influenced marine life on continental shelves *and* in the open ocean. But compelling answers remain a distant prospect. To date, hardly anyone has even asked the question.

The future

With biological revolutions in animals and phytoplankton, the modern biology, ecology, and biogeochemistry of skeletal organisms in the oceans was in place. On land, the modern pattern emerged somewhat later, with the mid-Tertiary expansion of grasses. Annual production of siliceous phytoliths now rivals that of marine diatom frustules (Conley 2002), a biological expansion of silica use perhaps recorded most clearly in the changing tooth morphology of mammalian grazers (Janis 1995).

Today, however, one last revolution appears set to change the cost/benefit balance of carbonate skeletons in tropical oceans. Since the beginning of the industrial revolution, carbon dioxide levels in the atmosphere and surface ocean have been rising, and they may be rising too fast for some skeleton formers to adapt. (The deleterious effects of environmental change nearly always hinge as much on rates as they do on magnitude.) Increased carbon dioxide leads to a decrease in carbonate concentration, and hence in the saturation state of seawater with respect to calcite and aragonite (Gattuso and Buddemeier 2000; Andersson et al. 2003). In light of this, it is not surprising that experimental growth of coccolithophorids, red algae, and symbiont-bearing corals at CO_2 levels predicted for the next century resulted in pronounced downturns in calcification (Gattuso et al. 1998; Kleypas et al. 1998; Riebesell et al. 2000; Langdon et al. 2000; Marubini et al. 2003). As P_{CO2} continues its rapid rise, the physiological cost of carbonate precipitation may once again become prohibitively high, especially for poorly buffered organisms like benthic algae and cnidarians—the same groups that suffered disproportionately at the end of the Permian Period. As they shed light on Earth's evolutionary past, then, experimental studies of growth in skeleton-forming organisms may illuminate the biological future that our grandchildren will inherit.

DISCUSSION AND CONCLUSIONS

From the proceeding sections, it should be clear that the fossil record is not simply a history of organisms preserved by virtue of their mineralized skeletons. The evolutionary fates of skeletonized animals, plants, and protists have actually been tied, in no small measure, to the physiological pathways by which those skeletons formed. At least three times in the past 600 million years, increases in predation pressure have fueled evolutionary changes in both the phylogenetic distribution and the functional biology of mineralized skeletons. Environmental perturbations have also decreased, at least temporarily, the saturation state of the oceans with respect to calcite and aragonite, challenging organisms that have only limited physiological control over the fluids from which they precipitate carbonate skeletons.

The evolutionary influence of marine chemistry may go beyond selective extinction. Two decades ago, Sandberg (1983) recognized large scale secular variations in the carbonate chemistry of the oceans. During the earliest Cambrian, the mid-Mississippian to Middle Jurassic, and the last 35 million years, non-skeletal carbonates (ooids, penecontemporaneous cements) have been precipitated largely as aragonite and Mg-calcite. During the intervening intervals (late Early Cambrian to mid-Mississippian and Late Jurassic to the Oligocene), non-skeletal carbonates formed predominantly as calcite. Sandberg (1983) ascribed this geological pattern to secular variations in P_{CO_2}, but Stanley and Hardie (1998) have argued instead that seawater Mg/Ca, itself governed by hydrothermal ridge activity, controls the mineralogy of carbonate precipitation in the oceans. Moreover, Stanley and Hardie (1998) noted that the skeletal mineralogy of major carbonate producers has varied through time in synchrony with the pattern of non-skeletal precipitation (Fig. 7). In the Paleozoic "calcite" sea, calcitic corals and sponges dominated reefs. During the ensuing "aragonitic" interval, however, major carbonate producers included aragonite-precipitating demosponges, algae, and scleractinian corals. Corals declined in distribution and ecological importance during the "calcitic" interval of the late Mesozoic and early Tertiary, but came back as principal reef builders when "aragonitic" conditions returned. In contrast, calcite use in coccoliths shows a polyphyletic decline from the Cretaceous to the present (Houghton 1991).

Some caveats are in order. Not all skeletal organisms appear to have noticed the oscillations in seawater chemistry. As Stanley and Hardie (1998) emphasized, it is specifically those organisms that precipitate massive skeletons under conditions of limited physiological control that show the predicted stratigraphic pattern. These of course, are the same groups that suffered disproportionately at the end of the Permian and which were so conspicuously absent from Late Cambrian and Early Triassic seafloors. For carbonate producers, physiology is destiny.

Note also that ocean chemistry correlates with the initial appearances of sponges, cnidarians, and algal taxa, but not necessarily with the last appearances of individual clades. In this regard, it is important to recognize that there is no evidence for a secular shift in mineralogy within skeleton-forming lineages. The groups highlighted by Stanley and Hardie (1998) all belong to larger clades that also include non-calcifying taxa. Thus, apparent shifts from calcite to aragonite skeletons in corals, in sponges, and in red and green algae actually reflect extinction followed by the *de novo* evolution of mineralized skeletons from unmineralized ancestors. It appears that in the recruitment of macromolecules to guide *de novo* carbonate precipitation, natural selection favored molecules that would minimize the physiological cost of precipitation. [Minor element chemistry does not seem to be so rigorously controlled. When Stanley et al. (2002) grew coralline red algae in simulated Cretaceous sea water, the algae precipitated skeletons of low Mg calcite, as predicted for the Cretaceous "calcite" sea, rather than the Mg-calcite formed today. In red algae, templating macromolecules appear to govern the overall mineralogy of carbonate precipitation, but they do not stringently control cation incorporation into precipitated minerals.]

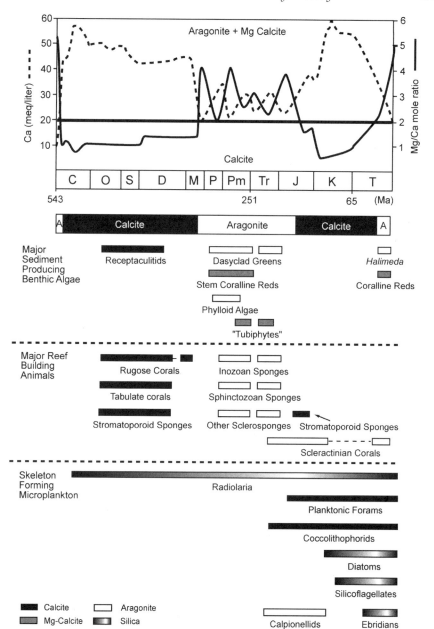

Figure 7. The stratigraphic relationship between ocean chemistry and skeletal mineralogy in major sediment producing benthic algae and animals, modified from Stanley and Hardie (1998). The figure shows Stanley and Hardie's (1998) estimates of Ca abundance and Mg/Ca in ancient oceans, as well as the principal time intervals dominated by "calcite" and "aragonite" seas. Aragonite and Mg-calcite are favored when the Mg/Ca mole ratio is above 2; calcite is favored at ratios below 2. Also shown are the time distributions of principal skeleton–forming protists; calpionellids are thought to be tintinnids with calcified tests (see text for references).

Harper et al. (1997) have claimed a role for seawater chemistry in the skeletal evolution of bivalved mollusks, but it is an influence of a different sort. Bivalves precipitate skeletal minerals from physiologically controlled fluids, so environmental influence on shell formation is not at issue. But for epifaunal bivalves, carbonate dissolution potentially weakens shells, thereby increasing the success rates of predators. Ancestral bivalves secreted aragonitic shells, but through time epifaunal bivalves show a pattern of increasing calcite use in shell laminae—a propensity that is potentially counterintuitive, given the superior mechanical strength of aragonite. Noting that infaunal bivalves show no comparable trend, Harper et al. (1997) proposed that the evolutionary pattern of increasing calcite incorporation was driven by calcite's greater resistance to dissolution, especially during the long Paleozoic and Mesozoic intervals of "calcite" oceans.

From the examples presented here, it can be seen that a number of key events in the history of life reflect the changing costs and benefits of skeletal biomineralization. Paleontologists have identified principal patterns of skeletal evolution, and biologists have provided functional explanations of this pattern, at least for larger organisms. The challenge—and the opportunity—for the future is to link paleontological insights more firmly to the molecular biology and physiology of skeletal biomineralization and to the physical history of the biosphere.

ACKNOWLEDGMENTS

Research leading to this paper was funded in part by NSF Biocomplexity Grant OCE-0084032, Project EREUPT. I thank Paul Falkowski and other members of the Project EREUPT team for helpful discussions; P. Westbroek and R. Bambach for a decade of stimulating conversations; S. Awramik for access to materials described by the late C. Allison; and R. Bambach, J. Payne, W. Fischer, P. Dove, and S. Weiner for constructive criticisms of my draft manuscript.

REFERENCES

Aizawa K, Miyachi S (1986) Carbonic anhydrase and carbon concentrating mechanisms in microalgae and cyanobacteria. FEMS Microbiol Rev 39:215–233

Allison CW, Hilgert JW (1986) Scale microfossils from the early Cambrian of northwest Canada. J Paleontol 60:973–1015

Amthor JE, Grotzinger JP, Schroder S, Bowring SA, Ramezani J, Martin MW, Matter A (2003) Extinction of *Cloudina* and *Namacalathus* at the Precambrian-Cambrian boundary in Oman. Geology 31:431–434

Andersson AJ, Mackenzie FT, Ver LM (2003) Solution of shallow-water carbonates: An insignificant buffer against rising atmospheric CO_2. Geology 31:513–516

Anning T, Nimer N, Merrett MJ, Brownlee C (1996) Costs and benefits of calcification in coccolithophorids. J Mar Systems 9:45–56

Baldauf SL (2003) The deep roots of eukaryotes. Science 300:1703–1706

Bambach RK (1993) Seafood through time: changes in biomass, energetics, and productivity in the marine ecosystem. Paleobiology 19:372–397

Bambach RK (1999) Energetics in the global marine fauna: a connection between terrestrial diversification and change in the marine biosphere. Geobios 32:131–144

Bambach RK, Knoll AH (in review) Physiological selectivity during the end-Permian mass extinction. Earth Planet Sci Lett

Bambach RK, Knoll AH, Sepkoski JJ (2002) Anatomical and ecological constraints on Phanerozoic animal diversity in the marine realm. Proc Nat Acad Sci USA 99:6854–6859

Barron JA (1993) Diatoms. *In:* Fossil Prokaryotes and Protists. Lipps J (ed) Blackwell Scientific, Oxford, p 155–167

Baud A, Cirilli S, Marcoux J (1997) Biotic response to mass extinction: the lowermost Triassic microbialites. Facies 36:238–242

Becker L, Poreda RL, Hunt AG, Bunch TE, Rampino M (2001) Impact event at the Permian-Triassic boundary: evidence from extraterrestrial gases in fullerenes. Science 291:1530–1533

Bengtson S (1994) The advent of animal skeletons. In: Early Life on Earth, Nobel Symposium 84. Bengtson S (ed) Columbia University Press, New York, p 421–425

Bengtson S, Conway Morris S (1992) Early radiation of biomineralizing phyla. In: Origin and Early Evolution of the Metazoa. Lipps JH, Signor PW (eds) Plenum, New York, p 447–481

Bengtson S, Runnegar B (1992) Origins of biomineralization in metaphytes and metazoans. In: The Proterozoic Biosphere: A Multidisciplinary Study. Schopf JW, Klein C (eds) Cambridge University Press, Cambridge, p 447–451

Bengtson S, Yue Z (1992) Predatorial borings in late Precambrian mineralized xoskeletons. Science 257:267–369

Bengtson S, Conway Morris S, Cooper BJ, Jell PA, Runnegar BN (1990) Early Cambrian fossils from South Australia. Mem Assoc Australas Palontol 9:1–364

Berner RA (2002) Examination of hypotheses for the Permo-Triassic boundary extinction by carbon cycle modeling. Proc Nat Acad Sci USA 99:4172–4177

Berner RA, Kothavala Z (2001) GEOCARB III: A revised model of atmospheric CO_2 over Phanerozoic time. Am J Sci 301:182–204

Bowring SA, Erwin DH, Jin YG, Martin MW, Davidek K, Wang, W (1998) U/Pb zircon geochronology and tempo of the end-Permian mass extinction. Science 280:1039–1045

Brasier M, Green O, Shields G (1997) Ediacaran sponge spicule clusters from southwestern Mongolia and the origins of the Cambrian fauna. Geology 25:303–306

Briggs DEG, Erwin DH, Collier FJ (1994) The Fossils of the Burgess Shale. Smithsonian Institution Press, Washington

Brocks JJ, Logan GA, Buick R, Summons RG (1999) Archean molecular fossils and the early rise of eukaryotes. Science 285:1033–1036

Budd GE, Jensen S (2000) A critical reappraisal of the fossil record of the bilaterian phyla. Biol Rev 75:253–295

Butterfield NJ (2000) *Bangiomorpha pubescens* n. gen., n. sp; implications for the evolution of sex, multicellularity, and the Mesoproterozoic/Neoproterozoic radiation of eukaryotes. Paleobiology 26:386–404

Butterfield NJ (2002) A *Vaucheria*-like fossil from the Neoproterozoic of Spitsbergen. Geol Soc Am Abstr Progr 34:169

Butterfield NJ, Knoll AH, Swett K (1994) Paleobiology of the Upper Proterozoic vanbergfjellet Formation, Spitsbergen. Fossils and Strata 34:1–84

Carter JG (ed) (1990) Skeletal Biomineralization, Patterns, Processes and Evolutionary Trends. Van Nostrand Reinhold and Company, NewYork

Chacón-Baca E, Beraldi-Campesi H, Knoll AH, Golubic S, Cevallos-Ferriz SR (2002) 70 million year old non-marine diatoms from northern Mexico. Geology 30:279–281

Chen J, Zhou G (1997) Biology of the Chengjiang fauna. Bull Nat Mus Nat Sci (Taiwan) 10:11–106

Clack JA, Farlow (2002) Gaining Ground: The Origin and Early Evolution of Tetrapods. Indiana University Press, Bloomington, Indiana

Cloud PE (1968) Pre-metazoan evolution and the origins of the Metazoa. In: Evolution and Environment. Drake T (ed) Yale University Press, New Haven, p 1–72

Conley DJ (2002) Terrestrial ecosystems and the global biogeochemical silica cycle. Global Biogeochem Cycles 16(4):1121, doi:10.1029/2002GB001894

Conway Morris S, Mattes BW, Chen M (1990) The early skeletal organism *Cloudina*: new occurrences from Oman and possibly China. Am J Sci 290A:245–260

Culver SJ (1993) Foraminifera. In: Fossil Prokaryotes and Protists. Lipps J (ed) Blackwell Scientific, Oxford, p 203–247

Debrenne F, Lafuste J, Zhuravlev A (1990) Coralomorphes et spongiomorphes á l'aube du Cambrien. Bull Mus Natn Hist Nat Paris, 4ᵉ Sér 12(sec C, nr 1):17–39

Dong X, Knoll AH (1996) Middle and Late Cambrian sponge spicules from Hunan, China. J Paleontol 70:173–184

Dong X, Knoll AH, Lipps J (1997) Late Cambrian radiolaria from Hunan, China. J Paleontol 71:753–758.

Donoghue PCJ, Forey PL, Aldridge RJ (2000) Conodont affinity and chordate phylogeny. Biol Rev 75:191–251

Erwin DH (2001) Lessons from the past: biotic recoveries from mass extinctions. Proc Nat Acad Sciences USA 98:5399–5403

Erwin DH, Bowring SA, Jin YG (2002) End-Permian mass extinctions: a review. Geol Soc Am Spec Papers 356: 363–383

Falkowski PG, Schofield O, Katz ME, van Schotteberge B, Knoll AH (in press) Why is the land green and the ocean red? *In*: Coccolithophores: From Molecular Process to Global Impact. Thierstein H, Young JR (eds) Springer Verlag, Heidelberg

Fensome RA, MacRae RA, Moldowan JM, Taylor FJR, Williams GL (1996) The early Mesozoic radiation of the dinoflagellates. Paleobiology 22:329–338

Flügel E (2002) Triassic reef patterns. *In*: Phanerozoic Reef Patterns, SEPM Special Publication 72. Kiessling W, Flügel, Golonka J (eds) SEPM Press, Tulsa, Oklahoma, p 391–463

Gattuso J-P, Buddemeier RW (2000) Calcifcation and CO_2. Nature 407:311–312

Gattuso J-P, Frankignoulle M, Bourge I, Romaine S, Buddemeier RW (1998) Effect of calcium carbonate saturation of seawater on coral calcification. Global Planet Change 18:37–46

German TN (1990) Organic World One Billion Year Ago. Nauka, Leningrad

Germs GJB (1972) New shelly fossils from Nama Group, South West Africa. Am J Sci 272:752–761

Gibbs PE, Bryan GW (1984) Calcium phosphate granules in muscle cells of *Nephtys* (Annelida, Polychaeta)—a novel skeleton? Nature 310:494–495

Giribet G (2002) Current advances in the phylogenetic reconstruction of metazoan evolution: a new paradigm for the Cambrian explosion? Mol Phylogenet Evol 23:345–357

Gotliv BA, Addadi L, Weiner S (2003) Mollusk shell acidic proteins: in search of individual functions. Chembiochem 4:522–529

Grant SWF (1990) Shell structure and distribution of *Cloudina*, a potential index fossil for the terminal Proterozoic. Am J Sci 290A:261–294

Grant SWF, Knoll AH, Germs GJB (1991) A probable calcified metaphyte from the latest Proterozoic Nama Group, Namibia. J Paleontol 65:1–18

Grey K, Williams IR (1990) Problematic bedding-plane markings from the Middle Proterozoic Manganese Subgroup, Bangemall Basin, Western Australia. Prec Res 46:307–328

Grotzinger JP (1989) Facies and evolution of Precambrian carbonate depositional systems: emergence of the modern platform archetype. *In*: Controls on Carbonate Platform and Basin Development. SEPM Special Publication 44. Crevello PD, Wilson JL, Sarg JF, Read JF (eds) SEPM Press, Tulsa, Oklahoma, p 79–106

Grotzinger JP, Knoll AH (1995) Anomalous carbonate precipitates: Is the Precambrian the key to the Permian? Palaios 10:578–596

Grotzinger JP, Knoll AH (1999) Proterozoic stromatolites: evolutionary mileposts or environmental dipsticks? Ann Rev Earth Planet Sci 27:313–358

Grotzinger JP, Watters W, Knoll AH (2000) Calcareous metazoans in thrombolitic bioherms of the terminal Proterozoic Nama Group, Namibia. Paleobiology 26:334–359

Hagadorn JW, Bottjer DJ (1997) Wrinkle structures: microbially mediated sedimentary structures in siliciclastic settings at the Proterozoic-Phanerozoic transition. Geology 25:1047–1050

Hargraves PE (2003) The ebridian flagellates *Ebria* and *Hermesinum*. Plankton Biol Ecol 49:9–16

Harper EM, Palmer TJ, Alphey JR (1997) Evolutionary response by bivalves to changing Phanerozoic seawater chemistry. Geol Mag 134:403–407

Harper HE, Knoll AH (1975) Silica, diatoms, and Cenozoic radiolarian evolution. Geology 3:175–177

Harwood DM, Nikolaev VA (1995) Cretaceous diatoms: morphology, taxonomy, biostratigraphy. *In*: Siliceous Microfossils, Short Courses in Paleontology 8. Blome CD, Whalen PM, Reed KM (eds) Paleontological Society, Tulsa OK, p 81–106

Horodyski RJ, Mankiewicz C (1990) Possible late Proterozoic skeletal algae from the Pahrump Group, Kingston Range, southeastern California. Am J Sci 290A:149–169

Houghton SD (1991) Calcareous nannofossils. *In*: Calcareous Algae and Stromatolites. Riding R (ed). Springer Verlag, Heidelberg, p 217–266

Huh Y, Edmond JM (1999) The fluvial geochemistry of the rivers of Eastern Siberia: III. tributaries of the Lena and Anabar draining the basement terrain of the Siberian Craton and the Trans-Baikal Highlands. Geochim Cosmochim Acta 63:967–987

Janis CM (1995) Correlations between craniodental morphology and feeding behavior in ungulates: reciprocal illumination between living and fossil taxa. *In:* Functional Morphology in Vertebrate Paleontology. Thomason JJ (ed) Cambridge University press, Cambridge, p 76–98

Javaux EJ, Knoll AH, Walter MR (2001) Ecological and morphological complexity in early eukaryotic ecosystems. Nature 412:66–69

Javaux E, Knoll AH, Walter MR (2003) Recognizing and interpreting the fossils of early eukaryotes. Origins Life Evol Biosphere 33:75-94

Jin YG, Wang Y, Wang W, Shang QH, Cao CQ, Erwin DH (2000) Pattern of marine mass extinction near the Permian-Triassic boundary in south China. Science 289:432–436

Johnson JH (1966) A review of Cambrian algae. Quart Colorado School Mines 61:1–162

Kaiho K, Kajiwara Y, Nakano T, Miura Y, Kawahata H, Tazaki K, Ueshima M, Chen ZQ, Shi GR (2001) End-Permian catastrophe by a bolide impact: Evidence of a gigantic release of sulfur from the mantle. Geology 29:815–818

Kennard JM, James NP (1986) Thrombolites and stromatolites: two distinct types of microbial structures. Palaios 1:492–503

Kershaw S, Guo L, Swift A, Fan J (2002) Microbialites in the Permian-Triassic boundary interval in central China: structure, age, and distribution. Facies 47:83–90

Kleypas JA, Buddemeier RW, Archer D, Gattuso J-P, Langdon C, Opdyke BN (1999) Geochemical consequences of increased atmospheric carbon dioxide on coral reefs. Science 284:118–120

Knoll AH (1994) Proterozoic and Early Cambrian protists: evidence for accelerating evolutionary tempo. Proc Nat Acad Sci USA 91:6743–6750

Knoll AH (2003) Life on a Young Planet: The First Three Billion Years of Evolution on Earth. Princeton University Press, Princeton NJ

Knoll AH, Semikhatov MA (1998) The genesis and time distribution of two distinctive Proterozoic stromatolite microstructures. Palaios 13:407–421

Knoll AH, Swett K (1990) Carbonate deposition during the late Proterozoic era: An example from Spitsbergen. Am J Sci 290A:104–132

Knoll AH, Bambach R, Canfield D, Grotzinger JP (1996) Comparative Earth history and late Permian mass extinction. Science 273:452–457

Knoll AH, Fairchild IJ, Swett K (1993) Calcified microbes in Neoproterozoic carbonates: implications for our understanding of the Proterozoic-Cambrian transition. Palaios 8:512–525

Kouchinsky A (2000) Shell microstructures in Early Cambrian molluscs. Acta Palaeontol Pol 45:119–150

Krasko, A, Lorenz B, Batel R, Schröder HC, Müller IM, Müller WEG (2000) Expression of silicatein and collagen genes in the marine sponge *Suberites domuncula* is controlled by silicate and myotrophin. Eur J Biochem 267:4878–4887

Kröger N, Sumper M (1998) Diatom cell wall proteins and the biology of silica biomineralization. Protist 149: 213–219

Kröger N, Deutzmann R, Bergsdorf C, Sumper M (2000) Species-specific polyamines from diatoms control silica morphology. Proc Nat Acad Sci USA 97:14133–14138

Kröger N, Lorenz S, Brunner E, Sumper M (2002) Self-assembly of highly phosphorylated silaffins and their function in silica morphogenesis. Science 298:584–586

Langdon C, Takahashi T, Sweeney C, Chipman D, Goddard J, Marubini F, Aceves H, Barnett H (2000) Effect of calcium carbonate saturation state on the calcification rate of an experimental coral reef. Global Biogeochem Cycles 14:639–654

Leadbetter BSC, Riding R (1986) Biomineralization in Lower Plants and Animals. The Systematics Association Special Volume 30. Clarendon Press, Oxford

Lehrmann DL, Payne JL, Felix SV, Dillett PM, Wang H, Yu Y, Wei J (2003) Permian-Triassic boundary sections from shallow-marine carbonate platforms of the Nanpanjiang Basin, south China: Implications for oceanic conditions associated with the end-Permian extinction and its aftermath. Palaios 18:138–152

Looy CV, Twitchett RJ, Dilcher DL, Konijnenburg-Van Cittert JHA, Visscher H (2001) Life in the end-Permian dead zone. Proc Nat Acad Sci USA 98:7879–7883

Lopez E, Vidal B, Berland S, Camprasse S, Camprasse G, Silve C (1992) Demonstration of the capacity of nacre to induce bone formation by human osteoblasts *in vitro*. Tissue and Cell 24:667–679

Lowenstam HA, Weiner S (1989) On Biomineraliztion. Oxford University Press, Oxford

Luo G, Ducy P, McKee MD, Pinero GJ, Loyer E, Behringer RR, Karsenty G (1997) Spontaneous self-calcification of arteries and cartilage in mice lacking matrix GLA protein. Nature 386:78–81

Maldonado M, Carmona MG, Uriz MJ, Cruzado A (1999) Decline in Mesozoic reef-building sponges explained by silicon limitation. Nature 401:785–788

MacLennan DH, Rice WJ, Green NM (1997) The mechanism of Ca^{2+} transport by sarco(endo)plasmic reticulum Ca^{2+}-ATPases. J. Biol Chem 272:28815–28818

Maliva R, Knoll AH, Siever R (1990) Secular change in chert distribution: a reflection of evolving biological participation in the silica cycle. Palaios 4:519–532

Marin F, Smith M, Isa Y, Muyzer G, Westbroek P (1996) Skeletal matrices, muci, and the origin of invertebrate calcification. Proc Nat Acad Sci USA 93:1554–1559

Marin F, Corstjens P, de Gaulejac B, de Vrind-de Jong EW, Westbroek P (2000) Mucins and molluscan calcification: molecular characterization of mucoperlin, a novel mucin-like protein from the nacreous layer of the fan mussel *Pinna nobilis* (Bivalvia, Pteriomorpha). J Biol Chem 275:20667–20675

Marubini F, Ferrier-Pages C, Cuif JP (2003) Suppression of skeletal growth in scleractinian corals by decreasing ambient carbonate-ion concentration: a cross-family comparison. Proc Royal Soc London B270:179–184

Maxwell MD (1992) Permian and Early Triassic extinction of non-marine tetrapods. Paleontology 35:571–583

McIlroy D, Green OR, Brasier MD (2001) Palaeobiology and evolution of the earliest agglutinated Foraminifera: *Platysolenites*, *Spirosolenites* and related forms. Lethaia 34:13–29

Midgley JJ, Midgley G, Bond WJ (2002) Why were dinosaurs so large? A food quality hypothesis. Evol Ecol Res 4:1093-1095

Miller AI (1977) Dissecting global diversity patterns: examples from the Ordovician radiation. Annu Rev Ecol Syst 28:85–104

Mishler BD, Lewis LA, Buchheim MA, Renzaglia KS, Garbary DJ, Delwiche CF, Zechman FW, Kantz TS, Chapman RL (1994) Phylogenetic relationships of the "green algae" and "bryophytes." Ann Missouri Bot Gard 81:451–483

Moldowan JM, Talyzina NM (1998) Biogeochemical evidence for dinoflagellate ancestors in the Early Cambrian. Science 281:1168–1170

Moldowan JM, Dahl J, Jacobson SR, Huizinga BJ, Fago FJ, Shetty R, Watt DS, Peters KE (1996) Chemostratigraphic reconstruction of biofacies: Molecular evidence linking cyst-forming dinoflagellates with pre-Triassic ancestors. Geology 24:159–162

Narbonne GM (1998) The Ediacaran biota: a terminal Proterozoic experiment in the evolution of life. GSA Today 8:1–7

Noffke N, Knoll AH, Grotzinger JP (2002) Sedimentary controls on the formation and preservation of microbial mats in siliciclastic deposits: a case study from the Upper Neoproterozoic Nama Group, Namibia. Palaios 17:533–544

Porter SM (in press) Closing the 'phosphatization' window: implications for interpreting the record of small shelly fossils. Paleobiology

Porter SM, Knoll AH (2000) Testate amoebae in the Neoproterozoic Era: evidence from vase-shaped microfossils in the Chuar Group, Grand Canyon. Paleobiology 26: 360–385

Porter SM, Meisterfeld R, Knoll AH (2003) Vase-shaped microfossils from the Neoproterozoic Chuar Group, Grand Canyon: a classification guided by modern testate amoebae. J Paleontol 77:205–225.

Quigg A, Finkel ZV, Irwin AJ, Rosenthal Y, Ho T-Y, Reinfelder JR, Schofield O, Morel FMM, Falkowski PG (2003) The evolutionary inheritance of elemental stoichiometry in marine phytoplankton. Nature 425:291-294

Racki G, Cordey F (2000) Radiolarian palaeoecology and radiolarites: is the present the key to the past? Earth Sci Rev 52:83–120

Rapp GS, Mulholland SC (1992) Phytolith Systematics: Emerging Issues. Plenum, New York

Raup DM (1979) Size of the Permo-Triassic bottleneck and its evolutionary implications. Science 206:217–218

Rasmussen B, Bengtson S, Fletcher IR, McNaughton NJ (2002) Discoidal impressions and trace-like fossils more than 1200 million years old. Science 296:1112–1115

Remane J (2000) International stratigraphic chart. International Union of Geosciences. IUGS Secretariat, Geological Survey of Norway, Trondheim, Norway

Reitner J, Wörheide, Lange R, Thiel W (1997) Biomineralization of calcified skeletons in three Pacific coralline demosponges—an approach to the evolution of basal skeletons. Cour Forch-Inst Senckenberg 201:371–383

Rhoades DR (1979) Evolution of plant chemical defense against herbivores. *In*: Herbivores: their interaction with secondary plant metabolites. Rosenthal GH, Janzen DH (eds) Academic Press, New York, p 3–54

Riding R, Zhuravlev AYu (1995) Structure and diversity of oldest sponge-microbe reefs: Lower Cambrian, Aldan River, Siberia. Geology 23:649–652

Riding R, Cope JCW, Taylor PD (1998) A coralline-like red alga from the lower Ordovician of Wales. Palaeontology 41:1069–1076

Riebesell U, Zondervan I, Rost B, Tortell PD, Zeebe RE, Morel FMM (2000) Reduced calcification of marine plankton in response to increased atmospheric CO_2. Nature 407:364–367

Rodland DL, Bottjer DJ (2001) Biotic recovery from the end-Permian mass extinction: behavior of the inarticulate brachiopod as a disaster taxon. Palaios 15: 95–101

Romano SL, Cairns SD (2000) Molecular phylogenetic hypotheses for the evolution of scleractinian corals. Bull Mar Sci 67:1043–1068

Rothpletz A (1896) Über die Flysch-Fucoiden und einige andere fossile Algen, sowie über liasische, Diatomeen führende Hörenschwämme. Deutsch Geol Ges 48:858–914

Rowland SM, Shapiro RS (2002) Reef patterns and environmental influences in the Cambrian and earliest Ordovician. *In*: Phanerozoic Reef Patterns, SEPM Special Publication 72. Kiessling W, Flügel, Golonka J (eds) SEPM Press, Tulsa, Oklahoma, p 95–128

Runnegar B (1985) Shell microstructures of Cambrian mollusks replicated by phosphate. Alcheringa 9:245–257

Runnegar B (1987) Rates and modes of evolution in the Mollusca. *In*: Rates of Evolution. Campbell KS, Day MF (eds) Allen and Unwin, London, p 39–60

Sandberg PA (1983) An oscillating trend in Phanerozoic non-skeletal carbonate mineralogy. Nature 305:19–22

Sanders D, Brownlee C, Harper JF (1999) Communicating with calcium. Plant Cell 11:691–706

Samankassou E (1996) Skeletal framework mounds of dasycladalean alga *Anthracoporella*, Upper Paleozoic, Carnic Alps, Austria. Palaios 13:297–300

Schubert JK, Bottjer DJ (1992) Early Triassic stromatolites as post-mass extinction disaster forms. Geology 20:883–886

Schubert JK, Bottjer DJ (1995) Aftermath of the Permian-Triassic mass extinction event: paleoecology of Lower Triassic carbonates in the western USA. Palaeogeogr Palaeoclimatol Palaeoecol 116:1–39

Senowbari-Daryan B, Flügel E (1993) *Tubiphytes* Maslov, an enigmatic fossil: classification, fossil record and significance through time, Part I: discussion of Late Paleozoic material. Soc Paleontol Ital Boll Spec Vol 1:353–382

Siesser W (1993) Calcareous nannoplankton. *In*: Fossil Prokaryotes and Protists. Lipps J (ed) Blackwell Scientific, Oxford, p 169–201

Sepkoski JJ (1982) A compendium of fossil marine families. Milwaukee Publ Mus Contr Biol Geol 51:1–125

Sepkoski JJ, Sheehan (1992) Diversification, faunal change, and community replacement during the Ordovician radiation. *In*: Biotic Interactions in Recent and Fossil Communities. Tevesz MJS, McCall PL (eds) Plenum, New York, p 673–717

Shu DG, Luo HL, Morris SC, Zhang XL, Hu SX, Chen L, Han J, Zhu M, Li Y, Chen LZ (1999) Lower Cambrian vertebrates from South China. Nature 402:42–46

Simkiss K, Wilbur KM (1989) Biomineralization: Cell Biology and Mineral Deposition. Academic Press, New York

Stanley GD (2003) The evolution of modern corals and their early history. Earth-Sci Rev 60:195–225

Stanley SM (1975) Adaptive themes in the evolution of the Bivalvia. Annu Rev Earth Planetary Sci 3:361–385

Stanley SM, Hardie LA (1998) Secular oscillations in the carbonate mineralogy of reef-building and sediment-producing organisms driven by tectonically forced shifts in seawater chemistry. Palaeogeogr Palaeoclimatol Palaeoecol 144:3–19

Stanley SM, Ries JB, Hardie LA (2002) Low-magnesium calcite produced by coralline algae in seawater of Late Cretaceous composition. Proc Nat Acad Sci USA 99:15323–15326

Sullivan BK, Miller CB, Peterson WT, Soeldner AH (1975) A scanning electron microscope study of the mandibular morphology of boreal copepods. Marine Biol 30:175–182

Summons RE, Walter MR (1990) Molecular fossils and microfossils of prokaryotes and protists from Proterozoic sediments. Am J Sci 290A:212–244

Thomas RDK, Reif WE (1993) The skeleton space: a finite set of organic designs. Evolution 47:341–360

Thomas RDK, Shearman RM, Stewart GW (2000) Evolutionary exploitation of design options by the first animals with hard skeletons. Science 288:1239–1242

Tinbergen N (1963) On aims and methods of ethology. Zeitschr Tierpsychol 20:410–413

Verma HC, Upadhyay C, Tripathi RP, Shukla AD, Bhandari N (2002) Evidence of impact at the Permian/Triassic boundary from Mössbauer spectroscopy. Hyperfine Interactions 141:357–360

Vermeij GJ (1977) The Mesozoic marine revolution: evidence from snails, predators and grazers. Paleobiology 3:245–258

Vermeij GJ (1995) Economics, volcanos, and Phanerozoic revolutions. Paleobiology 21:125–152

Wahlman GP (2002) Upper Carboniferous-Lower Permian reefs (Bashkiri-Kungarian) mounds and reefs. *In*: Phanerozoic Reef Patterns, SEPM Special Publication 72. Kiessling W, Flügel, Golonka J (eds) SEPM Press, Tulsa, Oklahoma, p 271–228

Walter MR, Du R, Horodyski RJ (1990) Coiled carbonaceous megafossils from the Middle Proterozoic of Jixian (Tinjian) and Montana. Am J Sci 290A:133–148

Wang DYC, Kumar S, Hedges SB (1999) Divergence time estimates for the early history of animal phyla and the origin of plants, animals and fungi. Proc Royal Soc London B266:163–171

Weidlich O (2002) Middle and Late Permian reefs—distributional patterns and reservoir potential. *In*: Phanerozoic Reef Patterns, SEPM Special Publication 72. Kiessling W, Flügel, Golonka J (eds) SEPM Press, Tulsa, Oklahoma, p 339–390

Weiner S, Addadi L (1997) Design strategies in mineralized biological materials. J Mater Chem 7:689–702

Westbroek P, Marin F (1998) A marriage of bone and nacre. Nature 392:861–862

Wignall PB, Morante R, Newton R (1998) The Permo-Triassic transition in Spitsbergen: $\delta^{13}C_{org}$ chemostratigraphy, Fe and S geochemistry, facies, fauna and trace fossils. Geol Mag 135:47–62

Williams A, Luter C, Cusack M (2001) The nature of siliceous mosaics forming the first shell of the brachiopod *Discinisca*. J Structur Biol 134:25–34

Wollast R, Mackenzie FT (1983) The global cycle of silica. *In*: Silicon Geochemistry and Biochemistry. Astor SR (ed) Academic Press, New York, p 39–76

Won MZ, Below R (1999) Cambrian radiolaria from the Georgina Basin, Queensland, Australia. Micropaleontology 45:325–363

Won MZ, Iams WJ (2002) Late Cambrian radiolarian faunas and biostratigraphy of the Cow Head Group, western Newfoundland. J Paleontol 76:1–33

Wood RA, Grotzinger JP, Dickson JAD (2002) Proterozoic modular biomineralized metazoan from the Nama Group, Namibia. Science 296:2383–2386

Wray JL (1977) Calcareous Algae. Elsevier, Amsterdam

Xiao S, Knoll AH (2000) Phosphatized animal embryos from the Neoproterozoic Doushantuo Formation at Weng'an, Guizhou Province, South China. J Paleontol 74:767–788

Xiao S, Knoll AH, Yuan X, Pueschel C (in press) Phosphatized multicellular algae in the Neoproterozoic Doushantuo Formation, China, and the early evolution of florideophyte red algae. Am J Bot

Yoon HS, Hackett JD, Pinto G, Bhattacharya D (2002) The single, ancient origin of chromist plastids. Proc Nat Acad Sci, USA 99:15507–15512

Young JR, Davis SA, Bown PR, Mann (1999) Coccolith ultrastructure and biomineralisation. J Struc Biol 126:195–215

Zhuravlev AY, Wood RA (1996) Anoxia as the cause of the mid-Early Cambrian (Botomian) extinction event. Geology 24:311–314

Zondervan I, Zeebe RE, Rost B, Riebesell (2001) Decreasing marine biogenic calcification: a negative feedback on rising atmospheric pCO$_2$. Global Biogeochem Cycles 15:507–516

Zurzolo C, Bowler C (2001) Exploring bioinorganic pattern formation in diatoms. A story of polarized trafficking. Plant Physiol 127:1339–1345

12 Biomineralization and Global Biogeochemical Cycles

Philippe Van Cappellen
Faculty of Geosciences, Utrecht University
P.O. Box 80021
3508 TA Utrecht, The Netherlands

INTRODUCTION

Biological activity is a dominant force shaping the chemical structure and evolution of the earth surface environment. The presence of an oxygenated atmosphere-hydrosphere surrounding an otherwise highly reducing solid earth is the most striking consequence of the rise of life on earth. Biological evolution and the functioning of ecosystems, in turn, are to a large degree conditioned by geophysical and geological processes. Understanding the interactions between organisms and their abiotic environment, and the resulting coupled evolution of the biosphere and geosphere is a central theme of research in biogeology. Biogeochemists contribute to this understanding by studying the transformations and transport of chemical substrates and products of biological activity in the environment.

Biogeochemical cycles provide a general framework in which geochemists organize their knowledge and interpret their data. The cycle of a given element or substance maps out the rates of transformation in, and transport fluxes between, adjoining environmental reservoirs. The temporal and spatial scales of interest dictate the selection of reservoirs and processes included in the cycle. Typically, the need for a detailed representation of biological process rates and ecosystem structure decreases as the spatial and temporal time scales considered increase.

Much progress has been made in the development of global-scale models of biogeochemical cycles. Although these models are based on fairly simple representations of the biosphere and hydrosphere, they account for the large-scale changes in the composition, redox state and biological productivity of the earth surface environment that have occurred over geological time. Since the Cambrian explosion, mineralized body parts have been secreted in large quantities by biota. Because calcium carbonate, silica and calcium phosphate are the main mineral phases constituting these hard parts, biomineralization plays an important role in the global biogeochemical cycles of carbon, calcium, silicon and phosphorus.

The chapter starts with introducing the basic concepts of global biogeochemical cycles. An overview of the main forcing mechanisms is presented, followed by a discussion of box models used to simulate global elemental cycles. Emphasis is on the variety of spatial and temporal scales over which biogeochemical cycles operate, and the resulting need to identify and describe processes and system variables that are relevant at the different spatio-temporal scales. The first part concludes with a short discussion of the carbon cycle, because of the central role of this element in biogeochemistry.

Next, the impact of biomineralization on the cycles of C, Ca, Si and P is reviewed. A comparison is made between elemental fluxes associated with soft and hard tissues in the oceans. An important conclusion is that calcareous and biosiliceous skeletal remains are the dominant forms under which C, Ca and Si are removed from the oceans. Burial of biogenic calcium phosphate is a relatively unimportant sink for P in the modern ocean, but this may have been different in the past during periods of ocean anoxia. The intense cycling of silicon in terrestrial ecosystems is briefly discussed. A geochemist's perspective on the implications of the "hard body part revolution" marking the end of the Precambrian is then given. While calcifying organisms essentially took advantage of the existing (abiotic) equilibria of the calcium carbonate-seawater system, silicifying organisms completely changed how the marine silicon cycle functions.

The last part of the chapter deals with the marine silicon cycle. A simple steady state analysis highlights the importance of geological and geochemical controls on the production of biogenic silica in the oceans. Then the upscaling of silicate dissolution rates to the global weathering flux of dissolved silicic acid is briefly discussed. The final section summarizes some recent work on biogenic silica preservation in marine sediments. In particular, the results point to aluminum as an important long-term regulator of marine biosiliceous production.

BIOGEOCHEMICAL CYCLES

Forcing mechanisms and time scales

Biogeochemical cycles operate on many different spatial and temporal scales (Fig. 1). At the level of individual ecosystems, biogeochemical fluxes are controlled by interactions between organisms and by external forcings that typically fluctuate on diurnal to decadal time scales. Interactions include competition for food and habitat, parasitism and predation, while examples of forcings are air temperature, rainfall, tidal inundation, wind stress and flooding events. At this level, biogeochemistry is closely linked to ecology, and the emphasis is on relating fluxes of chemical elements to the

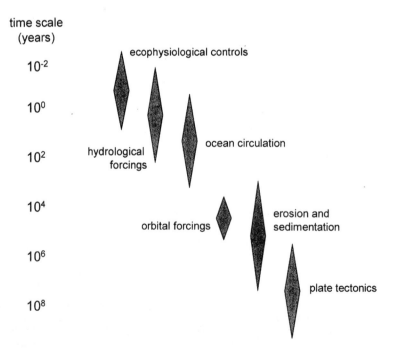

Figure 1. Examples of forcings of biogeochemical cycles and their characteristic time scales.

structure and functioning of the ecosystem. The monographs of De Angelis (1992) and Valiela (1995) illustrate this close integration of ecology and biogeochemistry.

At the other end of the spectrum, the biogeochemical dynamics of the global earth system respond to geophysical forcings on time scales of millions of years and longer. Plate tectonics continuously rearrange the physical setting in which the biosphere functions and evolves, by shaping the morphology of ocean basins and modifying the position and elevation of the continents. Equally important, plate tectonics control the intensity of chemical exchanges between the earth's surface environment and the underlying lithosphere (e.g., Berner 1990).

The major bioessential elements, including carbon and the nutrient elements N, P, S, Si, K and Ca, are efficiently recycled within terrestrial and marine ecosystems (Schlesinger 1997). As a consequence, on a yearly basis, only very small fractions of these elements escape through removal to the lithosphere. Therefore, on relatively short time scales, say, less than 1000 years, the lithosphere is a minor sink in most biogeochemical cycles. At these time scales, modeling efforts tend to focus on the redistribution of chemical constituents among the atmosphere, hydrosphere and biosphere. A good example is provided by global carbon cycle models used in predicting the future fate of anthropogenic CO_2 emissions to the atmosphere. How much of this CO_2 remains in the atmosphere is, in the short term, mainly determined by transfer of CO_2 from the atmosphere to the ocean, and to vegetation plus soils on land (e.g., Sarmiento and Gruber 2002).

With the passing of time, however, the cumulative loss of bioessential elements to the lithosphere becomes significant. For most major nutrient elements, the principal escape route is incorporation in marine sedimentary deposits. Unless somehow compensated, burial in ocean sediments would ultimately deplete the surface reservoirs of nutrients, resulting in the collapse of biological activity on earth. Marine turnover times (Equation 3) of limiting nutrients, relative to removal by sediment burial, provide rough estimates of the time scale over which such a collapse would take place. These turnover times are on the order of 10^4 to 10^5 years.

Fortunately, the loss of bioessential elements to the lithosphere is countered by their release by volcanic outgassing and chemical weathering of rocks. Hence, the latter processes are essential for the continued survival of life on geological time scales. From the point of view of the global cycles of bioessential elements, we can therefore distinguish between long and short time scales, depending on whether the lithosphere is a significant sink and source, or not. The distinction between short and long times is not a sharp one and varies from one element to the other. It is safe to state, however, that for time spans $\geq 10^4$ years chemical exchanges with the lithosphere become a key factor controlling biogeochemical cycles.

With the exception of N, the lithosphere is the largest reservoir of the major nutrient elements. Sediments and rocks contain orders of magnitude more C, P, S, Fe, Si, Ca, and K than the atmosphere, hydrosphere, pedosphere and biosphere combined (e.g., Garrels and Mackenzie 1971; Drever et al. 1988; Chameides and Perdue 1997; Reeburgh 1997; Mackenzie 1998). The transit times of these elements through the lithosphere are similarly orders of magnitude longer than their turnover times at the earth surface. On average, it takes several hundreds of millions of years for non-volatile materials buried in marine sediments to be exposed on land by plate tectonic processes. Once incorporated in the lithosphere, bioessential elements may follow a variety of different pathways, however (Fig. 2). As a result, transit times through the lithosphere exhibit a broad range of values.

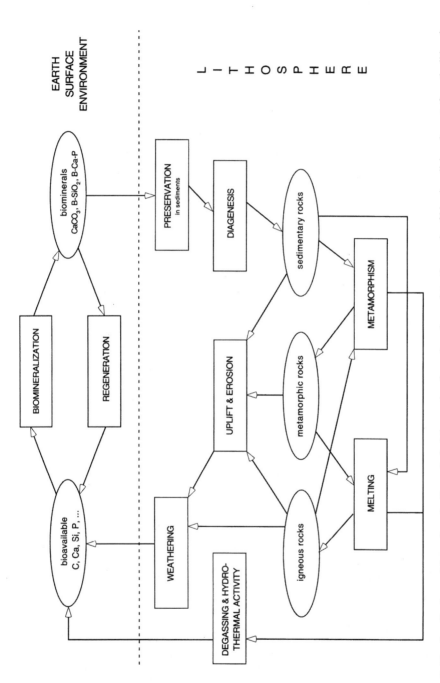

Figure 2. Biomineralization and the rock cycle. On geological time scales, the rock cycle regulates the supply of biomineralizing elements from the lithosphere, as well as their removal from the earth surface environment.

Pathways through the lithosphere not only determine how long a given element is removed from the earth surface environment, but also its chemical state and reactivity when it reappears. Marine sediments may be entrained deep into the mantle at subduction zones, and their chemical constituents return to the earth surface as part of igneous and metamorphic rocks, or they make their way back with volcanic emanations and hydrothermal discharge. Alternatively, sediments accumulating in sedimentary basins may never experience temperatures and pressures sufficient to cause melting and metamorphism. Instead, they resurface as sedimentary rocks, after having undergone variable degrees of diagenesis.

Biogenic constituents of sediments are particularly sensitive to diagenetic alteration. With advancing diagenesis, organic debris mature into kerogen, while calcareous and siliceous shell fragments are subject to recrystallization and cementation processes (e.g., Pettijohn 1957; Tegelaar et al. 1989). As diagenetic transformations generally increase the resistance of rocks to chemical weathering, carbon and nutrient elements tend to be more efficiently remobilized from younger sedimentary rocks that resurface after a relatively brief cycle of burial and uplift. For instance, sedimentary rocks younger than 250 Ma are estimated to release about three times as much dissolved silicic acid to the world's rivers as do older sedimentary rocks, although twice as much Si is locked up in the older rock reservoir (Garrels and Mackenzie 1972). In a similar vein, Berner (1987) proposes that more realistic simulations of the biogeochemical cycles of C and S over Phanerozoic time are obtained when splitting the sedimentary rock reservoir into young and old sub-reservoirs. Model results indicate that first-order rate constants for weathering of C and S are 10-100 times higher for rocks younger than 100 Ma, compared to older rocks (Berner and Canfield 1989).

On million-year time scales, earth surface reservoirs respond near-instantaneously to changes in the supply rates of bioessential elements from the lithosphere. The fast transit times through the earth surface environment, relative to those through the lithosphere, mean that the global sources and sinks of carbon and nutrient elements track each other closely, when integrated over 1 Ma or more. Hence, on these long time scales, biogeochemical cycles at the earth surface can be thought of as evolving from one quasi-steady state to another, under the influence of tectonically-driven variations in the intensity of exchanges with the lithosphere. Major transitions in the biogeochemical state and functioning of the earth surface system can therefore often be traced back to large-scale geological events, such as the opening or closing of ocean basins, massive lava eruptions (plateau basalts), and mountain building episodes.

There are additional forcings acting on global biogeochemical cycles, however. In particular, the evolution of the sun and the evolution of life itself impose their constraints on the biogeochemical dynamics of the earth surface environment (e.g., Caldeira and Kasting 1992; Van Andel 1994). Some of the most dramatic biogeochemical changes that have taken place on earth are related to the build-up in the atmosphere and oceans of O_2 produced by photosynthetic organisms. This build-up is a billion year-long story of complex interactions between changing solar luminosity, microbial evolution, weathering, ecosystem development, climate, sediment burial and volcanic activity (for a lively and up-to-date account, see Lane 2002). In the more recent geological past, the widespread appearance of mineralized body parts, and the colonization of continents by plants have markedly modified biogeochemical cycles at the earth's surface (e.g., Siever 1992; Berner 1992).

Besides being the principal energy source for biological activity, incoming solar radiation also drives the hydrological cycle, as well as atmospheric and oceanic circulation. On temporal scales of months to millennia, groundwater flow, river runoff, atmospheric transport and ocean circulation redistribute bioessential elements at the earth's surface. On time scales of thousands of years and more, rainfall, air temperature and thermohaline circulation play major roles in regulating rock weathering (e.g., Berner and Berner 1997), ocean fertility (e.g., Van Cappellen and Ingall 1994) and sealevel (Turekian 1996). As the time span of interest increases, the solar-driven forcings of biogeochemical cycles (and climate) increasingly overlap with plate tectonic forcings. Because of the multiple forcings acting over a wide range of time scales, as well as the numerous feedbacks and couplings within and among elemental cycles, the interpretation of the earth's biogeochemical workings in terms of simple cause and effect is often no longer possible. At this point, it is best to turn to models.

Models

Environmental conditions and the persistence of a biosphere at the earth's surface are dependent on the continuous flow of matter between different parts of the earth system. This flow is ultimately sustained by the supply of energy from the sun and by the release of heat deep within the earth. As matter is moved around it is also continuously transformed from one chemical form into another. The biogeochemical cycle of an element describes the combined transformation and transport pathways of that element in the earth system. A numerical biogeochemical model is an attempt to translate the conceptual understanding of an element's cycle into mathematical expressions that account for its transport and transformation.

An important aspect in the development of numerical biogeochemical models is discretization. One approach uses a continuous representation of space and time. This approach yields partial derivatives with respect to the spatial coordinates as well as time. The set of partial differential equations that describe mass conservation of the basis species considered are usually referred to as a reactive transport model. Reactive transport models are used when the detailed chemical structure of the system, and its evolution over time, are desired model outcomes. For example, reactive transport modeling may help predict the spreading of pollutants through a groundwater system. Various applications of reactive transport modeling in the earth and environmental sciences can be found in Steefel and Van Cappellen (1998). It is worth noting that reactive transport models have also been used to describe biomineralization processes at the cellular level (e.g., Wolf-Gladrow and Riebesell 1997).

Reactive transport models are mathematically and numerically demanding. When dealing with multicomponent systems, finite difference or finite element methods are needed to solve the equations (see, for example, Steefel and MacQuarry 1996). Therefore, a simpler approach to modeling biogeochemical cycles is frequently used. This so-called box model approach is based on reactor modeling concepts borrowed from chemical engineering. The biogeochemical system is subdivided in a finite number of *reservoirs*. In each reservoir, the relevant chemical, physical and biological properties are assumed to be (reasonably) uniform. *Fluxes* transfer matter from one reservoir to another. A flux into a reservoir is sometimes referred to as a *source*, a flux out of the reservoir as a *sink*. In box models of biogeochemical cycles, matter leaving a given reservoir may ultimately return to that reservoir, often via a number of alternative pathways.

In chemical engineering applications, the reservoir is a physically well-defined reacting system, i.e. the reactor. No attempt is made at describing the mixing processes inside the reactor; rather the extent of reaction is derived from the *residence time distribution* in the reactor. For a perfectly mixed reactor, the concentration of a reactant or product in the outflow is equal to the uniform concentration inside the reactor, C_{res}. The mass balance equation for a reacting species in a perfectly mixed, constant volume reactor with a single inlet and a single outlet, is

$$\frac{dC_{res}}{dt} = \frac{1}{\bar{t}_f}(C_{in} - C_{res}) + R \tag{1}$$

where C_{in} is the concentration of the species in the inflow, R is the rate, per unit volume, at which the species is produced in the reactor (note: when the species is being consumed R is negative), and $\bar{t}_f$ is the mean residence time (or transit time) of the carrier fluid in the reactor:

$$\bar{t}_f = \frac{V}{Q} \tag{2}$$

where V is the volume of the reactor and Q the volumetric flow in and out of the reactor. The residence time thus emerges naturally as a fundamental property in the governing mass balance equation. For reactors, the inflow and outflow are purely advective transport fluxes. As can be seen from Equation (1), governing equations for reactor models, and by extension those of biogeochemical box models, are ordinary differential equations with time as the only independent variable.

Although mathematically similar, box models of biogeochemical cycles differ from simple reactor models in at least two major ways. First, the reservoirs do not have to occupy a continuous physical space, nor do they necessarily correspond to a reacting system. Instead they refer to a collection of matter of a certain chemical type, within given physical and/or biological boundaries. For example, a key reservoir in the carbon cycle is living biomass of marine biota. Included in this reservoir are all organic carbon atoms of living marine organisms. The oceans, as a well-defined physical space, contain many more biogeochemical reservoirs, including other types of organic carbon, e.g., dissolved organic molecules and dead particulate organic matter. Similarly, marine biota encompass other biogeochemically relevant reservoirs, for example, calcium carbonate and nutrient elements such as N, P and Si.

Second, fluxes between reservoirs are not restricted to purely advective flows of matter. Most often, a flux in a biogeochemical box model combines physical transport and biogeochemical transformation processes. For example, the supply of silicic acid to the oceans by weathering of rocks on land may be represented by a flux linking the terrestrial rock reservoir to the marine dissolved silica reservoir. Clearly, this flux encompasses a variety of processes acting in concert, including, uplift and erosion of rocks on land, mineral dissolution and soil formation, continental drainage, and retention of silica by terrestrial and estuarine ecosystems. A large number of geological and environmental variables therefore affects the net flux of dissolved silicic acid to the oceans. Different variables may dominate at different time-scales. Hence, on relatively short time scales, say $\leq 10^3$ years, storage of reactive silica in vegetation and estuaries may be an important regulator of the supply of silicic acid to the oceans. On longer time scales, however, variations in the rate of uplift, rock lithology and climate become the major sources of variability in the supply flux.

The two main tasks facing the developer of a biogeochemical box model is the definition of the reservoirs and the parameterization of the fluxes. There are no magic guidelines, other than to clearly define the goals of the modeling effort and to start as simple as possible. Obviously, a model only provides direct information on reservoirs, fluxes and parameters that are explicitly represented. The reader is referred to the excellent textbook on biogeochemical box modeling by Chameides and Perdue (1997), for a step-by-step introduction to model building and scenario testing. A shorter overview can be found in Rodhe (1992).

It may sometimes be possible to simplify a model by lumping together adjoining reservoirs. The *turnover time* of the reservoirs can help decide whether this is reasonable or not. The turnover time is defined as

$$t_R = \frac{M}{F_{out}} \quad (3)$$

where M is the mass of the reservoir (say, the total number of moles of organic carbon in marine biota) and F_{out} is the total flux out of the reservoir (i.e., the sink). If t_R is much smaller than the time step of the simulations, then joining the reservoir to a larger one does not significantly modify the biogeochemical dynamics of the overall system. For example, the turnover time of reactive phosphorus in the surface ocean is about 2.6 years, while it is on the order of 50,000 years for the ocean as a whole (Reeburgh 1997). For simulations at a million year-time scale, surface ocean phosphorus can therefore be integrated into a single total oceanic phosphorus reservoir. Table 1 compares turnover times at the earth's surface and in the lithosphere of the major elements involved in biomineralization. Note that, at steady state, turnover times of these elements coincide with their (mean) residence times.

Once all the reservoirs are chosen and the fluxes between them identified, mathematical expressions must be derived to describe the fluxes. When dealing with the cycle of a single element, say carbon, a convenient starting point is to adopt a linear box model where the fluxes are assumed to be linearly proportional to the masses in the reservoirs from which they originate:

$$F_{ij} = k_{ij} M_i \quad (4)$$

where F_{ij} and k_{ij} are the flux and linear transfer constant from reservoir i to reservoir j, respectively, while M_i is the mass in reservoir i. If the masses M_i and the fluxes F_{ij} are known at some point in time, then the transfer constants can be derived from Equation (4). The parameterization of the transfer constants for a given element may, for example, be carried out using estimated reservoir sizes and fluxes during the recent geological past, prior to significant human interference.

The great advantage of linear box models is that there exists a general solution, $M_i(t)$, based on the eigenvector-eigenvalue method of matrix calculus (Lasaga 1981). Linear models are not always appropriate for representing biogeochemical systems, however (Lasaga 1981; Rodhe 1992). It has been shown, for instance, that the flux of CO_2 from

Table 1. Turnover times of biomineralizing elements at the earth surface and in the lithosphere. The earth surface environment comprises the atmosphere, hydrosphere and biosphere, plus soils and surficial sediments. The turnover times for the lithosphere are calculated using estimates of the elemental reservoir sizes of the continental and oceanic crust (i.e., the lower part of the lithosphere is excluded). Data used to derive the turnover times are from a variety of sources.

Element	Earth Surface (years)	Lithosphere (years)
C	1.2×10^5	2×10^8
O	3×10^6	—
Si	$1.5–2.8 \times 10^4$	$5–50 \times 10^8$
P	6×10^4	$>2 \times 10^8$
Ca	$\geq 10^6$	7×10^8

the oceans to the atmosphere depends in a highly non-linear fashion on the mass of total dissolved inorganic carbon (DIC) in the ocean surface layer (Rodhe 1992). This non-linearity derives from the fact that the actual species being transferred, CO_2, is only a minor form of DIC in seawater.

Furthermore, over long time spans, it may not be realistic to assume strictly constant proportionalities between fluxes and reservoir masses. For example, fluxes of carbon and nutrients from the lithosphere to the earth surface environment vary through time because of changes in plate tectonic regime, climate, and terrestrial vegetation. One way to deal with these multiple variables is to introduce dimensionless parameters that express relative effects on the fluxes as a function of time. The linear equations (4) are then modified to

$$F_{ij} = \{f_A(t)f_B(t)...f_K(t)\} \bullet k_{ij}M_i \qquad (5)$$

where $f_N(t)$ expresses the effect of variable N, normalized to the value at some point in time. Parameterization now requires one not only to derive values for the transfer constants, k_{ij}, but also to reconstruct the dimensionless functions $f_N(t)$ over the time span of interest (for further details, see, Berner 1990, 1994; Berner and Kothavala 2001).

The derivation of meaningful functions $f_N(t)$ over geological time scales is possible only through integration of data and knowledge from a wide range of scientific fields, from (geo)biology to geophysics and climatology. An example of a dimensionless function $f_N(t)$ in the long-term geochemical carbon cycle introduced by R. A. Berner is the so-called weathering feed-back function. This function expresses the dependence of weathering of silicate rocks on atmospheric CO_2 levels. In a nutshell, the idea is that an increase in atmospheric CO_2 increases global temperature (greenhouse effect), continental river runoff, and land plant productivity (CO_2 fertilization). All three factors accelerate silicate mineral weathering, which is a net sink for atmospheric CO_2. On geological time scales, this negative feedback helps guarding the atmospheric CO_2 concentration against unreasonably large fluctuations.

A feature that is not captured by linear box models is coupling between the cycles of different elements. A typical example is the coupling of photosynthetic production of organic carbon to the availability and regeneration of the limiting nutrient elements N and P. Mathematically, this means that some of the fluxes in the carbon cycle depend on reservoir sizes and fluxes of other elements. In that case, the mass balance equations of the different coupled cycles must be solved together, which requires numerical integration techniques. A good example is provided by Mackenzie et al. (2002), who use a coupled box model to simulate the effects of human activity on the carbon and nutrient cycles on century time scales.

Coupling of elemental cycles may also lead to feedbacks. Van Cappellen and Ingall (1994, 1996) have explored positive and negative feedbacks in the coupled cycles of phosphorus, carbon and oxygen, using a box modeling approach. Feedback mechanisms are a common feature of biogeochemical cycles, although many of the exact mechanisms are still poorly known. An important role of model simulations is to help uncover the existence of feedbacks and constrain their potential effectiveness. Nowadays, access to user-friendly software packages for dynamic modeling greatly facilitates experimentation with box models, even when dealing with coupled, non-linear biogeochemical cycles.

Carbon cycle

Table 2 presents an overview of reservoir sizes in the short and long term carbon cycles. The short term cycle is receiving a lot of attention, because of the potential climatic effects of the fast release of CO_2 to the atmosphere by humans, primarily as a result of fossil fuel burning and deforestation (e.g., Harvey 2000). The global carbon

Table 2. Reservoirs in the carbon cycle.

reservoir	carbon (g)
atmosphere	6×10^{17}
terrestrial biota	8×10^{17}
soils & detritus	1.5×10^{18}
marine biota	3×10^{15}
ocean surface water	10^{18}
deep ocean water	3.8×10^{19}
surface sediments	1.5×10^{17}
sediments & sedimentary rocks	5.5×10^{22}
basaltic oceanic crust	7×10^{20}
granitic continental crust	9×10^{21}

Sources: Garrels and Mackenzie (1971), Reeburgh (1997), Chameides and Perdue (1997).

balance on short time scales ($\leq$ 1000 years) is dominated by the relatively small reservoirs of the earth surface environment, including the atmosphere, the oceans, terrestrial and marine biota, soils and nearshore sediments. For a good introduction to the issues related to the short term redistribution of anthropogenic CO_2 among these reservoirs, the reader is referred to Sarmiento and Gruber (2002).

On short time scales, anthropogenic CO_2 may be sequestered by the oceans via the so-called *biological pump*, that is, the export to the deep ocean of organic matter synthesized in the surface ocean. Marine organisms, in particular calcifying algae, coccolithophores and foraminifera, also secrete calcium carbonate ($CaCO_3$). Part of the calcium carbonate precipitated in the surface ocean sinks to the deep ocean, together with the organic carbon. This so-called *carbonate pump* is an important component of the oceanic carbon cycle. Together, the export fluxes of particulate organic carbon (POC) and particulate inorganic carbon (PIC) to the deep ocean represent about 10 Gt of carbon annually (1 Gt = 10^{15} g), of which 20–40% is under the form of $CaCO_3$.

The role of $CaCO_3$ production and dissolution in the marine carbon cycle is rather complex, however (e.g., Archer and Maier-Reimer 1994). Although it is a net sink for dissolved inorganic carbon (DIC), calcification actually generates CO_2 while consuming alkalinity:

$$Ca^{2+}(aq) + 2HCO_3^-(aq) \Leftrightarrow CaCO_3(s) + H_2O(l) + CO_2(g) \qquad (6)$$

Thus, uptake of atmospheric CO_2 by surface ocean waters depends on the proportion of biomass production carried out by calcifying organisms (mainly coccolithophorids), versus that by non-calcifying organisms (mainly diatoms).

Calcium carbonate dissolution (i.e., reaction (6) in reverse direction) below the photic zone is a net sink for CO_2. Dissolution is usually assumed to take place at great depths, typically below 3–4 km, where seawater is undersaturated with respect to calcite and aragonite (Broecker and Peng 1982). This deep ocean sink should therefore affect atmospheric CO_2 levels on time scales dictated by the rate of deep water renewal, that is, on the order of 1000 years. Recent studies, however, suggest that substantial $CaCO_3$ dissolution may take place in the upper 1000 m of the water column, probably coupled to organic matter respiration (e.g. Wollast and Chou 1998; Milliman et al. 1999). If confirmed, shallow $CaCO_3$ dissolution may represent an important marine sink for anthropogenic CO_2 on time scales of tens of years.

Long term fluctuations in the level of atmospheric CO_2 are dominated by exchanges with the lithosphere (e.g., Berner 1990). The CO_2 sinks in the long-term carbon cycle are burial of organic carbon and $CaCO_3$ in marine sediments, and CO_2 consumption during weathering of silicate rocks. The sources are CO_2 produced by the thermal degradation of buried organic matter and carbonate minerals, plus the oxidative weathering of sedimentary organic matter after uplift.

The carbon sinks and sources are often represented by overall chemical reactions, such as

$$CO_2(g) + H_2O(l) \Leftrightarrow \text{"}CH_2O\text{"} + O_2(g) \qquad (7)$$

where "CH_2O" is a simplified representation for organic matter. The forward reaction in Equation (7) corresponds to net global photosynthesis, that is, photosynthesis minus respiration. Because the earth surface environment is at (quasi-)steady state, net global photosynthesis is equal to the burial rate of organic matter in sediments. The backward reaction represents total oxidative degradation of organic matter, either through weathering of sedimentary organic matter or the oxidation in the atmosphere of reduced carbon gases derived from the degradation of organic matter during late diagenesis, metamorphism or magmatic activity (Fig. 2). The backward reaction can thus be thought of as the global respiration of the geosphere.

Weathering of carbonate rocks on land does not affect the long term, average concentration of CO_2 in the atmosphere. This is because, on a million-year time scale, the carbonate weathering flux is balanced by an equal amount of carbonate precipitation and burial in the oceans. The latter returns the CO_2 consumed during carbonate weathering back to the atmosphere. The situation is different when marine carbonate minerals are formed with calcium (and magnesium) ions produced by weathering of silicate minerals. In that case, there is a net consumption of atmospheric CO_2, as shown schematically by the overall reaction (e.g., Berner 1990):

$$CaSiO_3(s) + CO_2(g) \Leftrightarrow CaCO_3(s) + SiO_2(s) \qquad (8)$$

where $CaSiO_3(s)$ represents calcium containing silicate minerals. The forward reaction corresponds to silicate mineral weathering coupled to carbonate mineral burial in ocean sediments, and the reverse reaction to the thermal decomposition of the carbonates deep within the lithosphere. Because of the long time span that separates deposition of carbonate minerals at the seafloor and the return of carbon to the earth's surface as volcanic CO_2, reaction (8) can be out of balance, resulting in a net source or sink of atmospheric CO_2.

In box models for the long term carbon cycle, the forward and backward rates of overall reactions such as those represented by Equations (7) and (8) correspond to fluxes linking carbon reservoirs at the earth surface and in the lithosphere. These fluxes obviously combine many different processes and are subject to a variety of controls. For example, the burial flux of organic carbon in sediments (i.e., the forward flux in reaction 7) depends not only on how much organic matter is produced globally, but also on the fraction of organic matter that survives degradation and is ultimately incorporated in the sedimentary column (e.g., Canfield 1989). In other words, the derivation of a mathematical expression for the burial flux of organic carbon must be based on a careful evaluation of the factors controlling both production and preservation of organic matter. The same is true for the burial fluxes of other biogenic materials, in particular the products of biomineralization.

BIOMINERALIZATION IN A GLOBAL CONTEXT

Biomineralization and biogeochemical cycles

The biogeochemical cycle of carbon is coupled to the cycles of nutrient elements, particularly N, P, through the production and degradation of organic matter. The Redfield ratios of marine phytoplankton biomass exemplify this coupling (Redfield et al. 1963). The ratios express the relative proportions in which N, P and other nutrient elements are used, relative to C, during primary production in the oceans. The concept of limiting nutrient(s) is a direct consequence of the stoichiometric constraints on nutrient uptake fluxes during synthesis of new biomass. The Redfield ratios also provide base line values against which the relative release fluxes of nutrients can be compared during degradation of organic matter. Preferential release or preservation of nutrient elements exerts a major control on the biological productivity of the oceans (e.g., Van Cappellen and Ingall 1994).

The production of skeletal hard parts creates additional coupling among elemental cycles. This is particularly true in the oceans, where vast amounts of calcium carbonate and biogenic opal (B-SiO_2) are secreted by organisms living in the surface waters (i.e. depths <200 m). Table 3 compares the rates of production of soft and mineralized tissues in the surface ocean. Photosynthetic fixation of carbon in organic molecules is estimated to be about 20 times higher than the gross production of calcium carbonate or biogenic opal (on a molar basis). Nearly 90% of the organic matter, however, is rapidly degraded and the carbon and nutrients are returned to the surface waters. In contrast, only about 50% of the biogenic silica dissolves in the upper 200 m of the water column (Tréguer et al. 1995). For calcium carbonate, this fraction is even smaller, because the surface ocean is supersaturated with respect to calcite and aragonite. As a result, net export of organic carbon from the surface ocean is only 2–5 times larger than the corresponding fluxes of B-SiO_2 and $CaCO_3$. The biogenic materials are removed from the surface waters, either by sinking into deeper waters or deposition in shallow water sediments.

Degradation of organic matter continues in the water column and after deposition at the seafloor. Ultimately, less than 1% of the organic matter originally synthesized in the surface waters is preserved in sediments. A significantly larger fraction of the biogenic $CaCO_3$ survives dissolution, however (Table 3). The end result is that about twice as much carbon is removed from the earth surface environment as biogenic $CaCO_3$ than as organic carbon. Because of highly undersaturated ocean waters, relatively more B-SiO_2 dissolves than $CaCO_3$. About 3% of the total amount of B-SiO_2 secreted by diatoms, silicoflagelates and radiolarians is buried in marine sediments. In terms of absolute magnitudes, the burial fluxes of organic carbon and biogenic silica are fairly similar (Table 3).

Table 3. Fluxes of organic matter and biogenic minerals in the oceans. C_{org} and P_{org} correspond to carbon and phosphorus in particulate organic matter. B–SiO_2 and B–CaP refer to biogenic silica and biogenic calcium phosphate. The latter is mainly associated with fish bones and scales; B-CaP does not include calcium phosphate minerals forming authigenically in sediments. All fluxes are in Tmol phosphorus per year (Tmol = 10^{12} mol)[1].

	C_{org}	P_{org}	$CaCO_3$	B-SiO_2	B-CaP^2
Production in surface ocean	4200	39.6	≥ 223	240	0.4
Export to deep ocean	497	4.7	213	120	0.34
Deposition at seafloor	74	0.32	74	30	?
Burial in sediments	8–11	0.02	17–24	6.7	< 10^{-3}

[1]***Sources:*** Tréguer et al. (1995); Berner and Berner (1996); Sarmiento and Gruber (2002); Klaas and Archer (2002); Slomp et al. (2003).

Most $CaCO_3$ and $B-SiO_2$ is produced by organisms from relatively low trophic levels, including the major marine primary producers, coccolithophorids and diatoms. Therefore, the fluxes of $CaCO_3$ and $B-SiO_2$ are large and intimately coupled to those of organic matter. As mentioned earlier, the relative proportion of siliceous versus calcareous production is of global climatic significance, because it regulates the uptake of CO_2 by marine surface waters. When silica-secreting organisms dominate, more CO_2 can be extracted from the atmosphere, because there is no release of CO_2 by reaction (6). In addition, it has been proposed that diatoms are responsible for much of the export of organic matter from the surface waters (e.g., Smetacek 1999). This is because large diatom blooms are usually followed by aggregation and rapid sinking of biomass out of the surface mixed layer. Diatom blooms could thus enhance the efficiency of the biological (organic) carbon pump in the oceans (for an opposing view, see, Klaas and Archer 2002).

Much oceanographic research is currently directed at understanding how circulation patterns, sea surface temperature, nutrient limitation, atmospheric dust input, and ecological factors affect the ratio of silica-secreting organisms to calcium carbonate-secreting organisms in the surface ocean. One particular area of focus is the Southern Ocean, where a significant fraction of available P and N in the surface waters is not being utilized for photosynthesis (e.g., Martin et al. 1990). It is now widely believed that diatoms in the Southern Ocean are limited by the availability of iron and, in some cases, silica. A related hypothesis is that, during glacial times, wind-blown dust alleviates iron limitation of diatom production and, hence, causes a drawdown of atmospheric CO_2.

Biogenic calcium phosphate (B-CaP) is mainly secreted by marine vertebrates, that is, organisms higher up the food chain. We may therefore expect the fluxes of B-CaP to be of lesser importance in global ocean biogeochemistry, compared to those of $CaCO_3$ and $B-SiO_2$. The estimated gross production of B-CaP in the surface ocean is indeed two orders of magnitude smaller than phosphorus fixation by photosynthesis (Table 3). However, the export fluxes of organic phosphorus and phosphorus associated with B-CaP only differ by one order of magnitude. This reflects the fast remineralization of organic matter in the surface ocean, but also relatively minor B-CaP dissolution in the upper water column. B-CaP consists primarily of hydroxapatite, a rather insoluble mineral phase. Furthermore, a large fraction of B-CaP is probably exported with relatively large, fast sinking fish debris.

As it settles through the water column, and after deposition at the seafloor, most B-CaP dissolves. The soluble phosphate is either recycled back to the surface ocean or it precipitates in sediments as fine-grained (authigenic) carbonate fluorapatite (Ruttenberg and Berner 1993). Globally, the contribution of B-CaP to total burial of reactive P in marine sediments is estimated at less than 1% (Berner and Berner 1996). Burial of B-CaP appears to be mainly associated with coastal upwelling areas that exhibit low bottom water oxygen concentrations. In sediments of the Arabian Sea located within the so-called Oxygen Minimum Zone (OMZ), 25-40% of the total burial flux of P is attributed to biogenic calcium phosphate (Schenau and de Lange 2001). In sediments of the OMZ along the Peru margin, the burial flux of B-CaP may exceed that of organic P (Suess 1981).

The observed enhanced B-CaP preservation under oxygen-depleted bottom waters is puzzling, and the responsible mechanisms are not yet understood. However, it could imply that burial of B-CaP may have been a more significant sink for oceanic phosphorus during so-called Ocean Anoxic Events (OAEs). These are periods during the geological past, for example during the mid-Cretaceous, when anoxic bottom waters were widespread in the oceans. These periods are characterized by the abundant accumulation of laminated, organic-rich shale deposits (e.g., Arthur et al. 1984; Wignall 1994). In many cases, these sediments have, upon deep burial, acted as petroleum source rocks.

Model simulations suggest that oxygen-dependent burial of phosphorus plays a major role in the development of ocean anoxia (Van Cappellen and Ingall 1994).

The biogeochemical cycles of P and Si in the oceans are completely dominated by biological uptake during organic matter synthesis and skeletal secretion in the surface waters, and regeneration in deeper waters and surface sediments (e.g., Broecker and Peng 1982). These elements are extremely efficiently recycled in the oceans and are often limiting nutrients for marine ecosystems (e.g., Ragueneau et al. 1996; Falkowski et al. 1998). Nearly all dissolved phosphate and silicate that enter the photic zone are used for the production of new soft tissue and skeletal hard parts. In contrast, C and Ca rarely limit primary production. Their reservoir sizes and oceanic turnover times are substantially larger than for P and Si. For comparison, the estimated turnover times of Si, P, C and Ca are 1.5×10^4, 5×10^4, 1.3×10^5 and 10^6 years, respectively. In essence, C and Ca are less biogenic than P and Si. Abiotic processes play a relatively larger role in their biogeochemical cycles, for example, air-sea gas exchange for C and seawater-basalt reaction for Ca.

So far, the focus has been on biomineralization processes in the oceans. On the continents, calcifying and silicifying organisms may play significant roles, albeit at a more local scale, in the biogeochemistry of aquatic environments, e.g., lakes, wetlands and large river systems. In terrestrial ecosystems, a recent assessment by Conley (2002) emphasizes the very intense biological cycling of silicon by land plants. He estimates that the gross uptake of dissolved silicate by terrestrial vegetation (60-200 Tmol Si yr^{-1}) is of the same order of magnitude as biosiliceous production in the oceans (240 Tmol Si yr^{-1}, Table 3). The plants deposit the silica as structural support elements, or phytoliths, in their tissues. Phytoliths accumulate together with other biogenic detritus in soils, which may therefore constitute a significant but, as yet, poorly quantified terrestrial pool of reactive silica.

While both the continents and oceans are sites of continuous cycling of biogenic elements, a major difference is the generally smaller storage capacity of the terrestrial biosphere plus soils, compared to the oceanic reservoir (for carbon this is illustrated in Table 2). This means that turnover times of these elements on land tend to be shorter than in the oceans. For example, the turnover time of reactive Si in forest ecosystems appears to be on the order of a few thousands of years (Conley 2002), compared to a marine turnover time on the order of $1-2 \times 10^4$ years. The available estimates, however, are based on a very limited data set. In particular, there is a lack of (steady state) mass balances of reactive Si for grasslands, which are known to produce large amounts of phytoliths. At present, the global terrestrial phytolith reservoir size is a major unknown in the global biogeochemical cycle of Si.

To summarize the discussion so far, the following, admittedly simplified, outline of the biogeochemical cycles of the major elements involved in biomineralization can be drawn. The continents are the main site of mobilization of C, Si, P and Ca from the lithosphere. The elements typically remain on land for a few thousands of years, during which they participate in the biogeochemical cycles of terrrestrial and aquatic ecosystems. They are then delivered to the oceans, where they reside for periods ranging from 10^4 to 10^6 years. After extensive recycling within the ocean-atmosphere system, they are ultimately removed by burial in sediments. Tectonic processes close the cycles by returning the elements to the earth surface, on time scales of tens to hundreds of millions of years.

Biomineralization through time

Secretion of mineralized body parts is a relatively recent phenomenon in geological history, with little evidence suggesting it originated much earlier than 600 Ma ago (Van Andel 1994). The Cambrian explosion truly marks the start of the radiation of organisms with hard parts, first in the oceans and, from about 400 Ma on, also on land. Given the large impact of biomineralization on biogeochemical cycles in today's world, one wonders how the adoption of mineralized body parts has altered the biogeochemical dynamics of the earth surface environment. Unfortunately, there is no easy answer to this question. The available geological records are not only incomplete, they also provide ample evidence that many of the dominant features of the earth surface environment in the past may not have modern analogues.

During the early days of the Cambrian (570–510 Ma B.P.), life probably experimented with a variety of different biominerals. Yet, calcium carbonate rapidly emerged as the dominant building material for shells and skeletons among invertebrates. The geological history of $CaCO_3$ biomineralization since the end of the Precambrian is fascinating, with periods of massive reef building, e.g., during the Ordovician, and the radiative expansion of calcareous plankton during the Mesozoic (Martin 1995). However, whether in shallow-water environments or in the open ocean, invertebrates have been producing large quantities of $CaCO_3$ throughout the entire Phanerozoic, with the possible exception of episodes of marine extinction (Martin 1998).

Today, precipitation and dissolution of biogenic $CaCO_3$ maintain the oceans close to thermodynamic equilibrium with calcite (e.g., Turekian 1976). The widespread availability of $CaCO_3$ buffers the pH of seawater to mildly alkaline values around 8. Even fairly large variations in the partial pressure of CO_2 therefore only induce relatively small shifts in the pH of ocean waters. It is likely that this buffering mechanism has been operational at least since the early Cambrian, and has effectively kept seawater pH within a relatively narrow range. Buffering of seawater pH by biogenic $CaCO_3$ may have helped create stable environmental conditions in which complex marine life forms could evolve during the Phanerozoic.

The biological control on seawater carbonate chemistry probably extends much further in time, however. Some of the oldest life forms in the geological record, cyanobacteria, are associated with calcareous formations called stromatolites. These have been around since the Archean, but are particularly abundant in sedimentary rocks from the Proterozoic (Van Andel 1994). The $CaCO_3$ of stromatolites is a by-product of the photosynthetic activity of the cyanobacteria, not an intracellular biomineralization process under direct genetic control. From the viewpoint of the carbonate chemistry of seawater, however, it makes little difference how $CaCO_3$ is produced. In fact, given enough time, supersaturated (undersaturated) seawater will precipitate (dissolve) $CaCO_3$, even in the absence of any biological involvement. Thus, to a first approximation, cyanobacteria and, later, calcifying invertebrates simply turned the existing chemical equilibria of the $CaCO_3$–seawater system to their advantage.

The story is different for silica-secreting organisms. During the Precambrian, prior to biological silicification, the marine silica cycle was probably dominated by sorption reactions of dissolved silicic acid to clay minerals, zeolites and organic matter (Siever 1992). There is no compelling evidence for large-scale accumulation of primary silica deposits. Compared to today, the oceans at the start of the Phanerozoic were highly enriched in dissolved silicic acid, with concentrations on the order of 1 mM (Siever 1992). Silica-secreting biota, in particular radiolarians and siliceous sponges, took advantage of this high abundance of silicic acid.

During the first half of the Phanerozoic, radiolarians were the dominant predators in the plankton food chain. The availability of dissolved silica, however, started to decline dramatically after the appearance of diatoms (Racki and Cordey 2000). It is believed that many of the evolutionary trends in biosiliceous biota are related to increased competition for a dwindling oceanic reservoir of silicic acid (e.g., Conley et al. 1994). Diatoms clearly won the competition and have been the predominant group of marine siliceous organisms for the last 60 Ma. They currently account for about half of total marine primary production (Nelson et al. 1995).

Today, the concentration of dissolved silicic acid in the deep ocean rarely exceeds 250 µM. In surface waters, the concentrations are typically below 30 µM. The oceanic silica cycle has thus fundamentally changed since the Precambrian. While at first silica was an abundant resource, the oceans are now silica-starved. In order to maintain high siliceous productivity, biogenic silica must be continuously recycled to silicic acid. Efficient recycling is only possible because diatoms keep seawater highly undersaturated with respect to biogenic silica. Silicon in the oceans has thus evolved from a non-biogenic to a highly biogenic element. In contrast, calcium and (inorganic) carbon have retained part of their non-biogenic character.

There have been suggestions that during the early Cambrian secretion of phosphatic shells may have been more common than that of calcareous ones (Van Andel 1994). Certainly some brachiopods opted for calcium phosphate mineralization. Nonetheless, the vast majority of phosphatic hard parts in the geological record have been produced by crustaceans and marine vertebrates. The reason why calcium phosphate was never used on a large scale by protists and most invertebrate groups may reflect the scarcity of phosphorus in the environment. Phosphorus is absolutely essential for the synthesis of soft tissue and, hence, indispensable to all organisms. Thus, even at the start of the Phanerozoic, phosphorus was probably in short supply, compared to Ca, C and Si. This would have put organisms building phosphatic hard parts in a competitive disadvantage, compared to those opting for silica or calcium carbonate. The disadvantage would have been most severe for organisms at the lowest trophic levels, which dominate the production of biomass.

In the end, only higher organisms, starting with fishes in the Silurian, have been producing phosphatic body parts to any considerable extent. The production and preservation of B-CaP was probably never a decisive factor in the oceanic phosphorus cycle, except during periods of prolonged ocean anoxia. Recent modeling results indicate that under conditions of reduced circulation and, hence, decreased ocean ventilation, B-CaP may become a significant phosphorus sink (Slomp et al. 2003). Enhanced burial of B-CaP under anoxia fits with the observation that black shales are often enriched in well-preserved fish debris, compared to bioturbated shales (Wignall 1994).

MARINE BIOGEOCHEMICAL CYCLE OF SILICON

Controls on biosiliceous production

With the exception of the earth's core, the geosphere consists primarily of silicon and oxygen atoms, with some cations thrown in between. Silicate compounds, under the form of minerals, rocks and melts, are therefore dominant players in the geological rock cycle. Yet, silicon is also a major biogenic element. The availability of dissolved silicate can dramatically affect the structure, health and productivity of marine, freshwater and terrestrial ecosystems (Brzezinski et al. 1990; Epstein 1994). Some organisms have an absolute requirement for silicon. These organisms use dissolved silicate to build structural elements of hydrated, amorphous silica, or biogenic opal. The overall reaction describing this biomineralization process is

$$H_4SiO_4(aq) \Leftrightarrow SiO_2 \cdot nH_2O(s) + (2-n)H_2O(l) \tag{9}$$

The productivity of the main group of primary producers in the oceans, diatoms, is directly dependent on the supply of dissolved silicic acid. For normal growth, diatoms need about as much silicon as nitrogen (on a molar basis). When nutrients are plenty, diatoms can displace other primary producers, because of their relatively high photosynthetic capacity and low maintenance energy requirements. Therefore, diatoms often dominate the early stages of open ocean blooms. They are also abundantly present in coastal seas and upwelling areas, where nutrient supplies are high. Accumulation rates of biogenic silica and the composition of diatom assemblages in sediments have been used as indicators of paleo-productivity and paleo-upwelling intensity (e.g., Koning et al. 1997, 2001).

Short-term fluctuations of biogenic silica production by diatoms in a particular oceanic area are controlled by factors such as light intensity, hydrographic conditions and nutrient status of upwelling waters. However, for periods of time on the order of, or longer than, the oceanic turnover time of H_4SiO_4 ($1-2\times10^4$ years), global biosiliceous production is regulated by the oceanic sources and sinks of silica. Because the turnover time of H_4SiO_4 is significantly longer than the characteristic time scale of mixing of the oceans, which is on the order of a few thousands of years, a simple box model can be constructed to represent the long-term marine silica cycle (Fig. 3).

The oceans are assumed to consist of two, internally homogeneous, reservoirs: dissolved (bioavailable) H_4SiO_4 and biogenic silica. Input of new H_4SiO_4 to the oceans is due to river inflow and hydrothermal venting. During seawater-basalt interactions at mid-oceanic ridges, H_4SiO_4 is released and transported to the water column by hydrothermal circulation. River inflow is by far the dominant supply route of new H_4SiO_4 to the oceans, however (Berner and Berner 1996). Biosiliceous production (i.e., the forward reaction in Eqn. 9) is sustained by the input of new H_4SiO_4, but also by H_4SiO_4 regenerated during the dissolution of biogenic silica (i.e., the backward reaction in Eqn. 9). Some of the biogenic silica produced by diatoms, but also radiolarians and silicoflagellates, escapes dissolution and is buried in sediments, below the zone of early diagenesis.

At the time scales considered ($\geq 10^4$ years), the oceanic silica cycle remains close to steady state. Hence, the following relationships exist between the fluxes shown in Figure 3:

$$IN = BUR \qquad (10)$$

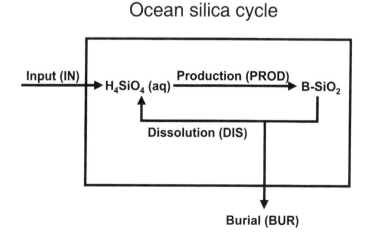

Figure 3. Box model representing the long-term ($\geq 10^4$ years) marine silica cycle. See text for discussion.

and
$$\text{PROD} = \text{IN} + \text{DIS} \tag{11}$$

Furthermore, we can define the dissolution (or recycling) efficiency

$$\alpha = \frac{\text{DIS}}{\text{PROD}} \tag{12}$$

where $0 \leq \alpha \leq 1$. In the modern ocean, α is on the order of 0.97, reflecting very efficient silica recycling. Combining Equations (11) and (12), we obtain

$$\text{PROD} = \frac{\text{IN}}{(1-\alpha)} \tag{13}$$

The denominator in Equation (13), $(1-\alpha)$, can also be viewed as the ocean-wide biogenic silica preservation efficiency.

According to Equation (13), for marine production of biogenic silica to increase either the external supply of H_4SiO_4 must increase or the preservation efficiency must decrease, or both. The numerator and denominator of the right-hand side of Equation (13) are related to processes such as silicate weathering, volcanic-seawater interaction, dissolution plus ageing of biosiliceous shells, and sediment burial. In other words, Equation (13) links the global rate of silica biomineralization in the oceans to geological and geochemical processes that regulate the oceanic sources and sinks of silica.

Weathering

The main source of new silica for the oceans is delivery by rivers of H_4SiO_4 produced by weathering of silicate rocks on the continents. Broadly speaking, silicate minerals consist of SiO_4^{4-} tetrahedra that are linked together by strong (covalent) $\equiv$Si–O–Si$\equiv$ (siloxane) bonds. The exception is olivine where the SiO_4^{4-} units are not linked, but instead separated from one another by cations, mainly Fe^{2+} and Mg^{2+}. Because its mineral structure is held together by much weaker electrostatic forces, olivine dissolves much faster than other silicates. Olivine is a major constituent of the basaltic oceanic crust, it is relatively rare in continental rocks. Therefore, the rate-controlling steps in the dissolution of silicate minerals commonly exposed on land are rather similar and involve the breaking, or hydrolysis, of $\equiv$Si–O–Si$\equiv$ and, in alumino-silicates, also $\equiv$Si–O–Al$\equiv$ bonds.

With the exception of olivine, dissolution rates of silicate minerals measured in the laboratory fall in a fairly narrow range. At room temperature and under near-neutral conditions the rates are typically on the order of 10^{-12} to 10^{-11} mol Si m^2 s^{-1} (e.g., Brady and Walther 1989). A long-standing problem in geochemistry is to relate these experimental dissolution rates to regional or global scale weathering rates (e.g., Velbel 1993). To illustrate the magnitude of the problem, a theoretical weathering flux of dissolved silicic acid is calculated by extrapolating the experimental dissolution rates. The calculation also illustrates the type of rough approximations that are often involved in making global flux estimates.

The main difficulty is to estimate the amount of silicate mineral surface area exposed on the continents. Let us start by assuming that limestones, which cover about 15% of the continental surface area (Garrels and Mackenzie 1971), do not contribute significantly to silicate weathering. Let us further assume that a 10 cm deep weathering layer covers 50% of the remaining continental area. The volume of the active weathering layer is then approximately 6.5×10^{12} m^3. This is likely a minimum estimate, as nearly 60% of the continental surface are assumed not to be subjected to chemical weathering. Using (low) estimates of the density (1 g cm^{-3}) and specific surface area (0.1 m^2 g^{-1}) of silicate minerals in the weathering layer, the total exposed silicate mineral surface area is then 6.5×10^{17} m^2. Combining this value with the dissolution rates given above (10^{-12} to 10^{-11} mol Si m^{-2} s^{-1}), the total flux of dissolved H$_4$SiO$_4$ produced by weathering should be comprised between 29 and 290 Tmol per year.

This extrapolation is fraught with uncertainties, yet it conveys a simple message. Even with low estimates of the exposed silicate mineral surface area, the laboratory dissolution rates predict weathering rates that are much higher than the total flux of H$_4$SiO$_4$ delivered by rivers to the oceans. The latter is on the order of 6 Tmol per year (Tréguer et al. 1995). There are a number of reasons for this discrepancy (Velbel 1993; Berner and Berner 1996). In contrast to experiments conducted in reactors, much of the available mineral surface area in soils may only be intermittently in contact with soil solution. This is particularly true in dry climates and in highly aggregated soils. Thus, the mineral surface area that is actually dissolving is only a fraction of the total available surface area. Soil formation also reduces the net production of silicic acid through the precipitation of secondary silicate minerals. The total river flux of H$_4$SiO$_4$ is further reduced because of hydraulic short-circuiting in watersheds, internal continental drainage, as well as biogenic silica retention in aquatic and terrestrial ecosystems.

Upscaling from laboratory dissolution rates to global weathering fluxes thus requires a combination of approaches (Fig. 4). In the same way that biogeochemical cycles respond to different forcings depending on the spatio-temporal scale of interest, the different approaches yield information on processes and variables acting at variable scales. Typically the interpretation of data and observations at a given scale relies on process-based knowledge from the underlying scales. For example, climate may affect continental weathering rates through the direct effect of temperature on mineral dissolution rates, or via its effects on vegetation, rainfall and mechanical erosion. A

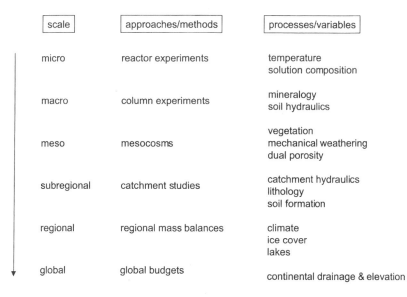

Figure 4. Approaches for studying weathering rates: different approaches yield information on controls acting at different scales, from the molecular to global scale. Typically, mechanistic understanding at one scale helps rationalize observations made at higher scales.

mechanistic understanding of the individual effects is therefore needed in order to derive meaningful mathematical representations of the net effect of climate on the global H_4SiO_4 weathering flux. This understanding can be obtained through laboratory experimentation and observations at the catchment level.

Preservation of biogenic silica

The high productivity of diatoms in the oceans requires a very efficient dissolution of diatom frustules after the organisms die, in order to regenerate H_4SiO_4. The high dissolution efficiency, in turn, is due to the high degrees of undersaturation of seawater with respect to biogenic silica. This is illustrated in Figure 5, which shows silica solubilities measured experimentally on fresh phytoplankton, cultured diatoms and sinking particulate matter collected at different depths of the ocean with so-called sediment traps. The general increasing tendency with depth reflects a moderate effect of pressure on the solubility of amorphous silica (Dixit et al. 2001). The measured

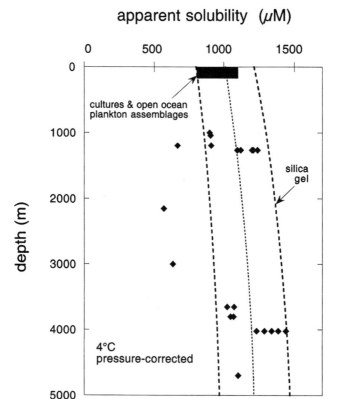

Figure 5. Silica solubilities in seawater, at 4°C. The data points are experimental solubilities measured in flow-through reactors on sediment trap samples (Gallinari 2002). The pressure correction proposed by Dixit et al. (2001) was used to account for differences in water depth. Also shown at zero depth is the range of silica solubilities reported for diatom cultures and open ocean siliceous plankton (Dixit et al. 2001). The thick broken lines are the pressure-corrected solubility of synthetic silica gel and the lower limit for plankton and diatom cultures. The thin broken line is the average biogenic silica solubility as a function of depth. The available data show that solubilities of siliceous materials sinking through the water column are consistent with solubilities measured on fresh diatoms and phytoplankton.

solubilities are all much higher than the concentrations of silicic acid observed in the oceans. In the water column, dissolved silicate concentrations rarely exceed 250 µM.

According to the results in Figure 5, all biogenic silica should dissolve in the oceans. Dissolution is prevented in living organisms by the presence of protective organic membranes. After death of the organisms, however, bacteria rapidly break down the organic membranes and dissolution starts immediately (Biddle and Azam 1999). Despite the high degree of thermodynamic disequilibrium of the oceans, some biogenic shell material nonetheless survives dissolution and accumulates at the seafloor. In its most spectacular form, this accumulation results in diatomaceous oozes that contain more than 85% SiO_2 by weight. The preservation of a fraction of the biogenic silica produced in the surface waters provides paleooceanographers with a precious record for the reconstruction of past patterns in ocean productivity, upwelling intensity and nutrient utilization (e.g., Berger and Herguera 1992; Koning et al. 2001; Brzezinski et al. 2002).

Preservation and burial of biogenic silica is explained in part by the build-up of silicic acid in the pore waters of marine sediments. Siliceous remains that manage to escape dissolution during their transit through the water column continue to dissolve after deposition. Because the solid matrix of the sediment limits turbulent solute transport, H_4SiO_4 released by opal dissolution can accumulate in the pore waters and the H_4SiO_4 concentration increases with depth. Typically, the pore water concentration stabilizes at a near constant value 5–10 cm below the sediment surface; this value is referred to as the asymptotic (or saturation) concentration, C_{sat}. It has been traditionally assumed that this concentration corresponds to the equilibrium solubility of the deposited biogenic opal. Any biogenic silica that reaches the depth of the asymptotic concentration is therefore preserved, because thermodynamic equilibrium has been reached.

A problem with this interpretation is that C_{sat} varies widely from sediment to sediment (Fig. 6). This is in sharp contrast to the fairly narrow range of silica solubilities observed in the water column (Fig. 5). In biosiliceous oozes of the Southern Ocean, C_{sat} approaches silica solubilities measured on water column samples, but in most other sediments C_{sat} is much lower than expected for equilibrium with biogenic silica (compare Figs. 5 and 6). Dixit et al. (2001) showed that the large variations in C_{sat} are related to interactions between deposited biogenic silica remains and lithogenic minerals. The lithogenic (or detrital) fraction of marine sediments consists mostly of oxides and aluminosilicate phases. Soluble aluminum produced by the slow dissolution of the lithogenic minerals interacts with biosiliceous remains, and thereby profoundly modifies their thermodynamic and kinetic properties.

Incorporation of aluminum into the surface layers of biosiliceous fragments decreases their solubility. This effect is seen even when relatively small amounts of detrital minerals are present. In batch reactor experiments where biogenic silica was mixed with model detrital phases (basalt or kaolinite), a measurable drop in apparent silica solubility was observed at mass ratios of detrital matter to biogenic silica (= %detrital:%opal) as low as 0.1 (Dixit et al. 2001). Incorporation of Al also affects the reactivity of biogenic silica. In experiments with cultured diatoms, the rate constant of dissolution decreased by nearly one order of magnitude when the molar Al/Si ratio in the frustules increased from 0.3–0.6×10^{-3} to 2–3×10^{-3} (Van Cappellen et al. 2002).

For open ocean environments, most aluminum uptake by biogenic silica occurs after deposition at the seafloor. In coastal waters, diatoms may also incorporate significant amounts of Al in their frustules during biomineralization, because of the much higher availability of aluminum in the ambient seawater (Van Beusekom 1991). In sediments with high %detrital:%opal ratios, the relatively high supply of soluble aluminum may

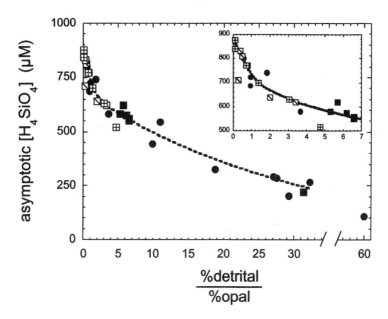

Figure 6. Asymptotic pore water concentration of silicic acid versus the detrital to opal ratio of the sediment. The different symbols correspond to a variety of ocean basins (for detailed site description and original references, see Dixit and Van Cappellen 2003). The observed inverse relationship is the result of early diagenetic interactions between biogenic silica remains and aluminum solubilized from detrital aluminosilicates.

ultimately result in the precipitation of new (authigenic) silicate minerals (Michalopoulos and Aller 1995). Formation of authigenic silicates, a process also known as "reverse weathering," is a largely overlooked sink for reactive silica in the oceans. Whether through Al incorporation into biogenic opal or via reverse weathering reactions, increased aluminum availability in the oceans decreases the silica recycling efficiency.

The early diagenetic interactions between Al and biogenic silica explain the observed inverse relationship between the asymptotic pore water concentration of silicic acid and the %detrital:%opal ratio of marine sediments (Fig. 6). This relationship provides a striking illustration of the strong coupling between the benthic cycles of silica and aluminum. This coupling has potentially far-reaching consequences for our understanding of regional patterns and global variability of biosiliceous production in the oceans. In particular, the supply of detrital aluminosilicates mobilized by soil erosion and mechanical rock weathering on land may modulate the preservation efficiency of biogenic silica in the oceans and, therefore, by virtue of Equation (13), the biosiliceous productivity of the oceans. It is important to note in this respect that the responses of mechanical erosion and chemical weathering to changes in tectonic regime, climate or vegetation may be very different. For instance, continental glaciation is likely to increase mechanical erosion, but decrease chemical weathering.

In addition to the role of Al, biogenic silica surfaces age with time (Van Cappellen 1996). This aging further protects the biosiliceous fragments from dissolution and, hence, enhances their preservation in the sedimentary record. Preliminary acid-base titrations of freshly cultured diatom shells and older sedimentary biosilicious oozes suggest that aging corresponds to a loss in the surface density of ionizable silanol groups (Dixit and Van Cappellen 2002). Clearly much work remains to be done to unravel the changes in the chemical structure of biogenic silica surfaces, from the moment they are first exposed to seawater until their ultimate burial in sediments. These molecular changes, however, are ultimately responsible for the preservation of biogenic silica in marine sediments and, therefore, influence how much silica biomineralization can be sustained in the oceans.

CONCLUSIONS

In this chapter, biomineralization is analyzed within the context of global biogeochemical cycles. Of particular importance are the magnitudes of elemental fluxes associated with the secretion of mineralized tissues, but also the efficiency with which nutrients are regenerated by dissolution of biominerals. Biomineralization and subsequent regeneration are major processes in the marine cycling of C, Ca, P and Si. Biogenic silica production by land plants may rival that by marine diatoms, however. On time scales $\geq 10^4$ years, geological and geochemical processes exert a major control on global biomineralization rates. This is illustrated for the marine silica cycle. The global average rate of biogenic silica production in the oceans is regulated by the intensity of chemical weathering on the continents and by early diagenetic interactions in marine sediments.

ACKNOWLEDGMENTS

The last section is based in large measure on the PhD thesis research of Suvasis Dixit. Caroline Slomp is thanked for providing estimates of B-CaP fluxes prior to publication. The author is grateful to the volume Editors for their patience and help. Financial support from the Netherlands Organisation of Scientific Research is acknowledged (NWO Pionier Grant).

REFERENCES

Archer DE, Maier-Reimer E (1994) Effect of deep-sea sedimentary calcite preservation on atmospheric CO_2 concentration. Nature 367:260-264
Arthur MA, Dean WE, Stow DAV (1984) Models for the deposition of Mesozoic-Cenozoic fine grained organic-carbon-rich sediment in the deep-sea. Geol Soc London Special Pub 26:527-562
Berger WH, Herguera JC (1992) Reading the sedimentary record of ocean's productivity. *In:* Primary Production and Biogeochemical Cycles in the Sea. Falkowski PG, Woodhead AD (eds) Plenum Press, p 455-486
Berner RA (1987) Models for carbon and sulfur cycles and atmospheric oxygen: Application to Paleozoic history. Am J Sci 287:77-196
Berner RA (1990) Atmospheric carbon dioxide levels over Phanerozoic time. Science 249:1382-1386
Berner RA (1992) Weathering, plants and the long term carbon cycle. Geochim Cosmochim Acta 56:3225-3231
Berner RA (1994) GEOCARB II: A revised model of atmospheric CO_2 over Phanerozoic time. Am J Sci 291:56-91
Berner RA, Berner EK (1996) Global Environment: Water, Air, and Geochemical Cycles. Prentice Hall
Berner RA, Berner EK (1997) Silicate weathering and climate. *In:* Tectonic Uplift and Climate Change. (Ruddiman WF) Plenum Press, p 353-365
Berner RA, Canfield DE (1989) A new model for atmospheric oxygen over Phanerozoic time. Am J Sci 289:333-361
Berner RA, Kothavala Z (2001) GEOCARB II: A revised model of atmospheric CO_2 over Phanerozoic time. Am J Sci 301:182-204
Biddle K, Azam F (1999) Accelerated dissolution of diatom silica by marine bacterial assemblages. Nature 397:508-512
Brady PV, Walther JV (1989) Controls on silicate dissolution rates in neutral and basic pH solutions at 25°C. Geochim Cosmochim Acta 53:2823-2830
Broeker WS, Peng TH (1982) Tracers in the Sea. Eldigio Press

Brzezinski MA, Olson RJ, Chisholm SW (1990) Silicon availability and cell-cycle progression in marine diatoms. Mar Ecology Progress Series 67:83-96

Brzezinski MA, Pride CJ, Franck VM, Sigman DM, Sarmiento JL, Matsumoto K, Gruber N, Rau GH, Coale KH (2002) A switch from $Si(OH)_4$ to NO_3^- depletion in the glacial Southern Ocean. Geophys Res Letters 29:doi:10.1029/2001GL014349

Caldeira K, Kasting JF (1992) The life span of the biosphere revisited. Nature 360:721-723

Canfield DE (1989) Sulfate reduction and oxic respiration in marine sediments: Implications for organic carbon preservation in euxinic environments. Deep-Sea Res 36:121-138

Chameides WL, Perdue EM (1997) Biogeochemical Cycles: A Computer-Interactive Study of Earth System Science. Oxford University Press, New York

Conley DJ (2002) Terrestrial ecosystems and the global biogeochemical silica cycle. Global Biogeochem Cycles 16: doi:10.1029/2002GB001894

Conley DJ, Zimba PV, Theriot E (1994) Silica content of freshwater and marine diatoms. *In:* Proceedings of the 11th International Diatom Symposium. Kociolek JP (ed) California Academy Sci Memoir 17:95-101

De Angelis DL (1992) Dynamics of Nutrient Cycling and Food Webs. Chapman and Hall

Dixit S, Van Cappellen P (2002) Surface chemistry and reactivity of biogenic silica. Geochim Cosmochim Acta 66:2559-2568

Dixit S, Van Cappellen P (2003) Predicting benthic fluxes of silicic acid from deep-sea sediments. J Geophys Res – Oceans (in press)

Dixit S, Van Cappellen P, van Bennekom AJ (2001) Processes controlling solubility of biogenic silica and pore water build-up of silicic acid in marine sediments. Mar Chem 73:333-352

Drever JI, Li Y-H, Maynard JB (1988) Geochemical cycles: the continental crust and the oceans. *In:* Chemical Cycles in the Evolution of the Earth. Gregor CB, Garrels RM, Mackenzie FT, Maynard JB (eds) Wiley, p 17-53

Epstein E (1994) The anomaly of silicon in plant biology. Proceedings of the National Academy of Sciences USA 91:11-17

Falkowski PG, Barber RT, Smetacek V (1998) Biogeochemical controls and feedbacks on ocean primary production. Science 281:200-206

Gallinari M (2002) Dissolution et Préservation de la Silice Biogénique dans les Sédiments Marins. PhD Thesis, Université de Bretagne Occidentale

Garrels RM, Mackenzie FT (1971) Evolution of Sedimentary Rocks. Norton

Garrels RM, Mackenzie FT (1972) A quantitative model for the sedimentary rock cycle. Mar Chem 1:27-41

Harvey LDD (2000) Global Warming: The Hard Science. Prentice Hall

Klaas C, Archer DE (2002) Association of sinking organic matter with various types of mineral ballast in the deep sea: Implications for the rain ratio. Global Biogeochem Cycles 16:doi:10.1029/2001GB001765

Koning E, Brummer G-J, van Raaphorst W, van Bennekom AJ, Helder W, Iperen J (1997) Settling, dissolution and burial of biogenic silica in the sediments off Somalia (northwestern Indian Ocean). Deep-Sea Research 44:1341-1360

Koning E, van Iperen JM, van Raaphorst W, Helder W, Brummer G-JA, Weering TCE (2001) Selective preservation of upwelling-indicating diatoms in sediments off Somalia, NW Indian Ocean. Deep-Sea Research 48:2473-2495

Lane N (2002) Oxygen: The Molecule that Made the World. Oxford University Press, New York

Lasaga AC (1981) Dynamic treatment of geochemical cycles: Global kinetics. Rev Mineral 8:69-109

Mackenzie FT (1998) Our Changing Planet: An Introduction to Earth System Science and Global Environmental Change. Prentice-Hall

Mackenzie FT, Ver LM, Lerman A (2002) Century-scale nitrogen and phosphorus controls of the carbon cycle. Chem Geol 190:13-32

Martin RE (1995) Cyclic and secular variation in microfossil biomineralization: Clues to the biogeochemical evolution of Phanerozoic oceans. Global Planetary Change 11:1-23

Martin RE (1998) Catastrophic fluctuations in nutrient levels as an agent of mass extinction: Upward scaling of ecological processes? *In:* Biodiversity Dynamics: Turnover of Populations, Taxa, and Communities. McKinney ML, Drake JA (eds) Columbia University Press, p 405-429

Martin JH, Fitzwater SR, Gordon RM (1990) Iron deficiency limits phytoplankton growth in Antarctic waters. Global Biogeochem Cycles 4:5-12

Michalopoulos P, Aller RC (1995) Rapid clay mineral formation in Amazon delta sediments: Reverse weathering and oceanic elemental cycles. Science 270:614-617

Milliman JD, Troy PJ, Balch WM, Adams AK, Li Y-H, Mackenzie FT (1999) Biologically mediated dissolution of calcium carbonate above the chemical lysocline? Deep-Sea Res 46:1653-1669

Nelson DM, Tréguer P, Brzezinski MA, Leynaert A, Quéguiner B (1995) Production and dissolution of biogenic silica in the ocean: Revised global estimates, comparison with regional data and relationship to biogenic silica sedimentation. Global Biogeochem Cycles 9:359-372
Pettijohn FJ (1957) Sedimentary Rocks. Harper and Row
Racki G, Cordey F (2000) Radiolarian paleoecology and radiolarites: Is the present the key to the past. Earth-Sci Reviews 52:83-120
Ragueneau O, Leynaerts A, Treguer P, DeMaster DJ, Anderson R (1996) Opal studied as a marker of paleoproductivity. EOS 77:491-493
Redfield AC, Ketchum BH, Richards FA (1963) The influence of organisms on the composition of seawater. *In:* The Sea. Hill MN (ed) John Wiley, p 12-37
Reeburgh WS (1997) Figures summarizing the global cycles of biogeochemically important elements. Bull Ecological Soc Am 78:260-267
Rodhe H (1992) Modeling biogeochemical cycles. *In:* Global Biogeochemical Cycles. Butcher SS, Charlson RJ, Orians GH, Wolfe GV (eds). Academic Press
Ruttenberg KC, Berner RA (1993) Authigenic apatite formation and burial in sediments from non-upwelling, continental margin environments. Geochim Cosmochim Acta 57:991-1007
Sarmiento JL, Gruber N (2002) Sinks for anthropogenic carbon. Physics Today 55:30-36
Schenau SJ, De Lange GJ (2001) Phosphorus regeneration versus burial in sediments of the Arabian Sea. Mar Chem 75:201-207
Schlesinger WH (1997) Biogeochemistry: An Analysis of Global Change. Academic Press
Siever R (1992) The silica cycle in the Precambrian. Geochim Cosmochim Acta 56:3265-3272
Slomp CP, Meile C, Van Cappellen P (2003) The global phosphorus cycle: Response to ocean anoxia (submitted)
Smetacek V (1999) Diatoms and the ocean carbon cycle. Protist 150:25-32
Steefel CI, MacQuarry KT (1996) Approaches to modeling reactive transport in porous media. Rev Mineral 34:83-129
Steefel CI, Van Cappellen P (eds) (1998) Reactive transport modeling of natural systems. J Hydrology 209:1-388
Suess E (1981) Phosphate regeneration from sediments of the Peru continental margin by dissolution of fish debris. Geochim Cosmochim Acta 45:577-588
Tegelaar EW, de Leeuw JW, Derenne S, Largeau C (1989) A reappraisal of kerogen formation. Geochim Cosmochim Acta 53:3103-3106
Tréguer P, Nelson DM, Van Bennekom AJ, Demaster DJ, Leynaerts A, Queguiner B (1995) The silica balance in the world ocean: A re-estimate. Science 268:375-379
Turekian KK (1976) Oceans. Prentice Hall
Turekian KK (1996) Global Environmental Change. Prentice Hall
Valiela I (1995) Marine Ecological Processes. Springer-Verlag, Berlin
Van Andel TH (1994) New Views on an Old Planet. Cambridge University Press, Cambridge
Van Beusekom JEE (1991) Weschelwirkungen zwischen Gelöstem Aluminium und Phytoplankton in Marinen Gewässern. PhD Thesis, University of Hamburg
Van Cappellen P (1996) Reactive surface area control of the dissolution kinetics of biogenic silica in deep-sea sediments. Chem Geol 132:125-130
Van Cappellen P, Ingall ED (1994) Benthic phosphorus regeneration, net primary production, and ocean anoxia: A model of the coupled marine biogeochemical cycles of carbon and phosphorus. Paleoceanography 9:677-692
Van Cappellen P, Ingall ED (1996) Redox stabilization of the atmosphere and oceans by phosphorus-limited marine productivity. Science 271:493-496
Van Cappellen P, Dixit S, van Beusekom J (2002) Biogenic silica dissolution in the oceans: Reconciling experimental and field-based dissolution rates. Global Biogeochem Cycles 16:1075:doi:10.1029/2001GB001431
Velbel MA (1993) Constancy of silicate-mineral weathering-rate ratios between natural and experimental weathering: Implications for hydrologic control of differences in absolute rates. Chem Geol 105:89-99
Wignall PB (1994) Black Shales. Clarendon Press
Wolf-Gladrow DA, Riebesell U (1997) Diffusion and reactions in the vicinity of microalgae: A refined model. Mar Chem 59:17-34
Wollast R, Chou L (1998) Distribution and fluxes of calcium carbonate along the continental margin in the Gulf of Biscay. Aquatic Geochem 4:369-393